MW00341375

STOCHASTIC MODELS IN OPERATIONS RESEARCH

Volume I

Stochastic Processes and Operating Characteristics

Daniel P. Heyman

Lincroft, New Jersey

Matthew J. Sobel

Weatherhead School of Management
Case School of Engineering
Case Western Reserve University

DOVER PUBLICATIONS, INC.
Mineola, New York

TO OUR PARENTS
Asenath and Jerome Heyman

and

Alice and Willard Sobel

Bibliographical Note

This Dover edition, first published in 2004, is an unabridged republication of the work originally published in the "McGraw-Hill Series in Quantitative Methods for Management" by McGraw-Hill Book Company, New York, in 1982. This Dover edition has been published by special arrangement with Lucent Technologies, Inc., 600 Mountain Avenue, Murray Hill, N.J. 07974.

There is a list of corrections to *Stochastic Models in Operations Research, Volumes I and II* at http://weatherhead.cwru.edu/orom/sobel/index.htm. This website also has information that instructors will find useful such as solutions to many problems and comments on each chapter.

International Standard Book Number: 0-486-43259-9

Manufactured in the United States of America
Dover Publications, Inc., 31 East 2nd Street, Mineola, N.Y. 11501

CONTENTS

v

NOTE: Sections marked with one or two asterisks may be skipped without interfering with the reader's understanding of the rest of the text.

PREFACE

This book covers stochastic processes and operating characteristics of stochastic systems. The latter includes queueing theory and comparable analyses of stochastic processes in models of many phenomena such as inventories, cash balances, computer systems, dams, and reservoirs. The book is intended primarily as a text at the graduate school level for students in operations research, management science, computer science, all branches of engineering, applied mathematics, statistics, and economics. The comprehensive coverage should make the book useful for practitioners, teachers, and researchers who want an up-to-date review of stochastic processes and their applications. Volume II is devoted to stochastic optimization, particularly Markov decision processes.

The principal objectives of the book are:

1. To give the central facts and ideas of stochastic processes, and show how they are used in various models which arise in applied and theoretical investigations
2. To demonstrate the interdependence of three areas of study that usually appear separately: stochastic processes, operating characteristics of stochastic systems, and stochastic optimization
3. To show the importance of structured models by formulating and analyzing models in many contexts
4. To provide a comprehensive treatment that emphasizes the practical importance, intellectual stimulation, and mathematical elegance of stochastic models

ORGANIZATION AND COVERAGE

This book is organized on the basis of generic properties of many models rather than on the types of applications which prompt models. This choice fosters the intellectual unity of the three areas that have usually been treated as disparate.

We believe in getting to examples as quickly as possible. We start Part A, "Stochastic Processes and Models," with Chapter 2, which presents deterministic analyses of six models from different disciplines. This sets the stage for the stochastic models in succeeding chapters.

Chapter 3, "Prologue to Stochastic Processes," defines stochastic processes and categorizes the types of stochastic processes that we emphasize. Then we give a complete treatment of birth-and-death processes (Chapter 4) to show how basic ideas are used in a simple, but nontrivial, stochastic process that has many applications. Many of the examples introduced in Chapter 4 are expanded in subsequent chapters.

The cornerstone for much of our analysis of stochastic processes is the idea of regeneration epochs. This makes Chapters 5 and 6, on renewal theory and regenerative processes, the core of the book. The notions and results in these chapters are exploited in discrete-time Markov chains (Chapter 7); continuous-time Markov chains (Chapter 8); semi-Markov processes, Markov renewal processes, random walks, and Brownian motion (Chapter 9); and stationary processes (Chapter 10).

Our goal of illustrating the interdependence of stochastic processes and operating characteristics has a major impact on the book's organization. Nearly all the usual content of a first course in queueing theory is contained in Part A. As a result, Part B, "Operating Characteristics of Stochastic Systems," which is couched in terms of queueing models, differs from extant books on queueing theory. Part B draws on results in queueing theory which are not straightforward consequences of the types of stochastic processes presented in Part A.

Part B emphasizes general properties, generic models, and techniques for analyzing a wide variety of specific models. We favor sample-path methods over transform methods, although the latter are not absent. The main features of Chapter 11 are proofs of "Poisson arrivals 'see' time averages" and the queueing formula $L = \lambda W$, as well as illustrations of how these results are used to draw more detailed conclusions in specific models. Chapter 12 introduces networks of queues and emphasizes systems with Poisson arrivals and exponential service times. Chapter 13 briefly explores bounds, diffusion process approximations, and ad hoc approximation methods for intractable queueing models.

An appendix on probability theory, transforms, and mathematical analysis is included so that the prerequisite facts can be looked up as needed.

MATHEMATICAL PREREQUISITES

The following are the mathematical prerequisites to read the book:

1. A rigorous calculus sequence or the usual calculus sequence followed by a course in advanced calculus
2. Elementary matrix algebra
3. A first course in probability theory with a calculus prerequisite
4. Mathematical maturity

"Mathematical maturity" is a state of grace that usually is attained through course work, beyond items 1, 2, and 3, in mathematics or rigorous applied mathematics. Most students will not have completed items 1 through 4 before the senior year of their undergraduate programs. We have found that students can satisfy the advanced calculus prerequisite by starting that course at the same time as beginning this book.

COURSE OUTLINES

The ideal way to go through this book is to cover the 13 chapters in order during a one-year sequence on stochastic models. Once Markov chains are reached (this could be after one quarter), Volume II could be started in a companion course on stochastic optimization.

We have designed the book for several courses:

1. Stochastic processes: one quarter, one semester, or two quarters
2. Queueing theory: one quarter, one semester, or two quarters
3. Advanced queueing theory: one quarter or one semester
4. Advanced stochastic models: one quarter or one semester

For the first option, we found that we can review probability theory and the material from birth-and-death processes through semi-Markov processes in one semester. In a one-quarter course, we omit continuous-time Markov chains and semi-Markov processes.

For the second option, we have covered Markovian queues (Sections 4-4, 4-8, 4-9, 8-6, and 8-8), queues with embedded Markov chains (Sections 6-5, 7-6, and 7-9), and most of Chapter 11 in one semester. This presentation assumes the frequently encountered prerequisite of a first course on stochastic processes which includes Poisson processes, renewal theory, and Markov chains.

For options 3 and 4 we selected topics from the starred sections in Chapters 4 through 10 and from all of Chapters 11 through 13. The choice is influenced by the background and interests of the instructor and the students.

ADVICE TO INSTRUCTORS

We found it useful to review basic probability theory (in Section A-1) at the start of a first course in stochastic processes. The material on the exponential distribution (Section A-2) usually is needed, especially conditioning arguments. We do not emphasize transform methods, but they are useful in renewal theory; we recommend giving a brief review before starting renewal theory. Except for the "little-oh" notation, facts from mathematical analysis (in Section A-5) should be covered as they come up. A concentrated dose will give the wrong idea of the thrust of the course.

When we use the book as a text, we find that (particularly in second courses) we skip around to accommodate students with different preparations and interests. We do not recommend that you try to cover everything in each chapter or section. We do not. The book contains material which we think is important and interesting, but the usual academic calendars make it impossible to cover it all.

We frequently leave proofs, or parts of proofs, to the exercises. An entire proof deferred to an exercise parallels a previous proof or is easy, or else hints are given. A portion of a proof is left as an exercise to avoid breaking the thread of an argument with tedious manipulations. We recommend that you assign these exercises.

We occasionally refer to mathematical ideas that are beyond the scope of the book. This is done for the benefit of those readers who know such things and to point out further directions of study.

ACKNOWLEDGMENTS

During the years we spent writing this book, we received valuable advice from many people. We may not always have recognized the wisdom of their counsel, so any flaws in the book should not be attributed to them. It is a pleasure to thank for their comments on preliminary drafts Drs. David Y. Burman, Shlomo Halfin, Richard Serfozo, Diane D. Sheng, Donald R. Smith, Donald M. Topkis, Ward Whitt, and Eric Wolman of the Bell Telephone Laboratories; Messrs. Glenn Cordingley and Bruce J. Linskey and Profs. Robert G. Jeroslow, George E. Monahan, Loren K. Platzman, and Jonathan E. Spingarn of the Georgia Institute of Technology; Prof. Mordecai Henig of the University of Tel Aviv; and Dr. Roy Mendelssohn of the National Marine Fisheries Service. Mr. Jagadeesh Chandramohan and Prof. Ralph L. Disney of the Virginia Polytechnic Institute and State University, Prof. Edward J. Ignall of Columbia University; Prof. William S. Jewell of the University of California at Berkeley; Prof. Edward P. C. Kao of the University of Houston; Prof. Steven Lippman of the University of California, Los Angeles; and Prof. Haim Mendelson of the University of Rochester read the entire manuscript, and their suggestions greatly improved its content and presentation.

The preparation of this book was supported by the Bell Telephone Laboratories, the Georgia Institute of Technology, the National Science Foundation, and Yale University. We are especially grateful to Bell Laboratories for offering us the opportunity to collaborate on this book.

The seemingly endless task of typing the manuscript (and its many revisions) was accomplished by Judy Boné, Lisa Giummo, Janet Greene, and the Word Processing Center supervised by Mrs. Viola Trenta.

Daniel P. Heyman
Matthew J. Sobel

ONE

INTRODUCTION

Uncertainty is an important characteristic in many settings where operations research is applied. In a warehouse, demands for an item may fluctuate from day to day. In a telephone company, requests for assistance from a team of operators may arrive in an irregular manner. A computer center often receives programs that require different amounts of processing time, which causes congestion. A power company's turbine sometimes needs emergency repair. In these settings, the exact times and magnitudes of the events cannot be predicted with certainty. *Uncertainty is pervasive.*

Applications of operations research focus on decision making. Stochastic models in operations research are used to compare alternative decisions when uncertainty is too important to be ignored. For the warehouse, we would compare alternative rules for ordering from the manufacturer. For the telephone company, we would compare the effects of alternative sizes for the team of operators. For the computer center, we would compare the congestion caused by alternative rules for operating the computer system. For the power company, we would compare alternative preventive maintenance rules. *Uncertainty is manageable.*

During the twentieth century, there has been a continual interplay between the formulation of stochastic models and mathematical developments that permit us to analyze the models more completely. The models and their attendant mathematics now compose an enormous literature. The motivating contexts are extraordinarily varied, and the models and their analysis often exhibit ingenuity and elegance. *Uncertainty is intellectually stimulating.*

Operations research is not, of course, the only applied science in which stochastic models play a major role. Other areas include engineering (particularly industrial engineering), computer science, most of the social sciences (particularly economics), and most of the biological sciences. Some of the subjects which ordinarily are not associated with operations research but where stochastic models developed in an operations research setting play a central role include reservoir management and animal population dynamics.

1-1 OVERVIEW OF THE BOOK

The modeling phase of operations research can be divided into four parts: understanding the operation, choosing a model, validating the appropriateness of the model, and drawing conclusions from the model. This book concentrates on descriptions of settings in which various models have been applied and on derivations of the properties of the models.

All branches of applied mathematics consist of formal mathematical theorems, intuitive results, and a collection of applications. Intuitive explanations of the theorems are essential for successful applications and for extensions of the theory. We emphasize both the intuitive basis of the theorems and the formal proofs.

We present fundamental ideas and facts that provide a background for future learning and are apt to be useful in new applications. The examples have been chosen to illustrate the large number of potential applications and the power of general methods of analysis.

This book, like ancient Gaul, is divided into three parts. Part A describes the basic stochastic processes and models used in operations research and allied fields. The stochastic processes are used in Part B to obtain the operating characteristics of a wide variety of models, most of them concerning congestion and storage. Part C, in Volume II, is devoted to the characterization and computation of optimal policies for controlling a stochastic process.

1-2 USING THE BOOK

This book is both a text and a reference. It assumes no prior knowledge of stochastic models; it does assume that the reader is familiar with the topics usually covered in a calculus-based probability course. The material in Part A builds sequentially. The sections marked with a single star treat special topics which use the context of queues to amplify the unstarred material. Some subsequent examples and motivational remarks may refer to these sections, but they may be skipped without loss of continuity. The three sections marked with a double star are interesting derivations of important (and hard) theorems, which may strike some readers as dry. They, too, may be skipped.

Part B assumes a knowledge of the material in Part A, which may have been obtained from another source. Except for some motivating material, Part C is independent of Part B. Part C is almost independent of Part A, requiring only an elementary knowledge of Markov chains.

The examples and exercises are essential for appreciating the content of the theorems and for developing intuition. Many important results, which are used in subsequent sections, are given in the exercises. An exercise that is a continuation of the exercise preceding it is marked "(Continuation)." Frequent questions have been placed in the text for you to check your understanding of earlier material.

Although we provide rigorous proofs, we do not believe that the main content of a proof is its maneuvering of epsilons and deltas and its justifications of limiting operations. These are details (albeit important ones) that must be mastered eventually. It is the conception, or what some call the "physics," of the proof that should be understood first.

Definitions, theorems, examples, exercises, and figures are numbered sequentially within each chapter. The sixth theorem in Chapter 2, for example, is numbered 2-6.

1-3 NOTATION

We have tried to denote random variables by uppercase Latin letters and their realizations by the corresponding lowercase letters. Since there are only 26 Latin letters, and some of them (such as E and P) have special uses, we were unable to always adhere to this convention. Lowercase Greek letters typically represent constants. Since matrices are always distinguished from random variables by context, they, too, are denoted by uppercase Latin letters. Vectors are denoted by boldface lowercase letters, and their elements by the same letter without boldface.

The following symbols are used:

$\triangleq$ equal by definition

$\equiv$ identically equal

$\doteq$ approximately equal

$f(t) \approx g(t)$ means $f(t)/g(t) \to 1$ as t approaches some specified limit

$I \triangleq \{0, 1, 2, \ldots\}$

$I_+ \triangleq \{1, 2, \ldots\}$

$(a)^+ = \max\{a, 0\}$

$(a)^- = \max\{-a, 0\}$

$s\uparrow t$ means s approaches t from below

$s\downarrow t$ means s approaches t from above

$f(t^+) = \lim\limits_{s\downarrow t} f(s)$

$f(t^-) = \lim\limits_{s\uparrow t} f(s)$

$\#\{\cdot\} = $ number of elements in the set $\{\cdot\}$

$\square = $ end of a proof or an example

The identity matrix is also denoted by I; it will be clear from the context which interpretation of I is intended. The empty set is denoted by $\emptyset$.

The abbreviations *r.v.* and *i.i.d.* stand for *random variable* and *independent and identically distributed*, respectively. When it is inconvenient to denote the density function of an r.v., X say, explicitly, we write Pd $\{X = x\}$ for the density function evaluated at x.

STOCHASTIC PROCESSES AND MODELS

TWO

POSTERITY ANALYSIS

2-1 INTRODUCTION

The models in this chapter concern maintenance, congestion, cash management, reservoir operations, inventory replenishment, and commercial fisheries management. Many applications of operations research have occurred in these settings; in each of these settings, most applications use stochastic models. Throughout both volumes we use models from these settings to illustrate general results. As disparate as the problem settings may be, the models and analyses in this chapter possess considerable unity. Indeed, the unity among the ways of modeling a real situation and among the approaches to analyzing a model is a major theme of both volumes.

None of the analyses in this chapter is probabilistic, so this book on stochastic models has a strange beginning. Instead, for several reasons we analyze typical "posterities," i.e., historical outcomes from now to the future. In later chapters, we endow elements of the models in this chapter with probabilities. Consequently, any particular posterity will be only one of many candidate posterities that might have been "realized," i.e., "outcomes" or "sample paths" that might have occurred. Each posterity will be weighted by the probability that *it* would have been the one to occur. Then weighted averages will be computed for various quantities. In fact, not one but *two* averages are computed. For example, the analysis of an inventory model might include computation of the average inventory level at each time t and then the averaging of this average over time.

For simplicity, suppose that the set of possible posterities has J elements and that $x_j(t)$ is the inventory level at time t in the jth of the J posterities. Let p_j be the probability that posterity j actually ensues, and suppose $[0, T)$ is the time interval of interest. Then one way to compute the mean inventory level x^{**} is to compute first the *average over posterities* (or paths) of the inventory level at each t,

$$x^*(t) \triangleq \sum_{j=1}^{J} p_j x_j(t)$$

and then the *time average* of $x^*(t)$,

$$x^{**} \triangleq \frac{1}{T} \int_0^T x^*(t) \, dt \tag{2-1}$$

An alternative is to compute first the *time average* for each posterity

$$x_j^* \triangleq \frac{1}{T} \int_0^T x_j(t) \, dt \tag{2-2}$$

and then the average over all posterities

$$x^{**} \triangleq \sum_{j=1}^{J} p_j x_j^* \tag{2-3}$$

This second approach sometimes has the important advantage that x_j^* is relatively easy to compute, so that (2-3) is a much simpler step than (2-1). However, the computation of x_j^* in (2-2) summarizes the properties of a particular posterity, namely the jth. The computations in this chapter are analogous to (2-2). They lay the groundwork for averages similar to (2-3) in the following chapters.

Another reason to analyze posterities before endowing them with probabilities is that important managerial insights often are obtained. A property that pertains to each posterity usually is true for the weighted averages after probabilities are assigned; many expositions of stochastic models present the insights in a probabilistic context. However, you should appreciate which properties are valid regardless of the manner in which probabilities are assigned, indeed which are valid for deterministic models.

Terminology

The term *posterity* is not standard. In other books you might find the notion of a sample space whose elements are functions of time. Each of these elements might be called a sample path in other books because the next step usually is to discuss probability spaces. A probability function assigns numbers to certain subsets of outcomes, so the term *sample path* is consistent with such a development.

This chapter does *not* emphasize that any particular outcome is merely one of many. We do not study the issue of the extent to which some particular outcome is representative of the ensemble. Instead, we elicit some major features of whatever particular outcome we are studying. We avoid the sample-path terminology which reminds the reader of the probabilistic background. Instead, we use the word *posterity* to focus on the future consequences of following a particular path.

2-2 MAINTENANCE OF A COMPUTER

Consider a major computer center whose main memory, or *central processing unit* (CPU), sporadically becomes defective and must be shut down to be repaired. Figures 2-1 and 2-2 exhibit typical posterities of the times at which repairs were completed (the CPU "went up") and the times at which the CPU became defective ("went down"). The first record shows the cumulative number of repairs as a function of time, and the second shows the cumulative number of failures as a function of time. These data would be recorded carefully because a new CPU might cost $1 million dollars or more. Therefore, it is very expensive for the CPU to be "down" and unusable. Maintenance processes are modeled in practice because facility managers wish to spend the least money on maintenance that is consistent with adequate service to computer users. The models focus on the trade-off between cost and service.

Define 0 as the *epoch*, i.e., the particular time, at which the computer was first usable following its installation, and let τ be the amount of time that has elapsed since 0. Also, let $r(\tau)$ denote the number of "failures," i.e., the number of times that the CPU went down, up to and including epoch τ. The jumps in $r(\tau)$ occur at epochs when a failure occurs. Let r_k be the epoch of the kth such failure.

In Figure 2-1, $a(\tau)$ is the cumulative number of repairs completed up to and including epoch τ. The epochs at which the repairs were completed are labeled t_2, t_3, ... It is convenient to label 0 as t_1 and to regard the installation of the CPU

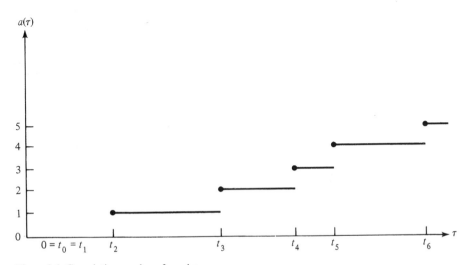

Figure 2-1 Cumulative number of repairs.

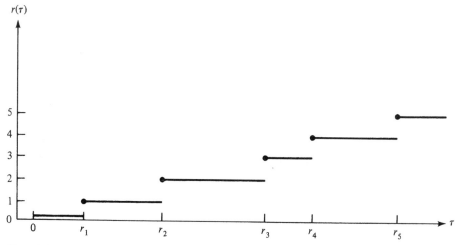

Figure 2-2 Cumulative number of CPU failures.

as the completion of the first repair. Then t_k is the epoch at which the $(k-1)$th repair was completed.

Observe that at every τ, either $a(\tau) = r(\tau)$ or $a(\tau) + 1 = r(\tau)$. If $r(\tau) - a(\tau) = 1$, then the CPU is down; and if $r(\tau) - a(\tau) = 0$, then the CPU is up. Therefore, the difference $x(\tau) \triangleq r(\tau) - a(\tau)$ indicates whether the CPU is up or down. When the CPU is down, $x(\tau)$ is 1; when the CPU is up, $x(\tau)$ is 0. This indicator function is graphed in Figure 2-3.

Epoch 0 ($=t_1$) initiates an interval during which the CPU is up. A *cycle* is the time between successive occurrences of this phenomenon. In Figure 2-3, the cycles are $[0, t_2), [t_2, t_3), [t_3, t_4), [t_4, t_5),$ and $[t_5, t_6)$. A cycle consists of an up period followed by an interval during which the CPU is being repaired. An up period is sometimes called a *period*. It is that part of the cycle when $x(\tau) = 0$. The periods in Figure 2-3 are $[0, r_1), [t_2, r_2), [t_3, r_3), [t_4, r_4),$ and $[t_5, r_5)$.

Let us examine the first few cycles and periods in greater detail. During the first cycle, the proportion of time that the CPU is down is $(t_2 - r_1)/t_2$. The same

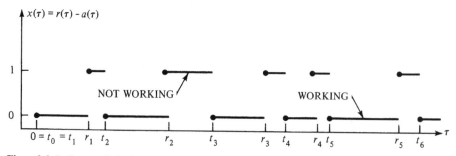

Figure 2-3 Indicator of whether CPU is working.

fraction during the first k cycles, labeled f_k, is

$$f_k = \sum_{i=1}^{k} \frac{t_{i+1} - r_i}{t_{k+1}} \tag{2-4a}$$

As a result, the proportion of time that the CPU is up is

$$1 - f_k = \frac{\sum_{i=1}^{k}(r_i - t_i)}{t_{k+1}} = \frac{k/t_{k+1}}{k/\sum_{i=1}^{k}(r_i - t_i)} = \frac{\lambda_k}{\mu_k} = \rho_k \tag{2-4b}$$

where λ_k and μ_k are the failure and repair rates discussed below and ρ_k is their ratio. The definitions are

$$\lambda_k \triangleq \frac{k}{t_{k+1}} \qquad \mu_k \triangleq \frac{k}{\sum_{i=1}^{k}(r_i - t_i)} \quad \text{and} \quad \rho_k \triangleq \frac{\lambda_k}{\mu_k} \tag{2-5}$$

Then

$$f_k = 1 - \rho_k \qquad k = 1, 2, \ldots \tag{2-6}$$

The *rate* of an activity is its average number of occurrences per unit time. Velocity, for example, is a rate of passage where an "occurrence" is the traversal of a unit distance. In (2-5), both λ_k and μ_k are rates. Table 2-1 presents the data on which Figures 2-1 and 2-2 are based and accompanying calculations for λ_k, μ_k, k/r_k, and ρ_k.

Inspection of Figure 2-1 and column 4 of Table 2-1 verifies that λ_k is the rate at which $a(\cdot)$ climbs during the first cycles. By comparison, μ_k is *not* the rate at which $r(\cdot)$ climbs during $[0, t_{k+1}]$. If it were, then the denominator of μ_k would be r_k, which is larger than $\sum_{i=1}^{k}(r_i - t_i)$ if $k > 1$. The difference $r_k - \sum_{i=1}^{k}(r_i - t_i) = \sum_{i=1}^{k-1}(t_{i+1} - r_i)$ is the down time during $[0, r_k]$. Thus μ_k is the rate of increase of $r(\cdot)$ during that portion of the first k cycles when the CPU is up. Compare columns 5 and 6 in Table 2-1. Finally, (2-6) asserts that $1 - f_k$, the fraction of time that the CPU is up, is the ratio of the rate at which $a(\cdot)$ climbs to the rate at which $r(\cdot)$ climbs when $x(\tau) = 0$.

Table 2-1 Data for Figures 2-1 and 2-2 and accompanying calculations

(1)	(2)	(3)	(4)	(5)	(6)	(7)
k	Times at which repairs are completed t_k	Times at which failures occur r_k	$\lambda_k = k/t_{k+1}$	$\mu_k = k/\sum_{i=1}^{k}(r_i - t_i)$	k/r_k	$\rho_k = \lambda_k/\mu_k$
1	0	2.5	0.333	0.400	0.400	0.833
2	3.0	6.2	0.250	0.351	0.323	0.712
3	8.0	11.0	0.261	0.345	0.273	0.757
4	11.5	13.1	0.305	0.388	0.305	0.786
5	13.5	17.0	0.294	0.362	0.294	0.812
6	17.5	—	—	—	—	—

Check the calculations by observing that $1 - f_k$ should be the fraction of time during $[0, t_{k+1})$ that the CPU is up. Let L_k denote this fraction. Then looking at $x(t)$, we have

$$L_k = \frac{1}{t_{k+1}} \int_0^{t_{k+1}} [1 - x(\tau)] \, d\tau$$

$$= \frac{1}{t_{k+1}} \sum_{i=1}^{k} (r_i - t_i) = \frac{k/t_{k+1}}{k/\sum_{i=1}^{k} (r_i - t_i)} = \rho_k$$

and so L_k does indeed equal $1 - f_k$:

$$L_k = \frac{\lambda_k}{\mu_k} = \rho_k \qquad k = 1, 2, \dots \tag{2-7}$$

It is reasonable to evaluate the long-run behavior of the CPU. The first few cycles may be unrepresentative because the computer center personnel are unfamiliar with the new model. As an immediate consequence of (2-6) and (2-7), we have Theorem 2-1.

Theorem 2-1 If λ_k and μ_k possess limits λ and μ as $k \to \infty$, and if $\mu > 0$, then L_k, f_k, and ρ_k have limits $L, f,$ and ρ and $\rho = \lambda/\mu$, with

$$L = 1 - f = \rho \tag{2-8}$$

The graphs of $a(\cdot)$, $r(\cdot)$, and $x(\cdot)$ can be regarded as two kinds of functions; it is apparent that they are functions of time, as is any posterity. However, our CPU may have been merely one of many CPUs of the same model made by the manufacturer. Peculiarities in materials and minute changes in workmanship might have led to somewhat different posterities if some other CPU had been delivered. Therefore, we might view $a(\cdot)$, $r(\cdot)$, and $x(\cdot)$ as functions, too, of the serial number, say, of the particular CPU that we happened to receive. This latter perspective views $a(\tau)$, $r(\tau)$, and $x(\tau)$ as random variables for each τ. Other sources of randomness might include the training and identity of the maintenance workers (which influence repair times), voltage fluctuations from the power company, and changes in business conditions that influence the size of the computing load and its composition.

Most of this book examines processes and models in which the principal variables are random. Then f_k, L_k, λ_k, μ_k, and ρ_k become random variables. However, in this chapter, we examine the relationships among these quantities for only the particular posterity that ensues from the delivery of *our* CPU.

EXERCISES

2-1 The data shown in the table refer to a computer center which shuts down for two days each weekend. The workday lasts from 8 a.m. to 6 p.m. Compute λ_k, μ_k, and ρ_k for $k = 1, \dots, 5$. Use one hour as the unit of time, count only work hours, and assume that repairs are made only on weekdays between 8 a.m. and 6 p.m. Is this assumption reasonable?

Day	Date	Event
Monday	April 2	Working all day
Tuesday	April 3	Repair starts at 10 a.m.
Tuesday	April 3	Repair finishes at 3 p.m.
Friday	April 6	Repair starts at 3 p.m.
Monday	April 9	Repair finishes at 9 a.m.
Wednesday	April 11	Repair starts at 4 p.m.
Thursday	April 12	Repair finishes at 10 a.m.
Monday	April 16	Repair starts at 11 a.m.
Monday	April 16	Repair finishes at 3 p.m.
Thursday	April 19	Repair starts at 3 p.m.
Friday	April 20	Repair finishes at 9 a.m.

2-2 Repeat Exercise 2-1, using its data, under the assumption that there are two shifts daily, from 8 a.m. to 4 p.m. and from 4 p.m. to midnight. Use one hour as the unit of time, count only work hours, and assume that repairs are made only on weekdays between 8 a.m. and midnight.

2-3 Let M_k denote the mean square variation of $1 - x(\tau) - L_k$ during the first k cycles:

$$M_k = \frac{1}{t_{k+1}} \int_0^{t_{k+1}} [1 - x(\tau) - L_k]^2 \, d\tau$$

(a) Prove that $M_k = (1 - \lambda_k/\mu_k)\lambda_k/\mu_k$, $k = 1, 2, \ldots$.
(b) How do you interpret M_k ?

2-3 CONGESTION AT A BANK

Tellers and their equipment constitute the major cost of operating a retail bank. More tellers means that customers wait less before being served. To see whether having more tellers might be worthwhile, banks sometimes model the congestion at their branches. The goal is to understand the trade-off between the customers' waiting times and the costs of altering the number of tellers or their equipment.

Consider a bank where someone is recording the epochs at which people arrive and depart. Suppose separate records are maintained for arrivals and departures. A typical posterity of arrivals is illustrated in Figure 2-4. The record of arrivals is exhibited as the cumulative number of "arrivals," i.e., people who have arrived as a function of time. Let us define 0 as the epoch after the bank was opened when the first customer arrived, and let $a(\tau)$ denote the number of people who arrived up to (and including) time τ. The jumps in $a(\cdot)$ occur at epochs when someone arrives; let t_k be the epoch of the kth such jump. By assumption, $t_1 = 0$. The corresponding segment of the departure observations also appears in Figure 2-4. There $r(\tau)$ is the cumulative number of departures up to and including time τ, and r_k is the epoch at which the kth departure occurs. It is convenient to define $r_0 = t_0 = 0$.

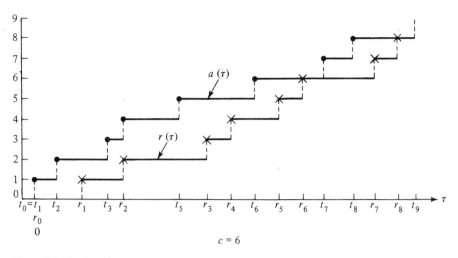

Figure 2-4 Number of arriving customers $a(\cdot)$ and number of departing customers $r(\cdot)$.

Customers do not necessarily leave the bank in the order of their arrival, so r_k is not necessarily the departure time of the customer who arrives at t_k. We discuss this point later.

The records of $a(\cdot)$ and $r(\cdot)$ are often the most easily observed characteristics of congested facilities, and fortunately several useful conclusions can be drawn from such data. Note first that the difference $x(\tau) \triangleq a(\tau) - r(\tau)$ is the number of people who are still inside the bank at epoch τ. This difference is graphed in Figure 2-5. Epoch 0 has been defined so that it marks the end of an interval during which the bank was empty. Label a cycle here as the time between successive occurrences of this phenomenon. In Figure 2-5 there is a cycle $[t_1, t_7)$ followed by a cycle $[t_7, t_9)$. A cycle encompasses an episode when there are customers in the bank (it is not empty) and the subsequent time while there are none (the bank is empty). A *busy period* is the "nonempty" portion of a cycle. In Figure 2-5, for example, the first busy period is $[t_1, r_6)$, and the second is $[t_7, r_8)$.

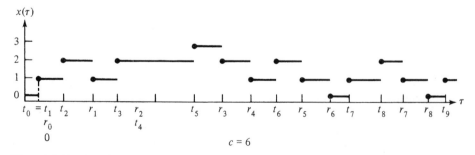

Figure 2-5 Number of people in the bank.

Analysis of a Typical Busy Period with Only One Teller

Now we examine a typical busy period, say the first one in Figure 2-5, in some detail. Let c denote the number of people who are served during the busy period; $c = 6$ in Figure 2-5. During the cycle, the proportion of time the bank was empty is

$$f \triangleq \frac{t_{c+1} - r_c}{t_{c+1}} \tag{2-9}$$

and the proportion of time it was occupied is $1 - f$. Other interesting operating characteristics during the cycle include various averages. Some of the averages are computed with respect to time; an example is L^*, the average number of people in the bank during the cycle:

$$L^* = \frac{1}{t_{c+1}} \int_0^{t_{c+1}} x(\tau) \, d\tau \tag{2-10}$$

It is also true that

$$L^* = \frac{1}{t_{c+1}} \int_0^z x(\tau) \, d\tau \qquad r_c \le z < t_{c+1}$$

The rates of arrival and of departure, i.e., the average number of arrivals and departures per unit time, are the reciprocals of the average times between successive arrivals and departures. The average time between successive arrivals t_{c+1}/c and the average time between successive departures r_c/c are examples of customer averages. The arrival rate is $\lambda^* \triangleq c/t_{c+1}$, and the departure rate is $\mu^* \triangleq c/r_c$. Let $\rho^* \triangleq \lambda^*/\mu^*$. Then the definitions of λ^*, μ^*, and f in (2-9) yield

$$1 - f = \frac{r_c}{t_{c+1}} = \frac{c/t_{c+1}}{c/r_c}$$

Hence

$$f = 1 - \frac{\lambda^*}{\mu^*} = 1 - \rho^* \tag{2-11}$$

The raw data are two sequences, $0 \triangleq t_0 = t_1 \le \cdots \le t_c \le t_{c+1}$ and $0 \triangleq r_0 \le r_1 \le r_2 \le \cdots \le r_c$ with the following properties:

(i) $t_j \le r_j$ for all j
(ii) $t_{j+1} < r_j$ if $1 \le j < c$
(iii) $t_{c+1} \ge r_c$

Property i asserts that at every epoch, the cumulative number of arrivals is greater than or equal to the cumulative number of departures. Properties ii and iii define r_c as the end of the first busy period and t_{c+1} as the end of the first cycle. The raw data do not permit us to identify whether the c individuals left the bank in the same order that they entered. Many permutations are possible, and the first to enter may have been the last to leave. Suppose, for the moment, that the order of departure was the same as the order of arrival. This particular queueing discipline is called *FIFO* (*first-in, first-out*).† Also, if there is only one teller who serves customers, sometimes it is called *FCFS* (*first-come, first-served*) and *HOL* (*head-of-the-line priority*). Under this egalitarian assumption, the ith arrival is inside the bank for an elasped time of $w_i \triangleq r_i - t_i$, and

$$W* \triangleq \frac{1}{c} \sum_{i=1}^{c} w_i \tag{2-12}$$

denotes the average waiting time inside the facility.

Analysis of a Bank with One Teller

If the bank is a branch office located in a small shopping center, it might well have only a single teller's window. A customer enters the bank, and if any other customer is inside, the new arrival goes to the end of a line, or queue. The new arrival reaches the head of the line and is served by the teller after all earlier arrivals have been served in their turn. In models of service processes such as this branch bank, sometimes it is convenient to transform the raw data, $0 \triangleq t_0 = t_1 \le t_2 < \cdots$ and $0 \triangleq r_0 \le r_1 \le r_2 \le \cdots$, as follows. The transformed data will be two sequences $u_1, u_2, \ldots$ and $v_1, v_2, \ldots$. Let $u_i = t_i - t_{i-1}$ for $i = 1$, $2, \ldots$, so that u_i is simply the time between the arrival of the ith and $(i-1)$th customers. The transformation is one-to-one because $t_i = \sum_{j=1}^{i} u_j$. The u_i are called *interarrival times*.

We want to construct the v_i's so that v_i is the time needed to serve customer i. This *service time* is the difference between the departure epoch r_i and the epoch at which customer i reaches the head of the line. However, $r_i - r_{i-1}$ would be a fallacious specification of v_i for some i. In Figure 2-5, for example, $r_7 - r_6$ includes the time $t_7 - r_6$ at whose end the seventh customer arrives. At that epoch, the facility is empty, so the seventh customer's service begins without delay and ends at epoch r_7. Therefore, the service time is $r_7 - t_7 < r_7 - r_6$. The trouble is that $v_i < r_i - r_{i-1}$ if r_{i-1} terminates a busy period, in which case $v_i = r_i - t_i$; otherwise, $v_i = r_i - r_{i-1}$. Let max $\{\cdots\}$ denote the largest number in the set $\{\cdots\}$. Therefore,

$$v_i = r_i - \max \{t_i, r_{i-1}\} \qquad \text{for all } i = 1, 2, \ldots \tag{2-13}$$

is adequate because $r_{i-1} \le t_i$ if and only if r_{i-1} terminates a busy period.

To state (2-9), (2-10), and (2-11) in terms of the u_i and v_i, first observe that $a(\cdot)$ counts the u_i. Specifically, $a(\tau)$ is the integer k with the property that the kth arrival occurred by time τ but the $(k+1)$th arrival was later than τ:

$$a(\tau) = \max \{k: \ t_k \le \tau\}$$

$$= \max \left\{ k: \ \sum_{i=1}^{k} u_i \le \tau \right\} \qquad \tau \ge 0$$

† The literature sometimes uses the label *FIFO* interchangeably with *FCFS* even if there is more than one server. In that case, customers do not necessarily leave in the same order as they arrived.

Similarly,

$$r(\tau) = \max \left\{ k : \sum_{i=1}^{k} v_i \le \tau \right\} \qquad 0 \le \tau \le r_c$$

Both $a(\,\cdot\,)$ and $r(\,\cdot\,)$ are called *counting functions*. Since $x(\tau)$ is $a(\tau) - r(\tau)$ and r_c is the earliest epoch after 0 that the bank is empty, the number of arrivals exceeds the number of departures until the busy period ends. Thus

$$c = \min \left\{ k : \; k \ge 1 \quad \text{and} \quad \sum_{i=2}^{k+1} u_i \ge \sum_{i=1}^{k} v_i \right\}$$

and

$$r_c = \sum_{i=1}^{c} v_i$$

The u_i and v_i data permit an alternative representation of (2-9), (2-10), and (2-11) and are useful in their own right. If the queueing discipline is FIFO and all the interarrival and service times are given, can we deduce the waiting times? From the construction of the u_i and v_i, the elapsed time which customer i spends in the bank, namely, the sum of the time on line (if any) and the time being served, is

$$w_i = r_i - t_i = v_i + \max \{t_i, r_{i-1}\} - t_i$$
$$= v_i + (r_{i-1} - t_i)^+$$

where $(z)^+$ denotes $\max \{z, 0\}$ for any number z. Since v_i is customer i's service time, $(r_{i-1} - t_i)^+$ must be customer i's time in line, or *queueing time*, if any. Let d_i denote customer i's queueing time:

$$d_i = (r_{i-1} - t_i)^+ \tag{2-14}$$

This expression makes sense; customer i has positive queueing time if i arrives before the previous customer departs.

From (2-13), (2-14), and $t_i = u_i + t_{i-1}$, we have

$$d_i = (r_{i-1} - t_i)^+ = (r_{i-1} - u_i - t_{i-1})^+$$
$$= (v_{i-1} + \max \{t_{i-1}, r_{i-2}\} - t_{i-1} - u_i)^+$$
$$= [(r_{i-2} - t_{i-1})^+ + v_{i-1} - u_i]^+ = (d_{i-1} + v_{i-1} - u_i)^+$$

and so

$$d_i = (d_{i-1} + v_{i-1} - u_i)^+ \tag{2-15}$$

Observe in Figure 2-6 that the smallest i for which $(d_{i-1} + v_{i-1} - u_i)^+ = 0$ is $i = 7$. Also, $d_1 = 0$, and so (by Exercise 2-4)

$$d_i = \sum_{j=1}^{i} (v_{j-1} - u_j) \qquad i \le c$$

We return to (2-15) in Chapter 9.

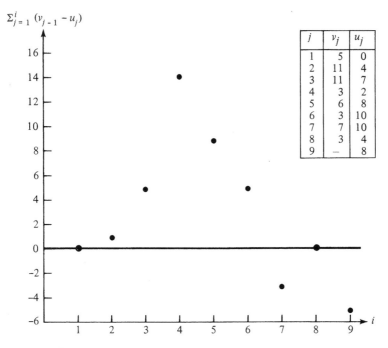

$\sum_{j=1}^{i} (v_{j-1} - u_j)$

j	v_j	u_j
1	5	0
2	11	4
3	11	7
4	3	2
5	6	8
6	3	10
7	7	10
8	3	4
9	—	8

Figure 2-6 $\sum_{j=1}^{i} (v_{j-1} - u_j)$ for the data on which Figures 2-4 and 2-5 are based.

A Conservation Theorem

The previous subsection exploited the assumption of a single teller via the simple relationship $v_i = r_i - \max\{t_i, r_{i-1}\}$. With more than one teller, this expression does not specify the service time of the ith customer to enter the bank (because the customer who arrives at t_i does not necessarily leave at r_i). However, in this subsection we do not rely on this expression, so the restriction to a single teller is abandoned.

The time average L^* and the customer averages λ^* and W^* are related by Theorem 2-2.

Theorem 2-2 During a cycle, regardless of the queueing discipline,

$$L^* = \lambda^* W^* \qquad t_{c+1} > 0 \tag{2-16}$$

Before proving Theorem 2-2, we present an example based on Figure 2-7.

Example 2-1 This example has $c = 4$ customers served during the busy period. First, suppose that customers are processed FIFO by a single server. The waiting times w_i labeled in Figure 2-7 are consistent with FIFO. In this example, $W^* = (w_1 + w_2 + w_3 + w_4)/4$ and $\lambda^* = 4/t_5$, so

$$\lambda^* W^* = (w_1 + w_2 + w_3 + w_4)/t_5$$

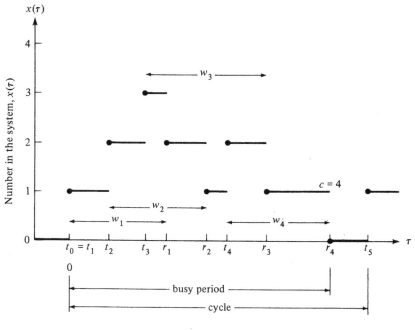

Figure 2-7 Example of a cycle with FIFO.

We verify graphically in Figure 2-8 that L^* has the same value as λ^*W^*. If customers were processed FIFO, then $w_i = r_i - t_i$, and so L^* would equal $(w_1 + w_2 + w_3 + w_4)/t_5$. In Figure 2-8, let A_i be the cumulative area of all boxes marked i. Then

$$A_1 = r_1 - t_1 = W_1$$
$$A_2 = (r_2 - r_1) + (r_1 - t_2) = r_2 - t_2 = W_2$$
$$A_3 = (r_3 - r_2) + (r_2 - r_1) + (r_1 - t_3) = r_3 - t_3 = W_3$$

and

$$A_4 = (r_4 - r_3) + (r_3 - t_4) = r_4 - t_4 = W_4$$

Hence

$$L^* \triangleq \frac{1}{t_{c+1}} \int_0^{t_{c+1}} x(\tau)\, d\tau = \sum_{i=1}^4 \frac{A_i}{t_5}$$
$$= \frac{\sum_{i=1}^4 W_i}{t_5} = \frac{4}{t_5}\left(\frac{\sum_{i=1}^4 W_i}{4}\right) = \lambda^*W^*$$

Thus L^* would equal λ^*W^* under the FIFO assumption.

How does the FIFO assumption affect this example? Suppose, now, that customers do not necessarily depart in the order of their arrival. Then

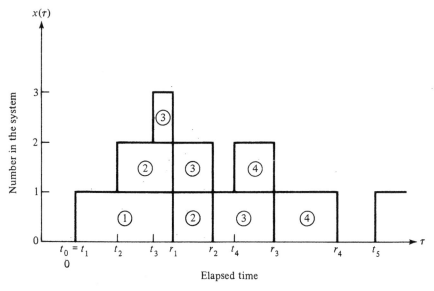

Figure 2-8 Calculation of L^* in the example.

$w_i \neq r_i - t_i$ is possible for some i. However, our graphical computation of L^* based on Figure 2-7 is still valid. We merely summed the area under $x(\cdot)$. Also, $\lambda^* = 4/t_5$ is unchanged. The following proof of Theorem 2-2 verifies $L^* = \lambda^* W^*$, so the numerical value of W^*, too, does not depend on FIFO. We make this assertion in Theorem 2-3. $\qquad \Box$

PROOF OF THEOREM 2-2 For $k = 1, \ldots, c$ and $0 \leq \tau \leq t_{c+1}$, let

$$I_k(\tau) = \begin{cases} 1 & \text{if} \quad t_k \leq \tau < r_k \\ 0 & \text{if} \quad \tau < t_k \quad \text{or} \quad \tau \geq r_k \end{cases}$$

Under the FIFO assumption, $I_j(\tau)$ indicates whether the jth customer to arrive is still inside the bank at time τ. Without depending on such an assumption,

$$x(\tau) = \sum_{k=1}^{c} I_k(\tau) \qquad 0 \leq \tau < t_c \tag{2-17}$$

To verify (2-17), let $\#\{\cdots\}$ denote the number of elements contained in the bracketed set. Then

$$
\begin{aligned}
x(\tau) = a(\tau) - r(\tau) &= \max \{k: \ t_k \leq \tau\} - \max \{k: \ r_k \leq \tau\} \\
&= \#\{k: \ t_k \leq \tau\} - \#\{k: \ r_k \leq \tau\} \\
&= \#(\{k: \ t_k \leq \tau\} \cap \{k: \ r_k > \tau\}) \\
&= \#\{k: \ I_k(\tau) = 1\} = \sum_{k=1}^{c} I_k(\tau)
\end{aligned}
$$

From the definition of $I_k(\tau)$, if $r_c \leq z \leq t_{c+1}$, then

$$\int_0^z I_k(\tau)\, d\tau = \int_0^{r_c} I_k(\tau)\, d\tau = r_k - t_k = w_k$$

With (2-17) this implies

$$\sum_{k=1}^c w_k = \int_0^{r_c} x(\tau)\, d\tau \qquad (2\text{-}18)$$

By using in (2-18) the definitions of L^* and W^* in (2-10) and (2-12), we get

$$cW^* = t_{c+1} L^*$$

or $\qquad\qquad L^* = \lambda^* W^* \qquad$ because $\qquad \lambda^* \triangleq \dfrac{c}{t_{c+1}} \qquad\qquad \square$

The formula $L^* = \lambda^* W^*$ was proved without depending on the FIFO assumption. What, then, is the interpretation of W^*? We shall see that W^* can be interpreted as the average waiting time, *regardless of the order in which customers are served inside the bank*, when the arrival and departure epochs are held fixed. Therefore, Theorem 2-2 relates time averages to customer averages in a service facility regardless of the queue discipline.

Example 2-2 (Continuation of Example 2-1) Table 2-2 displays data on which Figures 2-7 and 2-8 were based. Customers are processed FIFO in Table 2-2. However, Table 2-3 displays data on which Figures 2-7 and 2-8 could have been based except that the FIFO assumption is not valid.

Table 2-2 Data for Figure 2-7 with the FIFO assumption

Customer	Arrival time	Departure time	Difference
1	0	5	5
2	2	7	5
3	4	11	7
4	9	15	6
			$\overline{23} \div 4 = 5.75$

Table 2-3 Data for Figure 2-7 without the FIFO assumption

Customer	Arrival time	Departure time	Difference
1	0	7	7
2	2	5	3
3	4	15	11
4	9	11	2
			$\overline{23} \div 4 = 5.75$

In Table 2-2, customer 2 leaves sooner than customer 1, and customer 4 leaves before customer 3. Nevertheless, the mean wait, that is, W', is the same in both tables. This is shown by the fact that both tables have the same sum of the entries in their fourth columns. □

Suppose each customer is identified when he or she enters and leaves the service facility. If customers are labeled $1, \ldots, c$ according to their order of arrival, then the order of their departure is a permutation of $1, \ldots, c$. Let $M(\cdot)$ denote this permutation; i.e., the customer labeled 1 is the $M(1)$th to leave, customer 2 is the $M(2)$th to leave, etc. Therefore, customer k is inside the facility from the arrival epoch t_k until the departure epoch $r_{M(k)}$. The only restriction on the permutation $M(\cdot)$ is that $r_{M(k)} - t_k > 0$ (what does this mean?) for all $k = 1, \ldots, c$. The average elapsed time that customers spend inside the facility is

$$c^{-1} \sum_{k=1}^{c} (r_{M(k)} - t_k) = c^{-1} \left(\sum_{k=1}^{c} r_{M(k)} - \sum_{k=1}^{c} t_k \right)$$

$$= c^{-1} \left(\sum_{k=1}^{c} r_k - \sum_{k=1}^{c} t_k \right)$$

$$= c^{-1} \sum_{k=1}^{c} (r_k - t_k)$$

$$= c^{-1} \sum_{k=1}^{c} w_k = W^*$$

In the first equality above,

$$\sum_{k=1}^{c} r_{M(k)} = \sum_{k=1}^{c} r_k$$

because $M(\cdot)$ is a permutation, and so $\{1, 2, \ldots, c\} = \{M(1), M(2), \ldots, M(c)\}$. Therefore $L^* = \lambda^* W^*$ is generally meaningful because of Theorem 2-3.

Theorem 2-3 During a cycle, if a set of disciplines all yield $\{t_1, \ldots, t_{c+1}\}$ and $\{r_0, r_1, \ldots, r_c\}$ but differ with respect to $M(\cdot)$, then they have the same value of W^*.

Theorem 2-3 says that if two disciplines permute the order in which customers depart but leave the set of departure times unaltered, then the average waiting times are the same with both disciplines.

Comparison of the Computer Maintenance and Bank Congestion Models

How similar are the models of computer maintenance and bank congestion? A bank teller is like a repairperson. Both reduce the number of items needing attention. The "items" are customers in the bank model and failed CPUs in the maintenance model. A CPU failure is similar to a customer's entering the bank. The completion of a repair is similar to a customer's leaving the bank.

How do the analyses differ? First, the graphs of $x(\cdot)$ in Figures 2-7 and 2-3 show that $x(\tau) > 1$ is possible at the bank while $x(\tau) \le 1$ for the computer. Second, $L_k = \lambda_k/\mu_k$ in (2-7) whereas $L^* = \lambda^* W^*$ in Theorem 2-2. In both contexts, $x(\tau) = |a(\tau) - r(\tau)|$ and λ_k and λ^* are the rates at which $a(\cdot)$ increases. Also, in both contexts L_k and L^* are the average values of $x(\cdot)$. Therefore, the two models are fundamentally different unless W^* is analogous to μ_k^{-1}. This analogy is valid, as will be shown.

For the bank, $w_i = r_i - t_i$ and $W^* = \sum_{i=1}^{c} w_i/c$ by definition. For the computer CPU, from (2-5), $\sum_{i=1}^{k}(r_i - t_i)/k = \mu_k^{-1}$. Hence μ_k^{-1} is indeed analogous to W^* in Section 2-2. In other words, in Section 2-2, t_{k+1} denotes the epoch at which the kth repair is completed, and the installation of the CPU is regarded as the completion of the first repair. Therefore, the analysis of bank congestion can be viewed as a generalization of Section 2-2, where $x(t)$ is any nonnegative integer instead of only 0 or 1. The next subsection generalizes Theorem 2-1.

After the First Cycle

Banks experience long sequences of cycles rather than just one cycle. This is true also of other service facilities. Fortunately the preceding results can be extended to sequences of cycles.

The sequences of cycles are embedded in sequences of arrival epochs $0 \triangleq t_0 = t_1 \le \cdots$ and departure epochs $0 \triangleq r_0 \le r_1 \le r_2 \le \cdots$ with the properties $t_j \le r_j$ for all j. Episodes of emptiness occur when $r_k \le t_{k+1}$. Suppose the facility is left empty for the first time by the $c(1)$th departure, for the second time by the $c(2)$th departure, ..., and for the jth time by the $c(j)$th departure, $j = 1$, 2, ... Arrival number $c(j) + 1 \triangleq k(j)$ ends the jth cycle and starts the $(j + 1)$th cycle. Therefore, the length of the first cycle is $t_{k(1)} - t_1$, and generally the length of the jth cycle is $t_{k(j)} - t_{k(j-1)}$ [where $k(0) \triangleq 1$]. The first busy period extends from t_1 to $r_{c(1)}$ and the jth from $t_{k(j-1)}$ to $r_{c(j)}$. In Figures 2-4 and 2-5, $c(1) = 6$, $c(2) = 8$, $k(1) = 7$, and $k(2) = 9$.

Let L_i, W_i, and λ_i denote averages during the first i cycles:

$$\lambda_i = c(i)/t_{k(i)}$$

$$L_i = \frac{1}{t_{k(i)}} \int_0^{t_{k(i)}} x(z)\, dz$$

and

$$W_i = \frac{1}{c(i)} \sum_{j=1}^{k(i)} w_j$$

where $w_j = r_j - t_j$. In the notation of Section 2-2, $L^* = L_1$, $W^* = W_1$, and $\lambda^* = \lambda_1$.

Theorem 2-4 For each $i = 1, 2, \ldots$

$$L_i = \lambda_i W_i$$

PROOF As in the proof of Theorem 2-3,

$$x(\tau) = \sum_{j=1}^{k(i)} I_j(\tau) \qquad 0 \le \tau \le t_{k(i)}$$

and so

$$L_i = \frac{1}{t_{k(i)}} \int_0^{t_{k(i)}} \sum_{j=1}^{k(i)} I_j(\tau) \, d\tau$$

$$= \frac{1}{t_{k(i)}} \sum_{j=1}^{k(i)} (r_j - t_j) = \frac{1}{t_{k(i)}} \sum_{j=1}^{k(i)} w_j$$

Therefore, $t_{k(i)}L_i = c(i)W_i$, or $L_i = \lambda_i W_i$. $\qquad\square$

It is natural to consider unending service processes, i.e., the number of cycles $i \to \infty$. As a direct result of Theorem 2-2, we have Theorem 2-5.

Theorem 2-5 If L_i, W_i, and λ_i possess limits L, W, and λ, respectively, as $i \to \infty$, then

$$L = \lambda W \qquad (2\text{-}19)$$

PROOF This is left as Exercise 2-5. $\qquad\square$

The typical assumption in queueing theory is that λ_i has a limit. A stronger requirement is that $a(t)/t$ has a limit as $t \to \infty$ (why is this stronger?), i.e., that the arrival rate stabilizes. The intuitive basis of the following proposition is that the arrival rate should be the reciprocal of the average time between arrivals.

Proposition 2-1 For any finite $\lambda > 0$,

$$\lim_{\tau \to \infty} \frac{a(\tau)}{\tau} = \lambda \Leftrightarrow \lim_{n \to \infty} \frac{t_n}{n} = 1/\lambda \qquad (2\text{-}20)$$

PROOF ($\Rightarrow$) From the definition of $a(\cdot)$ and/or Figure 2-4, observe that

$$a(t_n) = n \qquad \text{and} \qquad t_{a(\tau)} \le \tau \le t_{a(\tau)+1}$$

for each $n \in I_+$ and $\tau \ge 0$. Hence

$$\frac{a(\tau)}{a(\tau)+1} \frac{a(t_{a(\tau)+1})}{t_{a(\tau)+1}} = \frac{a(\tau)}{a(\tau)+1} \frac{a(\tau)+1}{t_{a(\tau)+1}} \le \frac{a(\tau)}{\tau} \le \frac{a(\tau)}{t_{a(\tau)}} = \frac{a(t_{a(\tau)})}{t_{a(\tau)}} \qquad (2\text{-}21)$$

The existence of the limit on the left-side of (2-20) and $\lambda > 0$ imply $a(\tau) \to \infty$ as $\tau \to \infty$. Therefore,

$$\lambda = \lim_{\tau \to \infty} \frac{a(\tau)}{\tau} \le \lim_{\tau \to \infty} \frac{a(t_{a(\tau)})}{t_{a(\tau)}}$$

$$= \lim_{n \to \infty} \frac{a(t_n)}{t_n} = \lim_{n \to \infty} \frac{n}{t_n}$$

Taking limits on the left-side of (2-21) yields

$$\lambda \geq \lim_{\tau \to \infty} \frac{a(t_{a(\tau)+1})}{t_{a(\tau)+1}} = \lim_{\tau \to \infty} \frac{a(t_{n+1})}{t_{n+1}}$$

$$= \lim_{n \to \infty} \frac{n}{t_n}$$

because $a(\tau)/[a(\tau) + 1] \to 1$ as $\tau \to \infty$.

PROOF ($\Leftarrow$) The limit on the right-side of (2-20) exists and is positive, so $t_n \to \infty$ as $n \to \infty$. This divergence and $a(t_n) = n$ yield $a(\tau) \to \infty$ as $\tau \to \infty$. Then taking limits in (2-21) yields

$$\lim_{\tau \to \infty} \frac{a(\tau)}{\tau} \leq \lim_{\tau \to \infty} \frac{a(\tau)}{t_{a(\tau)}} = \lim_{n \to \infty} \frac{a(t_n)}{t_n} = \lambda$$

To establish the reverse inequality, taking limits in (2-21) also yields

$$\lim_{\tau \to \infty} \frac{a(\tau)}{\tau} \geq \lim_{\tau \to \infty} \frac{a(\tau)}{a(\tau) + 1} \frac{a(\tau) + 1}{t_{a(\tau)+1}} = \lim_{\tau \to \infty} \frac{a(\tau) + 1}{t_{a(\tau)+1}} = \lim_{n \to \infty} \frac{a(t_n) + 1}{t_{n+1}} = \lambda \quad \square$$

During the first cycle, the facility is empty for a fraction of time $f = 1 - \lambda^*/\mu^* = 1 - \lambda_1/\mu_1$. Let f_i denote the same fraction during the first i cycles. Then

$$f_i = \frac{1}{t_{k(i)}} \sum_{j=1}^{i} (t_{k(j)} - r_{c(j)})$$

because during the jth cycle the facility is empty from $r_{c(j)}$ until $t_{k(j)}$. Therefore,

$$1 - f_i = \frac{c(i)/t_{k(i)}}{c(i)/\sum_{j=1}^{i}(r_{c(j)} - t_{k(j-1)})} \tag{2-22}$$

where $k(0) \triangleq 1$. Let μ_i denote the right-hand denominator:

$$\mu_i = \frac{c(i)}{\sum_{j=1}^{i}(r_{c(j)} - t_{k(j-1)})}$$

which is the departure rate during the first i busy periods. Then (2-22) can be written as

$$f_i = 1 - \frac{\lambda_i}{\mu_i} = 1 - \rho_i \qquad \rho_i \triangleq \frac{\lambda_i}{\mu_i} \tag{2-23}$$

Letting $i \to \infty$ in (2-23) proves

Theorem 2-6 If $\lambda_i \to \lambda$ and $\mu_i \to \mu$ as $i \to \infty$, then $\rho \triangleq \lambda/\mu \leq 1$ and

$$f_i \to p_0 \triangleq 1 - \rho \tag{2-24}$$

The proof that $\rho \leq 1$ is left as Exercise 2-6.

A probabilistic version of Theorem 2-6 is given in Chapter 6.

Looking ahead to models with random variables, if the interarrival times $u_1 \triangleq t_1, u_2 \triangleq t_2 - t_1, \ldots$ are independent and identically distributed random variables (i.i.d. r.v.'s) with $1/\lambda$ as their common mean, in Chapter 5 we show that $a(t)/t \to \lambda$ (with probability 1). In Chapter 6 we also show that L_i and W_i possess limits in many cases of interest.

Also it is typically assumed that the service times are i.i.d. and independent of the arrival times; let $1/\mu$ denote their common mean. When the facility contains a single server, the interdeparture times during a busy period are the service times. A result in Chapter 5 will establish $\mu_i \to \mu$ (with probability 1).

EXERCISES

2-4 Reformulate (2-12) for any order of service. Show that FIFO minimizes the mean square wait [that is, $\sum_{j=1}^{c(i)} w_j^2 / c(i)$ for each $i \in I_+$]. *Hint*: Suppose FIFO does not minimize the wait; then obtain a contradiction by interchanging the order of service of two customers.

2-5 Prove Theorem 2-5.

2-6 Prove $\rho_i \le 1$ for each i and hence $\rho \le 1$ under the assumptions of Theorem 2-6.

2-7 In the analysis of a typical busy period with a single teller, we claimed

$$d_i = \sum_{j=1}^{i-1} (v_j - u_j) \qquad i \le c$$

Verify this claim.

2-8 Let M_k denote the mean square variation of $1 - x(\tau) - L_k$ during the first k cycles:

$$M_k = \frac{1}{t_{k+1}} \int_0^{t_{k+1}} [1 - x(\tau) - L_k]^2 \, d\tau$$

Exercise 2-3 asserts that

$$M_k = \frac{(1 - \lambda_k/\mu_k)\lambda_k}{\mu_k} \tag{2-25}$$

for the CPU model. Either prove that (2-25) is valid for the bank congestion model or give a numerical counterexample.

2-4 CASH MANAGEMENT

The first objective of this section is to introduce a model whose stochastic versions are widely used. Many applications of operations research concern the management of financial assets, and often stochastic models are used. Another objective of the section is to show that a posteriority which is inappropriate for the analyses in Sections 2-2 and 2-3 sometimes can be transformed to a new posteriority to which our previous analyses can be applied.

Large and small businesses find it difficult to predict their exact cash needs in future weeks and months. Therefore, they establish "lines of credit" with commercial banks. Sometimes a credit is much more than offset by a large balance in

Table 2-4 Transactions in a cash account

Elapsed days	Amount deposited, $	Amount withdrawn, $	Balance, $
0			60,000
4		30,000	30,000
6		20,000	10,000
10	30,000		40,000
13		20,000	20,000
15		20,000	0
16	20,000		20,000
18		10,000	10,000
21	30,000		40,000
23		20,000	20,000
26	40,000		60,000

a longer-term account. Here we suppress the important issue of the management of the longer-term assets and focus on an account from which cash disbursements (including payroll checks) are made and into which cash receipts are deposited. Receipts include transfers of funds from longer-term accounts to the shorter-term one. Disbursements include transfers of funds from the shorter-term account to the longer-term ones. Businesses use models to reduce the costs of their cash management decisions.

Table 2-4 is a record of receipts deposited in the short-term account, or "cash account," and disbursements or withdrawals from it. Figure 2-9 displays the consequent posterity of balances during the same period. The balance at time τ is labeled $x(\tau)$. The balance $x(\tau)$ is the algebraic sum of the initial balance $x(0)$ and the changes—deposits minus withdrawals—that occurred during $[0, \tau]$. In Figure 2-9, τ_k is the epoch at which the kth deposit occurs, and s_k is the epoch at which the kth withdrawal occurs. Also, z_k and y_k are the respective amounts of the kth deposit and withdrawal. We assume that all the z_k's and y_k's are positive integers (for example, an integral number of cents) and begin the analysis by converting

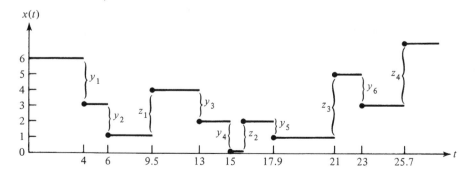

Figure 2-9 Balances in a cash account.

the data to the form used in earlier sections of this chapter. The conversion task is necessary because $x(\cdot)$ in the models for the CPU and bank congestion could jump up or down only one unit at a time.

In the present context, let

$$t_1 = t_2 = \cdots = t_{z_1} = \tau_1 \quad \text{and} \quad t_{z_1+1} = t_{z_1+2} = \cdots = t_{z_1+z_2} = \tau_2$$

For the general case, let $j(n) = \sum_{i=1}^{n} z_i$ and $t_{j(n-1)+1} = t_{j(n-1)+2} = \cdots = t_{j(n)} = \tau_n$ with $j(0) = 0$. Then t_i is the epoch at which the ith unit of revenue is deposited [not counting the initial balance of $x(0)$ units]. Also the interarrival time $u_i = t_i - t_{i-1}$ separates the deposits of the ith and $(i + 1)$th units.

Similarly, for the withdrawals, let $k(n) = \sum_{i=1}^{n} y_i$. Then

$$r_1 = r_2 = \cdots = r_{y_1} = s_1, \, r_{y_1+1} = r_{y_1+2} = \cdots = r_{y_1+y_2} = s_2$$

and generally $r_{k(n-1)+1} = r_{k(n-1)+2} = \cdots = r_{k(n)} = s_n$ with $k(0) = 0$. In this notation, the ith unit withdrawn leaves at epoch r_i. Define $0 = r_0 = t_0$.

The notation is the same as in Sections 2-2, 2-3, and 2-4, and so the derived quantities L_i, W_i, λ_i, μ_i, ρ_i, and f_i can be computed here. Then $L_i = \lambda_i W_i$ and $f_i = 1 - \rho_i$. But do these expressions have useful interpretations in the context of cash management? The rate at which cash enters the account is λ_i, and the average balance is L_i. But what is W_i? We might think of it as the average time between deposit and withdrawal of a monetary unit—and then agree that this notion is awkward for cash management. However, we also consider cash to be an important and expensive intermediate commodity for a firm. Wishing to invest as little in it as is needed, we seek a high *turnover rate* for cash, just as we do for all expensive stocks of goods. A high turnover rate for money in the cash account means that, on the average, money does not stay in the account too long, that is, W_i is small. Then $L_i = \lambda_i W_i$ asserts that for λ_i constant, the turnover rate can be increased, that is, W_i can be reduced, only by reducing L_i.

What is the meaning of $f_i = 1 - \rho_i$? Suppose, for example, that the bank automatically transfers funds to the cash account whenever the balance dips below $50,000 and that "dips" are the cumulative effect of issuing payroll checks which do not exceed, say, $7,000. Then $f_i = 0$ because the balance, that is, $x(\cdot)$, never drops to zero. Alternatively, suppose the account is a *debit account* used only to issue, say, payroll checks and that the bank transfers funds into the account at the end of each day so that the balance $x(\cdot)$ is raised to zero. Then f_i is the average fraction of the banking day when the day's first check has not yet cleared. For a very large firm, this average fraction is close to zero. Is the relationship $f_i = 1 - \rho_i$ useful here?

The answer, of course, is yes! Would we lead you astray so early in the book? Consider the task of computing the proportion of time that the balance is less than or equal to, say, $20,000 in the account in Figure 2-9. Let $x^{\#}(\tau) = [x(\tau) - 2]^{+}$, so that $x(\tau) \leq 2$ if and only if $x^{\#}(\tau) = 0$. Then the proportion of time that $x(\tau) \leq 2$ is the value of f_i computed for the $x^{\#}(\cdot)$ record. Figure 2-10 is the $x^{\#}(\cdot)$ record that corresponds to Figure 2-9. The times, magnitudes, and direc-

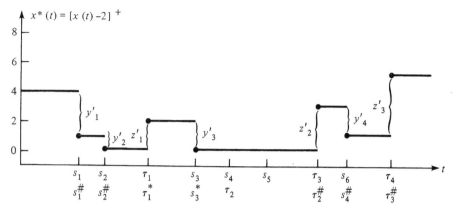

Figure 2-10 Amounts by which cash balances exceed $20,000.

tions (up or down) are the same as in Figure 2-9 if $x(\tau) > 2$. Finally, values of the r_i and t_i can be computed from the values of $\tau_i^\#$, $s_i^\#$, $z_i^\#$, and $y_i^\#$ in the $x^*(\cdot)$ record; from these, the values of λ_i and μ_i, hence ρ_i and f_i, can be computed.

In summary, if a posterity does not exhibit $x(\tau) = 0$ for a positive fraction of the values of τ, then add or subtract a suitable constant. The new posterity is amenable to the analyses and useful interpretations in the preceding sections.

EXERCISES

2-9 Suppose $x(\cdot)$ is the balance of a debit cash account; that is, $x(\tau) \le 0$ for all τ. Explain how to use the records of deposits and withdrawals to compute the proportion of time that the debit is $50,000 or more.

2-10 Suppose c is computed in Exercise 2-9 in the same way as in Section 2-3. What is its interpretation?

2-5 RESERVOIR REGULATION

The first objective of this section is to expand on the idea of transforming a posterity so that it is suited to the analyses and interpretations in Section 2-3. Here we transform the same posterity in two ways, and each one provides useful results. The second objective is to introduce a simple water resource model. Stochastic generalizations of this model have been used in many applications of operations research.

Reservoirs are valuable because they contain both water and space. Water can be used for drinking, and space can be used to prevent floods. These uses and others often are served by the same reservoir, so the trade-offs can be complicated. Water resource models are used to identify "good" trade-offs.

Consider a reservoir whose capacity is Q volumetric units of water and whose inflows and discharges are expressed as integer quantities. (Is this a limitation? Imagine measurements made in liters at Lake Superior or Lake Powell!) Suppose the contents $x(0)$ of the reservoir at the epoch when records are first kept is an integer. Then the content $x(\tau)$ at any time t afterward also is an integer. Similarly, the *freeboard*, i.e., the residual space in the reservoir, $Q - x(\tau)$ also is an integer for all $\tau \geq 0$.

Suppose that the cumulative inflows and discharges during $[0, \tau]$ are $a(\tau)$ and $r(\tau)$, respectively. Must it be true that $x(\tau) = x(0) + a(\tau) - r(\tau)$? The answer depends on the manner in which $a(\cdot)$ and $r(\cdot)$ are constructed. Consider the occurrence of flooding and subsequent overflow. If $a(\cdot)$ includes all the water that would have flowed into the reservoir had the capacity been great enough to hold the water, then $x(\tau) = x(0) + a(\tau) - r(\tau)$ overstates the actual contents by the amount of water which uncontrollably spilled over the dam. Similarly, suppose an extended drought causes the reservoir to run dry. If $r(\cdot)$ includes all requests for discharges, regardless of whether there was enough water to make the discharges, then $r(\tau)$ overstates the cumulative amount discharged. In this case, $x(\tau) = x(0) + a(\tau) - r(\tau)$ understates the actual contents by the amount of unsatisfied requests for discharges.

In this chapter, we resolve the ambiguity by using $a(\cdot)$ and $r(\cdot)$ for the actual inflows and discharges, rather than the ones that would have occurred had Q and $x(0)$ been virtually infinite. In subsequent chapters we use more detailed models that begin with raw inflows and demands, regardless of capacity or storage quantity, and then adjust these quantities in response to the actual capacity and the available storage.

Figures 2-11 and 2-12 graph the contents $x(\tau)$ and the freeboard $Q - x(\tau)$ for a certain time. Observe that both posterities have the features needed to compute quantities such as L_i, ρ_i, f_i, etc. in earlier sections:

1. The level at epoch 0 is a nonnegative integer.
2. The amounts of upward and downward jumps are positive integers.
3. Cycles and periods occur; i.e., the level drops to zero now and then.

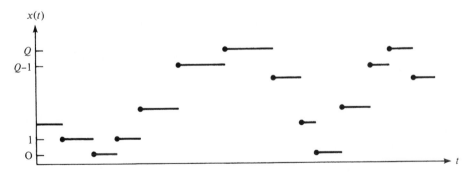

Figure 2-11 Reservoir contents.

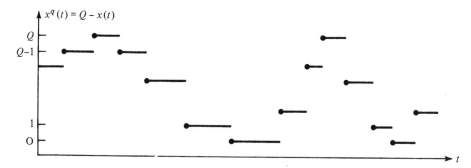

Figure 2-12 Reservoir freeboard.

Cycles in the graph of $x(\cdot)$ separate epochs of dryness, whereas they separate epochs of overflow in the graph of $Q - x(\cdot)$. Therefore, the quantities L_i, ρ_i, f_i, etc. have related but markedly different interpretations in the two graphs.

Let $x^q(\tau) = Q - x(\tau)$; let L_i, ρ_i, f_i, and μ_i pertain to $x(\cdot)$; and let L_i^q, ρ_i^q, f_i^q, and μ_i^q pertain to $x^q(\cdot)$. Then L_i is the average amount of water in the reservoir, and L_i^q is the average freeboard; f_i is the proportion of time that the reservoir is dry, and f_i^q is the proportion of time that it is full and has no freeboard; λ_i is the rate at which water flows in, and λ_i^q is the rate at which water is discharged. (What do μ_i and μ_i^q signify?) Also W_i is important for some water quality considerations, namely, the average age of stored water.

Is the relationship $0 = \lambda_i - \mu_i^q = \lambda_i^q - \mu_i$ true or false? Why or why not?

Reservoirs often are managed to avoid overflow and emptiness so that the cycles and periods in $x(\cdot)$ and $x^q(\cdot)$ may be lacking in some cases. Then L_i, f_i, ρ_i, etc. and L_i^q, f_i^q, ρ_i^q, etc. are ill-defined. As in Section 2-4, such cases invite analysis of transformed records. Suppose the reservoir is regarded as being "perilously low" if $x(\tau) \leq 0.1Q$ and "dangerously high" if $x(\tau) \geq 0.92Q$. Suppose that these events occur from time to time. Let $x_1(\tau) = [x(\tau) - 0.1Q]^+$ and $x_2(\tau) = [0.92Q - x(\tau)]^+$, so cycles occur in both $x_1(\cdot)$ and $x_2(\cdot)$. Then a version of ρ_i can be computed for each of the transformed records. For $x_1(\cdot)$, ρ_i is the proportion of time that the reservoir is perilously low, and for $x_2(\cdot)$ it is the fraction of time that the reservoir is dangerously high.

The simple transformations in Section 2-4 and this section are related to powerful methods for the statistical analysis of simulation outputs. We explain this connection briefly in Chapter 10.

EXERCISES

2-11 Suppose Q is so large that overflow never occurs (perhaps because discharges are increased to forestall overflow). Restate this section's results in this case.

2-12 Either prove $0 = \lambda_i - \mu_i^q = \lambda_i^q - u_i$ or give a numerical counterexample.

2-6 INVENTORY REPLENISHMENT

Inventory processes are modeled with the dual objective of reducing operating costs and improving services. Firms and public agencies invest enormous sums in inventories of primary, intermediate, and finished goods. A Western Electric Co. warehouse (distributing house) that supplies a local telephone company, for example, might contain 10,000 different kinds of items worth $25,000,000. A stockroom for the New Haven, Connecticut, municipal housing authority's maintenance staff might include 2,500 different kinds of items worth $100,000. A U.S. Navy supply depot might stock 50,000 different kinds of items. Frequently inventory managers are pressured to reduce their operating costs *and* the frequency with which they are out of stock.

In none of these three activities would it be prudent to construct a detailed separate model for each kind of item that is stocked. Instead, it is standard practice to group similar kinds of items and to use just one model as the basis of replenishment decisions for all kinds of items in the group. We defer detailed cost models to Volume II, but here, as preparation, we introduce policies.

A *policy* is a contingency plan that stipulates which action will be taken in the future depending on the prevailing state of affairs. In the context of inventory management, consider the limited task of deciding whether to order additional stock of a particular item and, if so, how much to order. Then a policy is a rule that specifies, for each time τ and each conceivable posterity during $[0, \tau]$, the size of the order (perhaps 0) placed at τ.

Suppose that customers request goods throughout the day and that they immediately are issued their requested amounts if the inventory level is high enough for their needs to be met. Otherwise, the amount by which their requests exceed supply is backordered and issued to them when additional stock is next delivered. Let $r(\tau)$ denote the aggregate *demand*, i.e., the number of units requested from the beginning of the first day up to the *end* of the τth day. Suppose that requests always are made for an integer number of units, so that $r(\cdot)$ is nondecreasing and integer-valued and has jumps only on the discrete set $\{0, 1, 2, \ldots\}$.

For simplicity, suppose that a replenishment decision is made at the beginning of each day and additional stock which is ordered (if any) is delivered virtually immediately. Let $x(\tau)$ denote the amount of stock on hand at the end of day τ, and let $z(\tau)$ denote the amount of additional stock procured at the start of day τ. This is a *discrete-time model* because τ takes values only in the discrete set of epochs $\{1, 2, \ldots\}$. In Sections 2-2 through 2-5, we presented *continuous-time models*. There τ could take values in the continuum $[0, \infty)$.

Suppose that the initial inventory level $x(0)$ is known and an integer. Then a simple relationship connects successive inventory levels: Today's level equals yesterday's level plus the amount added to stock minus the amount issued to satisfy demand. In symbols,

$$x(\tau) = x(\tau - 1) + z(\tau) - [r(\tau) - r(\tau - 1)]$$

where $r(0) \triangleq 0$. Alternatively, if $a(\tau)$ denotes the aggregate addition to stock by time τ, then

$$a(\tau) = \sum_{i=1}^{\tau} z(i) \qquad \text{for} \quad \tau \in I_+$$

where I_+ denotes the set of positive integers $\{1, 2, \ldots\}$. Let $a(0) = 0$. Then we obtain the expression (which should be familiar by now)

$$x(\tau) = x(0) + a(\tau) - r(\tau) \qquad \tau = 0, 1, 2, 3, \ldots \tag{2-26}$$

The quantity $r(\tau)$ is the cumulative amount requested, whether or not inadequate inventory levels caused some delays in satisfying the demands. Therefore, $x(\tau) < 0$ is possible for some values of τ. If $x(\tau) < 0$, then $-x(\tau)$ is interpreted as the number of previously requested units that are owed to consumers whose needs were backordered.

Now we can pose questions analogous to those in earlier sections. What proportion of days begin by being out of stock or in a backorder condition? What is a cycle here? What is the meaning of $f_i = 1 - \rho_i$? We resolved a similar ambiguity in the cash management model by transforming $x(\cdot)$. Here let $x_1(\tau) = [x(\tau)]^+$, and suppose $x(\tau) < 0$ occurs sporadically. Then $x_1(\cdot)$ has well-defined cycles, and f_i computed for $x_1(\cdot)$ is the proportion of time that $x(\cdot) \le 0$.

The values of f_i and other operating characteristics depend on three factors: the initial inventory level $x(0)$, the demand posterity $r(\cdot)$, and the inventory replenishment policy that is used. Consider a policy that can be administered easily: Each day it replenishes the amount of the preceding day's demand. Then $z(1) = 0$, so $x(1) = x(0) - r(1)$ and $z(2) = r(1)$. Therefore, $x(2) = x(0) + z(2) = x(0) - [r(2) - r(1)]$. Generally, for $\tau \in I_+$, let $r_\tau = r(\tau) - r(\tau - 1)$, so that $z(\tau) = r_\tau$ and

$$x(\tau) = x(0) - r_\tau \tag{2-27}$$

with this simple policy.

Another policy, called a *base stockage policy*, is specified by

$$z(\tau) = \begin{cases} S - x(\tau - 1) & \text{if } x(\tau - 1) < S \\ 0 & \text{if } x(\tau - 1) \ge S \end{cases}$$

or

$$z(\tau) = [S - x(\tau - 1)]^+ \tag{2-28}$$

where S is a parameter selected by the inventory manager. Let T be the length of time until cumulative demand draws the stock level down to S:

$$T = \min\{\tau: \quad r(\tau) \ge [x(0) - S]^+\}$$

And so $T = 0$ if $x(0) \le S$. It follows from (2-28) and (2-26) that

$$x(\tau) = S - r_\tau \qquad \tau > T \tag{2-29}$$

which is the same as (2-27) when $x(0) = S$. In other words, the base stockage policy does not order until T; after that, it specifies a replenishment $z(\tau)$ equal to the preceding day's demand.

Ordinarily a significant expense is incurred each time that $z(\tau) > 0$. The expenses include preparation of a purchase order and a delivery report, shelving of the delivered material, issuing a check to the vendor, and the clerical effort (even if automated) to update the inventory record to reflect the delivered quantity. These expenses provide an incentive to pool several days' demands into one replenishment order and, as a consequence, to replenish (positive quantities) less often. Various kinds of policies have been devised to behave this way. Consider an (s, S) *policy*:

$$z(\tau) = \begin{cases} S - x(\tau - 1) & \text{if } x(\tau - 1) \le s \\ 0 & \text{if } x(\tau - 1) > s \end{cases} \tag{2-30}$$

where $s < S$ are two parameters. Comparison of (2-29) with (2-30) leads to the observation that a base stockage policy is an (s, S) policy with $s = S - 1$. It follows from (2-27) and (2-30) that an (s, S) policy induces

$$x(\tau) = \begin{cases} S - r_\tau & \text{if } x(\tau - 1) \le s \\ x(\tau - 1) - r_\tau & \text{if } x(\tau - 1) > s \end{cases} \tag{2-31}$$

Inventory theory compares the effectiveness of different kinds of policies, for example, base stockage and (s, S). The theory describes the operating characteristics of different policies and contains algorithms to compute parameters of policies that are optimal with respect to various criteria. Many results in inventory theory are contained in both volumes.

EXERCISES

2-13 Consider an (s, S) inventory policy used on $r(\tau) = \omega\tau$, $\tau \ge 0$, for ω an integer such that $m \equiv (S - s)/\omega$ is an integer, too. Let $x_2(\tau) = [x(\tau) - s]^+$, and suppose $x(0) = x$. What are $\lambda_i, \mu_i, \rho_i, f_i, L_i$, and W_i for $x_2(\cdot)$ when $i = 1, 2, \ldots$? If these quantities possess limits as $i \to \infty$, what are they?

2-14 Suppose $r(\tau) \le \omega\tau$, an (s, S) policy is being used, and $r(\tau) > 0$ for all $\tau > 0$. Show that the long-run fraction of days on which orders occur is at most $\omega/(S - s)$. *Hint*: As i gets larger, ρ_i in Exercise 2-13 is an upper bound here.

2-7 FISHERY HARVESTS

A *fishery* is a subpopulation of a marine species whose members are more or less in the same geographical area. Examples include cod on the Grand Banks (off Newfoundland), the northern anchovy (southern California and Baja California), and lobsters on the New England and New Brunswick coasts. We also talk about the fishery of a more or less homogeneous subpopulation of a migratory species. An example is the Columbia River salmon.

From an economist's perspective, a fishery is a *renewable resource* because the resource can be prudently depleted and afterward its size self-regenerated. In plain words, we can catch plenty of fish and if the residual population is not too low, the population will increase eventually to its former size. However, larger catches (i.e., the aggregate caught by all boats) in a season often have two consequences of managerial importance. First, there is a higher risk that the population will "crash" and never regenerate. Second, even if the population perseveres, it will take longer to return to its former size than if the catch were smaller. Therefore, a prudent manager balances a sure immediate profit, a high one at that, with the impaired prospects for future profits.

This balance is a central issue in the management of other biological systems such as cattle ranches. Also it characterizes *capital accumulation* programs. Consider an individual's investment management and savings program. The task is to balance the satisfaction of immediately consuming (i.e., spending) part of the assets with the diminished base from which future earnings and capital growth stem. Investment portfolio managers and regional fishery councils regularly use stochastic models to identify a proper balance. Now we see that renewable resources can be identified in many contexts, but the discussion here uses terms from the fishery context. In both volumes, we introduce separate and more detailed models for biological resource management and capital accumulation.

Let $x(\tau)$ denote a fishery's size at time τ, measured in units of biomass (i.e., aggregate weight). Let $a(\tau)$ and $r(\tau)$ denote the cumulative increments and decrements, respectively, to population size during $[0, \tau]$. By conservation,

$$x(\tau) = x(0) + a(\tau) - r(\tau) \qquad (2\text{-}32)$$

where $r(\tau)$ encompasses both natural mortality and fishing effort. Suppose that changes in population size depend only on the size of population. This simplistic assumption ignores such important details as the age and sex distribution of the population. The assumption is represented by

$$\frac{dx(\tau)}{d\tau} = h[x(\tau)] \qquad \tau \ge 0 \qquad (2\text{-}33)$$

in which we ignore any restriction of $x(\,\cdot\,)$ to integer values and assume $h(\,\cdot\,)$ to be continuous.

One example that has been investigated repeatedly is the *logistic model*, in which $h(\,\cdot\,)$ is the function $h(y) = y[\beta(\alpha - y) - \delta y]$, so that

$$\frac{dx(\tau)/d\tau}{x(\tau)} = \beta[\alpha - x(\tau)] - \delta x(\tau) \qquad x(\tau) > 0 \qquad (2\text{-}34a)$$

where $\beta[\alpha - x(\tau)]$, the per capita (i.e., per unit biomass) birthrate, diminishes as the population rises to α. The per capita death rate $\delta x(\tau)$ is proportional to the size of the population. This model is motivated by the contention that each fishery has a natural limitation to its size, a *carrying capacity*, namely α. As $x(\tau)$ nears α, the fish are crowded together, and so predators have easier pickings and larvae less successfully compete with one another for nourishment.

We rewrite (2-34a) as

$$\frac{dx(\tau)}{d\tau} = \gamma_1 x(\tau) - \gamma_2 x(\tau)^2 \qquad (2\text{-}34b)$$

where $\gamma_2 = \beta + \delta$ and $\gamma_1 = \beta\alpha$. The solution of (2-33) and (2-34a and b) is known to be

$$x(\tau) = \frac{(\gamma_1/\gamma_2)e^{\gamma_1 \tau}}{[\gamma_1/\gamma_2 - x(0)]/x(0) + e^{\gamma_1 \tau}} \qquad 0 < x(0) \qquad (2\text{-}35)$$

Suppose, now, that mortality due to fishing is a constant proportion p of the population size, in addition to the natural mortality which alone occurs at a per capita rate $\delta[x(\tau) - \alpha]$. Then

$$\frac{dx(\tau)/d\tau}{x(\tau)} = \beta[\alpha - x(\tau)] - \delta x(\tau) - p \qquad (2\text{-}36)$$

and so, instead of (2-34b),

$$\frac{dx(\tau)}{d\tau} = \gamma_1^* x(\tau) - \gamma_2 x(\tau)^2$$

where $\gamma_1^* = \gamma_1 - p$. Therefore, (2-35) can be used to investigate the effects on population size of various catch rates p simply by replacing γ_1 with $\gamma_1 - p$. For example, in (2-35),

$$\lim_{\tau \to \infty} x(\tau) = \frac{\gamma_1}{\gamma_2} = \frac{\alpha}{1 + \delta/\beta} \qquad (2\text{-}37)$$

Therefore, when (2-36) describes the *population dynamics*, we have

$$\lim_{\tau \to \infty} x(\tau) = \frac{\beta\alpha - p}{\beta + \delta}$$

and so, in the limit, the catch accrues at a rate $p(\beta\alpha - p)/(\beta + \delta)$. This rate is maximized by $p = \beta\alpha/2$, which causes $\lim_{\tau \to \infty} x(\tau) = (\alpha/2)/(1 + \delta/\beta)$ which is half the value of (2-37). Observe that $p > 1$ is possible for appropriate values of β and α. This would be physically feasible in (2-36), for example, if β were, say, 4 and δ were small, say, 1. Then the population would approximately triple itself per unit time interval if $x(\tau)$ were small. Hence, by the end of the interval it would be feasible to harvest 150 percent, say, of the fish population at the start of the unit time interval.

EXERCISE

2-15 Suppose

$$x(\tau) = \begin{cases} (1 - p)\tau & \text{if } \tau < p^{-1} \\ 0 & \text{if } \tau \geq p^{-1} \end{cases}$$

where $0 < p \leq 1$ is a fishing rate. This is a simple model of a population crash that occurred because the population overwhelmed the carrying capacity of its habitat. Suppose $e^{-\gamma \tau}$ is the present value at time 0 of a unit quantity caught at time τ ($\gamma \geq 0$). The problem is to choose p to maximize the present value of the entire catch, i.e., to maximize

$$\int_0^{1/p} px(\tau)e^{-\gamma \tau}\, d\tau = \int_0^{1/p} p(1-p)e^{-\gamma \tau}\, \tau\, d\tau$$

Let $p(\gamma)$ denote the dependence on the parameter γ of an optimal value of p. Must $p(\gamma) \to 0$ as $\gamma \to 0$?

THREE

PROLOGUE TO STOCHASTIC PROCESSES

This chapter introduces some fundamental concepts of stochastic processes. These concepts are used here primarily to develop a technical vocabulary. Chapters 4 to 10 describe the analysis and applications of those stochastic processes that appear most often in operations research models.

3-1 DEFINITIONS AND BASIC CONCEPTS

We begin by defining a stochastic process.

> **Definition 3-1** A *stochastic process* is a collection of random variables that are all defined on the same probability space and indexed by a real parameter.

We denote a stochastic process by $\{X(t); t \in \mathcal{T}\}$, where $\mathcal{T}$ is a set of numbers that indexes the random variables $X(t)$. Often it is appropriate to interpret t as time and $\mathcal{T}$ as the range of times being considered. In this context, $X(t)$ is the value of the process at time t. Occasionally we refer to $\{X(t); t \in \mathcal{T}\}$ as the *X-process* and $\mathcal{T}$ is understood from the context. If $\mathcal{T}_1$ and $\mathcal{T}_2$ are different index sets, then $\{X(t); t \in \mathcal{T}_1\}$ and $\{X(t); t \in \mathcal{T}_2\}$ are different processes.

We call t the *parameter of the process* and $\mathcal{T}$ the *parameter set*. We always consider t as a scalar and $\mathcal{T}$ as a set of scalars, but the general theory of

38

stochastic processes does not require this. When $\mathcal{T}$ is finite or countably infinite, we say that $\mathcal{T}$ is *discrete*; otherwise, $\mathcal{T}$ is called *continuous*. Let $\mathcal{S}$ be the set of possible values which the $X(t)$'s can assume; it is called the *state space* of the process. When $\mathcal{S}$ contains a finite or countably infinite number of points, it is called *discrete*; otherwise, it is called *continuous*.

Two prevalent choices for $\mathcal{T}$ are $\{0, 1, 2, \ldots\} \triangleq I$ and $[0, \infty)$. The former is the natural choice when the process evolves in discrete time, and the latter is the natural choice for continuous time. Sometimes we wish to emphasize that time "zero" is actually picked to be a reference point chosen after the process has been evolving for a long time; usually this is intended to mean that the process is in the "steady state." In this case, we may want to write $\mathcal{T}$ as $\{0, \pm 1, \pm 2, \ldots\}$ or as $(-\infty, \infty)$ for discrete and continuous time, respectively.

When $\mathcal{S}$ is discrete, it is conventional to label the states by nonnegative integers. When the number of states is $m + 1$, we take $\mathcal{S} = \{0, 1, \ldots, m\}$; when the number of states is infinite (but countable), we take $\mathcal{S} = \{0, 1, \ldots\} \triangleq I$. When $\mathcal{T}$ is a set of integers, typically we write the random variables (r.v.'s) with subscripts: for example, X_n.

For any state $i \in \mathcal{S}$ and any time $t \in \mathcal{T}$, if $X(t) = i$, we say that the process is in state i at time t. Sometimes we substitute the term *stage* or *epoch* for *time*; the former is used primarily for discrete time and the latter for continuous time.

You will see how this notion is used (and useful) many times in this text. To give you an idea of its use, recall the CPU example described in Section 2.2. Suppose the times between failures are r.v.'s, and let R_k be the epoch of the kth failure. Then $\{R_k; k = 1, 2, \ldots\}$ is a discrete-parameter stochastic process with state space $[0, \infty)$. Let the r.v. $X(t)$ be 1 if the CPU is down at time t and 0 otherwise. Then $\{X(t); t \geq 0\}$ is a continuous-parameter stochastic process with the discrete state space $\{0, 1\}$.

Finite-Dimensional Distributions

What information is needed to specify a stochastic process? Let $A_n = \{t_1, t_2, \ldots, t_n\}$ be a set of n points in $\mathcal{T}$ and

$$F_{A_n}(x_1, \ldots, x_n) = P\{X_{t_1} \leq x_1, \ldots, X_{t_n} \leq x_n\}$$

We call $F_{A_n}(\cdot)$ a *finite-dimensional distribution function* (abbreviated *fidi*) of the process. The collection of all such distribution functions (DFs) as $\mathcal{A}_n$ ranges over all the finite subsets of $\mathcal{T}$ is called the *family* of fidi's.

Any probabilistic statement about a discrete parameter process can be expressed in terms of the fidi's, and we say that *the family of fidi's determines a discrete-parameter process*. Unfortunately, things are not so simple for a continuous-parameter process: The family of fidi's is not sufficient to completely describe the process. To see the latter, let $\mathcal{T} = [0, 2]$, and suppose we want to know the probability that $X(t) \geq 0$ for all $0 \leq t \leq 1$. Obtaining this probability requires simultaneous knowledge of the process at an uncountable number of points, information that cannot be obtained only from the fidi's. This fact is one reason why continuous-parameter stochastic processes often are more difficult to treat rigorously.

Suppose we use data analysis, theory, and experience to choose a family of fidi's that specifies the stochastic process $\{R_k\,;\ k = 1, 2, ...\}$ which are the failure epochs in the CPU model of Section 2-2. Is there a stochastic process with these fidi's? The answer is yes provided a consistency condition is satisfied.

A family of fidi's is *compatible* if whenever A_n and A_m are two finite subsets of $\mathscr{T}$ with A_n contained in A_m, then $F_{A_n}(\,\cdot\,) = F_{A_m}(\,\cdot\,)$ with the appropriate arguments set at infinity. For example, with $A_1 = \{t_1\}$ and $A_2 = \{t_1, t_2\}$, we must have

$$F_{A_1}(x_1) = F_{A_2}(x_1, \infty)$$

A fundamental theorem of stochastic processes asserts that for any given compatible family of fidi's, there exists a stochastic process having that family for its fidi's.

Sample Paths

Now let us consider another way to describe a stochastic process. We start with a simple example, two independent flips of a fair coin. Let $X_n = 1$ if the nth flip yields a head and let $X_n = 0$ otherwise, $n = 1, 2$. The stochastic process $\{X_n\,;\ n = 1, 2\}$ is determined by its fidi's, which are

$$P\{X_1 = 1\} = P\{X_1 = 0\} = P\{X_2 = 1\} = P\{X_2 = 0\} = \tfrac{1}{2} \qquad (3\text{-}1a)$$

and

$$P\{X_1 = 0, X_2 = 0\} = P\{X_1 = 0, X_2 = 1\} = P\{X_1 = 1, X_2 = 0\}$$
$$= P\{X_1 = 1, X_2 = 1\} = \tfrac{1}{4} \qquad (3\text{-}1b)$$

Each realization of this process is a pair (X_1, X_2), where X_n is the outcome of the nth flip, $n = 1, 2$. Thus, (3-1a and b) give the probability of each possible realization of the process. This is an intuitive way of describing the probabilistic structure of a stochastic process.

To obtain this description in general, we choose a set Ω that represents the possible realizations of the process. Each element ω of Ω is called a *sample path*, or *sample function*, of the process. Then we choose an appropriate collection of subsets of Ω to which to assign probabilities; next we assign the probabilities. In this framework, $X(t, \omega)$ is the value of the ωth function at time t, which is a number. For each fixed ω, $X(t, \omega)$ is a function of t and is a sample path of the process. For this reason, stochastic processes are sometimes called *random functions* because the trajectory of the function depends on which ω is chosen. For each fixed t, $X(t, \omega)$ is a function of ω and so is a random variable, as before.

The probabilities assigned to subsets of Ω must be consistent with the fidi's. The probability of every *discrete-interval* event of the form $\{\omega\colon\ a_j < X(t_j, \omega) \le b_j\}$ for each $j = 1, 2, ..., n$ is obtained from the fidi's. It may be shown that

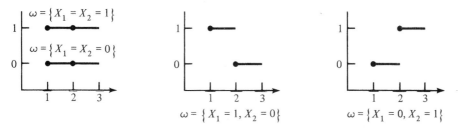

Figure 3-1 Sample paths for two flips of a coin.

these probabilities uniquely determine a probability measure on Ω for all sets in the smallest† σ-field containing all the discrete intervals. These concepts are illustrated in Figure 3-1 by using the coin-flipping example. The points have been converted to line segments for visual clarity. Notice that each sample function has probability $\frac{1}{4}$ of occurring and for each t, $1 \le t < 3$, $P\{X(t) = 1\} = \frac{1}{2}$.

The preceding discussion emphasizes that a stochastic process can be viewed both as a collection of random variables and as a kind of generalized random variable whose values are functions of time. Each of the deterministic processes described in Chapter 2 can serve as a sample path of a stochastic model. The results in Chapter 2 hold on any given sample path; one of the tasks in the next eight chapters is to ascertain how these results change (if at all) when stochastic features are introduced.

Pitfalls to Avoid

It is possible to specify a stochastic process by making assumptions about the fidi's and about the sample paths. For example, a diffusion process (defined in Chapter 9) is a Markov process (defined in Section 3-2) with continuous sample paths. When a stochastic process is specified in this way, care must be taken to ensure that it is well defined.

Example 3-1 Let $\mathcal{T} = [0, 1]$, and suppose we insist that the sample paths $X(\cdot, \omega)$ all be continuous on $[0, 1]$ and for any n, the r.v.'s $X(t_1), \ldots, X(t_n)$ are i.i.d. and normally distributed, where $0 \le t_1 < \cdots < t_n \le 1$. Let us see why such a process cannot exist. For any number δ and points t and s in $[0, 1]$,

$$P\{X(t) > \delta, X(s) < -\delta\} = \left(\int_\delta^\infty \frac{1}{\sqrt{2\pi}} e^{-u^2/2} \, du \right)^2 \tag{3-2}$$

However, the continuity of the sample paths implies that for each ω and positive integer m, the sets

$$A_m = \{\omega : \quad X(t, \omega) > \delta, X(t + 1/m, \omega) < -\delta\}$$

† See Section A-1 for a definition of a σ-field and its role in probability theory.

must become empty as $m \to \infty$. That is, given $\varepsilon > 0$,

$$P\{X(t) > \delta, \ X(t + 1/m) < -\delta\} < \varepsilon \tag{3-3}$$

when m is large enough. Clearly (3-2) and (3-3) are incompatible, so no such process exists.

The reason that such a process does not exist is not mysterious. The requirement that the sample paths be continuous implies that when $|t - s|$ is small, the probability that $X(t)$ is close to $X(s)$ must be large. However, the requirement that $X(t)$ and $X(s)$ be independent and have normal distributions implies that the probability that $X(t)$ is close to $X(s)$ is small. $\qquad\square$

Since a stochastic process is specified by its fidi's, which can consider only a finite number of points simultaneously, strange things may happen when infinitely many points are considered simultaneously. The next example illustrates what can happen.

Example 3-2 Choose $\mathcal{T} = [0, 1]$, let $\omega \in [0, 1]$ with $P\{\omega \le y\} = y$ for $0 \le y \le 1$, and set

$$X(t, \omega) = \begin{cases} 0 & \text{if } t \ne \omega \\ 1 & \text{if } t = \omega \end{cases}$$

for $0 \le t \le 1$. Thus, the sample paths have a uniform distribution on $[0, 1]$, and the trajectories are functions that are zero except at the single epoch $t = \omega$. For each fixed t, $P\{\omega = t\} = 0$ and hence

$$P\{X(t_1) = X(t_2) = \cdots = X(t_n) = 0\} = 1$$

for any positive integer n and times $0 \le t_1 < t_2 < \cdots < t_n \le 1$. However,

$$P\{ \max_{0 \le t \le 1} X(t) = 0\} = P\{X(t) = 0 \text{ for all } t \in [0, 1]\} = 0$$

Notice that $\{X(t); \ 0 \le t \le 1\}$ has the same fidi's as $\{Y(t); \ 0 \le t \le 1\}$, which is constructed in the same manner as the X-process, except that $Y(t, \omega) \equiv 0$. This certainly shows that more than one process may share the same family of fidi's. $\qquad\square$

These issues do not appear explicitly in the subsequent chapters. For some continuous-time processes there are allusions to the fact that these issues exist and have been resolved.

3-2 SPECIAL CLASSES OF STOCHASTIC PROCESSES

Conclusions of a practical nature are obtained when some structure is imposed on the general features of a stochastic process. This structure often takes the form of specifying (1) how the r.v.'s of the process depend on one another and (2) what restrictions are placed on the form of their joint distributions. In this section we define three classes of stochastic processes: The first two are studied extensively in Chapters 4 through 10, while the third is mentioned only in passing.

Markov Processes

In a Markov process, the future depends on only the present state. The history of the process prior to its present state has no influence on the subsequent evolution of the process. Here is the formal definition.

> **Definition 3-2: Markov process** The stochastic process $\{X(t); t \in \mathcal{T}\}$ is a *Markov process* if for any positive integer n, time points (in $\mathcal{T}$) $t_1 < t_2 < \cdots < t_n < t_{n+1}$, and any set of states A,
>
> $$P\{X(t_{n+1}) \in A \mid X(t_1) \cdots X(t_n)\} = P\{X(t_{n+1}) \in A \mid X(t_n)\} \qquad (3\text{-}4)$$
>
> holds true. We call (3-4) the *Markov property*. A Markov process is called *homogeneous* if
>
> $$P\{X(t + s) \in A \mid X(t)\}$$
>
> does not depend on t.

Given $X(t_n)$, $X(t_{n+1})$ is independent of $X(t_1), \ldots, X(t_{n-1})$. If we call t_n the present, we say that (3-4) asserts that the future of the process is *conditionally independent* of the past, given the present.

Markov processes are good models for many situations. They are analytically tractable and are widely used in many applied fields. Chapters 4, 7, 8, and 9 are devoted exclusively to Markov processes.

Stationary Processes

A stationary stochastic process has the property that the joint distributions do not depend on the time origin.

> **Definition 3-3: Stationary process** The stochastic process $\{X(t); t \in \mathcal{T}\}$ is *stationary* if whenever $t_i \in \mathcal{T}$ and $t_i + s \in \mathcal{T}$, $i = 1, 2, \ldots, n$ (n is any positive integer), then $\{X(t_1), \ldots, X(t_n)\}$ and $\{X(t_1 + s), \ldots, X(t_n + s)\}$ have the same joint distribution.

Choosing $n = 1$ shows that, in particular, the distribution of $X(t)$ is the same for all $t \in \mathcal{T}$.

Stationary processes typically arise in operations research models as a description of processes that have evolved for a long time. Chapter 10 is devoted to stationary processes.

Martingales

A *martingale* is a mathematical model of a sequence of fair gambles.

Definition 3-4: Martingale The stochastic process $\{X_n ; n \in I\}$ is a *martingale* if

$$E(|X_n|) < \infty \qquad n \in I$$

and

$$E(X_{n+1} | X_0, \ldots, X_n) = X_n \qquad n \in I$$

If the equality in the above equation is replaced by $\geq$ ($\leq$), the process is called a *supermartingale* (*submartingale*).

Martingales are important for many theoretical investigations; as yet, they rarely arise in operations research (an exception is noted in Section 11-2). Some of the processes studied in this book are martingales, but that is not their essential feature (for example, the symmetric simple random walk described in Section 7-8). Usually they have more structure than martingales. A discussion of martingales is given in Karlin and Taylor (1975, chap. 6).

3-3 NOTATION FOR MODELS OF QUEUES

Queueing models are important in their own right, and they are used as examples of various types of stochastic processes. The following notation for models of queues is used throughout this book:

$a \triangleq \lambda/\mu =$ offered load

$A(t) =$ number of arrivals during $(0, t]$

$B_i =$ length of ith busy period

$B(\cdot) =$ common distribution function of each B_i

$c =$ number of servers

$C_i =$ number of customers served in ith busy period

$D_n =$ delay (time in queue until service starts) of nth arriving customer

$F(\cdot) =$ distribution function of interarrival times

$G(\cdot) =$ distribution function of service times

$Q(t) =$ number in queue at time t

$R_n = n$th departure epoch

$T_n = n$th arrival epoch

$U_n = n$th interarrival time ($U_1 = T_1$, $U_n = T_n - T_{n-1}$ for $n \geq 2$)

$V_n = n$th service time

W_n = waiting time of customer n

$W(t)$ = work in system at time t

$\lambda \triangleq \lim_{t \to \infty} A(t)/t$ = arrival rate

$\mu \triangleq 1/E(V_1)$ = service rate

$v_G \triangleq E(V_1)$ = mean service time

$\rho \triangleq \lambda/(c\mu)$ = traffic intensity (when $c = 1$, $\rho = a$ and ρ is used for their common value)

Other symbols are defined as needed.

D. G. Kendall introduced a compact notational scheme to describe queueing models. The Kendall notation for a queueing model is $(\,\cdot\,)/(\,\cdot\,)/c/K$: The first position describes the arrival process, the second position describes the service time distribution, c is the number of servers, and K is the maximum number of customers that can be in queue or in service. It is always true that $K \geq c$; when $K = \infty$, it is often omitted. The following symbols are used to describe i.i.d. interarrival or service time distributions: M for exponential, D for deterministic, E_k for Erlang-k, and H_k for hyperexponential of order k.

When the service times have a general distribution, the symbol G is used; the symbol G is used, too, for interarrivals when they are not necessarily independent. When the interarrival times are i.i.d. with a general distribution, the symbol GI is used.

Unless an assumption to the contrary is explicitly stated, it is always assumed that the arrival times and service times are independent of one another.

BIBLIOGRAPHIC GUIDE

There are many good books on stochastic processes at all levels of mathematical sophistication. Here is a list of a few that explore some of the concepts covered in this chapter.

Intermediate level (Non-measure-theoretic)

Çinlar, Erhan: *Introduction to Stochastic Processes*, Prentice-Hall, Englewood Cliffs, N.J. (1975).
Cox, D. R., and H. D. Miller: *The Theory of Stochastic Processes*, Methuen, London (1965).
Karlin, Samuel, and Howard M. Taylor: *A First Course in Stochastic Processes*, 2d ed., Academic Press, New York (1975).
Parzen, Emanuel: *Stochastic Processes*, Holden-Day, San Francisco (1962).

Advanced level (Measure-theoretic)

Cramer, Harold, and M. R. Leadbetter: *Stationary and Related Stochastic Processes*, Wiley, New York (1967).
Doob, J. L.: *Stochastic Processes*, Wiley, New York (1953).
Feller, William: *An Introduction to Probability Theory and Its Applications*, vol. II, 2d ed., Wiley, New York (1971).
Wong, Eugene: *Stochastic Processes in Information and Dynamical Systems*, McGraw-Hill, New York (1971).

FOUR

BIRTH-AND-DEATH PROCESSES

4-1 INTRODUCTION

The birth-and-death process is a simple and very useful stochastic process. It is used extensively to represent queueing and inventory systems and machine and facility maintenance processes. Among its many other applications, as its name suggests, it is a model of a biological population whose size can both increase (there are births) and decrease (there are deaths). The analysis of specific birth-and-death processes illustrates many of the concepts encountered in more complicated stochastic processes. Some assumptions lead to easily obtained solutions with a simple form, while other assumptions require more advanced solution methods. Still other assumptions make the solution so complicated that it is not used for practical problems.

> **Example 4-1** As an example of the phenomena that are modeled with birth-and-death processes, consider the computer CPU in Section 2-2. It experiences an alternating sequence of up and down periods, namely, periods of satisfactory operation followed by repairs. Suppose the lengths of the ups are $U_1, U_2, \ldots$, and the lengths of the downs are $V_1, V_2, \ldots$. Assume that the U's and the V's are mutually independent exponential random variables with
>
> $$E(U_k) = \lambda^{-1} \qquad E(V_k) = \mu^{-1} \qquad k = 1, 2, \ldots \qquad (4\text{-}1)$$

Let $X(t) = 0$ if the CPU is up at time t and $X(t) = 1$ if it is down at time t. [Think of $X(t)$ as the number of CPUs being repaired at time t.] Suppose $X(0) = 0$. We wish to know such quantities as

$$P\{X(t) = 0 \mid X(0) = 0\}$$

$$\lim_{t \to \infty} P\{X(t) = 0 \mid X(0) = 0\}$$

and

$$\lim_{t \to \infty} [P\{X(t) = 0 \mid X(0) = 1\} - P\{X(t) = 0 \mid X(0) = 0\}]$$

These are, respectively, the probability that the CPU is up at time t, the "long-run probability" that the CPU is up, and the effect on this long-run probability of the initial status $X(0)$. We shall see that all three quantities are easy to compute because this example is a simple birth-and-death process. $\square$

In general, the sample path, or outcome, of a birth-and-death process has two notable features. First, at any epoch, the state is an element of $I = \{0, 1, \dots\}$. Second, transitions from one state occur only to neighboring ones; i.e., from $i \in I$ it is possible to move only to $i + 1$ and (if $i > 0$) $i - 1$. Every sample path in Example 4-1 has these features because either $X(t) = 1$ or $X(t) = 0$, so $\{0,1\} \subset I$ is the set of states and 0 and 1 are neighbors. Other salient properties of the birth-and-death process are deduced from a formal definition.

4-2 THE DEFINITION OF A BIRTH-AND-DEATH PROCESS

There are several equivalent definitions of a birth-and-death process. By *equivalent* we mean that a process which satisfies any one of the definitions can be shown, as a consequence, to satisfy the others. The following definition emphasizes the probabilistic basis of transition functions.

From Definition 3-2, a stochastic process $\{X(t); t \geq 0\}$ with state space $S \subset I$ is a *homogeneous Markov process* if for any $n \in I$, states $i_0, i_1, \dots, i_n, i,$ and j, and times $0 \leq \tau_0 < \tau_1 < \cdots < \tau_n < s,$

$$P\{X(t + s) = j \mid X(\tau_0) = i_0, \dots, X(\tau_n) = i_n, X(s) = i\}$$
$$= P\{X(t + s) = j \mid X(s) = i\} = P\{X(t) = j \mid X(0) = i\} \qquad (4\text{-}2)$$

Definition 4-1: Birth-and-death process The stochastic process $\{X(t); t \geq 0\}$ is a *birth-and-death process* with state space S if the following axioms are satisfied:

1. $S \subset I = \{0, 1, 2, \dots\}$.
2. It is a homogeneous Markov process.

3. There are nonnegative constants λ_j and μ_j, $j = 0, 1, \ldots$ with $\mu_0 = 0$, such that for $s > 0$ and $t \geq 0$, the following equations† hold:

$$P\{X(t + s) = j + 1 \mid X(t) = j\} = \lambda_j s + o(s) \qquad j \in I \qquad (4\text{-}3)$$

$$P\{X(t + s) = j - 1 \mid X(t) = j\} = \mu_j s + o(s) \qquad j \in I \qquad (4\text{-}4)$$

$$P\{X(t + s) = j \mid X(t) = j\} = 1 - (\lambda_j + \mu_j)s + o(s) \qquad j \in I \qquad (4\text{-}5)$$

We interpret $X(t)$ as the population size at time t. The constant λ_j is called the *birth rate* in state j because from (4-3), the probability that a population of size j will increase by 1 during a short time is essentially proportional to λ_j. Similarly, μ_j is called the *death rate* in state j. The boundary condition $\mu_0 = 0$ is natural because populations of size zero should not experience deaths. (Depending on your theology, you might not want to allow births in a population of size zero; it may be more apt to interpret λ_0 as an immigration rate to an empty population.)

We label the states by nonnegative integers for convenience. This restriction is not essential. For example, suppose each "birth" represents the birth of identical twins who also die together. In order to conform to the first axiom of Definition 4-1, the state of the population is the number of pairs of twins in the population (i.e., the total size of the population divided by 2). Any countable state space that is not naturally formulated as I can be transformed to I by a suitable relabeling of the states.

Observe that a birth-and-death process with a finite state space, e.g., Example 4-1, can be formally cast as a process with an infinite state space by setting $\lambda_j = 0$ and $\mu_j = 0$ for $j \notin S$ *and* $\lambda_m = 0$, where m is the largest state in S. For example, if $S = \{0, 1\}$, setting $\lambda_j = 0$ for $j \in I_+$ in (4-2) shows that states outside S cannot be entered. Since most results about birth-and-death processes are the same for finite and infinite state spaces, we wish to treat them simultaneously. Toward that end, we adopt this convention:

Convention When S is the finite set $\{0, 1, \ldots, m\}$, set $\lambda_j = 0$ for $j \geq m$ and $\mu_j = 0$ for $j \geq m + 1$ and interpret $\lambda_j / \mu_j = 1, j \geq m + 1$.

One property of birth-and-death processes follows immediately from the definition: By adding (4-3), (4-4), and (4-5) and recalling that $-o(s)$ is also $o(s)$, we have

$$P\{X(t + s) \in [j - 1, j + 1] \mid X(t) = j\} = 1 - o(s)$$

or $\qquad (4\text{-}6)$

$$P\{|X(t + s) - X(t)| \leq 1\} = 1 - o(s)$$

and so the probability that the population change exceeds unity during a short time is negligible.

---

† See Section A-5 for an explanation of the notation $o(s)$.

An important cog in the analysis of birth-and-death processes is the probability

$$P_{ij}(t) \triangleq P\{X(t) = j \mid X(0) = i\} \qquad t \geq 0$$

which is termed the *transition function*. The homogeneity part of the second axiom is equivalent to the following assertion:

$$P_{ij}(t) = P\{X(s + t) = j \mid X(s) = i\} \qquad s \geq 0, t > 0 \tag{4-7}$$

Therefore, (4-6) asserts

$$P_{i, i-1}(t) + P_{ii}(t) + P_{i, i+1}(t) = 1 - o(t) \qquad i \in I_+ \cap S$$

$$P_{00}(t) + P_{01}(t) = 1 - o(t)$$

and $\qquad P_{i, i+1}(t) = 0 \qquad i + 1 \notin S$

The unconditional probability of being in state j at time t is obtained from the transition function via

$$P_j(t) \triangleq P\{X(t) = j\} = \sum_{i \in S} P\{X(0) = i\} P_{ij}(t) \qquad t > 0 \tag{4-8}$$

The state at time zero is frequently known exactly, but it may be an r.v. in some applications. For example, at the start of a measurement period, the size of a geographically dispersed population may be subject to large measurement errors and modeled as a random variable.

Whenever a birth-and-death process changes its state, we say that a *transition* has occurred.

4-3 EXAMPLES

Example 4-2: Example 4-1 continued. The model of CPU reliability in Section 4-1 has up and down intervals distributed as independent exponential random variables. Recall $X(t)$ equals 0 or 1 to indicate whether the CPU is up or down. Now we show that $\{X(t); t \geq 0\}$ satisfies the axioms of a birth-and-death process with

$$\lambda_0 = \lambda \qquad \mu_0 = 0 \qquad \lambda_1 = 0 \qquad \mu_1 = \mu$$

We already observed that $X(t) \in \{0, 1\} \subset I$ for all t, so the first axiom of Definition 4-1 is valid. The second axiom is verified by appealing, as follows, to the memoryless property of exponential random variables (Section A-2).

Choose a time $s > 0$. The assumption that up and down periods are independent exponential random variables has the following implication. If $X(s) = 0$, then regardless of the lengths of up and down intervals prior to epoch s, and regardless of the elapsed length of the up interval in whose midst we find the process at epoch s, the remaining time until the CPU is down next has distribution function $1 - e^{-\lambda x}$ for $x \geq 0$. The entire lengths of subsequent up intervals have the same distribution. Also the future down intervals have distribution function $1 - e^{-\mu x}$, for $x \geq 0$, no matter what sample path has been realized during $[0, s]$. Similarly, if $X(s) = 1$, then the same argument is valid and the process is in the midst of a down interval whose residual length has distribution function $1 - e^{-\mu x}$ for $x \geq 0$. Thus, the process is Markov; since s was chosen arbitrarily, it is homogeneous.

Toward verifying the third axiom of Definition 4-1, let U and V be adjacent (indeed, any pair of) up and down intervals. Since U and V are independent,†

$$P\{U + V \leq s\} = \int_0^s P\{V \leq s - x\} \text{Pd}\{U = x\} \, dx$$

$$= \int_0^s (1 - e^{-\mu(s-x)})\lambda e^{-\lambda x} \, dx$$

$$= 1 + \frac{\lambda e^{-\mu s}}{\mu - \lambda} - \frac{\mu e^{-\lambda s}}{\mu - \lambda}$$

From the definition of the exponential function,

$$e^{-\mu s} \triangleq \sum_{n=0}^{\infty} \frac{(-\mu s)^n}{n!} = 1 - \mu s + s^2 \mu^2 \sum_{n=0}^{\infty} \frac{(-s\mu)^n}{(n+2)!}$$

The sum in the rightmost term approaches $\frac{1}{2}$, and $s^2/s \to 0$ as $s \downarrow 0$, hence

$$e^{-\mu s} = 1 - \mu s + o(s) \tag{4-9}$$

A similar result obviously holds for $e^{-\lambda s}$, hence

$$P\{U + V \leq s\} = \frac{1 + \lambda[1 - \mu s + o(s)]}{\mu - \lambda} - \frac{\mu[1 - \lambda s + o(s)]}{\mu - \lambda}$$

$$= o(s) \tag{4-10}$$

Since the probability that two or more transitions occur in a time interval of length s is no larger than the probability that exactly two transitions occur, it is $o(s)$.

In the verification of the second axiom, it was shown that if $X(t) = 0$, then the distribution function of the residual up time is $1 - e^{-\lambda x}$. In order that $X(t + s) = 0$ when $X(t) = 0$, either the same up is in effect at $t + s$ or at least two transitions occur during $(t, t + s)$. By (4-10), the probability of the latter is $o(s)$. Therefore,

---

† Recall that the notation $\text{Pd}\{U = x\}$ represents the probability density of the random variable U evaluated at x.

$$P\{X(t + s) = 0 \mid X(t) = 0\} = e^{-\lambda s} + o(s)$$

$$= 1 - \lambda s + o(s) \qquad (4\text{-}11)$$

$$= 1 - (\lambda_0 + \mu_0)s + o(s)$$

because $\lambda_0 = \lambda$ and $\mu_0 = 0$; and (4-11) conforms to (4-5).

Similarly, if $X(t) = 1$, then $X(t + s) = 1$ if, and only if, either the down in effect at t lasts at least until $t + s$ or an even number of transitions occurs during $(t, t + s]$. The probability of the former is $e^{-\mu s}$ (by the memoryless property and the definition of the distribution function of an exponential random variable), and the probability of the latter is $o(s)$ [by (4-10)]. Therefore,

$$P\{X(t + s) = 1 \mid X(t) = 1\} = e^{-\mu s} + o(s) = 1 - (\lambda_1 + \mu_1)s + o(s) \quad (4\text{-}12)$$

because $\lambda_1 = 0$ and $\mu_1 = \mu$. Taken together, (4-11) and (4-12) verify that (4-5) holds for this model.

If $X(t) = 0$, then $X(t + s) = 1$ if, and only if, either the subsequent down is in effect at $t + s$ or an odd number of transitions occurs during $(t, t + s]$. Let U' be the residual up time at t and V be the length of the subsequent down time. Then from (4-10) and $\lambda_0 = \lambda$,

$$P\{X(t + s) = 1 \mid X(t) = 0\} = \int_0^s P\{V > s - x\}\mathrm{Pd}\{U' = x\} \, dx + o(s)$$

$$= \int_0^s e^{-\mu(s-x)}\lambda e^{-\lambda x} \, dx + o(s)$$

$$= \frac{\lambda(e^{-\lambda s} - e^{-\mu s})}{\mu - \lambda} + o(s)$$

$$= \lambda_0 s + o(s)$$

which verifies (4-3).

If $X(t) = 1$, then $X(t + s) = 0$ only if either the subsequent up is in effect or an odd number of completions of ups and downs has occurred, so (4-10) and $\mu_1 = \mu$ yield

$$P\{X(t + s) = 0 \mid X(t) = 1\} = \int_0^s \mu e^{-\mu x}e^{-\lambda(s-x)} \, dx + o(s)$$

$$= \mu_1 s + o(s)$$

which verifies (4-4). All three axioms are valid. $\qquad \square$

Example 4-3: M/M/1 queue Recall the bank congestion model described in Section 2-3. There the time between the arrival of the ith and $(i + 1)$th customers was denoted u_i and the service time of the ith customer was v_i. Suppose for each i that u_i and v_i are outcomes of random variables U_i and V_i, that $U_1, U_2, \ldots, V_1, V_2, \ldots$ are mutually independent, and that

$$P\{U_i \leq a\} = 1 - e^{-\lambda a} \qquad P\{V_i \leq a\} = 1 - e^{-\mu a} \qquad a \geq 0, i \in I_+$$

Only for the moment, suppose that an arbitrarily large waiting room is available and that the queue discipline is FIFO (i.e., service in order of arrival). Let $X(t)$ denote the number of customers in the bank at time t. You should verify that the three axioms are satisfied with $\lambda_j = \lambda$ for all $j \in I$, $\mu_j = \mu$ for all $j \in I_+$, and $\mu_0 = 0$.

In this example, it is natural to interpret the number of customers present as the size of the population, the arrival of customers as births, and the service completions as deaths. The conditions $\mu_0 = 0$ and $\lambda_0 = \lambda > 0$ occur naturally because service cannot occur without a customer being present but a customer can arrive when the system is empty. □

Example 4-4: Machine repair model Example 4-2 is a simple version of a maintenance process that has one machine and one repairer. What if there are several machines and several repairers? For example, suppose the computer system has m identical input-output devices, say teletypewriters (TTYs), and that each device occasionally breaks down. After a device breaks down, it is repaired when a mechanic is free to work on it; afterward, it is put in service again.

Suppose newly repaired TTYs run for independent and identical exponentially distributed lengths of time before breaking down; let $1/\lambda$ be their common mean. We assume that there are n mechanics and that the times to repair TTYs depend on neither a mechanic's identity nor the TTY "serial number." Specifically, we assume that the repair times are i.i.d. exponential random variables with common mean $1/\mu$. Of course, if $m > n$, then occasionally a TTY may break down when n or more devices are already defective so all n mechanics are busy making repairs. In that case, the repair time of the newly defective TTY refers to the time required for a mechanic to fix it and does not include the delay until a mechanic is free to repair the TTY. We assume that mechanics, as they become free, work on failed TTYs in the same chronological order as the breakdowns occurred.

Let $X(t)$ be the number of inoperative TTYs at time t (i.e., being repaired or waiting to be repaired), with $X(0)$ specified. You should verify that $\{X(t), t \geq 0\}$ is a birth-and-death process. Here we specify the parameters $\{\lambda_j, \mu_j\}$. A TTY is a candidate for a breakdown only if it is running, so

$$\lambda_j = \lambda(m - j) \qquad j = 0, 1, \ldots, m$$

because $m - j$ devices are working when j devices are not working, and each working device fails at rate λ. Only those nonworking devices that are attended by mechanics can be repaired, and every repaired machine decreases the number of nonworking machines by 1, so

$$\mu_j = \mu \min \{j, n\} \qquad \qquad □$$

Example 4-5 The previous model also can be interpreted as a description of the use of a timeshared computer system. Suppose m terminals (TTYs) are attached to a computer that can simultaneously handle n active users; the interesting (and typical) situation is that $m > n$. If you log on and request to be processed when n or more other users already have requested service, then your request is held in a buffer until it can receive service. Here we interpret $X(t)$ as the total number of requests that are either being processed by the computer or in the buffer at time t. For the jth TTY, consider the set of times between completion of processing a request for a TTY and placement of its next request. We assume that the union of these sets, for $j = 1, 2, \ldots, m$, consists of i.i.d. exponential random variables with mean $1/\lambda$. Similarly, we assume that the processing times are i.i.d. exponential random variables with mean $1/\mu$.

In Example 4-4, the primary interest would be in the distribution of the number of operative TTYs, namely $P\{X(t) = m - j\}$ ($j = 0, 1, \ldots, m$), which is the probability that j machines are working. In this model, suppose you attempt to log on at time t. A quantity of primary interest is the probability that the computer is fully loaded, which is

$$\sum_{j=n}^{m} P\{X(t) = j\}. \qquad \square$$

Example 4-6: Population model The population dynamics in the fishery model in Section 2-7 are quite general and moderately informative at best. We begin a march to greater specificity with a relatively simple stochastic model for the number of fish in the central subpopulation of the northern anchovy (off Baja and southern California). Each fish, whatever its age (post-larva), is assumed to have probability $\mu \, \Delta t + o(\Delta t)$ of dying during the next Δt units of time; each female is assumed to have probability $\lambda \, \Delta t + o(\Delta t)$ of producing a single offspring during the next Δt units of time. Suppose that the proportion of females in the population is constant, and denote it by α. We assume that the population receives immigrants from other subpopulations at random times, the immigrants arrive one by one, and the times between them are i.i.d. exponential random variables with mean $1/\theta$. Finally, let K denote the *carrying capacity* of the anchovy's habitat.

Let $X(t)$ be the size of the population at time t. The probabilistic assumptions of the model can be used, as in Example 4-2, to verify that $\{X(t); t \geq 0\}$ is a birth-and-death process. The method of analysis used in Example 4-4 can be employed to obtain

$$\lambda_j = \theta + j\alpha\lambda \qquad j = 0, 1, \ldots, K$$

and $\lambda_j = 0$ if $j > K$; also,

$$\mu_j = \mu j \qquad j \in I_+$$

and $0 = \mu_0$. The qualitative structure of the parameters in a birth-and-death process influences the type of results one seeks. If $\theta > 0$, then $\lambda_0 > 0$, so a defunct population is regenerated. It is tempting to seek $P\{X(t) = j \mid X(0) = j\}$ for all $t > 0$ or at least for large values of t. How will the population evolve in the short run and what are prudent long-run predictions? If $\theta = 0$, then $\lambda_0 = 0$ and extinction is possible. Now one seeks the probability that the population will become extinct and the distribution of the time at which extinction occurs. $\square$

You probably noticed that in all the examples, the memoryless property of exponential r.v.'s is used to verify the Markov property. Surprisingly many random phenomena are adequately described by the exponential distribution function. A theoretical explanation is presented in Section 5-8.

Example 4-7: Pure birth process When $\mu_n = 0$ for all $n \in I_+$, deaths are impossible, so $\{X(t); t > 0\}$ counts the number of births and is called a *pure birth process*. When $\lambda_n = n\lambda$, $n \in I_+$, and $X(0) > 0$, the process is called a *Yule process*; when $\lambda_n \equiv \lambda$, it is called a *Poisson process*. The latter is one of the most widely used stochastic processes and describes many different phenomena. We refer to it many times in this book as we explore its many properties.

In a pure birth process there are two important quantities. One is the distribution of the number of births in a time interval; it is determined by $\{P_{ij}(t)\}$. The other is the distribution of time until the population reaches a given size; this is called the *first-passage-time* problem. Formally, we define the *first-passage time* from i to j $(j > i)$, T_{ij} by

$$T_{ij} = \inf\{t : X(t) = j \mid X(0) = i\}$$

Let $G_{ij}(t) = P\{T_{ij} \le t\}$; it can be expressed in terms of $P_{ij}(t)$ by the following device. Set $\lambda_n = 0$ for $n \ge j$, so there will be no more births once the population reaches size j. Then the events $\{T_{ij} \le t \mid X(0) = i\}$ and $\{X(t) = j \mid X(0) = i\}$ are equivalent, so that $G_{ij}(t) = P_{ij}(t)$. $\square$

The *pure death process* in which $\lambda_n = 0$ (for all n) is analogous to the pure birth process. A pure death process can be analyzed by constructing a pure birth process whose births occur in the same stochastic way as the deaths in the original process. Then the distribution of the number of deaths by time t in the original process will be the same as the distribution of the number of births by time t in the new process. To be specific, suppose $\{X(t); t \ge 0\}$ is a pure death process with parameters $\{\mu_n\}$ and $X(0) = i > 0$. (If $i = 0$, the process is trivial.) Define the pure birth process $\{Y(t); t \ge 0\}$ with parameter set $\{\lambda_n\}$, where $\lambda_0 = \mu_i$, $\lambda_1 = \mu_{i-1}, \ldots, \lambda_i = \mu_0 = 0$, and $Y(0) = 0$. The birth rate of the Y-process in state n is equal to the death rate of the X-process in state $i - n$. These rates determine the transition functions, so

$$P\{X(t) = j \mid X(0) = i\} = P\{Y(t) = i - j \mid Y(0) = 0\} \qquad 0 \le j \le i$$

EXERCISES

4-1 Verify that the stochastic processes described in Examples 4-3, 4-4, and 4-6 are birth-and-death processes.

4-2 Suppose $X = 0$ in Example 4-6, and let T_i denote the (random) epoch at which the population becomes extinct when $X(0) = i$. Is $P\{T_i \le t\} = P\{X(t) = 0 \,|\, X(0) = i\}$? Let T_{ij} denote the epoch at which the population achieves size j for the first time, $j > i$. Assume that $P\{T_i \le t\}$ can be found for any choice of birth rate and death rate. Show how such a result can be used to find $P\{T_{ij} \le t\}$.

4-3 The M/M/c *queue* is a congestion model in which customers arrive one at a time with the times between arrivals being i.i.d. exponential random variables having mean $1/\lambda$, say. There are c "servers," and each server processes one customer at a time. The service times, regardless of customer or server identity, are i.i.d. exponential random variables with mean $1/\mu$, say. Let $X(t)$ denote the number of customers who are either waiting in the queue or being processed by a server. Assume that no customer waits while a server is idle.

  (a) Show that $\{X(t); t \ge 0\}$ is a birth-and-death process, and specify the birth rate and death rate.

  (b) Let $Y(t)$ denote the number of customers waiting in the queue. Is $\{Y(t); t \ge 0\}$ a birth-and-death process? If so, specify its parameters.

4-4 A simple model of the central processor of a computer can be constructed from the M/M/1 queue. Let the interarrival times be generated as before; we interpret "arrivals" as programs needing computation. Each program will be processed for a random length of time, called a quantum, as in the M/M/1 queue. At the end of each quantum, either the program will be finished with probability p, in which case it leaves the system, or it will require more computation and joins the end of the queue. This procedure ensures that one long program will not delay much shorter ones. Let $X(t)$ be the number of programs in the system (waiting or being processed) at time t. Show that $\{X(t); t \ge 0\}$ is a birth-and-death process, and specify its parameters. What is the distribution of a program's number of passes through the central processor? Assuming exponentiality and independence, what data are needed to specify the birth and death parameters?

4-5 The motor pool of a company contains c cars. Users of the cars need them for random lengths of time. Potential users who arrive when all c cars are out use an outside rent-a-car agency. Construct a birth-and-death process to describe this operation. What distributional and independence assumptions are needed? If you were the manager of the motor pool, how would you define "the solution" of the model? How could you use this solution? How would you estimate the required parameters and check that the model "works"?

4-6 A small bank branch has a single teller. Arriving customers join the line in front of the teller with probability $p(j)$ when j customers are already in the line; otherwise, they deem the line too long and leave. Once on line, a customer waits until served or until his or her patience runs out. We assume that all customers are patient for amounts of time that can be viewed as i.i.d. exponential random variables with mean $1/\gamma$. Let N be the number of potential customers (i.e., people with accounts at the bank) and $X(t)$ the number of people at the bank at time t.

  What further assumptions are needed for $\{X(t); t \ge 0\}$ to be a birth-and-death process? What are the parameters? What changes if there are two tellers? Assume that your model adequately describes the operation of the bank and that you can calculate $P\{X(t) = j\}, j \in I$. What uses can be made of the model?

4-7 M/M/1/K **Queue** Consider a more realistic version of Example 4-3 in which the waiting room is finite, say K. Here we mean that the total of the number of people in the queue plus the customer being served cannot exceed K. Therefore, if the waiting room is full and someone arrives, then the arrival is turned away. Construct a birth-and-death process to describe this model.

4-8 Suppose that, at some transition epoch, $n > 0$ customers are present in an M/M/1 queue. Show that the time until the next transition occurs has an exponential distribution, and specify its mean. Explain why it is independent of the lengths between previous transitions and of the states at these transitions. What changes if $n = 0$? Choose another model in this section and repeat.

4-4 TRANSITION FUNCTION

The transition function $P_{ij}(\,\cdot\,)$ describes the stochastic evolution of a birth-and-death process, so the first task is to characterize it as much as possible. We obtain two systems of differential equations that are satisfied by $P_{ij}(\,\cdot\,)$ and conditions for these systems to have a unique solution.

First we observe that axiom 1 (Definition 4-1) yields

$$\{X(t + s) = j\} = \bigcup_{k \in S} \{X(t) = k, \, X(t + s) = j\}$$

so axiom 2 implies

$$P_{ij}(t + s) = \sum_{k \in S} P\{X(t + s) = j \mid X(0) = i, \, X(t) = k\}P\{X(t) = k \mid X(0) = i\}$$

or
$$P_{ij}(t + s) = \sum_{k \in S} P_{kj}(s)P_{ik}(t) = \sum_{k \in S} P_{ik}(s)P_{kj}(t) \qquad i, j \in S \qquad (4\text{-}13)$$

Equation (4-13) is called the Chapman-Kolmogorov equation for a birth-and-death process. Its interpretation is that the probability that the process goes from state i to state j by time $t + s$ is the probability that the process goes from i to some state k in time s and then goes from state k to state j in time t. The derivation of (4-13) uses only axiom 1 and axiom 2.

Backward Kolmogorov Equations

We may use (4-13) to obtain a system of equations that are satisfied by the transition function. To find $P_{ij}(t + s)$, divide the interval $(0, t + s]$ into two parts, $(0, s]$ and $(s, t + s]$, where s is small and $s, t > 0$. Write (4-13) as

$$P_{ij}(t + s) = \sum_{k: \, |k-i| \leq 1} P_{ik}(s)P_{kj}(t) + \sum_{k: \, |k-i| \geq 2} P_{ik}(s)P_{kj}(t) \qquad i, j \in S$$

From (4-5), because $P_{kj}(t) \leq 1$, we have

$$\sum_{k: \, |k-i| \geq 2} P_{ik}(s)P_{kj}(t) \leq \sum_{k: \, |k-i| \geq 2} P_{ik}(s) = o(s) \qquad i, j \in S$$

Using axiom 3, we expand the first term of $P_{ij}(t + s)$ and obtain [for $i \notin S$ set $P_{ij}(t) \equiv 0$ for all j and set $\mu_0 = 0$]

$$
\begin{aligned}
P_{ij}(t + s) &= [\mu_i s + o(s)]P_{i-1,\,j}(t) + [1 - (\lambda_i + \mu_i)s + o(s)]P_{ij}(t) \\
&\quad + [\lambda_i s + o(s)]P_{i+1,\,j}(t) + o(s) \\
&= \mu_i s P_{i-1,\,j}(t) + [1 - (\lambda_i + \mu_i)s]P_{ij}(t) + \lambda_i s P_{i+1,\,j}(t) \\
&\quad + o(s)
\end{aligned}
$$

for all pairs $i, j \in S$. Subtraction of $P_{ij}(t)$ from both sides and division by s gives

$$\frac{P_{ij}(t+s) - P_{ij}(t)}{s} = \mu_i P_{i-1,j}(t) - (\lambda_i + \mu_i)P_{ij}(t)$$

$$+ \lambda_i P_{i+1,j}(t) + \frac{o(s)}{s} \qquad (4\text{-}14a)$$

You should repeat this procedure on the intervals $(0, s]$ and $(s, t - s]$ and verify that

$$\frac{P_{ij}(t) - P_{ij}(t-s)}{s} = \mu_i P_{i-1,j}(t-s) - (\lambda_i + \mu_i)P_{ij}(t-s)$$

$$+ \lambda_i P_{i+1,j}(t-s) + \frac{o(s)}{s} \qquad (4\text{-}14b)$$

As $s \downarrow 0$, the left-sides of (4-14a) and (4-14b) become the left and right derivatives of $P_{ij}(t)$. Since the right-sides have the same limit, the derivative exists for all $t > 0$, and

$$\frac{d}{dt} P_{ij}(t) = \mu_i P_{i-1,j}(t) - (\lambda_i + \mu_i)P_{ij}(t) + \lambda_i P_{i+1,j}(t) \qquad i, j \in S \quad (4\text{-}15)$$

This is a system of *differential-difference equations* because the equation containing $P_{ij}(t)$ and $dP_{ij}(t)/dt$ also contains $P_{i-1,j}(t)$ and $P_{i+1,j}(t)$. These equations are called the *backward Kolmogorov equations*.

Forward Kolmogorov Equations

The forward Kolmogorov equations are similar to the backward equations. We begin by deriving an equation like (4-14a) that yields a right derivative. For (4-14a), the interval $(0, t + s]$ was divided into $(0, s]$ and $(s, t + s]$, with s small. Here we use the partition $(0, t]$ and $(t, t + s]$, with $s, t > 0$ and s small. From (4-13) and axiom 3,

$$P_{ij}(t+s) = \sum_{k:\,|k-j|\,\leqslant 1} P_{ik}(t)P_{kj}(s) + \sum_{k:\,|k-j|\,\geqslant 2} P_{ik}(t)P_{kj}(s)$$

As in the derivation of (4-15), we would like to assert that the second sum is $o(s)$, but it is not necessarily true without further assumptions. Since $P_{ik}(t) \leq 1$,

$$\sum_{k:\,|k-j|\,\geqslant 2} P_{ik}(t)P_{kj}(s) \leq \sum_{k:\,|k-j|\,\geqslant 2} P_{kj}(s)$$

Although, by axiom 3, each term of the sum is $o(s)$, the crux of the matter is that a denumerable number of $o(s)$ terms may not have a sum which is also $o(s)$ (see Exercise 4-1). [Why was this not an issue in the derivation of (4-14a)?] It can be shown† that a sufficient condition for the sum to be $o(s)$ is the following assumption:

Assumption For any $t > 0$, the probability that infinitely many transitions occur during $[0, t]$ is zero.

A necessary and sufficient condition for this assumption to hold is given in Theorem 4-1.

This assumption and axiom 3 yield, for $t, s > 0$,

$$P_{ij}(t + s) = P_{i, j-1}(t)[\lambda_{j-1}s + o(s)] + P_{ij}(t)[1 - (\lambda_j + \mu_j)s + o(s)]$$

$$+ P_{i, j+1}(t)[\mu_{j+1}t + o(s)] + o(s) \qquad i, j \in S$$

Subtraction of $P_{ij}(t)$ from both sides, collection of the $o(s)$ terms, and division by s yield

$$\frac{P_{ij}(t + s) - P_{ij}(t)}{s} = \lambda_{j-1}P_{i, j-1}(t) - (\lambda_j + \mu_j)P_{ij}(t)$$

$$+ \mu_{j+1}P_{i, j+1}(t) + \frac{o(s)}{s} \qquad i, j \in S$$

As $s \downarrow 0$, the left-side becomes the right derivative of $P_{ij}(t)$. An expression for the left derivative, comparable to (4-14), can be derived by partitioning $[0, t)$ into $[0, t - s)$ and $[t - s, t)$, with $t > 0$, $s > 0$, and s small (see Exercise 4-2). The left and right derivatives have the same value. This yields the *forward Kolmogorov equations*

$$\frac{d}{dt} P_{ij}(t) = \lambda_{j-1}P_{i, j-1}(t) - (\lambda_j + \mu_j)P_{ij}(t) + \mu_{j+1}P_{i, j+1}(t) \qquad i, j \in S \qquad (4\text{-}16)$$

The backward Kolmogorov equations are obtained by considering the ramifications of the first transition after time zero. The forward Kolmogorov equations are obtained by considering the ramifications of the last transition prior to t. If the Assumption did not hold, the last transition prior to t would not be well defined and the derivation would break down.

Digraph Representation and the Flow-of-Probability Argument

The forward Kolmogorov equations have an interpretation that is often useful in conceptualizing the transitions of a birth-and-death process. Suppose containers are marked 0, 1, 2, ..., one for every state, and a unit of fluid is placed in one of the containers at time zero. This fluid-flow system is portrayed in Figure 4-1. The circles represent containers, and the arcs represent pipes. The arrows on the pipes indicate the directions of flow, and the labels λ_j and μ_j are the rates (in units of, say, gallons per second per gallon of fluid) at which the contents of container j flow through the pipes leaving container j. The technical term for the object

---

† See Theorem 4 in Section 2.17 of Kai Lai Chung, *Markov Chains with Stationary Transition Probabilities*, Springer-Verlag, New York (1967).

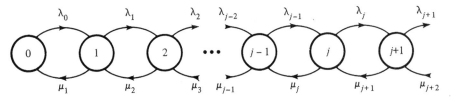

Figure 4-1 Digraph of a birth-and-death process.

, shown in Figure 4-1 is a *directed graph*, commonly abbreviated *digraph*. At time t, the rate of flow from container j to container $j + 1$ is the product of the amount of fluid in container j and λ_j. Similarly, the rate of flow from j to $j - 1$ is the product of the amount of fluid in container j and μ_j.

Let $P_{ij}(t)$ denote the amount of fluid in container j at epoch t when all the fluid is in container i at epoch 0. The rate of change of the content of container j is $dP_{ij}(t)/dt$, and this equals the rate of inward flow minus the rate of outward flow. Thus

$$\frac{d}{dt} P_{ij}(t) = \lambda_{j-1} P_{i,\,j-1}(t) + \mu_{j+1} P_{i,\,j+1}(t) - (\lambda_j + \mu_j) P_{ij}(t)$$

which is precisely (4-16). Thus, if we think of probability as a fluid, the forward Kolmogorov equations describe the "flow of probability," and they assert that the rate of change of the amount of this fluid equals the input rate minus the output rate. This observation and the digraph provide a heuristic method for obtaining the forward equations for particular birth-and-death processes. The process in Example 4-2 is portrayed simply by

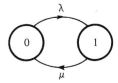

The pure birth process is represented by

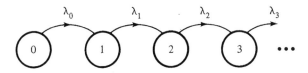

Existence and Uniqueness of Solutions

We choose the initial conditions for both (4-15) and (4-16) as

$$P_{ii}(0) = 1 \qquad P_{ij}(0) = 0 \qquad \text{for } j \neq i \tag{4-17}$$

The intuitive reason behind (4-17) is self-evident, and we know of no practical models where (4-17) is inappropriate. Since $P_{ij}(t)$ is a probability,

$$P_{ij}(t) \geq 0 \qquad t \geq 0 \tag{4-18}$$

The final condition imposed on the transition function is based on the observation that $\sum_{j=0}^{\infty} P_{ij}(t) = P\{X(t) \in I \mid X(0) = i\}$, which is unity unless the population may become infinite by time t. Thus,

$$\sum_{j=0}^{\infty} P_{ij}(t) \leq 1 \qquad t \geq 0 \tag{4-19}$$

The probabilistic evolution of a birth-and-death process is determined by the solution of (4-15) to (4-19). When you study differential equations, it is a particularly good idea to check that the equation has a unique solution (providing it has a solution at all) because pathologies are legion. The reward for this exploration is that multiple solutions of (4-15) through (4-19) can occur when the state space is infinite and the birth-and-death rates permit infinitely many births in a finite time. Then "state infinity" may be reached, and because no conditions on (4-15) and (4-16) are imposed "at infinity," multiple solutions are possible. If $\mathcal{S}$ is finite, as in Example 4-4, pathologies do not arise.

When $X(t; \omega) = \infty$, we say that the sample path has *exploded*. Although the possibility of explosions causes analytical difficulties, you should not a priori impose a finite state space on all models to avoid the difficulties. Real populations do explode (for all practical purposes); for example, the gypsy moth population in the United States has expanded rapidly because it lacks native natural predators. A reasonable model of this population probably ought not to preclude explosions.

Proposition 4-4 in Section 4-11 asserts that (w.p.1) an explosion occurs at t if, and only if, infinitely many transitions occur by t. Thus, the Assumption made earlier—for any $t > 0$, the probability that infinitely many transitions occur during $[0, t]$ is zero—holds if and only if (4-19) is an equality for all $t \geq 0$. The conditions needed for this to occur are given by Theorem 4-1.

Theorem 4-1 For any birth-and-death process with birth rates $\{\lambda_j\}$ and death rates $\{\mu_j\}, j \in I$, with $\mu_0 = 0$,

$$\sum_{j=0}^{\infty} P_{ij}(t) = 1 \qquad t \geq 0 \tag{4-19'}$$

(equivalently, $P\{$infinitely many transitions occur by $t\} = 0$) if, and only if,

$$\sum_{j=0}^{\infty} [b_j + (\lambda_j b_j)^{-1}] = \infty \tag{4-20}$$

where

$$b_0 = 1 \qquad b_j = \frac{\lambda_0 \lambda_1 \cdots \lambda_{j-1}}{\mu_1 \mu_2 \cdots \mu_j} \qquad j \in I_+ \tag{4-20'}$$

PROOF See Section 4-11.   □

Equation (4-20) has another important consequence.

Theorem 4-2 There is a unique solution of (4-15) through (4-19) if, and only if, (4-20) holds. Furthermore, the solution satisfies (4-19) with equality holding.

PROOF See Karlin and McGregor (1957a).   □

To our knowledge, there have been no applications which failed to satisfy (4-20). Karlin and McGregor (1957a) also constructed the general solution to (4-15) through (4-19), but an exposition of their method and the solution is beyond the scope of this book. Observe that if a unique solution of (4-15), [alternatively (4-16)] subject to (4-17), (4-18), and (4-19) is obtained, then the uniqueness part of Theorem 4-2 implies that this is a solution of (4-16) [alternatively (4-15)]. The solution in some special cases is not too difficult; perhaps the simplest solution is shown in Example 4-8.

Examples

Example 4-8: Examples 4-1 and 4-2 continued Recall the computer CPU which occasionally breaks down. The lengths of the time intervals when it operates properly are i.i.d. exponential random variables with mean $1/\lambda$. The repair times are i.i.d. exponential random variables with mean $1/\mu$.

Recall that $X(t) = 0$ means that the computer is working at time t and that $X(t) = 1$ means that it is being repaired at time t. Let 0 be the epoch at which the computer is installed and first ready to work, so that $P\{X(0) = 0\} = 1$. We know that $\{X(t); t \geq 0\}$ is a birth-and-death process with parameters $\lambda_0 = \lambda$, $\mu_1 = \mu$, and all other λ's and μ's zero. For convenience, use the notation $P_j(t) = P_{0j}(t)$. The forward equations (4-16) for this process are

$$\frac{d}{dt} P_0(t) = -\lambda P_0(t) + \mu P_1(t) \qquad t > 0$$

$$\frac{d}{dt} P_1(t) = -\mu P_1(t) + \lambda P_0(t) \qquad t > 0$$

Since $P_1(t) = 1 - P_0(t)$, the first equation becomes

$$\frac{d}{dt} P_0(t) = \mu - (\lambda + \mu)P_0(t) \qquad t > 0$$

with initial condition $P_0(0) = 1$. The solution of this differential equation is (see Example A-1)

$$P_0(t) = \frac{\mu}{\lambda + \mu} + \frac{\lambda}{\lambda + \mu} e^{-(\lambda + \mu)t} \qquad t \geq 0 \qquad (4\text{-}21a)$$

Use $P_1(t) = 1 - P_0(t)$ to obtain

$$P_1(t) = \frac{\lambda}{\lambda + \mu}(1 - e^{-(\lambda + \mu)t}) \qquad t \geq 0 \qquad (4\text{-}21b)$$

Equations (4-21a) and (4-21b) specify the state probabilities when $X(0) = 0$. For any time t, you can calculate the probability that the computer is working. A graphical version of (4-21a) and (4-21b) that shows how the state probabilities change over time when $\lambda = \frac{2}{3}$ and $\mu = \frac{1}{3}$ is shown in Figure 4-2. Notice that $P_0(t)$ and $P_1(t)$ each approach a limiting value as $t \to \infty$, and the approach occurs exponentially with time constant $\lambda + \mu$. The theory developed by Karlin and McGregor for the general birth-and-death process shows that this behavior is typical. □

The Poisson process is introduced briefly in Example 4-7 as a version of a pure birth process. It is used here to introduce a method of solving the forward Kolmogorov equations that is employed repeatedly in other chapters.

Example 4-9: Poisson process As described in Example 4-7, a Poisson process is a pure birth process with the same birth rate in all states, that is, $\mu_n \equiv 0$ and $\lambda_n \equiv \lambda$ for all $n \in I$. In this process, $X(t)$ is the number of births during $(0, t]$, and the object of interest is $P_j(t) = P\{X(t) = j \mid X(0) = 0\}$.

From the forward equation (4-16),

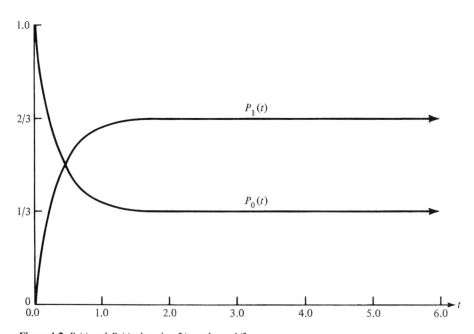

Figure 4-2 $P_0(t)$ and $P_1(t)$ when $\lambda = 2/\mu$ and $\mu = 1/3$.

$$\frac{d}{dt} P_j(t) = \lambda P_{j-1}(t) - \lambda P_j(t) \qquad j \in I_+ \tag{4-22a}$$

and

$$\frac{d}{dt} P_0(t) = -\lambda P_0(t) \tag{4-22b}$$

Equation (4-17) provides the initial condition

$$P_0(0) = 1 \qquad P_j(0) = 0 \qquad \text{for all } j \in I_+ \tag{4-22c}$$

The solution of (4-22b) and (4-22c) is

$$P_0(t) = e^{-\lambda t} \qquad t \geq 0 \tag{4-23}$$

To solve (4-22a), we introduce the generating function

$$\hat{P}(z; t) = \sum_{j=0}^{\infty} z^j P_j(t) \qquad 0 \leq z \leq 1$$

For the moment, assume that

$$\sum_{j=0}^{\infty} z^j \frac{d}{dt} P_j(t) = \frac{\partial}{\partial t} \hat{P}(z; t) \qquad 0 \leq z < 1 \tag{4-24}$$

is valid. For $z < 1$, multiply the jth equation in (4-22a) by z^j and sum from 0 to ∞ [include $j = 0$ in the sum and use (4-22b)] to obtain

$$\frac{\partial}{\partial t} \hat{P}(z; t) = \lambda \sum_{j=1}^{\infty} z^j P_{j-1}(t) - \lambda \sum_{j=0}^{\infty} z^j P_j(t)$$

$$= \lambda z \hat{P}(z; t) - \lambda \hat{P}(z; t) = \lambda(z - 1)\hat{P}(z; t) \tag{4-25}$$

One can use Theorem 4-1 to show that (4-19) holds as an equality when $\lambda < \infty$ (see also Theorem 4-13 in Section 11), so

$$\hat{P}(1; t) = \sum_{j=0}^{\infty} P_j(t) = 1$$

The solution of (4-25) with this initial condition is

$$\hat{P}(z; t) = e^{\lambda t(z-1)} = e^{-\lambda t} e^{\lambda t z} \tag{4-26}$$

Expansion of the second term on the right-side in Maclaurin's series yields

$$\hat{P}(z; t) = e^{-\lambda t} \sum_{j=0}^{\infty} \frac{(\lambda t z)^j}{j!}$$

Since $P_j(t)$ is the coefficient of z^j in $\hat{P}(z; t)$,

$$P_j(t) = \frac{e^{-\lambda t}(\lambda t)^j}{j!} \qquad j \in I \tag{4-27}$$

The fact that (4-27) is the Poisson distribution gives this process its name.

We use Propositions A-5 and A-7 to prove that (4-24) is valid. Let $f_n(t) = \sum_{j=0}^{n} z^j P_j(t)$. From (4-22a) and (4-22b) we obtain, as in the derivation of (4-25),

$$\left| \frac{d}{dt} f_n(t) \right| \triangleq \left| \sum_{j=0}^{n} z^j \frac{d}{dt} P_j(t) \right|$$

$$= |\lambda(z-1)[P_0(t) + zP_1(t) + \cdots + z^{n-1}P_{n-1}(t)] - \lambda z^n P_n(t)|$$

$$\leq |\lambda(z-1)(1 + z + \cdots + z^{n-1})| + |\lambda z^n|$$

$$= 2\lambda z^n$$

From Proposition A-5, this shows that $f_n(t)$ is uniformly convergent. Hypotheses 1 and 2 of Proposition A-7—that $f_n(\cdot)$ is differentiable and converges for some value of t—are easy to verify in this example, and we have just shown that hypothesis 3 holds; therefore (4-24) is valid. □

EXERCISES

4-9 Give an example of a sequence of functions $\{f_j(s)\}$ each of which is $o(s)$ but where $\sum_{j=1}^{\infty} f_j(s)$ is not $o(s)$.

4-10 Starting from the Chapman-Kolmogorov equation, derive the equation for

$$[P_{ij}(t) - P_{ij}(t-s)]/s$$

and complete the derivation of (4-15).

4-11 In Example 4-8, use a symmetry argument to obtain $P_{10}(t)$ and $P_{11}(t)$. Check your answer by showing that it satisfies the backward Kolmogorov equations.

4-12 (a) Solve (4-22) by Laplace transform methods. *Hint:*

$$\text{Show } \tilde{P}_0(s) = 1/(s + \lambda) \quad \text{and} \quad (s + \lambda)P_j(s) = \lambda \tilde{P}_{j-1}(s) \text{ for } j \in I_+$$

Iterate on j, show $\tilde{P}_j(s) = \lambda^j/(s + \lambda)^{j+1}, j \in I$, and invert by inspection.

(b) Solve (4-22) directly. *Hint:* Use the fact that the solution of the differential equation

$$\frac{d}{dt} y(t) + \lambda y(t) = \frac{\lambda(\lambda t)^{j-1} e^{-\lambda t}}{(j-1)!} \quad \text{with} \quad y(0) = 0$$

is $y(t) = (\lambda t)^j e^{-\lambda t}/j!$.

4-13 Animals arrive in a section of a national park one at a time, and the times between the arrival epochs have an exponential distribution. Assume that the section of the park is large enough that there is no effective bound on the number of animals that can be in it at any given time. Suppose each animal stays in this part of the park for a length of time that has an exponential distribution and is independent of the comings and goings of the other animals. How would you denote this model with the queueing terminology in Section 3-3? Let $\hat{P}(z; t) = \sum_{j=0}^{\infty} P_{0j}(t)z^j$; derive a partial differential equation for $\hat{P}(z; t)$ and solve it if you can. (To solve it, you should know Lagrange's method of characteristics, which is explained in an early chapter of any introduction to partial differential equations.)

4-14 Draw the digraphs for the birth-and-death processes described in Examples 4-2 to 4-4, and write the corresponding forward and backward equations. Under what conditions do these equations have a unique solution?

4-15 Construct parameter sets $\{\lambda_n\}$ and $\{\mu_n\}$ such that the Kolmogorov equations do *not* have a unique solution.

4-16 Let $\{X(t); t \geq 0\}$ be a birth-and-death process with $X(0) = 1$ and parameters $\lambda_n = n\lambda$ and $\mu_n = n\mu, n \in I$. Define $M(t) = E(X(t) | X(0) = 1)$; show that $M(t) = e^{(\lambda - \mu)t}, t \geq 0$.

4-5 STEADY-STATE PROBABILITIES

In many applications we are interested primarily in the long-run behavior of a birth-and-death process. For instance, in Example 4-2, suppose that order-of-magnitude estimates of $1/\lambda$ and $1/\mu$ are one week and one hour, respectively. Typically, the model would be used to find the proportion of time the computer works during a year. Indeed, even when we are not interested primarily in long-run results, the Kolmogorov equations may not yield tractable solutions for the short-run (or *transient*) state probabilities. In these situations, we get what we can from the model, and we always can calculate the long-run (or *asymptotic* or *steady-state* or *limiting*) probabilities. Often the long-run state probabilities are a useful surrogate for the transient state probabilities because the latter frequently approach the former exponentially, as illustrated in Example 4-2.

Definitions

Specifically, what is a steady-state probability? Intuitively, a system is "in the steady state" either when the initial conditions cease to influence its behavior or when the rate of change of its descriptors is zero. We adopt the former definition and derive the latter as a consequence.

Definition 4-2 The *steady-state* (or *limiting*) *probability* of state j is p_j if

$$p_j = \lim_{t \to \infty} P_{ij}(t) \qquad \text{for all } i \in \mathcal{S} \tag{4-28}$$

The set $\{p_j\}$ is called the *steady-state distribution*.

We interpret p_j as the asymptotic probability of being in state j.

Recall the Chapman-Kolmogorov equation:

$$P_{ij}(s + t) = \sum_{k \in \mathcal{S}} P_{ik}(s)P_{kj}(t) \tag{4-29}$$

Fix t and let $s \to \infty$ on both sides of (4-29); then when (4-28) is valid (Exercise 4-17)

$$p_j = \sum_{k \in \mathcal{S}} p_k P_{kj}(t) \qquad j \in \mathcal{S} \tag{4-30}$$

for each $t \geq 0$.

Suppose we are modeling a process that has been evolving for a long time. For example, in the model of CPU reliability in Example 4-2, suppose the CPU was installed several years ago and has experienced many up and down intervals. We may choose to model the current state $X(0)$ as a random variable with $P\{X(0) = j\} = p_j$.

For any birth-and-death process where (4-28) is valid, if $P\{X(0) = i\} = p_i$, then (4-30) and (4-8) yield

$$P\{X(t) = j\} = \sum_{i \in \mathcal{S}} p_i P_{ij}(t) = p_j \qquad t \geq 0 \tag{4-31}$$

Thus, with this choice of initial probabilities, $P\{X(t) = j\}$ is stationary in time, and we say that the steady-state probabilities are also *stationary probabilities*.

Here is another version of stationarity.

Proposition 4-1 If the limits in (4-28) exist and are independent of the initial state, then

$$\lim_{t \to \infty} \frac{d}{dt} P_{ij}(t) = 0 \qquad \text{for all } i, j \in \mathcal{S} \tag{4-32}$$

PROOF Take limits on both sides of the backward equations (4-15) to obtain

$$\lim_{t \to \infty} \frac{d}{dt} P_{ij}(t) = \mu_i p_j - (\lambda_i + \mu_i)p_j + \lambda_i p_j = 0 \qquad \text{for all } i, j \in \mathcal{S} \qquad \square$$

Flow Interpretations

Applying (4-32) to the forward equations (4-16) yields

$$(\lambda_j + \mu_j)p_j = \lambda_{j-1}p_{j-1} + \mu_{j+1}p_{j+1} \qquad j \in \mathcal{S} \tag{4-33}$$

This equation has an intuitive explanation in terms of flows. The left-side is the rate of flow of probability out of state j, the right-side is the rate of flow of probability into state j, and (4-33) asserts that these rates of flow are equal in the steady state.

If the rate of flow into state j equals the rate of flow out of state j, it seems reasonable that the net flow rate between adjacent states must be zero; see Figure 4-3. That is, the net flow rate across each dotted line in Figure 4-3 should be zero. Algebraically, this is

$$\lambda_j p_j = \mu_{j+1}p_{j+1} \qquad j \in \mathcal{S} \tag{4-34}$$

Proposition 4-2 Equations (4-33) and (4-34) are equivalent.

PROOF Assume (4-33) is true. Summing both sides of (4-33) over $j = 0, 1, \ldots,$ k for any $k \in \mathcal{S}$ yields

$$\sum_{j=0}^{k} (\lambda_j + \mu_j)p_j = \sum_{j=0}^{k-1} \lambda_j p_j + \sum_{j=1}^{k+1} \mu_j p_j$$

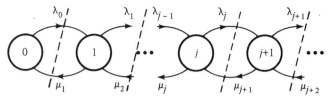

Figure 4-3 Flow diagram of Equation (4-34).

Canceling like terms on both sides leaves

$$\lambda_k p_k = \mu_{k+1} p_{k+1} \qquad k \in \mathcal{S}$$

which, except for the name of the dummy variable, is (4-34). Now assume (4-34) is true. We add $\mu_j p_j$ to both sides, obtaining

$$(\lambda_j + \mu_j) p_j = \mu_{j+1} p_{j+1} + \mu_j p_j$$

Evaluating (4-34) at $j - 1$ gives $\mu_j p_j = \lambda_{j-1} p_{j-1}$, and (4-33) is obtained. $\square$

Explicit Formula and Existence Conditions

From (4-34), $\mu_{j+1} = 0$ implies $p_j = 0$ and $\lambda_j = 0$ implies $p_{j+1} = 0$, and the limiting probabilities of a pure birth or a pure death process are obtained immediately. These processes are excluded in the remainder of this section.

Proposition 4-2 provides an easy way to obtain $\{p_j\}$ when the forward equations are valid. The solution to (4-34) is obtained in terms of p_0. From (4-34), with $j = 0$,

$$p_1 = \frac{p_0 \lambda_0}{\mu_1} = b_1 p_0$$

where b_1 is defined by (4-20a). Setting $j = 1$ in (4-34) yields

$$p_2 = \frac{p_1 \lambda_1}{\mu_2} = b_2 p_0$$

A simple induction proves

$$p_j = b_j p_0 \qquad j \in \mathcal{S} \tag{4-35}$$

where b_j is given by (4-20'). From (4-35) we obtain (recall that $b_0 = 1$)

$$\sum_{j \in \mathcal{S}} p_j = p_0 \sum_{j \in \mathcal{S}} b_j$$

and

$$p_j = b_j \frac{\sum_{k \in \mathcal{S}} p_k}{\sum_{k \in \mathcal{S}} b_k} \qquad j \in \mathcal{S} \tag{4-36}$$

If

$$\sum_{j \in \mathcal{S}} p_j = 1 \tag{4-37}$$

is valid, then (4-36) shows that

$$p_j = \frac{b_j}{\sum_{k \in \mathcal{S}} b_k} \qquad j \in \mathcal{S} \tag{4-38}$$

Furthermore, we must have $\sum_{k \in \mathcal{S}} b_k < \infty$; otherwise, $p_j \equiv 0$, which violates the hypothesis that $\sum_{j \in \mathcal{S}} p_j = 1$. However, if $\sum_{k \in \mathcal{S}} b_k < \infty$, then (4-38) satisfies (4-35) and (4-37). Thus we have established Theorem 4-3.

Theorem 4-3 If the limits $\{p_j\}$ given by (4-28) exist, then they are given by (4-38). Moreover, they form an honest probability distribution if, and only if,

$$\sum_{j \in \mathcal{S}} b_j < \infty \tag{4-39}$$

There still is the question of the existence of limiting probabilities. The answer is provided by Theorem 4-4.

Theorem 4-4 The limits $\{p_j\}$ given by (4-28) exist if, and only if,

$$\sum_{j \in \mathcal{S}} (\lambda_j b_j)^{-1} = \infty \tag{4-40}$$

PROOF See Karlin and McGregor (1957*b*). □

Combining Theorems 4-3 and 4-4 yields Corollaries 4-4*a* and *b*.

Corollary 4-4*a* The limits $\{p_j\}$ given by (4-28) exist and form an honest probability distribution if, and only if, (4-39) and (4-40) both hold.

It is not always necessary to check that (4-40) holds because of Corollary 4-4*b*.

Corollary 4-4*b* The limits $\{p_j\}$ given by (4-28) exist and form an honest probability distribution if (4-39) holds and $\{\lambda_j\}$ is bounded.

PROOF See Exercise 4-29. □

Combining Theorems 4-2, 4-3, and 4-4 yields the following: If (4-39) and (4-40) hold, then (1) the Kolmogorov equations have a unique solution, (2) the steady-state probabilities exist, and (3) they form an honest probability distribution.

Example 4-10: M/M/c queue The M/M/c queue is described in Exercise 4-3. It is a birth-and-death process with $\lambda_n \equiv \lambda$ and $\mu_n = n\mu$ for $0 \leq n \leq c$ and $\mu_n = c\mu$ for $n \geq c$. Hence $b_j = \lambda^j/(j!\mu^j)$ for $0 \leq j \leq c$, and $b_j = \lambda^j/(c!c^{j-c}\mu^j)$ for $j \geq c$. And so

$$\sum_{j=0}^{\infty} (\lambda_j b_j)^{-1} = \sum_{j=0}^{c-1} \frac{j!\mu^j}{\lambda^{j+1}} + \lambda^{-1}c!c^{-c} \sum_{j=c}^{\infty} \left(\frac{c\mu}{\lambda}\right)^j$$

The first sum is obviously finite, and the second sum diverges if, and only if, $c\mu/\lambda \geq 1$, so $\lambda < c\mu$ is necessary and sufficient for the limiting probabilities to exist. This condition has the intuitive meaning that the system behaves nicely when the mean time between arrivals $(1/\lambda)$ is larger than the mean time between services $[1/(c\mu)]$ when all the servers are working. This condition traditionally is expressed by defining $\rho = \lambda/(c\mu)$ and requiring $\rho < 1$. We call ρ the *traffic intensity* of the queue.

To verify (4-39), compute

$$\sum_{j=0}^{\infty} b_j = \sum_{j=0}^{c-1} \frac{a^j}{j!} + \frac{c^c}{c!} \sum_{j=c}^{\infty} \left(\frac{a}{c}\right)^j$$

where $a = \lambda/\mu$. The first sum is finite, and the second sum is finite if, and only if, $a/c < 1$, so again $\rho < 1$ is necessary and sufficient for nice behavior. Compute the geometric sum above and use (4-38) to obtain

$$p_0 = \frac{1}{\sum_{j=0}^{\infty} b_j} = \left[\sum_{j=0}^{c-1} \frac{a^j}{j!} + \frac{ca^c}{c!(c-a)} \right]^{-1} \tag{4-41}$$

and

$$p_j = \begin{cases} \dfrac{p_0 a^j}{j!} & 1 \leq j \leq c \\[2mm] \dfrac{p_0 a^j}{c^{j-c}c!} & j \geq c \end{cases} \tag{4-42}$$

Observe that the limiting probabilities depend on λ and μ only through their ratio a, which is the only quantity that needs to be measured. From (4-42) we can compute the average number of customers in the system, L say, from $L = \sum_{j=1}^{\infty} jp_j$. This is just an exercise in manipulating geometric sums and is left to you (see Exercise 4-23).

It is more interesting to compute the probability that all the servers are busy. From (4-42) this probability depends on only a and c, so we denote it by $C(c, a)$:

$$C(c, a) = \sum_{j=c}^{\infty} p_j = \frac{p_0 a^c}{c!} \sum_{j=c}^{\infty} \left(\frac{a}{c}\right)^{j-c}$$

$$= \frac{p_0 ca^c}{c!(c-a)} \tag{4-43}$$

Expression (4-43) is called the *Erlang delay formula* in honor of A. K. Erlang, who did the earliest significant work in queueing theory.

From (4-43), the average number of working servers, denoted by L_b, is

$$L_b = \sum_{j=0}^{c-1} jp_j + cC(c, a)$$

$$= p_0\left[\sum_{j=1}^{c-1} \frac{a^j}{(j-1)!} + \frac{ca^c}{(c-1)!(c-a)}\right]$$

$$= ap_0\left[\sum_{j=0}^{c-2} \frac{a^j}{j!} + \frac{ca^{c-1}}{(c-1)!(c-a)} + \frac{a^{c-1}}{(c-1)!} - \frac{a^{c-1}}{(c-1)!}\frac{c-a}{c-a}\right]$$

$$= ap_0\left[\sum_{j=0}^{c-1} \frac{a^j}{j!} + \frac{ca^c}{c!(c-a)}\right] = a \qquad (4\text{-}44)$$

with the last equality following from (4-41). Because of (4-44), a is called the *load* on the queue. □

EXERCISES

4-17 Derive (4-30) from (4-28) and (4-29) when the state space is finite. What difficulties arise when the state space is infinite?

4-18 Show that (4-30) is satisfied by the solution to Example 4-2.

4-19 For Examples 4-2, 4-3, and 4-4, (a) state what conditions (if any) have to be placed on the parameters so that a steady-state distribution exists, and (b) obtain the steady-state distributions.

4-20 Show that if $\lambda_j/\mu_j \leq 1$, $j \in I_+$, then (4-40) holds, so that a steady-state distribution exists. Construct an example to demonstrate that the steady-state distribution need not be proper, even if $\lambda_j/\mu_j \leq \rho < 1$, where ρ is some positive number.

4-21 Construct parameter sets $\{\lambda_n\}$ and $\{\mu_n\}$ such that (4-39) holds but (4-40) does not. This will show that values for the limiting probabilities can be obtained when the limits do not exist, so Theorem 4-3 is important.

4-22 For the M/M/1 queue, use (4-41) and (4-42) to show that $p_j = (1-\rho)\rho^j$, $L = \sum_{j=1}^{\infty} jp_j = \rho/(1-\rho)$. What are the distribution and mean value of the number in the queue alone (i.e., do not count the customer in the server, if any)? Assume that $\rho < 1$, so that the quantities discussed above exist.

4-23 Let p_j be given by (4-42) and $L = \sum_{j=1}^{\infty} jp_j$. Show that $L = a + p_0(a^{c+1}/c)/[c!(1-\rho)^2]$, where $\rho = a/c$. Calculate $\sum_{j=c}^{\infty}(j-c)p_j$ in terms of $C(c, a)$. What does this quantity represent?

4-24 Suppose we modify the description of the M/M/c queue so that customers who arrive when all the servers are busy are rejected. This system is denoted M/M/c/c and is called *Erlang's loss model*. It was used by Erlang to describe the operation of a telephone switchboard where it is possible to have up to c calls connected simultaneously but calls cannot wait in line for a server. The births represent requests for a connection, and the deaths represent call completions. A customer who tries to place a call when c calls are in progress receives a busy signal, and the state of the process is the number of calls in progress. This model arises in many contexts, and it appeared in Exercise 4-5. Find the limiting probabilities $p_0, p_1, \ldots, p_c$. For which values of a are these formulas valid?

4-25 (Continuation) The probability that all servers are busy is especially significant and is called *Erlang's loss formula*, denoted by $B(c, a)$. The solution to Exercise 4-24 gives

$$B(c, a) = \frac{a^c/c!}{\sum_{k=0}^{c} a^k/k!}$$

It is inefficient to program a computer to calculate $B(c, a)$ directly from this formula, and numerical problems are apt to arise because $k!$ can be so large that overflows occur. To overcome these difficulties, a recursion is used. Show that $B(c, a)$ satisfies the recursion

$$B(c, a) = \frac{aB(c - 1, a)}{c + aB(c - 1, a)} \qquad c = 1, 2, \dots ; B(0, a) = 1$$

4-26 A simple way to compute $C(c, a)$ is to compute $B(c, a)$ recursively and then use the identity

$$C(c, a) = \frac{aB(c - 1, a)}{aB(c - 1, a) + c - a}$$

Establish this identity and use it to show that $C(c, a) > B(c, a)$. Give an intuitive argument for this inequality.

4-27 How would you compute p_j in the M/M/c queue?

4-28 (Continuation of Exercise 4-7) Find steady-state probabilities for the single-server queueing model in Exercise 4-7 whose waiting room is finite, that is, $X(t) \le k$, where $X(t)$ is the number of customers inside the facility. Under what circumstances do the model's parameters satisfy (4-40)? Show that your answer satisfies

$$\rho = 1 - p_0 + \rho p_n \qquad \rho = \frac{\lambda}{\mu}$$

Using a flow argument, derive this equation directly, and give an interpretation of it.

4-29 Prove Corollary 4-4b.

4-30 Let system 1 be an M/M/1 queue with arrival rate λ and mean service time $1/\mu$; and let system 2 be an M/M/2 queue with arrival rate λ and mean service time $2/\mu$, $\lambda/\mu < 1$. Which system has a smaller mean number of customers present in the steady state? What about the mean number of customers in the queue? Justify your answer intuitively and by calculation. Make the analogous comparisons between an M/M/1 queue and an M/M/c queue.

4-6* START-UP AND SHUT-DOWN IN QUEUES

Birth-and-death processes occupy a central position in the armamentarium of stochastic models. This status is due partly to usefulness of the methods developed for birth-and-death processes in other kinds of models. In this section we analyze a queueing model in which a simple policy is used to decide when service is to be provided. The model is a generalization of Example 4-3, but it is not a birth-and-death process.

Consider a repair facility in which the single repairer, i.e., "server," can be productive at other tasks when not performing repairs. The economics of such situations sometimes lead to an operating regime for which Example 4-3 is *not* a good model. What happens, instead, is that the number of items awaiting repair is permitted to grow for a while before the repairer starts to repair them. Thereafter, the repairer continues to make repairs until no items still await attention. Then the repairer engages in other tasks until sufficiently many repairable items have arrived to warrant starting up the repair process, the repair process resumes, and the cycle repeats itself.

Let U_i denote the time between the arrivals of the ith and $(i + 1)$th items needing repairs, and let V_i denote the time needed to repair the ith item to arrive

(V_i does not include the delay until the repairer finally starts to repair item i). Suppose that U_1, U_2, ..., V_1, V_2, ... are mutually independent exponential random variables and that $E(U_i) = \lambda^{-1}$ and $E(V_i) = \mu^{-1}$ for all $i = 1, 2, ... (\lambda > 0$ and $\mu > 0$).

Let s be the number of waiting items that warrants initiation of the repair activity, let $X(t)$ denote the number of items in the facility at time t, and let $Z(t)$ indicate whether a repair is being made at time t. Specifically, $Z(t) = 1$ if a repair is being made at time t, and $Z(t) = 0$ otherwise. We think of the pair $(X(t), Z(t))$ as the *state* of the repair facility. Note that $X(t)$ alone is inadequate to specify probabilities for $X(t + s)$, $s > 0$. For example, if $s = 10$ and $X(t) = 5$, then it is possible that the facility was recently empty, so $Z(t) = 0$ and the repairer is engaged elsewhere until $s = 10$ items have arrived. In this case, $X(\cdot)$ is temporarily behaving like a Poisson process because no "deaths" can occur. However, if $Z(t) = 1$ and the repairer is striving to empty the facility, then $X(\cdot)$ is temporarily behaving like Example 4-3, which is a bona fide birth-and-death process.

It is instructive to draw a fluid-flow diagram for the states $\{(X(\cdot), Z(\cdot))\}$. Comparison of Figures 4-4 and 4-1 leads to the observation that the current model is a birth-and-death process except for the complication at state $(s, 1)$. Specifically, the state space here is denumerable, so it can be mapped into I. Also, we have a homogeneous Markov process, so axioms 1 and 2 in Section 4-2 are

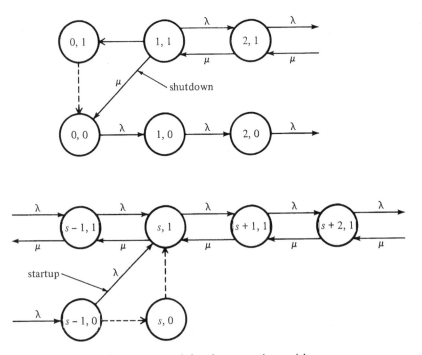

Figure 4-4 Diagraph for a start-up and shut-down queueing model.

satisfied. In place of axiom 3, there are

$$P\{X(t + r) = j + 1, \ Z(t + r) = z \mid X(t) = j, \ Z(t) = z\} = \lambda r + o(r)$$

$$\text{if } z = 1 \text{ and } j \in I \qquad \text{or if } z = 0 \text{ and } j < s - 1 \quad (4\text{-}45)$$

$$P\{X(t + r) = j - 1, \ Z(t + r) = 1 \mid X(t) = j, \ Z(t) = 1\}$$

$$= \mu r + o(r) \qquad \text{if } j > 1 \quad (4\text{-}46)$$

$$P\{X(t + r) = j, \ Z(t + r) = z \mid X(t) = j, \ Z(t) = z\}$$

$$= \begin{cases} 1 - \lambda r + o(r) & \text{if } j \leq s - 1 \text{ and } z = 0 \\ 1 - (\lambda + \mu)r + o(r) & \text{if } j > 0 \text{ and } z = 1 \end{cases} \quad (4\text{-}47)$$

$$P\{X(t + r) = s, \ Z(t + r) = 1 \mid X(t) = s - 1, \ Z(t) = 0\} = \lambda r + o(r) \quad (4\text{-}48)$$

$$P\{X(t + r) = Z(t + r) = 0 \mid X(t) = Z(t) = 1\} = \mu r + o(r) \quad (4\text{-}49)$$

Our derivation of the properties of $\{(X(\cdot), Z(\cdot))\}$ parallels Section 4-4 and is only sketched. Let

$$P_{ij}^{yz}(t) = P\{X(t) = j, \ Z(t) = z \mid X(0) = i, \ Z(0) = y\}$$

Of course, $y = 1$ necessarily unless $i \leq s - 1$. (Why?) Only axioms 1 and 2 were used to derive (4-13), the Chapman-Kolmogorov equations. Similarly, since axioms 1 and 2 apply to the present model, for $r > 0$ and $t \geq 0$,

$$P_{ij}^{yz}(t + r) = \sum_{m=0}^{1} \sum_{k=0}^{\infty} P_{kj}^{mz}(t) P_{ik}^{ym}(r) \quad (4\text{-}50)$$

To obtain backward equations such as (4-15), partition $(0, t + r]$ into $(0, r]$ and $(r, t + r]$, where r is small. Then use (4-45) through (4-49) in (4-50) as (4-2) through (4-4) were used in (4-13), to find that the right-hand derivative satisfies

$$\frac{d}{dt} P_{ij}^{0z}(t) = \lambda P_{i+1, j}^{0z}(t) - \lambda P_{ij}^{0z}(t) \qquad i < s - 1$$

$$\frac{d}{dt} P_{s-1, j}^{0z}(t) = \lambda P_{sj}^{1z}(t) - \lambda P_{s-1, j}^{0z}(t)$$

$$(4\text{-}51)$$

$$\frac{d}{dt} P_{ij}^{1z}(t) = \lambda P_{i+1, j}^{1z}(t) + \mu P_{i-1, j}^{1z}(t) - (\lambda + \mu) P_{ij}^{1z}(t) \qquad i > 1$$

$$\frac{d}{dt} P_{1j}^{1z}(t) = \lambda P_{2j}^{1z}(t) + \mu P_{0j}^{0z}(t) - (\lambda + \mu) P_{1j}^{1z}(t)$$

for $t > 0$. Partition $(0, t]$ into $(0, r]$ and $(r, t]$ to verify that the left-hand derivative satisfies (4-51). Hence, (4-51) is valid. Similarly, by analogy with the forward equations (4-16), it can be shown (Exercise 4-33) that

$$\frac{d}{dt}\, P_{ij}^{y0}(t) = \lambda P_{i,\,j-1}^{y0}(t) - \lambda P_{ij}^{y0}(t) \qquad\qquad 1 \le y \le s-1$$

$$\frac{d}{dt}\, P_{ij}^{y1}(t) = \lambda P_{i,\,j-1}^{y1}(t) + \mu P_{i,\,j+1}^{y1}(t) - (\lambda + \mu)P_{ij}^{y1}(t) \qquad j \neq s,\,1$$

$$\frac{d}{dt}\, P_{i0}^{y0}(t) = \mu P_{i1}^{y1}(t) - \lambda P_{i0}^{y0}(t) \qquad\qquad\qquad\qquad (4\text{-}52)$$

$$\frac{d}{dt}\, P_{i1}^{y1}(t) = \mu P_{i2}^{y1}(t) - (\lambda + \mu)P_{i1}^{y1}(t)$$

$$\frac{d}{dt}\, P_{is}^{y1}(t) = \lambda P_{i,\,s-1}^{y1}(t) + \mu P_{i,\,s+1}^{y1}(t)$$

$$\qquad\qquad + \lambda P_{i,\,s-1}^{y0}(t) - (\lambda + \mu)P_{is}^{y1}(t)$$

If the limit

$$p_j^z = \lim_{t \to \infty} P_{ij}^{yz}(t)$$

exists and is the same for all i and y, then

$$\lim_{t \to \infty} \frac{d}{dt}\, P_{ij}^{yz}(t) = 0$$

for all i, y, j, and z (Exercise 4-34). Use of this result with (4-52) yields

$$0 = \lambda p_{j-1}^0 - \lambda p_j^0 \qquad 1 \le j \le s-1 \qquad\qquad (4\text{-}53a)$$

$$0 = \lambda p_{j-1}^1 + \mu p_{j+1}^1 - (\lambda + \mu)p_j^1 \qquad j \neq 1,\,s \qquad (4\text{-}53b)$$

$$0 = \mu p_1^1 - \lambda p_0^0 \qquad\qquad\qquad\qquad\qquad\qquad (4\text{-}53c)$$

$$0 = \mu p_2^1 - (\lambda + \mu)p_1^1 \qquad\qquad\qquad\qquad\qquad (4\text{-}53d)$$

$$0 = \lambda p_{s-1}^1 + \mu p_{s+1}^1 + \lambda p_{s-1}^0 - (\lambda + \mu)p_s^1 \qquad (4\text{-}53e)$$

From (4-53a),

$$p_j^0 = p_0^0 \qquad 1 \le j \le s-1 \qquad\qquad\qquad (4\text{-}54)$$

From (4-53c) and (4-53d), using $\rho = \lambda/\mu$, we have

$$p_1^1 = \rho p_0^0 \qquad \text{and} \qquad p_2^1 = (1 + \rho)p_1^1 = (\rho + \rho^2)p_0^0$$

Start an induction with $j = 2$ in (4-53b) to verify

$$p_j^1 = p_0^0 \sum_{i=1}^{j} \rho^i \qquad 1 \le j \le s \qquad\qquad (4\text{-}55)$$

where (4-55) establishes the result for $j = 1$ and $j = 2$. A combination of (4-53e) with (4-54) and (4-55) yields

$$p^1_{s+1} = (1 + \rho)p^1_s - \rho p^1_{s-1} - \rho p^0_{s-1} = p^0_0 \sum_{j=2}^{s+1} \rho^j \qquad (4\text{-}56)$$

Finally, an induction that starts with $j = s + 1$ in (4-53b) establishes

$$p^1_{s+k} = p^0_0 \sum_{j=k+1}^{s+k} \rho^j = p^0_0 \frac{\rho^{k+1}(1 - \rho^s)}{1 - \rho} \qquad k \geq 1 \qquad (4\text{-}57)$$

A combination of (4-54) through (4-57) yields

$$\sum_{j=0}^{s-1} p^0_j + \sum_{j=1}^{\infty} p^1_j = p^0_0\left(s + \sum_{j=1}^{s} \sum_{i=1}^{j} \rho^i + \sum_{k=1}^{\infty} \sum_{j=k+1}^{s+k} \rho^j \right) \qquad (4\text{-}58)$$

However,

$$1 \geq \sum_{j=0}^{s-1} p^0_j + \sum_{j=1}^{\infty} p^1_j$$

Therefore, if $p^0_0 > 0$, then $\rho < 1$ is necessary for the rightmost sum in (4-58) not to diverge. If the steady-state probabilities sum to 1, then $p^0_0 = (1 - \rho)/s$, so

$$p^1_j = \frac{\rho - \rho^{j+1}}{s} \qquad\qquad 1 \leq j \leq s$$

$$p^1_{s+k} = \rho^{k+1}\frac{1 - \rho^s}{s} \qquad\qquad k \geq 1 \qquad (4\text{-}59)$$

$$p^0_j = \frac{1 - \rho}{s} \qquad\qquad 0 \leq j \leq s - 1$$

If $p^0_0 = 0$, then $p^z_j = 0$ for all j and z. It can be shown that $\rho < 1$ is sufficient as well as necessary for p^z_j to exist, for all j and z, and for the steady-state probabilities to sum to 1.

EXERCISES

4-31 Derive (4-50).

4-32 Verify the backward equations (4-51).

4-33 Verify the forward equations (4-52). What are the initial conditions?

4-34 Prove that $dP^{yz}_{ij}(t)/dt \rightarrow 0$ as $t \rightarrow \infty$, for all $i, j, y,$ and z, if for each j and $z, \lim_{t \rightarrow \infty} P^{yz}_{ij}(t)$ exists and is the same for all i and y.

4-35 Perform the inductions for (4-55) and (4-57).

4-36 Derive p^0_0 in (4-59) by setting the left-side of (4-58) equal to unity.

4-37 In Figure 4-2, every state has flow, at most, to and from its immediate neighbors, where the "neighbor" property is induced by the usual ordering of the integers. Figure 4-4 cannot be redrawn to have this property. Instead each state in Figure 4-2 has flow to, at most, two other states, but these need not be neighbors. Therefore, some states may receive flow from more than two states. The following process has these characteristics.

Let $\{Y(t); t \geq 0\}$ be a stationary Markov process with denumerable state space I that satisfies

$$P\{Y(t+r) = h_j \mid Y(t) = j\} = \lambda_j r + o(r) \qquad j \in I$$

$$P\{Y(t+r) = d_j \mid Y(t) = j\} = \mu_j r + o(r) \qquad j \in I \qquad (4\text{-}60)$$

$$P\{Y(t+r) = j \mid Y(t) = j\} = 1 - (\lambda_j + \mu_j)r + o(r) \qquad j \in I$$

where for each j, $h_j \in I \sim \{j\}$, $d_j \in I \sim \{j\}$, $\lambda_j \geq 0$, and $\mu_j \geq 0$. Verify that $\{(X(t), Z(t)); t \geq 0\}$ is a Y-process. Specify its parameters $\{\lambda_j, \mu_j, h_j, d_j\}$.

4-38 Derive the Chapman-Kolmogorov equations for the general Y-process in Exercise 4-37.

4-39 Derive the backward and forward equations for the general Y-process in Exercise 4-37.

4-40 (Continuous production with an (s, S) policy) Consider a production process whose output is inventoried to satisfy demands. Suppose that the times between successive demands are i.i.d. exponential random variables $U_1, U_2, \ldots$ and that each demand is for a single unit of output. Suppose that the times taken to produce successive units are i.i.d. exponential random variables $V_1, V_2, \ldots$. Assume that the U_j's and V_j's are mutually independent and $E(U_1) = \gamma$ and $E(V_1) = \beta$. Finally, any demand that arrives while the inventory level is zero is turned away (lost-sales assumption).

Suppose that an (s, S) policy is used to decide when to produce additional units for the inventory. In words, when the inventory level drops to s and if production is not occurring, then production starts up. Production continues until the inventory level has been built up to S. Then production ceases until the inventory level drops to s, which causes the cycle to be repeated.

Let $X'(t)$ denote the inventory level at time t; let $Z'(t) = 1$ if production is occurring at time t and $Z'(t) = 0$ if production is not occurring at time t. The description above suggests the following specifications. Suppose $\{(X'(t), Z'(t)); t \geq 0\}$ is a stationary Markov process with state space $\{(x, 1): x \in I$ and $0 \leq x \leq S\} \cup \{(x, 0): x \in I$ and $s < x \leq S\}$. In place of (4-45) through (4-49), we specify

$$P\{X'(t+r) = j+1, Z'(t+r) = 1 \mid X'(t) = j, Z'(t) = 1\} = \beta r + o(r) \qquad j < S$$

$$P\{X'(t+r) = j, Z'(t+r) = 1 \mid X'(t) = j, Z'(t) = 1\} = 1 - (\beta + \gamma)r + o(r) \qquad j < S$$

$$P\{X'(t+r) = j-1, Z'(t+r) = z \mid X'(t) = j, Z'(t) = z\} = \gamma r + o(r)$$

$$\text{if } z = 0 \text{ and } S \geq j > s+1 \text{ or if } z = 1 \text{ and } S \geq j > 0 \qquad (4\text{-}61)$$

$$P\{X'(t+r) = s, Z'(t+r) = 1 \mid X'(t) = s+1, Z'(t) = 0\} = \gamma r + o(r)$$

$$P\{X'(t+r) = S, Z'(t+r) = 0 \mid X'(t) = S-1, Z'(t) = 1\} = \beta r + o(r)$$

(a) Draw the digraph for this process.

(b) Show that it is another case of the general Y-process in Exercise 4-37.

(c) Show that this process is equivalent to an altered start-up and shut-down queuing process in which (4-45) is asserted only for $j < K$, (4-46) only for $j \leq K$, and (4-47) only for $j < K$. To (4-47) we add

$$P\{X(t+r) = K, Z(t+r) = 1 \mid X(t) = K, Z(t) = 1\} = 1 - \mu r + o(r)$$

This alteration of the queuing process corresponds to a waiting room of size K (Exercise 4-35). Assume $K > s$. *Hint*: Compare your digraph in (a) with an alteration of Figure 4-4 to include the finite waiting room feature.

(d) Suppose that the steady-state probabilities exist. Is any restriction on γ and β necessary for the probabilities to sum to 1? Specify the probabilities under the assumption that they exist and sum to 1.

(e) Under the assumptions of (d), what are the expressions for

(i) $\lim_{t \to \infty} E[X'(t) \mid X'(0) = S, Z'(0) = 0]$

(ii) $\lim_{t \to \infty} P\{Z'(t) = 1 \mid X'(0) = S, Z'(0) = 0\}$

4-7 SAMPLE-PATH PROPERTIES

If $\{X(t); \ t \geq 0\}$ is a birth-and-death process, its realizations are step functions that start at $X(0)$ and have upward jumps of size 1 when a birth occurs and downward jumps of size 1 when a death occurs. A transition is a birth or a death, so the transition epochs are precisely those times where $X(t)$ changes its value. In this section we obtain results about the spacings between transition epochs.

Preliminaries

Let S_n denote the time of the nth transition epoch with $0 = S_0 \leq S_1 \leq S_2 \leq \ldots$, and let $\tau_n = S_n - S_{n-1}$, $n \geq 1$; τ_n is the time between the $(n-1)$th and nth transitions. Arguing intuitively, we know that if for some sample path ω there are infinitely many transitions within a finite time interval, then the epochs $\{S_n(\omega)\}$ are not well defined. To preclude this pathology, we appeal to Theorem 4-1 and assume (4-20) holds.

In general, the birth rates and death rates depend on the state of the process, so the distribution of τ_n should depend on the state of the process $X(S_{n-1})$ at the *random* time S_{n-1}. This raises a difficulty. Axiom 2 of the birth-and-death process [cf. Eq. (4-2)] applies to *fixed* times s. Fortunately, there is Theorem 4-5.

Theorem 4-5 Let $\{X(t); \ t \geq 0\}$ be a birth-and-death process and S_n be the epoch of the nth transition, $n \in I$. If (4-20) holds, then

$$P\{X(S_n + t) = j \mid X(u); \ 0 \leq u \leq S_n\} = P\{X(S_n + t) = j \mid X(S_n)\} \quad (4\text{-}62)$$

for any $t \geq 0$ and $n \in I$.

PROOF This is a special case of Theorem 9-5 (whose proof is not presented in this book). $\qquad\square$

From (4-62) and the definition of a homogeneous Markov process, Eq. (4-2),

$$P\{X(S_n + t) = j \mid X(u), \ 0 \leq u < S_n, \ X(S_n) = i\} = P_{ij}(t) \qquad i, j \in \mathbb{S} \quad (4\text{-}63)$$

All the finite-dimensional distributions are needed to describe a stochastic process, and (4-63) considers just one value of t at a time. Thus (4-63) is not sufficient to claim that at each epoch when state i is entered, the future evolution of the process is probabilistically the same as the evolution of a process that starts in state i. The truth of the claim is asserted by Proposition 4-3.

Proposition 4-3 For each $i \in \mathbb{S}$ given S_n and $X(S_n) = i$, the fidi's of $\{X(S_n + t); \ t \geq 0\}$ are the same as the fidi's of $\{X(t); \ t \geq 0\}$ with $X(0) = i$.

PROOF This is a special case of Proposition 9-8 (whose proof is not presented in this book). $\qquad\square$

The assertions of Theorem 4-5 and Proposition 4-3 *seem* to follow directly from the Markov property. They do not, and their proofs require considerable analysis.

In the examples in Section 4-3, the times between transitions are all independent exponential r.v.'s. Theorem 4-6 shows that this always occurs in a birth-and-death process.

Theorem 4-6 In a birth-and-death process with birth rates $\{\lambda_i\}$ and death rates $\{\mu_i\}$, for any $n \in I$ and $i \in S$,

$$P\{\tau_{n+1} > t \mid S_1, \ldots, S_n, X(0), X(S_1), \ldots, X(S_n) = i\}$$

$$= P\{\tau_1 > t \mid X(0) = i\} = e^{-(\lambda_i + \mu_i)t} \qquad t \geq 0 \qquad (4\text{-}64)$$

PROOF When $S_1, \ldots, S_n, X(0), \ldots, X(S_n)$ are known, $X(u)$ is determined for all $0 \leq u \leq S_n$. Thus Proposition 4-3 establishes the first equality. To establish the second equality, let $G_i^c(t) = P\{\tau_1 > t \mid X(0) = i\}$. Since $\tau_1 > t + r$ precisely when no transitions occur during both $(0, t]$ and $(t, t + r]$, Equations (4-5), (4-6), and (4-62) yield

$$G_i^c(t + r) = G_i^c(t)[1 - (\lambda_i + \mu_i)r + o(r)] \qquad r > 0 \qquad (4\text{-}65)$$

Proceeding in the same manner as in the derivation of the Kolmogorov equations, we obtain

$$\frac{d}{dt} G_i^c(t) = -(\lambda_i + \mu_i) G_i^c(t) \qquad t > 0 \qquad (4\text{-}66)$$

By Theorem 4-1, $P\{\tau_1 > 0\} = 1$ (why?), so

$$G_i^c(0) = 1 \qquad (4\text{-}67)$$

The solution of (4-66) with boundary condition (4-67) is

$$G_i^c(t) = e^{-(\lambda_i + \mu_i)t} \qquad t \geq 0 \qquad (4\text{-}68)$$

and (4-64) is proved. $\qquad\qquad\qquad\qquad\qquad\qquad\qquad\qquad\qquad\square$

Corollary 4-6 In a Poisson process with birth rate λ in each state, $\tau_1, \tau_2, \ldots$ are i.i.d. with density function $\lambda e^{-\lambda t}$.

PROOF This is left to Exercise 4-44. $\qquad\qquad\qquad\qquad\qquad\qquad\qquad\square$

Choosing the Next State

If the process is currently in state i, then the next transition moves the process to either state $i + 1$ or state $i - 1$. Let h_i denote the probability that the next state is $i + 1$. Theorem 4-7 shows how to compute h_i and that the next state visited is *independent* of the time between transitions.

Theorem 4-7 Let $h_i(k) = P\{X(S_{k+1}) = i + 1 \mid X(S_k) = i\}$. Then

$$h_i(k) = h_i = \frac{\lambda_i}{\lambda_i + \mu_i} \qquad k \in I_+, i \in S \tag{4-69}$$

and

$$P\{X(S_{k+1}) = i + 1, \tau_{k+1} > t \mid X(u), 0 \le u < S_k, X(S_k) = i\}$$
$$= h_i e^{-(\lambda_i + \mu_i)t} \qquad t \ge 0, k \in I_+ \tag{4-70}$$

PROOF From Proposition 4-3,

$$P\{X(S_{k+1}) = i + 1, \tau_{k+1} > t \mid X(u), 0 \le u \le S_k, X(S_k) = i\}$$
$$= P\{X(S_1) = i + 1, \tau_1 > t \mid X(0) = i\} \tag{4-71}$$
$$= P\{\tau_1 > t \mid X(0) = i\} \ P\{X(\tau_1) = i + 1 \mid \tau_1 > t, X(0) = i\}$$

The first term is given by (4-64). To obtain the second term, we observe that the events $\{X(0) = i, \tau_1 > t\}$ and $\{X(u) = i, u \le t\}$ are equivalent and that the memoryless property of the exponential distribution implies that if $\tau_1 > t$, then $\tau_1 = t + \tau'$, where the distribution of τ' is the unconditional distribution of τ_1. Thus,

$$P\{X(\tau_1) = i + 1 \mid \tau_1 > t, X(0) = i\} = P\{X(t + \tau') = i + 1 \mid X(u) = i, u \le t\}$$
$$= P\{X(\tau_1) = i + 1 \mid X(0) = i\} = h_i \tag{4-72}$$

where (4-63) is used to obtain the second equality, and (4-71) and (4-72) prove (4-70).

The derivations of (4-71) and (4-72) establish the first equality in (4-69). To obtain the second equality, observe that $h_i = \lim_{t \downarrow 0} P\{X(t) = i + 1 \mid X(t) \ne i, X(0) = i\}$. For small values of t, axiom 3 and (4-6) are used to obtain

$$P\{X(t) = i + 1 \mid X(t) \ne i, X(0) = i\}$$
$$= \frac{P\{X(t) = i + 1, X(t) \ne i \mid X(0) = i\}}{P\{X(t) \ne i \mid X(0) = i\}}$$
$$= \frac{\lambda_i t + o(t)}{(\lambda_i + \mu_i)t + o(t)}$$

Dividing the numerator and denominator of the last expression by t and letting $t \downarrow 0$ completes the proof. $\qquad \Box$

Theorem 4-7 provides a method of constructing a birth-and-death process. First, $X(0)$ is chosen according to the appropriate probability function. Suppose $X(0) = i$. Then the time to the first transition is obtained by taking a random sample of a population with an exponential distribution having a mean of

$1/(\lambda_i + \mu_i)$. The effect of this transition is obtained by flipping a coin which lands heads with probability $\lambda_i/(\lambda_i + \mu_i)$. If the coin lands heads, then the transition is to state $i + 1$; otherwise, it is to state $i - 1$. When the new state is entered, these experiments are repeated with the appropriate parameter adjustments, and so on.

This procedure generates most birth-and-death processes of interest. The only ones it cannot generate are those where (4-20) fails and infinitely many transitions need to be generated in a finite time with positive probability.

Theorem 4-7 has some consequences which may be surprising. At an epoch when the process has a positive value, let τ be the elapsed time to the next transition, λ and μ the birth and death parameters (which may depend on the state), respectively, and $\{U\}$ and $\{D\}$, respectively, the events that the subsequent transition is to the next higher or lower states. Since (4-70) asserts that given the current state, the identity of the next state visited is independent of the time required to get there, we have

$$P\{\tau > t, U\} = \lambda(\lambda + \mu)^{-1} e^{-(\lambda + \mu)t} \quad \text{and} \quad P\{U\} = \frac{\lambda}{\lambda + \mu}$$

and so

$$P\{\tau > t \mid U\} = e^{-(\lambda + \mu)t} = P\{\tau > t\} = P\{\tau > t \mid D\} \tag{4-73}$$

where the last equality follows by similar reasoning.

Example 4-11 Applying (4-73) to the $M/M/1$ queue, we interpret it as showing that the time to the next transition, given that it is an arrival, has the same distribution as the unconditional random variable. This result usually is greeted with surprise because when λ is much smaller than μ, the mean time until the next arrival is much greater than the mean time until the next service completion. Then $P\{\tau > t \mid U\}$ "must be" greater than $P\{\tau > t\}$, which "must be" greater than $P\{\tau > t \mid D\}$. These statements are false because the time until the next transition does not stochastically depend on which transition occurs. □

EXERCISES

4-41 Let ω be a sample path of a birth-and-death process and $N(t; \omega)$ be the number of transitions by time t. Prove that (4-20) implies that $N(t; \omega) \to \infty$ as $t \to \infty$ (w.p.1).

4-42 Suppose $X(0) = 1$, $\lambda_n = n\lambda$, and $\mu_n = n\mu$. Let K be the transition where the first death occurs. Find $P\{K = k\}$, $k \in I_+$.

4-43 Suppose you have at your disposal computer programs that flip coins and draw samples from exponential distributions. Explain how you could use the programs to generate a sample path of an $M/M/c$ queue. Do the same for the model of Exercise 4-6. Be specific about the mathematical details and their justification, but do not specify the computer program in detail.

4-44 Prove that the times between transitions of a Poisson process are i.i.d. exponential random variables with mean $1/\lambda$.

4-45 State and prove versions of Theorems 4-6 and 4-7 and Corollary 4-6 for the start-up and shut-down queueing model in Section 4-6.

4-46 Repeat Exercise 4-43 for the start-up and shut-down queueing model in Section 4-6.

4-47 Repeat Exercises 4-45 and 4-43 for the model of continuous production with an (s, S) policy in Exercise 4-40.

4-8* RELATIONSHIPS AMONG AVERAGES

Recall Example 4-4, the machine breakdown and repair model. Section 4-5 contains theorems that enable us to obtain

$$p_j \triangleq \lim_{t \to \infty} P\{X(t) = j\}$$

for a birth-and-death process $\{X(t); t \geq 0\}$. Exercise 4-19 asked you to find $\{p_j\}$ for Example 4-4. Theorem 4-8 below states that p_j is the long-run proportion of time which the process spends in state j. In the context of Example 4-4, p_j is the asymptotic *proportion of time j* teletypewriters (TTYs) are inoperative; it is useful for determining how many TTYs to buy and how many repairers to hire.

Suppose you are using a TTY and it breaks down. You are now interested in how long it will be before your TTY is fixed, and that depends, in part, on how many other failed TTYs there are at the epoch your TTY failed. Is p_j the probability there are j of them? Let Y_n denote the number of machines being repaired or idle at the time of the nth failure. To be formal about this definition, let t_n be the epoch of the nth failure and

$$Y_n \triangleq X(t_n^-) \triangleq \lim_{s \downarrow 0} X(t_n - s) \qquad n \in I_+ \tag{4-74}$$

Define

$$\pi_j(n) = P\{Y_n = j\} \qquad \text{and} \qquad \pi_j = \lim_{n \to \infty} \pi_j(n)$$

Theorem 4-8 asserts that π_j is the asymptotic *proportion of failure epochs* where the number of failed TTYs was j (and then became $j + 1$).

We expect that π_j is related to p_j, but $\pi_j \neq p_j$ in general. In our current example, the solution to Exercise 4-19 shows that the probability that no machines are working is positive, that is, $p_M > 0$; even without doing that exercise, this result is intuitively correct. But $\pi_M = 0$ because no failure can occur when all the machines are down; hence $\pi_M \neq p_M$ for this process.

Now let $\{X(t); t \geq 0\}$ be any birth-and-death process where $\{p_j\}$ exists with $p_j > 0$ for each $j \in S$ and define Y_n by (4-74). Since Y_n is the state at the epoch of the nth birth, we call Y_n the *state seen by the nth birth*. In Chapter 6 we prove the following theorem (see Example 6-9).

Theorem 4-8 Let $\{X(t); t \geq 0\}$ be a birth-and-death process that has a limiting distribution $\{p_j\}$, and define Y_n by (4-74). For each sample path ω let

$$I_j(t; \omega) = \begin{cases} 1 & \text{if } X(t; \omega) = j \\ 0 & \text{otherwise} \end{cases}$$

and

$$J_j(n; \omega) = \begin{cases} 1 & \text{if } Y_n(\omega) = j \\ 0 & \text{otherwise} \end{cases}$$

Then for each $j \in S$,

$$p_j = \lim_{T \to \infty} \frac{1}{T} \int_0^T I_j(t; \omega) \, dt \tag{4-75}$$

and

$$\pi_j = \lim_{N \to \infty} \frac{1}{N} \sum_{n=1}^{N} J_j(n; \omega) \tag{4-76}$$

hold (with probability 1).

This theorem states that if we observe a birth-and-death process for a long time, then the proportion of time we observe that the process is in state j approaches p_j and the proportion of births that see state j approaches π_j. The right-sides of (4-75) and (4-76) are averages on a particular sample path; the left-sides are limiting probabilities. Theorems that equate sample-path averages to limiting probabilities attach a useful meaning to limiting probabilities. Theorems of this type are called *ergodic* theorems; many ergodic theorems are proved, and applied, in the sequel.

Relation between π_j and p_j

Theorem 4-8 is used now to derive the relation between π_j and p_j. First we give a *heuristic* explanation. The proportion of births that see state j should be proportional to both the proportion of time the process is in state j (that is, p_j) and the rate at which births occur when the process is in state j (that is, λ_j). If we set $\pi_j = k\lambda_j p_j$ for some unknown constant k, then $1 = \sum_{j \in S} k\lambda_i p_i$, and so

$$\pi_j = \frac{\lambda_j p_j}{\sum_{i \in S} \lambda_i p_i} \qquad j \in S \tag{4-77}$$

Theorem 4-9 If (4-38) and (4-39) are valid, then π_j exists and is given by (4-77) for all $j \in S$.

PROOF Without loss of generality, assume that $\lambda_j > 0$ and $\mu_j > 0$ for all $j \in S$. When (4-38) and (4-39) are valid, then p_j exists and is positive for each $j \in S$. Let $N(t; \omega)$ be the number of transitions that occur by time t; since (4-39) implies (4-20) is valid, $N(t; \omega) \uparrow \infty$ as $t \to \infty$, by Exercise (4-41). For notational convenience, let

$$C_j(T; \omega) = \int_0^T I_j(t; \omega) \, dt$$

and

$$b_j(T; \omega) = \sum_{n=1}^{N(T, \omega)} J_j(n; \omega) \qquad j \in \mathcal{S}$$

Hence

$$p_j = \lim_{T \to \infty} \frac{C_j(T; \omega)}{T} > 0 \tag{4-78}$$

and

$$B(T; \omega) \triangleq \sum_{j \in \mathcal{S}} b_j(T; \omega)$$

is the number of births by epoch T, and

$$\pi_j = \lim_{T \to \infty} \frac{b_j(T; \omega)}{B(T; \omega)} \tag{4-79}$$

Let $V_j(T; \omega)$ be the number of transitions that cause the process to enter state j by time T; call this the number of *visits* to state j by time T. Let $\tau_{jk}(\omega)$ denote the length of the kth visit to state j.† (From now on, we suppress the arguments ω and k *when they are apparent in the context*.)

According to Theorem 4-6, $\tau_{j1}, \tau_{j2}, \ldots$ are i.i.d. with mean $(\lambda_j + \mu_j)^{-1}$. And $C_j(T)$ is the sum of the first $V_j(T)$ τ_j's except (possibly) for the remaining part of the last one; this occurs only if $X(T) = j$. Let $\varepsilon_j(T)$ denote the length of that part of the last visit which extends beyond T, and observe that it is finite. Thus we have

$$C_j(T) = \tau_{j1} + \cdots + \tau_{jv(T)} - \varepsilon_j(T) \tag{4-80}$$

Since (4-78) implies $C_j(T) \uparrow \infty$ as $T \to \infty$, use (4-78) and (4-80) to obtain

$$P_j = \lim_{T \to \infty} \frac{C_j(T)}{V_j(T)} \frac{V_j(T)}{T} = \lim_{T \to \infty} \frac{C_j(T)}{V_j(T)} \lim_{T \to \infty} \frac{V_j(T)}{T}$$

$$= \frac{\lim_{T \to \infty} V_j(T)/T}{\lambda_j + \mu_j} \tag{4-81}$$

where we used the fact that $\varepsilon_j(T)$ is finite and the strong law of large numbers to calculate $\lim_{T \to \infty} C_j(T)/V_j(T)$. A consequence of (4-80) is that $V_j(T) \to \infty$ as $T \to \infty$.

A birth sees state j if and only if it is the first transition following an entrance to state j. (Why?) Theorem 4-7 shows that when the process is in state j, the conditional probability that the next transition is a birth is $\lambda_j/(\lambda_j + \mu_j)$ and that this is independent of what transpired on all previous visits to state j. This means that $b_j(T)$ is a binomial random variable with param-

---

† When the kth visit to state j occurs at the nth transition epoch, τ_{jk} is precisely τ_n as defined in Section 4-7.

eters $\lambda_j/(\lambda_j + \mu_j)$ and $V_j(T)$, so the strong law of large numbers yields

$$\lim_{T \to \infty} \frac{b_j(T)}{V_j(T)} = \frac{\lambda_j}{\lambda_j + \mu_j} \qquad \text{(w.p.1)} \qquad (4\text{-}82)$$

Combining (4-81) and (4-82), we obtain

$$\lim_{T \to \infty} \frac{b_j(T)}{T} = \lim_{T \to \infty} \frac{b_j(T)}{V_j(T)} \frac{V_j(T)}{T}$$

$$= \lambda_j(\lambda_j + \mu_j)^{-1}(\lambda_j + \mu_j)p_j = \lambda_j p_j \qquad (4\text{-}83)$$

Together (4-79) and (4-83) yield the final result

$$\pi_j = \lim_{T \to \infty} \frac{b_j(T)/T}{\sum b_i(T)/T} = \frac{\lambda_j p_j}{\sum \lambda_i p_i}$$

Interchange of the limit and the summation can always be rigorously justified. $\qquad \Box$

An immediate consequence of Theorem 4-9 is Corollary 4-9.

Corollary 4-9 For all $i \in \mathcal{S}$, $\lambda_i \equiv \lambda \Leftrightarrow \pi_i = p_i$.
   When the births occur according to a Poisson process, we indeed have $\lambda_i \equiv \lambda$, and we verbalize Corollary 4-9 by saying, "Poisson births see time averages."

Examples

Example 4-12: Waiting times in the M/M/1 queue In the M/M/1 queue, you showed, in Exercise 4-22, that $p_j = (1 - \rho) \rho^j$, where $\rho = \lambda/\mu < 1$. In this model, we interpret Y_n as the number of customers present at the nth arrival epoch. From Corollary 4-9 it follows that the steady-state probability that there are j other patrons in the system when a customer arrives is $\pi_j = p_j$.
   Let D_n denote the delay (the length of the time interval that starts when the customer arrives and ends when the customer begins service) of the nth customer to arrive. We want to find $P\{D_n \leq t\}$. If those, and only those, customers who arrive when the server is idle can obtain service immediately, we have

$$\{D_n = 0\} \Leftrightarrow \{Y_n = 0\}$$

Hence,

$$D(0) \triangleq \lim_{n \to \infty} P\{D_n = 0\} = \lim_{n \to \infty} \sum_{j=0}^{\infty} P\{D_n = 0 \mid Y_n = j\} P\{Y_n = j\}$$

$$= \lim_{n \to \infty} \pi_0(n) = p_0 = 1 - \rho \qquad (4\text{-}84)$$

Suppose that customers are served in order of arrival. When $Y_n > 0$, there is a customer in service when customer n arrives, and the memoryless property of the exponential distribution and (4-73) together assert that the remaining service time of this customer has an exponential distribution with mean $1/\mu$. When $Y_n = j > 0$, D_n is the sum of the remaining service time of the customer in service at customer n's arrival epoch and the service times of the $j - 1$ customers ahead of her or him in the queue. We have, for each n,

$$\{D_n \mid Y_n = j\} = \{Z_1 + Z_2 + \cdots + Z_j\} \qquad j \in S$$

where the Z's are i.i.d. with density function $\mu e^{-\mu t}$. Since $Z_1 + \ldots + Z_j$ has the gamma distribution with shape parameter j and scale parameter μ,

$$d_n(t) \triangleq \frac{d}{dt} P\{D_n \le t\} = \sum_{j=1}^{\infty} \frac{\mu(\mu t)^{j-1}}{(j-1)!} e^{-\mu t} \pi_j(n) \qquad t > 0$$

In the steady state,

$$d(t) \triangleq \lim_{n \to \infty} d_n(t) = \sum_{j=1}^{\infty} \frac{\mu(\mu t)^{j-1}}{(j-1)!} e^{-\mu t} p_j$$

$$= (1 - \rho)\lambda e^{-\mu t} \sum_{j=0}^{\infty} \frac{(\lambda t)^j}{j!}$$

$$= \lambda(1 - \rho)e^{-(\mu - \lambda)t} \qquad t > 0 \qquad (4\text{-}85)$$

Integrating (4-85) and combining the result with (4-84) yields the representation

$$D(t) = D(0) + \int_0^t d(s) \, ds = 1 - \rho + \rho(1 - e^{-(\mu - \lambda)t}) \qquad t \ge 0 \qquad (4\text{-}86)$$

Thus, the sequence of waiting times $\{D_n ; n \in I_+\}$ converges to a random variable D with distribution function $D(t)$ given by (4-86), in the sense that the associated sequence of distribution functions converges to $D(t)$. This mode of convergence for random variables is called *convergence in distribution*. Equation (4-86) shows that asymptotically the delay distribution has a mass of $1-\rho$ at zero and has an exponential density for positive arguments.

The waiting time of a customer is the delay plus the service time. Let W_n denote the waiting time of the nth customer to arrive, so

$$W_n = D_n + V_n \qquad n \in I_+$$

where V_n is the service time of customer n. Now D_n and V_n are independent because D_n depends on only the service times of those customers present when customer n arrives, and those times are independent of V_n.

Let $W_n(t) = P\{W_n \le t\}$ for each n:

$$W_n(t) = \int_0^t D_n(t - s)\mu e^{-\mu s} \, ds$$

It now follows from (4-86) that $W(t) = \lim_{n \to \infty} W_n(t)$ exists and is given by

$$W(t) = \mu \int_0^t e^{-\mu s} D(t-s)\, ds = 1 - e^{-(\mu-\lambda)t} \qquad t \geq 0 \qquad (4\text{-}87)$$

Since (4-87) shows that $W(t)$ has an exponential distribution, the mean waiting time in the steady state is

$$E(W) = \frac{1}{\mu - \lambda} \qquad (4\text{-}88)$$

where $W = \lim_{n \to \infty} W_n$ (again we interpret this limit to mean convergence in distribution). Using the formula for the mean number in the system L developed in Exercise 4-22 and Equation (4-88), we obtain

$$L = \frac{\lambda}{\mu} \frac{\mu}{\mu - \lambda} = \lambda E(W) \qquad (4\text{-}89)$$

This is a specific case of an important theorem which is discussed in Chapter 11. Sample-path (or posterity) versions of (4-89) were proved in Section 2-3. □

Example 4-13: Losses in the M/M/c/c queue In Exercise 4-25 you obtained a formula for the steady-state probability [denoted $B(c, a)$] that all servers are busy in an M/M/c/c queue. From (4-75), $B(c, a)$ is the long-run proportion of time that all the servers are busy. From Corollary 4-9, $B(c, a)$ is the probability that an arrival finds all the servers busy, i.e., the probability that the arrival is lost; and from (4-76), it is the long-run proportion of customers who are lost. In the context of providing enough servers so that the proportion of lost customers is not excessive, the last interpretation of $B(c, a)$ is needed. □

EXERCISES

4-48 Find π_j for each j in the machine repair model (Example 4-4).

4-49 Parameterize the machine repair models by M, the number of machines. Let $p_j(M)$ and $\pi_j(M)$ be the time-average and customer-average steady-state probabilities, respectively. Prove that for each $M \geq 1$, $\pi_j(M) = p_j(M-1)$, $j = 0, 1, \dots, M-1$.

4-50 Give a precise formulation of "the state left behind by a death." Show that asymptotically this has the same distribution as the state seen by a birth. *Hint:* When k births have seen state j, how many deaths left state j behind?

4-51 Suppose the number of cars in the lot of a rent-a-car agency can be described by a birth-and-death process. Explain why the processes are the same when the cars are assigned to new customers on a first-in, first-out basis, according to a last-in, first-out rule, or in random order. Explain, further, why the distribution of the time that a car spends in the lot differs among these three cases. Interpret (4-89) in the context of this model.

4-52 A simple model for determining the required size of an input buffer to a machine, e.g., a computer, can be constructed from the M/M/1 queue by imposing a *queue limit*, K say, so that customers (births) who arrive when $X(t)$, the number in the system (including the customer in service), equals K are rejected and depart without affecting $X(t + s)$ in any way, $s \geq 0$. The queue limit is the same as the finite waiting room. Use the notation of this section and find (a) p_j and π_j for all relevant values of j, (b) the probability that a customer is rejected, (c) the proportion of time the server is busy, (d) L and $E(W)$. Suppose $\rho = 0.9$ and no more than one customer out of a million, on the average, is to be rejected. What is the smallest value of K that can be used?

4-53 In an M/M/c queue, suppose customers enter service in order of arrival and that $\lambda < c\mu$. Let $D_F(t) = \lim_{n \to \infty} P\{D_n \leq t\}$. Show that

$$D_F(t) = \sum_{i=0}^{c-1} p_i + \frac{c\mu p_c (1 - e^{-(c\mu - \lambda)t})}{c\mu - \lambda} \qquad t \geq 0$$

and verify that $D_F(\infty) = 1$.

4-9* FIRST-PASSAGE TIMES

In this section the mean, variance, and distribution of the first-passage time between any pair of states is obtained when $\lambda_i \equiv \lambda$ and $\mu_i \equiv \mu$. For general $\{\lambda_i\}$ and $\{\mu_i\}$, only the mean of the first-passage time is obtained.

In Example 4-7, the first-passage time between distinct states i and j is denoted by T_{ij} and defined by†

$$T_{ij} = \inf\{t : X(t) = j \mid X(0) = i\} \qquad i \neq j \tag{4-90}$$

Axiom (2) for a birth-and-death process (that it is a homogeneous Markov process), Theorem 4-6, and the memoryless property of the exponential distribution combine to prove that

$$\inf\{t : X(s + t) = j \mid X(s) = i, \; X(u), \; 0 \leq u \leq s\}$$

has the same distribution as T_{ij}. This means that for obtaining distributional results, T_{ij} can always be considered as being defined by (4-90). Axiom (2) also implies that for each ordered pair of states (i, j), the successive first-passage times from i to j are independent r.v.'s.

Conditioning on the First Event

A first passage from i to j can be divided into two parts, namely, the first transition out of state i (to state k, say) followed by a first passage from k to j, as shown in Figure 4-5. Let S_1 be the time of the first transition; in Section 4-5 it is shown that S_1 is independent of $X(S_1)$ and that S_1 has an exponential distribution with mean $1/(\lambda_i + \mu_i)$. By *conditioning on the first event* we obtain

$$T_{ij} = S_1 + \begin{cases} T_{i+1, j} & \text{if } X(S_1) = i + 1 \\ T_{i-1, j} & \text{if } X(S_1) = i - 1 \end{cases} \tag{4-91}$$

---

† Note that formally (4-90) gives $T_{ii} \equiv 0$ for any $i \in I$; we adopt this as a definition of T_{ii}.

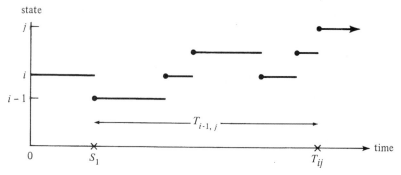

Figure 4-5 Conditioning on the first event.

Let $G_{ij}(t) = P\{T_{ij} \le t\}$; then (4-91) and Theorem (4-7) yield

$$G_{ij}(t) = \lambda_i \int_0^t G_{i+1, j}(t - x)e^{-(\lambda_i + \mu_i)x}\, dx$$

$$+ \mu_i \int_0^t G_{i-1, j}(t - x)e^{-(\lambda_i + \mu_i)x}\, dx \qquad i \in I \qquad (4\text{-}92)$$

Recall that $\mu_0 = 0$, so the second term on the right-side of (4-92) vanishes when $i = 0$. Let $\tilde{G}_{ij}(\cdot)$ be the Laplace-Stieltjes transform (LST) of $G_{ij}(\cdot)$; taking LSTs on both sides of (4-92), using the convolution property, yields

$$\tilde{G}_{ij}(s) = \frac{\lambda_i \tilde{G}_{i+1, j}(s) + \mu_i \tilde{G}_{i-1, j}(s)}{s + \lambda_i + \mu_i} \qquad i \in I \qquad (4\text{-}93)$$

Observe that (4-93) is a recursion that expresses $\tilde{G}_{i+1, j}(s)$ in terms of $\tilde{G}_{ij}(s)$ and $\tilde{G}_{i-1, j}(s)$.

Suppose $i < j$; since changes of state have unit magnitude in a birth-and-death process,

$$T_{ij} = T_{ik} + T_{kj} \qquad i < k < j \qquad (4\text{-}94)$$

as shown in Figure 4-6. Iterating (4-94) for $k = i + 1, i + 2, \ldots, j - 1$ yields

$$T_{ij} = \sum_{k=i}^{j-1} T_{k, k+1} \qquad i < j \qquad (4\text{-}95a)$$

Similarly, when $i > j$,

$$T_{ij} = T_{i, i-1} + T_{i-1, i-2} + \cdots + T_{j+1, j}$$

$$= \sum_{k=j}^{i-1} T_{k+1, k} \qquad i > j \qquad (4\text{-}95b)$$

The Markov property (axiom 2) implies that the summands in (4-95a) and (4-95b)

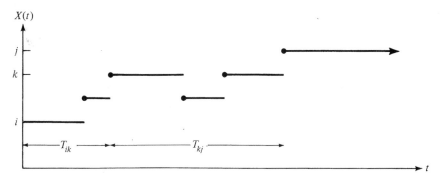

Figure 4-6 Some first-passage times.

are independent. Hence

$$\tilde{G}_{ij}(s) = \prod_{k=i}^{j-1} \tilde{G}_{k,k+1}(s) \qquad i < j \tag{4-96a}$$

and

$$\tilde{G}_{ij}(s) = \prod_{k=j}^{i-1} \tilde{G}_{k+1,k}(s) \qquad i > j \tag{4-96b}$$

Thus, in order to obtain $G_{ij}(\,\cdot\,)$, it is sufficient to find the first-passage times between the relevant adjacent states. This is particularly easy when the birth and death rates do not vary with the state and $i > j$.

Constant Birth Rates and Death Rates

Lemma 4-1 For $\lambda = \lambda_0 = \lambda_1 = \cdots$ and $\mu = \mu_1 = \mu_2 = \cdots$,

$$G_{k+1,k}(\,\cdot\,) = G_{10}(\,\cdot\,) \qquad k = 0, 1, \ldots$$

PROOF Now, $T_{k+1,k}$ is the time that elapses before the cumulative number of deaths first exceeds the cumulative number of births when $X(0) = k + 1$. Under the hypothesis given, Theorem 4-7 implies that magnitudes and epochs of the transitions prior to a visit to state zero do not depend on $X(0)$. $\qquad\square$

Lemma 4-1 indicates that $G_{10}(\,\cdot\,)$ is particularly important, and it assists in the derivation of an explicit formula for $G_{10}(\,\cdot\,)$.

Theorem 4-10 Let $\lambda = \lambda_0 = \lambda_1 = \cdots$, $\mu = \mu_1 = \mu_2 = \cdots$, $G(t) = P\{T_{10} \le t\}$, and $g(t) = dG(t)/dt$. Then

$$\tilde{G}(s) = \frac{s + \lambda + \mu - \sqrt{(s + \lambda + \mu)^2 - 4\lambda\mu}}{2\lambda} \tag{4-97}$$

and

$$g(t) = e^{-(\lambda + \mu)t}I_1(2t\sqrt{\lambda\mu})\frac{\sqrt{\mu/\lambda}}{t} \tag{4-98}$$

where $I_1(\ \cdot\ )$ is the *Bessel function* of imaginary argument and first order.†

PROOF Lemma 4-1 and Equation (4-96b) yield

$$\tilde{G}_{20}(s) = [\tilde{G}(s)]^2 \tag{4-99}$$

Set $i = 1$ and $j = 0$ in (4-93), and use (4-99) to obtain

$$\lambda[\tilde{G}(s)]^2 - (s + \lambda + \mu)\tilde{G}(s) + \mu = 0 \tag{4-100}$$

Equation (4-100) has the two solutions

$$\tilde{G}_+(s) = \frac{s + \lambda + \mu + [(s + \lambda + \mu)^2 - 4\lambda\mu]^{1/2}}{2\lambda}$$

and

$$\tilde{G}_-(s) = \frac{s + \lambda + \mu - [(s + \lambda + \mu)^2 - 4\lambda\mu]^{1/2}}{2\lambda}$$

but only one of them can be $\tilde{G}(s)$ because Laplace-Stieltjes transforms are unique. The selection is made by using the fact that $G(\ \cdot\ )$ is a distribution function, which implies that $\tilde{G}(s) \leq 1$ for s real and positive. For $s > 0$,

$$(s + \lambda + \mu)^2 - 4\lambda\mu > (\lambda + \mu)^2 - 4\lambda\mu = (\lambda - \mu)^2$$

so $G_+(s) > 1$. This establishes (4-97).

If (4-97) is a bona fide Laplace-Stieltjes transform, it is the Laplace transform of $g(\ \cdot\ )$. Recall that $\mathcal{L}\{e^{-ct}f(t)\} = \tilde{f}(s + c)$. Set $p = s + \lambda + \mu$, and write (4-97) as

$$\tilde{G}(p) = \frac{p - \sqrt{p^2 - 4\lambda\mu}}{2\lambda}$$

The numerator is the Laplace transform of $2(\lambda\mu)^{1/2}I_1(2t\sqrt{\lambda\mu})/t$; hence $\tilde{G}(s)$ is the Laplace transform (4-98). □

Equation (4-97) can be used to obtain moments.

---

† Bessel functions are among the most frequently used higher transcendental functions. They occur in connection with partial differential equations and are especially important in mathematical physics. In particular,

$$I_1(z) = \sum_{m=0}^{\infty} \frac{(\frac{1}{2}z)^{2M+1}}{M!(M+1)!} \qquad z = y\sqrt{-1},\ y \text{ is real}$$

When $y > 0$, $I_1(z)$ is real-valued. An important fact is $\mathcal{L}\{I_1(at)/t\} = (s + \sqrt{s^2 - a^2})/a$.

Corollary 4-10 With the hypothesis of Theorem 4-10,

$$P\{T_{i0} < \infty\} = 1 \quad \text{and} \quad E(T_{i0}) = \frac{i}{\mu - \lambda} \quad \mu \geq \lambda \quad (4\text{-}101a)$$

where $i/0$ is interpreted as infinity, and

$$P\{T_{i0} < \infty\} = \left(\frac{\mu}{\lambda}\right)^i \quad \text{and} \quad E(T_{i0}) = \infty \quad \mu < \lambda \quad (4\text{-}101b)$$

PROOF Since $P\{T_{10} < \infty\} = G(\infty) = \lim_{s\downarrow 0} \tilde{G}(s)$, Equation (4-97) yields

$$P\{T_{10} < \infty\} = \frac{\lambda + \mu - |\mu - \lambda|}{2\lambda}$$

$$= \begin{cases} 1 & \text{if } \mu \geq \lambda \\ \dfrac{\mu}{\lambda} & \text{if } \mu < \lambda \end{cases}$$

In the latter case, necessarily $E(T_{10}) = \infty$; in the former case,

$$E(T_{10}) = -\frac{d}{ds} \tilde{G}(s)\bigg|_{s=0} = \frac{1}{\mu - \lambda}$$

Use (4-96b) with $j = 0$ and Lemma 4-1 to obtain

$$\tilde{G}_{i0}(s) = [\tilde{G}(s)]^i$$

Hence

$$P\{T_{i0} < \infty\} = (P\{T_{10} < \infty\})^i \quad i \in I_+$$

and

$$E(T_{i0}) = iE(T_{10}) \quad i \in I_+$$

$\square$

Example 4-14: Busy period of the M/M/1 queue The busy period of an M/M/1 queue with arrival rate λ and service rate μ is precisely T_{10}, which is characterized completely by Theorem 4-10. Observe that the mean length of the busy period is finite if, and only if, the steady-state probabilities exist (cf. Example 4-8).

Let T' denote the length of a (generic) busy cycle; then $T' = T_{01} + T_{10}$. Since T_{01} is the time until the next arrival, it has an exponential d.f. with mean $1/\lambda$. Hence

$$E(T') = \frac{1}{\lambda} + \frac{1}{\mu - \lambda} = \frac{\mu}{\lambda(\mu - \lambda)} \qquad \square$$

The qualitative behavior of T_{ij} is different for $i < j$ than for $i > j$. Since T_{01} has an exponential d.f. with mean $1/\lambda$, it is finite (w.p.1), has a finite mean whenever $\lambda > 0$, and the situation where $\lambda = 0$ is completely uninteresting. It is less trivial to analyze T_{12}.

Setting $i = 1$ and $j = 2$ in (4-93) yields

$$\tilde{G}_{12}(s) = \frac{\lambda \tilde{G}_{22}(s) + \mu \tilde{G}_{02}(s)}{s + \lambda + \mu} \tag{4-102}$$

By definition, $G_{22}(s) = 1$, and a consequence of (4-94) is

$$\tilde{G}_{02}(s) = \tilde{G}_{01}(s)\tilde{G}_{12}(s) = \lambda \tilde{G}(s)/(s + \lambda)$$

Substituting these expressions into (4-102) yields

$$\tilde{G}_{12}(s) = \frac{\lambda + \mu \lambda G_{12}(s)/(s + \lambda)}{s + \lambda + \mu}$$

and hence

$$\tilde{G}_{12}(s) = \frac{\lambda(s + \lambda)}{s(s + \lambda + \mu) + \lambda(s + \lambda)} \tag{4-103}$$

This transform can be inverted and $g_{12}(t)$ obtained explicitly (Exercise 4-57). We use (4-103) to deduce the qualitative behavior of T_{12}. From (4-103),

$$P\{T_{12} < \infty\} = \lim_{s \downarrow 0} \tilde{G}_{12}(s) = 1$$

and

$$E(T_{12}) = \lim_{s \downarrow 0} \frac{d}{ds} \tilde{G}_{12}(s) = \frac{1}{\lambda} + \frac{\mu}{\lambda^2}$$

so T_{12} never exhibits the nonterminating behavior that T_{21} can. The reason is that during a first passage from state 1 to state 2, the process is constrained to be smaller than 2 and be nonnegative, whereas during a first passage headed toward† state 1 from state 2, the value of the process is unbounded above and the process can "reach infinity" and never get to state 1. However, it is the lack of an upper bound that leads to the closed-form expression for $\tilde{G}_{ij}(s)$ for $i > j$, and the presence of the lower bound makes the computation of $\tilde{G}_{ij}(s)$ cumbersome when $i < j$. In the latter case, $\tilde{G}_{ij}(s)$ is obtained from (4-93) by recursion.

Mean Values

Now consider the problem of obtaining $E(T_{ij})$ when the birth rates and death rates may depend on the state of the process. In particular, consider $E(T_{10})$. To develop a conjectured formula for $E(T_{10})$, suppose the birth rates and death rates do not depend on the state. For $\mu > \lambda$, write (4-101a) as

† This informal term is used to emphasize that state 1 may never be visited.

$$E(T_{10}) = \frac{(\lambda/\mu) + (\lambda/\mu)^2 + \cdots}{\lambda} = \sum_{i}^{\infty} \frac{b_j}{\lambda_0} \qquad (4\text{-}104)$$

where $b_j = (\lambda_0 \lambda_1 \cdots \lambda_{j-1})/(\mu_1 \mu_2 \cdots \mu_j)$ for $j \geq 1$ and $b_0 = 1$, as in Section 4-4. Exercise 4-61 shows that (4-104) is valid when λ_0, λ_1, λ_2, μ_1, and μ_2 are arbitrary nonnegative numbers and $\lambda_i \equiv \lambda$ and $\mu_i \equiv \mu$ for $i \geq 3$, with $\lambda/\mu < 1$. It is reasonable to conjecture that (4-104) is valid for any birth-and-death process. The truth of the conjecture is a special case of Theorem 4-11.

Theorem 4-11 For any birth-and-death process

$$E(T_{i,\,i-1}) = \frac{1}{\lambda_{i-1}} \left(\frac{\lambda_{i-1}}{\mu_i} + \frac{\lambda_{i-1}\lambda_i}{\mu_i \mu_{i+1}} + \cdots \right)$$

$$= (\lambda_{i-1} b_{i-1})^{-1} \sum_{j=i}^{\infty} b_j \qquad i \in I_+ \qquad (4\text{-}105)$$

If $\sum_1^{\infty} b_i$ diverges, interpret (4-105) as $E(T_{i,\,i-1}) = \infty$.

PROOF Let $\gamma_i = E(T_{i,i-1})$, $i \in I_+$, and set $\gamma_0 = 0$. The solution of Exercise 4-60 and Equation (4-91) yield

$$\gamma_i = \frac{1 + \lambda_i \gamma_{i+1}}{\mu_i} \qquad i \in I_+ \qquad (4\text{-}106)$$

Substituting (4-105) into (4-106) yields an identity, so (4-105) is a solution of (4-106). The most general solution of the difference equation (4-106) is†
(4-105) plus the solution of

$$\alpha_i = \frac{\lambda_i}{\mu_i} \alpha_{i+1}, \qquad i \in I$$

Hence the most general solution of (4-106) is

$$\gamma_i = (\lambda_{i-1} b_{i-1})^{-1} \sum_{j=1}^{\infty} b_j + \frac{\alpha_1 \lambda_0}{b_i \mu_i} \qquad (4\text{-}107)$$

where α_1 is an arbitrary constant. Since (4-107) must agree with (4-104) when $\lambda_j \equiv \lambda$, $\mu_j \equiv \mu$ with $\lambda < \mu$ and $i = 1$, we can conclude that $\alpha_1 = 0$. ☐

Corollary 4-11 Define $T' = T_{01} + T_{10}$. Then

$$E(T') = \frac{1}{\lambda_0} \sum_{j=0}^{\infty} b_j \qquad (4\text{-}108)$$

PROOF Use (4-105) with $i = 1$ and the fact that $E(T_{01}) = 1/\lambda_0$. ☐

---

† See, for example, Francis B. Hildebrand, *Finite-Difference Equations and Simulations*, Prentice-Hall, Englewood Cliffs, N.J. (1968), section 1.6.

Observe that T', as defined above, is the time between successive entries into state zero and that Theorem 4-4 asserts that the limiting probabilities, if they exist, form an honest probability distribution if, and only if, $E(T') < \infty$. The connection among Theorems 4-3 and 4-4 and Corollary 4-4a is explored more fully in Section 6-4.

Example 4-15: M/M/c queue For the M/M/c queue, $E(T_{10})$ represents the expected length of a busy period in the sense that no servers are busy right before T_{10} starts and right after T_{10} ends and at least one server is busy during T_{10}. With the birth rates and death rates associated with the M/M/c queue, (4-105) yields

$$E(T_{10}) = \frac{\sum_{i=1}^{c} (a^i/i!) + \sum_{i=c+1}^{\infty} (a^i/c^{i-c})/c!}{\lambda}$$

$$= \frac{\sum_{i=1}^{c-1} (a^i/i!) + \rho^c/[c!c^c(1-\rho)]}{\lambda} \qquad \rho < 1$$

and $E(T_{10}) = \infty$ when $\rho \geq 1$. $\qquad\qquad\qquad\qquad\qquad\qquad \square$

The analog of Theorem 4-11 for upward transitions is Theorem 4-12.

Theorem 4-12 For any birth-and-death process, let $\delta_i \triangleq E(T_{i,i+1})$, $i \in I$. Then δ_i obeys the recursion

$$\delta_i = \frac{1 + \mu_i \delta_{i-1}}{\lambda_i} \tag{4-109}$$

$$= (\lambda_i b_i)^{-1} \sum_{j=0}^{i} b_j \qquad i \in I \tag{4-110}$$

PROOF Condition on the next event, and use (4-95b) to obtain

$$\delta_i = \frac{1}{\lambda_i + \mu_i} + \frac{\mu_i}{\lambda_i + \mu_i} (\delta_{i-1} + \delta_i)$$

Solving for δ_i yields (4-109). Iterating (4-109) for $i = 0, 1, \ldots$ yields (4-110); or, just verify that (4-110) satisfies (4-109). The details are left as Exercise 4-63. $\qquad\qquad \square$

EXERCISES

4-54 Use (4-93) to show that $G_{ij}(\cdot)$ satisfies the backward Kolmogorov equations (4-15).

4-55 Derive the backward Kolmogorov equations by conditioning on the first event.

4-56 Establish that $P_{ij}(t) = \int_0^t G_{ik}(x) \, dP_{kj}(t-x)$, $i \leq k \leq j$, and hence that $\tilde{G}_{ik}(s) = \tilde{P}_{ij}(s)/\tilde{P}_{kj}(s)$.

4-57 Show that $g_{12}(t) \triangleq dG_{12}(t)/dt$ is a mixture of two exponential densities, and display $g_{12}(t)$ explicitly.

4-58 Derive a formula for $\tilde{G}_{23}(s)$.

4-59 Sketch a procedure to obtain $\tilde{G}_{36}(s)$.

4-60 By conditioning on the next event, derive a recursive equation for determining $E(T_{ij})$.

4-61 Show that if λ_0, λ_1, and μ_1 are arbitrary but $\lambda_i \equiv \lambda$ and $\mu_i \equiv \mu$ for $i \geq 2$ with $\lambda < \mu$, then $E(T_{10})$ is given by (4-104). Do the same when λ_2 and μ_2 are arbitrary.

4-62 Argue, without making any calculations, that (4-104) implies (4-106).

4-63 Obtain (4-110) from (4-109). Check the result by substituting (4-110) into (4-109).

4-64 Find the mean first-passage time between states 0 and $i > 0$ in an M/M/1 queue. What is the LST of the time to return to state 0? Find the mean and LST of a busy cycle (i.e., the time between successive visits to state 0).

4-65 In an M/M/c queue, $T_{c,c-1}$ is a length of time during which all servers are busy. Find its LST. *Hint*: This can be done without further calculation via Example 4-14.

4-66 In an M/M/c queue, suppose customers are served LIFO (last-in, first-out). Explain why the LST of the delay in queue of a customer who arrives when all servers are busy is given by the solution of Exercise 4-65. Let p_i be the limiting probability that i customers are in the system and $D_L(t)$ be the limiting probability that a customer is delayed in queue no more than t. Show that when $\lambda < c\mu$,

$$D_L(t) = \sum_{i=1}^{c-1} p_i + \left(1 - \sum_{i=1}^{c-1} p_i\right) B(t) \qquad t \geq 0$$

where $B(\cdot)$ is the d.f. of the length of a busy period in an M/M/1 queue with arrival rate λ and mean service time $1/(c\mu)$.

4-67 In Example 4-6, set $\theta = 0$ and find the mean time to extinction. Prove that $P\{\text{population becomes extinct}\} = 1$.

4-10** Transient Analysis of the M/M/1 Queue

The purpose of this section is to show how the forward Kolmogorov equations for the M/M/1 queue can be solved. The transient probabilities may be used to study the effects of a surge in the arrival process (see Exercise 4-69).

Differential-Difference Equations

The number of customers present in an M/M/1 queue (see Example 4-3) is a birth-and-death process with $\lambda_j \equiv \lambda$ and $\mu_j \equiv \mu$. It is easily verified that (4-20) holds for this process, so the transition function satisfies the forward Kolmogorov equations (4-16), with initial conditions (4-17), and this system of differential-difference equations satisfies (4-18) and (4-19). The equations to be solved are (a prime indicates differentiation)

$$P'_{i0}(t) = -\lambda P_{i0}(t) + \mu P_{i1}(t) \qquad (4\text{-}111a)$$

$$P'_{ij}(t) = -(\lambda + \mu)P_{ij}(t) + \lambda P_{i,j-1}(t)$$
$$+ \mu P_{i,j+1}(t) \qquad i \in I_+ \qquad (4\text{-}111b)$$

$$P_{ii}(0) = 1 \qquad P_{ij}(0) = 0 \qquad i \neq j \qquad (4\text{-}111c)$$

The method of solving this system of equations treats (4-111a) and (4-111c) as boundary and initial conditions, respectively, for the system (4-111b).

Solution of the Equations

A standard method of solving differential equations is to make a change of variables that transforms the given equation to an equation whose solution is known. Certain transformations are known to be useful in making simplifications.

The key observation to make about (4-111b) is that if i is treated as a parameter, then (4-111b) is similar to

$$2I'_n(t) = I_{n-1}(t) + I_{n+1}(t) \tag{4-112}$$

The solution of (4-112) is the Bessel function of the first kind with an imaginary argument, given by

$$I_n(z) \triangleq \left(\frac{z}{2}\right)^n \sum_{k=0}^{\infty} \frac{(z^2/4)^k}{k!\,(n+k)!} \qquad n = 0, \pm 1, \pm 2, \ldots \tag{4-113}$$

A useful property of these functions is

$$I_n(z) = I_{-n}(z) \qquad n \in I \tag{4-114}$$

The bare facts about $I_n(z)$ can be obtained from Chapter 9 of Abramowitz and Stegun (1964).

To transform (4-111b) to (4-112), define the functions $F_{ij}(\,\cdot\,)$ by

$$P_{ij}(t) = \beta^j e^{\gamma t} F_{ij}(\alpha t) \qquad t \geq 0 \tag{4-115}$$

for some constants α, β, and γ which are chosen below. An immediate consequence of (4-115) is

$$P'_{ij}(t) = \beta^j e^{\gamma t}[\alpha F'_{ij}(\alpha t) + \gamma F_{ij}(\alpha t)] \tag{4-116}$$

Substituting (4-115) and (4-116) into (4-111b) yields

$$\begin{aligned} P'_{ij}(t) &= \beta^j e^{\gamma t}[\alpha F'_{ij}(\alpha t) + \gamma F_{ij}(\alpha t)] \\ &= -(\lambda + \mu)\beta^j e^{\gamma t} F_{ij}(\alpha t) \\ &\quad + \gamma \beta^{j-1} e^{\gamma t} F_{i,\,j-1}(\alpha t) + \mu \beta^{j+1} F_{i,\,j+1}(\alpha t) \end{aligned} \tag{4-117}$$

Choosing $\gamma = -(\lambda + \mu)$ will eliminate $F_{ij}(\alpha t)$ from both sides of (4-117). Multiplying both sides by $2\beta^{-j} e^{-\gamma t}/\alpha$ (assume $\alpha\beta \neq 0$) yields

$$2F'_{ij}(\alpha t) = \frac{2\lambda}{\alpha\beta} F_{i,\,j-1}(\alpha t) + \frac{2\mu\beta}{\alpha} F_{i,\,j+1}(\alpha t) \tag{4-118}$$

Equation (4-118) is of the form (4-112) when α and β are chosen to satisfy

$$\frac{2\lambda}{\alpha\beta} = \frac{2\mu\beta}{\alpha} = 1$$

This is accomplished by choosing

$$\alpha = 2\sqrt{\lambda\mu} \quad \text{and} \quad \beta = \sqrt{\frac{\lambda}{\mu}} \tag{4-119}$$

Substituting (4-119) into (4-118) and making the change of variable $s = \alpha t$ yield

$$2F'_{ij}(s) = F_{i,\,j-1}(s) + F_{i,\,j+1}(s) \tag{4-120b}$$

Performing the same operations on (4-111a) and (4-111c) yields

$$2F'_{i0}(s) = \beta^{-1}F_{i0}(s) + F_{i1}(s) \tag{4-120a}$$

and

$$F_{ii}(0) = \beta^{-i} \qquad F_{ij}(0) = 0 \qquad \text{for} \qquad i \neq j \tag{4-120c}$$

Since (4-112) holds for any n, the general solution of (4-120b) is

$$F_{ij}(s) = \sum_{n=-\infty}^{\infty} c_n I_{n+j}(s) \tag{4-121}$$

where $\{c_n\}_{-\infty}^{\infty}$ is any set of constants. These constants are to be chosen to satisfy (4-120a) and (4-120c). From (4-113),

$$I_0(0) = 1 \qquad I_n(0) = 0 \qquad \text{for} \qquad n \neq 0 \tag{4-122}$$

Substituting (4-121) into (4-120c) and using (4-122) yield

$$c_{-j} = \beta^{-i} \quad \text{and} \quad c_{-j} = 0 \qquad \text{for} \qquad j \neq i, i \in I \tag{4-123}$$

Substituting (4-121) into (4-120a) and using (4-112) provide

$$2 \sum_{n=-\infty}^{\infty} c_n I_n'(s) = \sum_{n=-\infty}^{\infty} c_n [I_n(s) + I_{n+1}(s)]$$

$$= \beta^{-1} \sum_{n=-\infty}^{\infty} c_n I_n(s) + \sum_{n=-\infty}^{\infty} c_n I_{n+1}(s)$$

Hence

$$\sum_{n=-\infty}^{\infty} c_{n+1} I_n(s) = \beta^{-1} \sum_{n=-\infty}^{\infty} c_n I_n(s)$$

Using (4-114) to equate the coefficients of $I_n(s)$, $n \in I$, we obtain

$$c_{n+1} + c_{-n+1} = \beta^{-1}(c_n + c_{-n}) \qquad n \in I$$

Iterating for $n \in I_+$ yields

$$c_{n+1} = \beta^{-1}c_{-n} + (\beta^{-2} - 1)(c_{-n+1} + \beta^{-1}c_{-n+2} + \cdots + \beta^{-n-1}c_0) \tag{4-124}$$

Combining (4-123) and (4-124) shows

$$0 = c_0 = c_1 = \cdots = c_i \tag{4-125a}$$

$$c_{i+1} = \beta^{-1}c_{-i} = \beta^{-i-1} = \rho^{-(i+1)/2} \tag{4-125b}$$

$$c_{i+k} = \beta^{-2} - 1 = (\beta^{-2} - 1)\beta^{-i-k+2}$$

$$= (1 - \rho)\rho^{-(i+k)/2} \qquad k = 2, 3, \ldots \tag{4-125c}$$

Substituting (4-125) into (4-121), and then into (4-115), and simplifying yield the final result:

$$P_{ij}(t) = \rho^{(j-i)/2}e^{-(\lambda+\mu)t}\left[I_{i-j}(2t\sqrt{\lambda\mu}) + \rho^{-1/2}I_{i+j+1}(2t\sqrt{\lambda\mu}) \right.$$

$$\left. + (1-\rho)\sum_{k=2}^{\infty}\rho^{-k/2}I_{i+j+k}(2t\sqrt{\lambda\mu}) \right] \qquad i, j \in I, t \geq 0 \qquad (4\text{-}126)$$

This formula for $P_{ij}(t)$ is not easy to work with analytically, and it is not trivial to use for computational purposes. It does illustrate the point that transient solutions are difficult to obtain and to use, and it should motivate the search for approximations undertaken in Chapter 13.

EXERCISES

4-68 Let $X(t)$ denote the number of customers present at time t in an M/M/1 queue, and $L_i(t) = E(X(t)|X(0) = i), t \geq 0$. Give an intuitive argument why

$$L_i(t) = \lambda t - \mu \int_0^t [1 - P_{i0}(x)] \, dx + i \qquad (4\text{-}127)$$

Provide a rigorous proof of (4-127).

4-69 (Continuation) Suppose $t = 0$ represents an epoch just after an abnormally large surge of customers have arrived. How can you use (4-127) to describe the evolution of the mean number of customers in the system?

4-70 Let $V_i(t) = \text{Var}(X(t)|X(0) = i), t \geq 0$. Derive a differential equation for $V_i(\cdot)$.

4-71 Equation (4-126) is valid for any ρ. From Exercise 4-22, when $\rho < 1, \lim P_{ij}(t) = (1-\rho)\rho^j$ for each $j \in \mathcal{S}$ and any i. This fact is not immediately apparent from (4-126). Use the fact [see, e.g., Abramowitz and Stegun (1964)] that as $z \to \infty$,

$$I_n(z) = \frac{e^z}{\sqrt{2\pi z}} + o\left(\frac{1}{z}\right)$$

to show

$$P_{ij}(t) \to (1-\rho)\rho^j$$

4-11** EXPLODING BIRTH-AND-DEATH PROCESSES

The sample path $X(\omega; t)$ of the birth-and-death process $\{X(t); t \geq 0\}$ is said to *explode* at t if $X(t; \omega) > j$ for any positive integer j; roughly, $X(t; \omega) = \infty$. This can occur only if the state space is infinite; so assume $\mathcal{S} = I$. An explosion at t occurs with positive probability if

$$\sum_{j=0}^{\infty} P_{ij}(t) < 1 \qquad (4\text{-}128)$$

when $X(0) = i$. Observe that if (4-128) holds for some i, it holds for all i (Exercise 4-72).

Let S_k be the epoch of the kth transition. If an explosion occurs, it is clear that infinitely many transitions occur by a finite time. It is true, but not obviously so, that the reverse statement holds.

Proposition 4-4 Let $S_\infty(\omega) = \lim_{k \to \infty} S_k(\omega)$. Then

$$X(s; \omega) = \infty \text{ for some } s \leq t \Leftrightarrow S_\infty(\omega) \leq t \qquad (\text{w.p.1})$$

and

$$\sum_{j=0}^{\infty} P_{ij}(t) < 1 \Leftrightarrow P\{S_\infty \leq t\} > 0$$

PROOF Since the second statement is a direct consequence of the first and the $\Rightarrow$ part of the first is obvious, it suffices to prove the $\Leftarrow$ part of the first statement. Suppose

$$S_\infty(\omega) \leq t \qquad \text{and} \qquad X(s; \omega) < \infty \qquad 0 \leq s \leq t \qquad (4\text{-}129)$$

were true. Since there are infinitely many transitions and only finitely many states are visited, some state (n say) must be visited infinitely often. Let $\tau_{mn}(\omega)$ be the length of the mth visit to state n; fix n and suppress it for notational convenience. From Theorem 4-6, $\tau_1, \tau_2, \ldots$ are i.i.d. and have an exponential distribution with mean $1/(\lambda_n + \mu_n) < \infty$. The strong law of large numbers asserts that

$$\lim_{M \to \infty} \sum_{m=1}^{M} \tau_m(\omega)/M = \frac{1}{\lambda_n + \mu_n} \qquad (\text{w.p.1})$$

which implies $\lim_{M \to \infty} \sum_{m=1}^{M} \tau_m(\omega) = \infty$, which contradicts (4-129). $\qquad \square$

Pure Birth Process

The first step toward obtaining a necessary and sufficient condition for

$$\sum_{j=0}^{\infty} P_{ij}(t) = 1$$

to hold for a general birth-and-death process is to find such a condition for the pure birth process. The basic idea is that $1/\lambda_n$ is the mean time between the nth and the $(n + 1)$th birth. Then $S' \triangleq \sum_{n=0}^{\infty} (1/\lambda_n)$ is the mean time until infinitely many births occur. If S' is finite, there should be infinitely many births by epoch S', while if S' is infinite, there should be only finitely many births in any finite interval of time. This reasoning is made precise in the proof of Theorem 4-13.

Theorem 4-13: Feller-Lundberg Theorem Let $\{X(t); t \geq 0\}$ be a pure birth process with $X(0) = i$, and let $P_n(t) = P\{X(t) = n\}$. Then

$$\sum_{n \geq i} P_n(t) = 1 \text{ for all } t > 0 \Leftrightarrow \sum_{n \geq i} \frac{1}{\lambda_n} = \infty \tag{4-130}$$

PROOF Without loss of generality, choose $i = 0$ and suppress i in the notation. For any $t > 0$, define

$$P_j^c(t) \triangleq P\{X(t) > j\} = \sum_{n > j} P_n(t) \qquad Q_j(t) \triangleq 1 - P_j^c(t) = P\{X(t) \leq j\}$$

and $P^c(t) = \lim_{j \to \infty} P_j^c(t)$. The limit $P^c(t)$ must exist because $P_j^c(t)$ is decreasing with j and bounded by zero. Define $\lambda_{-1} = 0$, and obtain the forward equations for this process from (4-16):

$$\frac{d}{dt} P_n(t) = \lambda_{n-1} P_{n-1}(t) - \lambda_n P_n(t) \qquad n \in I$$

Summation from $n = 0$ to $n = j$ yields

$$\frac{d}{dt} Q_j(t) = -\lambda_j P_j(t) \qquad j \in I \tag{4-131}$$

Since the initial state is zero,

$$Q_j(0) = 1 \tag{4-132}$$

is the boundary condition for (4-131).

The solution to (4-131) and (4-132) is

$$1 - Q_j(t) \triangleq P_j^c(t) = \lambda_j \int_0^t P_j(u) \, du \tag{4-133}$$

Since $P_j^c(t) \leq 1$ and decreases with j for any fixed $t > 0$, (4-133) implies

$$\frac{P^c(t)}{\lambda_j} \leq \int_0^t P_j(u) \, du \leq \frac{1}{\lambda_j} \tag{4-134}$$

Summation of (4-134) from $j = 1$ to $j = n$ yields

$$P^c(t) \sum_{j=0}^n \frac{1}{\lambda_j} \leq \int_0^t Q_n(u) \, du \leq \sum_{j=0}^n \frac{1}{\lambda_j} \tag{4-135}$$

Since $Q_n(u) \leq 1$, the middle term in (4-135) is no larger than t. If

$$\lim_{n \to \infty} \sum_{j=0}^n (1/\lambda_j) = \infty$$

the first inequality in (4-135) is valid only if $P^c(t) = 0$, which proves the $\Leftarrow$ part of (4-130). If $\lim_{n \to \infty} \sum_{j=0}^n (1/\lambda_j) < \infty$, then $\int_0^\infty Q_n(u) \, du < \infty$, so $Q_n(t) < 1$ for at least some values of t. The proof is completed by verifying that if $Q_n(t) < 1$ for some $t > 0$, then $Q_n(t) < 1$ for all $t > 0$. This result is true for all birth-and-death processes (Exercise 4-73). $\qquad \square$

General Case

Let T_{ij} be the first-passage time from state i to state j. Set $T_i \triangleq T_{i,\,i+1}$ and let $\delta_i = E(T_i)$. A generalization of Theorem 4-13 to general birth-and-death process is found in Theorem 4-14.

Theorem 4-14

$$P\{S_\infty < \infty\} = 1 \Leftrightarrow \sum_{n=0}^{\infty} \delta_n < \infty$$

PROOF Let τ_{1n} be the length of the first visit to state n. For each sample path ω, $\tau_{1n}(\omega) \le T_{n,\,n+1}(\omega)$ holds. (Why?) The mean of an r.v. can be finite only if the r.v. is finite, so $\sum_{n=0}^{\infty} \delta_n < \infty \Rightarrow P\{\sum_{n=0}^{\infty} T_n < \infty\} = 1$. Manifestly,

$$\sum_{n=0}^{\infty} \delta_n \ge \sum_{n=0}^{\infty} E(\tau_{1n}) \qquad \text{and} \qquad P\left\{\sum_{n=0}^{\infty} \tau_{1n} < \infty\right\} \ge P\left\{\sum_{n=0}^{\infty} T_n < \infty\right\}$$

$$(4\text{-}136)$$

From Theorem 4-13, $\sum_{n=0}^{\infty} E(\tau_{1n}) = \infty \Rightarrow P\{\sum_{n=0}^{\infty} \tau_{1n} < \infty\} < 1$. Hence

$$\sum_{n=0}^{\infty} E(\tau_{1n}) < \infty \Leftarrow P\left\{\sum_{n=0}^{\infty} T_n < \infty\right\} = 1 \qquad (4\text{-}137)$$

Since $E(\tau_{1n}) = 1/(\lambda_n + \mu_n)$, if $\inf_n(\lambda_n + \mu_n)^{-1} = 0$, then $\sum_{n=0}^{\infty} E(\tau_{1n}) = \infty$, and (4-136) and (4-137) show that $\sum_{n=0}^{\infty} \delta_n = \infty$ and $P\{\sum_{n=0}^{\infty} T_n < \infty\} < 1$. Restate this, with the aid of Proposition 4-4, as

$$P\{S_\infty < \infty\} = 1 \Leftarrow \sum_{n=0}^{\infty} \delta_n < \infty \qquad (4\text{-}138)$$

If $\inf_n(\lambda_n + \mu_n)^{-1} = \beta > 0$, let $i_0, i_1, \ldots, i_m$ be the states visited during a first passage from n to $n + 1$; that is, $i_0 = n$, $i_m = n + 1$, and $i_j \notin \{n, n + 1\}$ for $i \le j \le m - 1$, $m \ge 1$. Let V_j be the length of the visit to state i_j, $0 \le j \le m$. By Theorem 4-6, $V_0, V_1, \ldots, V_m$ are independent, $E(V_j) = (\lambda_{i_j} + \mu_{i_j})^{-1}$ and $\text{Var}(V_j) = (\lambda_{i_j} + \mu_{i_j})^{-2} \le \beta(\lambda_{i_j} + \mu_{i_j})^{-1}$. Since

$$\sum_{j=0}^{m} \text{Var}(V_j) \le \beta \sum_{j=0}^{m} E(V_j)$$

for any $m \in I_+$,

$$\text{Var}(T_n) \le \beta E(T_n) \triangleq \beta \delta_n$$

Suppose

$$P\left\{\sum_{n=0}^{\infty} T_n < \infty\right\} = 1 \qquad \text{and} \qquad \sum_{n=0}^{\infty} \delta_n = \infty \qquad (4\text{-}139)$$

both hold. Then for any $\xi \in (0, 1)$, as $N \to \infty$,

$$P\left\{ \left| \sum_{n=0}^{N} (T_n - \delta_n) \right| > \xi \sum_{n=0}^{N} \delta_n \right\} \to 1 \qquad (4\text{-}140)$$

by the strong law of large numbers. But (4-139) and Chebyshev's inequality applied to the left-side of (4-140) yield

$$P\left\{ \left| \sum_{n=0}^{N} (T_n - \delta_n) \right| > \xi \sum_{n=0}^{N} \delta_n \right\} \le \frac{\beta}{\xi \sum_{n=0}^{N} \delta_n} \qquad (4\text{-}141)$$

If (4-139) were true, letting $N \to \infty$ in (4-141) would contradict (4-140), so (4-139) must be false and

$$P\{S_\infty < \infty\} = 1 \Rightarrow \sum_{n=0}^{\infty} \delta_n < \infty \qquad \square$$

The proof of Theorem 4-1 follows from Theorems 4-14 and 4-12, as shown below. Theorem 4-1 says that for any birth-and-death process,

$$\sum_{j=0}^{\infty} P_{ij}(t) = 1 \qquad t \ge 0 \qquad (4\text{-}19')$$

if, and only if,

$$\sum_{j=0}^{\infty} \left[b_j + (\lambda_j b_j)^{-1} \right] = \infty \qquad (4\text{-}20)$$

where

$$b_0 = 1 \qquad b_j = \frac{\lambda_0 \lambda_1 \cdots \lambda_{j-1}}{\mu_1 \mu_2 \cdots \mu_j} \qquad j \in I_+ \qquad (4\text{-}20')$$

PROOF Proposition 4-4 and Theorem 4-14 imply that (4-19') holds if, and only if, $\sum_{n=0}^{\infty} \delta_n = \infty$. Summing both sides of (4-110) over all $i \in I$ yields

$$\sum_{i=0}^{\infty} \delta_i = \sum_{i=0}^{\infty} (\lambda_i b_i)^{-1} \sum_{j=0}^{i} b_j$$

Hence, (4-19') holds if, and only if,†

$$\sum_{i=0}^{\infty} (\lambda_i b_i)^{-1} \sum_{j=0}^{i} b_j = \sum_{j=0}^{\infty} b_j \sum_{i=0}^{j} (\lambda_i b_i)^{-1} = \infty \qquad (4\text{-}142)$$

---

†Some authors state Theorem 4-14 with (4-142) in place of (4-20).

The proof is completed by establishing (4-142) $\Leftrightarrow$ (4-20), as follows. If (4-20) holds, either $\sum_0^\infty b_j$ or $\sum_0^\infty (\lambda_i b_i)^{-1}$ diverges. Assume the former sum diverges. Since $(\lambda_i b_i)^{-1} > 0$ for all $i \in I$,

$$\sum_{j=0}^{\infty} b_j \sum_{i=0}^{j} (\lambda_i b_i)^{-1} \ge (\lambda_0 b_0)^{-1} \sum_{j=0}^{\infty} b_j = \infty$$

and (4-142) holds. If the latter sum diverges,

$$\sum_{i=0}^{\infty} (\lambda_i b_i)^{-1} \sum_{j=0}^{i} b_j \ge b_0 \sum_{i=0}^{\infty} (\lambda_i b_i)^{-1} = \infty$$

and hence (4-20) $\Rightarrow$ (4-142). If (4-20) is false, we obtain

$$\sum_{j=0}^{\infty} b_j \sum_{i=0}^{j} (\lambda_i b_i)^{-1} \le \sum_{j=0}^{\infty} b_j \sum_{i=0}^{\infty} (\lambda_i b_i)^{-1} < \infty$$

Hence (4-142) is false; that is, (4-142) $\Rightarrow$ (4-20). $\qquad\square$

Theorem 4-1 holds formally for finite state spaces by setting $\lambda_i = 0$ for $i \in I - S$.

EXERCISES

4-72 Prove that if (4-128) holds for some i, then it holds for all $i \in I$.

4-73 For a birth-and-death process, prove that if $P_{ij}(t) > 0$, then $P_i^c(t) \triangleq \sum_{j=0}^{\infty} P_{ij}(t)$ is either identically 1 or $P_i^c(t) < 1$ for all $t > 0$. *Hint*: Use the Chapman-Kolmogorov equation.

4-74 Give a probabilistic derivation of (4-133), using the flow-of-probability concept.

BIBLIOGRAPHIC GUIDE

Queueing models were studied by the methods described in this chapter before the theory of birth-and-death processes was developed. The books about queueing theory listed in the bibliographic guide in Chapter 11 contain similar descriptions of the use of birth-and-death processes in the analysis of queues. Among the more notable papers that develop the theoretical aspects of birth-and-death processes are Lederman and Reuter (1954) and Karlin and McGregor (1957a, 1957b).

Most of the material presented in this chapter is standard fare in queueing theory. Proposition 4-2 does not seem to be widely known; it is a special case of the theorem in Morris and Wolman (1961). Section 4-10 follows Riordan (1962, Section 4.2), and Section 4-11 is inspired by Breiman (1968, Section 15.7).

Formulas and tables of Bessel functions, as well as many other important functions, can be found in Abramowitz and Stegun (1964).

Abramowitz, Milton, and Irene A. Stegun, eds., *Handbook of Mathematical Functions*, U.S. Department of Commerce, National Bureau of Standards, Washington, D.C. (1964).

Breiman, Leo, *Probability*, Addison-Wesley, Reading, Mass. (1968).

Karlin, S., and J. L. McGregor, "The Differential Equations of Birth-and-Death Processes, and the Stieltjes Moment Problem," *Trans. Am. Math Soc.* **85**: 489–546 (1957*a*).

——, and ——, "The Classification of Birth and Death Processes," *Trans. Am. Math. Soc.* **86**: 366–400 (1957*b*).

Lederman, W., and G. E. Reuter, "Spectral Theory for the Differential Equations of Simple Birth and Death Processes," *Phil. Trans. Roy. Soc. London* **A246**: 387–391 (1954).

Morris, Robert, and Eric Wolman, "A Note on 'Statistical Equilibrium,'" *Oper. Res.* **9**: 751–753 (1961).

Riordan, John, *Stochastic Service Systems*, Wiley, New York (1962).

RENEWAL THEORY

In Chapter 4 we defined the Poisson process (Example 4-7) and showed that it has the property that the times between transitions are i.i.d. with an exponential distribution. One way to express these results is to define the i.i.d. random variables X_1, X_2, ... and let X_i represent the time between the $(i - 1)$th and ith transitions of the stochastic process $\{N(t); t \geq 0\}$ which counts the number of transitions that occur by time t. When $F(t) = P\{X_1 \leq t\} = 1 - e^{-\lambda t}$, then $\{N(t); t \geq 0\}$ is a Poisson process. Renewal theory generalizes this structure by allowing $F(\cdot)$ (which is common to all the X_i's) to be *any* distribution function corresponding to a nonnegative random variable, and the objective of the theory is to discover the properties of $\{N(t); t \geq 0\}$, which is the *renewal process*.

Renewal theory originally was applied in the context of reliability studies of complex equipment. A typical problem was to use the known distributions for the time to failure of each component to find the mean number of failures of the entire system in an interval of length t. Renewal theory subsequently has become a powerful tool for analyzing stochastic processes which, on the surface, have nothing to do with the subject matter of renewal theory. For example, we use some of the theorems of this chapter in our analysis of Markov chains (Chapter 7), and in this chapter we show how renewal theory is used in the analysis of queues and inventories.

5-1 EXAMPLES AND A NONINTUITIVE RESULT

Example 5-1: A replacement model Many electronic devices, e.g., a large digital computer, are built out of smaller devices called *circuit packs*. A circuit pack is a collection of electronic components put together to perform a specific job; for example, a single circuit pack may contain all the circuitry required to be an amplifier. When a circuit pack fails, it is removed from its slot and replaced by another one which is assumed to have been manufactured under identical conditions.

Let us focus our attention on a single slot. Let X_i denote the length of time that the ith circuit pack worked and $N(t)$ denote the number of spare circuit packs installed by time t. It is reasonable to assume that $X_1, X_2, \ldots$ are i.i.d. random variables, so $\{N(t); t \geq 0\}$ is a renewal process. In order to make plans for providing spare parts, we might want to know $P\{N(t) = n\}$ and $E[N(t)]$. More subtle random variables arise in this context; at some fixed time t, we may ask for the length of time until the next failure and the length of time since the last failure. Characterizing these quantities is the primary goal of renewal theory. □

Example 5-2: Renewal-reward processes Continuing with the setting of Example 5-1, we may consider the cost associated with each machine failure and compare maintenance strategies. The machine lives $X_1, X_2, \ldots$ induce the actual costs of maintenance and failure. A possible strategy for repairing the machine is to fix it when, and only when, it fails. Another strategy is to do preventive maintenance periodically, at a low cost, and to make emergency repairs when the machine fails (which should be less often than with the first strategy). Call these two strategies 1 and 2, respectively, and let $C_i(t)$ be the cumulative costs incurred by strategy i by time t; we call $\{C_i(t); t \geq 0\}$ a renewal-reward process. By comparing $\{C_1(t); t \geq 0\}$ and $\{C_2(t); t \geq 0\}$ according to some criterion, we can decide which strategy is preferable. In Volume II we study how to choose the best strategy from a given class of strategies without explicitly computing a measure of cost for the renewal-reward process associated with each strategy. □

An extensive treatment of renewal-reward processes is given in Chapter 6.

We described the renewal-reward process as a way of attaching costs to failures in a natural setting for such a model. The use of renewal-reward processes goes beyond such settings, and they are a powerful way to obtain important probabilistic results. This notion is explained more fully in Example 5-31.

Example 5-3: $GI/G/1$ queue The $GI/G/1$ queue is a generalization of the $M/M/1$ queue where the exponential distributions are replaced by arbitrary distributions. Specifically, we assume that customer i arrives at time T_i, $i \in I$, with $0 = T_0 \leq T_1 \leq T_2 \leq \ldots$, and $U_i = T_{i+1} - T_i$, $i \in I_+$, are i.i.d. In other words, the number of arrivals in time t is a renewal process. Associated with customer i is a service time V_i, and these are i.i.d. and independent of the arrival times. There is a single server to process the arrivals.

Assume the system is empty at time zero, let T_{i_j} be the epoch of the jth customer to arrive when the system is empty, and define $X_j = T_{i_{j+1}} - T_{i_j}$.

You should be able to argue that $X_1, X_2, \ldots$ are i.i.d., and so the number of epochs, before t, when a customer arrives and finds the server idle is a renewal process. This observation leads to a powerful method for proving limit theorems and to a useful way of computing steady-state probabilities and analyzing simulation experiments. □

In these examples the X's and t had the dimension of time. Example 5-4 shows another interpretation.

Example 5-4 Suppose we want to store a list of names in a computer. The number of bits needed to represent a name is proportional to the number of letters in the name. When we do not look at the list before storing it in the computer, it is appropriate to model the number of bits in the ith name as a random variable, X_i say. The bits are stored in the computer on tracks; each track can hold a fixed number of bits, T say. The number of names stored in the first t bits of a track is a random variable, $N(t)$ say, and $N(T)$ is the number of names stored on a track. Clearly $\{N(t); t = 1, 2, \ldots, T\}$ is a renewal process. In the context of Example 5-1, the X's correspond to the times between failures, and $N(t)$ corresponds to the number of failures by the discrete time t.

In this model we are particularly interested in finding $P\{N(T) = k\}$, but other problems are important too. If the data are organized so that an incomplete name is continued on the next track, we want to determine how many bits would be left over from the truncated name at the end of a track. If an incomplete name is not continued on the next track, but is started over again at the start of the next track, we want to determine how much of a track is not used. In terms of Example 5-1, the former is the time until the next failure and the latter is the time since the last failure. □

Throughout this chapter we call $X_1, X_2, \ldots$ the *governing sequence* for the renewal process $\{N(t); t \geq 0\}$. Let X denote a generic interrenewal time, and let $F(t) = P\{X \leq t\}$, with $P\{X < 0\} = 0$. When it exists, we let $f(\cdot)$ be the density function of $F(\cdot)$. We also let $\nu = E(X)$ and $\sigma^2 = \text{Var}(X)$. To avoid trivialities, we assume $F(0) < 1$.

A Nonintuitive Result

To show that renewal theory provides qualitative conclusions that are not obvious, we start by giving an informal derivation of a result that is usually regarded as counterintuitive.

Suppose we observe a renewal process at a random time long after the process started (i.e., in the steady state) and measure the time between the last renewal and the next renewal. For example, suppose you walk into your office some day and decide to find out how long it has been since a certain light bulb has been replaced. Then you measure how long it takes for this light bulb to fail.

When $f(\cdot)$ is the density function of the time between failures, one would (naively) expect the density for the lifetime of the light bulb chosen above to be $f(\cdot)$. But this is not so! Let us see why.

Let $f_S(t)$ be the density of the lifetime of the sampled bulb. Since you are twice as likely to observe a bulb of lifetime $2t$ as one of lifetime t and the proportion of bulbs with lifetime t is $f(t)\,dt$, the probability that the lifetime of the sampled bulb is of length t is proportional to $tf(t)$. Hence,

$$f_S(t) = ctf(t)$$

where c is an appropriate constant. But

$$\int_0^\infty f_S(t)\,dt = 1$$

because $f_S(\cdot)$ is a probability density function (p.d.f.), so c must equal $1/v$. Thus,

$$f_S(t) = \frac{tf(t)}{v}$$

which is not equal to $f(t)$ unless all the lifetimes are a constant. We can also calculate

$$v_S = \int_0^\infty tf_S(t)\,dt = v + \frac{\sigma^2}{v}$$

so $v_S > v$ for any nondegenerate lifetime distribution because $\sigma^2 > 0$. For the exponential density with mean $1/\mu$, $\sigma^2 = 1/\mu^2$ and $v_S = 2/\mu = 2v$, so the sampled lifetime is twice as long, on the average, as an ordinary lifetime.

Once we have the result, it is easy to see why it is so. By sampling at random, one is more likely to observe long lifetimes than short ones. For this reason, $f_S(\cdot)$ is called the *length-biased p.d.f.*, and the fact that $f_S(\cdot) \neq f(\cdot)$ is called the *inspection paradox*.

EXERCISES

5-1 Is there a pure birth process other than the Poisson process which is a renewal process? Explain.

5-2 Use the memoryless property of the exponential distribution to obtain the inspection paradox for a Poisson process.

5-2 ELEMENTARY PROPERTIES OF A RENEWAL PROCESS

Formal Definition

For a given governing sequence, let $N(t)$ denote the number of renewals by time t, $t \geq 0$. To define $N(t)$ precisely, first define S_n by

$$S_0 = 0 \qquad S_n = X_1 + X_2 + \cdots + X_n \qquad n \in I_+ \tag{5-1}$$

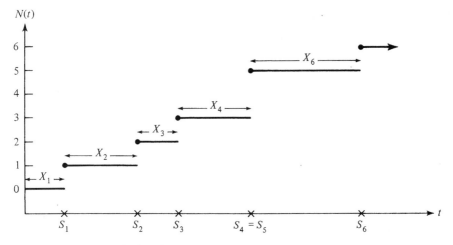

Figure 5-1 A partial realization of a renewal process.

For $n \geq 1$, S_n represents the epoch at which the nth renewal occurs; $\{S_n; n \in I_+\}$ is the set of renewal epochs. Since exactly n renewals occur by time t if, and only if, the epoch of the nth event is no later than t *and* the epoch of the $(n + 1)$th event is later than t, we have

$$\{N(t) = n\} \Leftrightarrow \{S_n \leq t, S_{n+1} > t\} \tag{5-2}$$

Hence, $N(t)$ is the largest index n such that $S_n \leq t$. The formal representation of this statement is given in Definition 5-1.

Definition 5-1 Let $X_1, X_2, \ldots$ be i.i.d. nonnegative r.v.'s and $\{S_n; n \in I\}$ be defined by (5-1). The *renewal process* $\{N(t); t \geq 0\}$ is defined by

$$N(t) = \sup \{n: S_n \leq t\} \qquad t \geq 0 \tag{5-3}$$

These concepts are illustrated in Figure 5-1. Note that S_0 is not a renewal epoch, although some authors treat it as such. This is just a matter of convention and results in increasing $N(t)$ by 1.†

Since $N(t)$ is a counting process, its sample paths are nondecreasing step functions that are continuous from the right. If $F(0) = 0$, the magnitudes of the jumps are 1 (w.p.1); if $F(0) > 0$, the magnitude of a jump is j when exactly j consecutive X's take the value zero; this occurs with probability $[F(0)]^j [1 - F(0)]$. Note that in Figure 5-1 there is a jump of size 2 at epoch S_4, consequently, for this sample path there is no t such that $N(t) = 4$.

---

† Some authors define the renewal process as $\{S_n; n = 0, 1, \ldots\}$ and call $\{N(t); t \geq 0\}$ the renewal *counting* process. When you refer to other works on renewal theory, be sure to check if they count a renewal at time zero.

Distribution of the Number of Renewals

Let

$$p_n(t) = P\{N(t) = n\} \quad \text{and} \quad F_n(t) = P\{S_n \le t\}$$

Since $X_1, X_2, \ldots$ are i.i.d., it follows from (5-1) that $F_n(\cdot)$ is the n-fold convolution of $F(\cdot)$ with itself,

$$F_n(t) = F^{*(n)}(t) \qquad n = 0, 1, \ldots \tag{5-4}$$

where we use the convention that $F_0(t) = 1$ for $t > 0$ and $F_0(t) = 0$ otherwise. It follows directly from (5-3), or Figure 5-1, that

$$\{N(t) < n\} \Leftrightarrow \{S_n > t\} \tag{5-5}$$

i.e., the number of renewals that occur by t is less than n if, and only if, the epoch of the nth renewal is after t. Therefore,

$$
\begin{aligned}
p_n(t) &= P\{n \le N(t) < n + 1\} = P\{N(t) < n + 1\} - P\{N(t) < n\} \\
&= P\{S_{n+1} > t\} - P\{S_n > t\} \\
&= 1 - F_{n+1}(t) - [1 - F_n(t)] \\
&= F_n(t) - F_{n+1}(t) \qquad t \ge 0, n \in I
\end{aligned}
\tag{5-6}
$$

Thus, $p_n(t)$ is determined, in principle, by (5-4) and (5-6). A more detailed characterization of $p_n(t)$ can be obtained by transform methods. (See Section A-3 for an explanation of the properties of transforms we will use.) We use a tilde to denote the Laplace-Stieltjes transform (LST), for example

$$\tilde{F}(s) = \int_0^\infty e^{-st} \, dF(t)$$

Theorem 5-1 For $p_n(\cdot)$ given by (5-6), $|z| \le 1$ and $s \ge 0$,

$$\sum_{n=0}^\infty z^n \tilde{p}_n(s) = \frac{1 - \tilde{F}(s)}{1 - z\tilde{F}(s)} \tag{5-7}$$

PROOF From the convolution property of LSTs, $\tilde{F}_n(s) = [\tilde{F}(s)]^n$. Taking LSTs on both sides of (5-6) yields

$$\tilde{p}_n(s) = [\tilde{F}(s)]^n [1 - \tilde{F}(s)] \qquad n \in I \tag{5-7'}$$

For each n multiply both sides of this equation by z^n and then sum over n to obtain (5-7). □

If we could invert (5-7) or (5-7'), we would obtain a formula for $p_n(t)$. Inversion is usually very difficult to do analytically, but is straightforward in a particularly important case.

Example 5-5 Let $F(t) = 1 - e^{-\lambda t}$, so $\tilde{F}(s) = \lambda/(s + \lambda)$. From (5-7'),

$$\tilde{p}_n(s) = \frac{s\lambda^n}{(s + \lambda)^{n+1}} \qquad n \in I$$

Since $\int_0^\infty e^{-st} p_n(t)\, dt = \tilde{p}_n(s)/s$, the Laplace transform of $p_n(t)$ is given by

$$\mathscr{L}\{p_n(t)\} = \frac{\lambda^n}{(s + \lambda)^{n+1}} \qquad n \in I$$

Inverting by inspection (see Table A-2) yields

$$p_n(t) = \frac{e^{-\lambda t}(\lambda t)^n}{n!} \qquad n \in I \qquad (5\text{-}8)$$

Note that (5-8) is identical to (4-26), which was obtained from different assumptions. Since a Poisson process is a renewal process (Exercise 5-7), it is the renewal process where the times between events have an exponential distribution. $\qquad\square$

EXERCISES

5-3 Let Y be a nonnegative random variable and $G^c(y) = P\{Y > y\}$. Prove that

$$E(Y) = \int_0^\infty G^c(y)\, dy$$

More generally, show that if $k \geq 0$, then

$$\int_0^\infty y^k G^c(y)\, dy = \frac{E(Y^{k+1})}{k + 1}$$

5-4 Use the result of Exercise 5-3 and Equation (5-5) to show that $E[N(t)] = \sum_{n=1}^\infty F_n(t)$.

5-5 Let $\gamma_k(t)$ be the gamma density with shape parameter k and scale parameter λ; that is, $\gamma_k(t) = (\lambda t)^{k-1} \lambda e^{-\lambda t}/(k - 1)!$. Use renewal theory to show

$$\sum_{j=k}^\infty \frac{e^{-\lambda t}(\lambda t)^j}{j!} = \int_0^t \gamma_k(x)\, dx$$

This is a relationship between the tail of the Poisson distribution and the gamma probability density function. What are the implications of this relationship for numerical computations?

5-6 Let $\{N(t); t \geq 0\}$ be a renewal process. Construct two new stochastic processes $\{N_1(t); t \geq 0\}$ and $\{N_2(t); t \geq 0\}$ as follows. At each renewal epoch of the original process, place the renewal in process 1 with probability p or in process 2 with probability $1 - p$; the placements of successive renewals are done independently. Show that the new processes are renewal processes, and find the distribution of the interevent times for process 1. Are the two new renewal processes independent? [That is, for each t, is $N_1(t)$ independent of $N_2(t)$?]

5-7 Explain why a Poisson process must be a renewal process. (That is, which theorems in Chapter 4 prove this assertion?)

5-8 In Exercise 5-6, let $\{N(t); t \geq 0\}$ be a Poisson process. Show that $\{N_1(t); t \geq 0\}$ and $\{N_2(t); t \geq 0\}$ are Poisson processes and that for each t, $N_1(t)$ and $N_2(t)$ are independent.

5-9 The superposition of two renewal processes is formed by interleaving the renewal epochs of the two processes to form a single set of epochs. Show that the superposition of two independent Poisson processes is another Poisson process. What is the distribution of the times between renewals?

5-10 Show by example that the superposition of two independent renewal processes is, in general, not a renewal process.

5-11 The *interrupted Poisson process* is formed by looking at two types of intervals, on-intervals and off-intervals. We start with an on-interval, and during it events are generated by a Poisson process with rate λ. The length of the on-interval has an exponential distribution with mean α^{-1}. At the end of an on-interval, an off-interval commences; its length has an exponential distribution with mean β^{-1}. During an off-interval, no events are generated. At the end of an off-interval, another on-interval commences, and so forth indefinitely. All off- and on-intervals are i.i.d. and independent of one another. Show that the interrupted Poisson process is a renewal process with interevent times distributed according to a hyperexponential distribution. Find the parameters of the latter.

5-3 ASYMPTOTIC RESULTS FOR $N(t)$

Even though we usually cannot invert (5-7), we can extract some useful information from it. We use it to find the mean and variance of $N(t)$ for large values of t. Then these are utilized to show that asymptotically $N(t)$ has a normal distribution.

Asymptotic Moments

Let

$$M(t) = E[N(t)] \quad \text{and} \quad \tilde{M}(s) = \int_0^\infty e^{-st}\, dM(t)$$

From (5-7)

$$
\begin{aligned}
\tilde{M}(s) &= \sum_{n=0}^\infty n\tilde{p}_n(s) = \lim_{z\uparrow 1} \frac{\partial}{\partial z} \sum_{n=0}^\infty z^n \tilde{p}_n(s) \\
&= [1 - \tilde{F}(s)] \lim_{z\uparrow 1} [1 - z\tilde{F}(s)]^{-2} \tilde{F}(s) \\
&= \frac{\tilde{F}(s)}{1 - \tilde{F}(s)}
\end{aligned}
\tag{5-9}
$$

Expand $\tilde{F}(s)$ in a Maclaurin's series. Assuming that $F(\,\cdot\,)$ has a finite third moment, we obtain [see (A-29)]

$$\tilde{F}(s) = 1 - vs + \frac{v_2 s^2}{2} + o(s^2) \tag{5-10}$$

where v_2 is the second moment of $F(\,\cdot\,)$. Substituting (5-10) into (5-9) yields

$$
\begin{aligned}
\tilde{M}(s) &= \frac{1 - vs + v_2 s^2/2 + o(s^2)}{vs - v_2 s^2/2 + o(s^2)} \\
&= \frac{1}{v}\frac{1}{s} + \frac{\sigma^2 - v^2}{2v^2} + o\left(\frac{1}{s}\right)
\end{aligned}
\tag{5-11}
$$

A formal inversion of (5-11) yields (see Table A-2)

$$M(t) = \frac{t}{v} + \frac{\sigma^2 - v^2}{2v^2} + o(1) \tag{5-12}$$

which shows that asymptotically $M(t)$ grows linearly with slope $1/v$ and has a bias term which is positive if the coefficient of variation (σ/v) of $F(\,\cdot\,)$ is larger than 1.

Now do the same thing to obtain the second factorial moment of $N(t)$, $M_2(t) \triangleq E\{N(t)[N(t) - 1]\}$. Let $\tilde{M}_2(s)$ be the Laplace-Stieltjes transform of $M_2(t)$. Then

$$\begin{aligned}
\tilde{M}_2(s) &= \lim_{z \uparrow 1} \frac{\partial^2}{\partial z^2} \sum_{n=0}^{\infty} z^n \tilde{p}_n(s) \\
&= \frac{2[\tilde{F}(s)]^2}{[1 - \tilde{F}(s)]^2}
\end{aligned} \tag{5-13}$$

If we assume that the fourth moment of $F(\,\cdot\,)$ is finite and denote the third moment by v_3, expand both numerator and denominator of (5-13) in Maclaurin's series, and divide, we obtain (after some algebra)

$$\tilde{M}_2(s) = \frac{2}{s^2} \frac{1}{v^2} + 2 \frac{\sigma^2 - v^2}{v^2} \frac{1}{s} \frac{1}{v^2} + \frac{2c_2}{v^2} + o\left(\frac{1}{s}\right)$$

where

$$c_2 = \frac{\sigma^2}{2} + \frac{3}{4}\left(v^2 + \frac{\sigma^4}{v^2}\right) - \frac{v_3}{3v}$$

Inverting term by term yields

$$M_2(t) = \frac{t^2}{v^2} + 2 \frac{\sigma^2 - v^2}{v^2} t + \frac{2c_2}{v^4} + o(1) \tag{5-14}$$

Let $V(t) = \mathrm{Var}[N(t)]$. Then

$$\begin{aligned}
V(t) &= M_2(t) + M(t) - M^2(t) \\
&= \frac{\sigma^2}{v^3} t + c_v + o(1)
\end{aligned} \tag{5-15}$$

where

$$c_v = \frac{2\sigma^2}{v^2} + \frac{3}{4} + \frac{5}{4} \frac{\sigma^4}{v^4} - \frac{2}{3} \frac{v_3}{v^3}$$

Notice that asymptotically $V(t)$ grows linearly with t, that the proportionality constant depends on the first two moments of $F(\,\cdot\,)$, and that the bias term depends on the first three moments of $F(\,\cdot\,)$.†

---

† Smith (1958) shows that an analogous property holds for the nth *cumulant* of $N(t), n \in I_+$. The nth cumulant of a random variable is the coefficient of $z^n/n!$ in the logarithm of its generating function. The first two cumulants are the mean and the variance, respectively.

By using the formulas for the mean and variance of the Poisson distribution, it is easy to see that, for the Poisson process, (5-12) and (5-15) hold for all values of t, with the bias and error terms equal to zero.

Asymptotic Normality of $N(t)$

Now we show that asymptotically $N(t)$ has a normal distribution. First, observe that since $S_r = X_1 + \cdots + X_r$ is a sum of i.i.d. random variables, the central limit theorem asserts that

$$\lim_{r \to \infty} P\left\{\frac{S_r - rv}{\sigma\sqrt{r}} \leq y\right\} = \Phi(y) \qquad (5\text{-}16)$$

where

$$\Phi(y) = \frac{1}{\sqrt{2\pi}} \int_{-\infty}^{y} e^{-u^2/2}\, du$$

is the standard normal distribution function. We want to find the probability that $N(t)$ is less than its mean plus y standard deviations when t is large. Let

$$r_t = \frac{t}{v} + y\sigma\sqrt{\frac{t}{v^3}}$$

From (5-12) and (5-15) r_t is roughly the mean plus y standard deviations of $N(t)$ when t is large. We want to prove that for large values of t,

$$P\{N(t) < r_t\} \doteq \Phi(y) \qquad (5\text{-}17)$$

This is established by Theorem 5-2.

Theorem 5-2: Central-limit theorem for renewal processes For a renewal process with $v < \infty$ and $\sigma^2 < \infty$, define

$$N^*(t) = \frac{N(t) - t/v}{\sqrt{\sigma^2 t/v^3}} \qquad t \geq 0$$

Then

$$\lim_{t \to \infty} P\{N^*(t) < y\} = \Phi(y) \qquad (5\text{-}18)$$

PROOF Observe that

$$P\{N(t) \geq r\} = P\left\{N^*(t) \geq \frac{r - t/v}{\sqrt{\sigma^2/v^3}}\right\}$$

$$= P\left\{N^*(t) \geq -\frac{t - rv}{\sigma\sqrt{r}}\sqrt{\frac{vr}{t}}\right\}$$

and that (5-5) implies

$$P\{N(t) \geq r\} = P\left\{\frac{S_r - rv}{\sigma\sqrt{r}} \leq \frac{t - rv}{\sigma\sqrt{r}}\right\}$$

Fix y and let r and t approach infinity in such a way that

$$\frac{t - rv}{\sigma\sqrt{r}} \to y$$

This implies that $t/\sqrt{r} \to \infty$ and hence $vr/t \to 1$; thus

$$\frac{t - rv}{\sigma\sqrt{r}} \sqrt{\frac{vr}{t}} \to y$$

Hence

$$\lim_{r, t \to \infty} P\{N(t) \geq r\} = \lim_{t \to \infty} P\{N^*(t) \geq -y\}$$

$$= \lim_{r \to \infty} P\left\{\frac{S_r - rv}{\sigma\sqrt{r}} \leq y\right\}$$

$$= \Phi(y)$$

where the last equality follows from the central limit theorem. Recalling that $\Phi(y) = 1 - \Phi(-y)$ will complete the proof. ☐

Strong Law of Large Numbers for Renewal Processes

Now we prove that for large t, $N(t)$ grows linearly with t at rate $1/v$, which conforms to our intuition. The proof illustrates a general method of obtaining bounds on $N(t)$.

Theorem 5-3: Strong law of large numbers for renewal processes Let $\{N(t); t \geq 0\}$ be a renewal process in which v is the mean time between renewals. Then for $v < \infty$,

$$\lim_{t \to \infty} \frac{N(t)}{t} = \frac{1}{v} \qquad \text{(w.p.1)}$$

PROOF Referring to Figure 5-1 we see that

$$S_{N(t)} \leq t < S_{N(t)+1} \qquad (5\text{-}19)$$

Therefore,

$$\frac{N(t)}{S_{N(t)+1}} < \frac{N(t)}{t} \leq \frac{N(t)}{S_{N(t)}}$$

Since $N(t) \to \infty$ as $t \to \infty$ (see Exercise 5-16), the strong law of large numbers implies that

$$\lim_{t \to \infty} \frac{N(t)}{S_{N(t)}} = \lim_{n \to \infty} \left(\frac{1}{n} \sum_{i=1}^{n} X_i \right)^{-i} = \frac{1}{\nu} \qquad \text{(w.p.1)}$$

By the same reasoning,

$$\frac{N(t)}{S_{N(t)+1}} = \frac{S_{N(t)}}{S_{N(t)+1}} \frac{N(t)}{S_{N(t)}} \to \frac{1}{\nu} \qquad \text{(w.p.1)}$$

Combining these results with (5-19) proves the theorem. $\qquad \square$

Summary

This ends our discussion of $N(t)$. We were able to obtain (5-7), which is the Laplace-Stieltjes transform of the generating function of $P\{N(t) = n\}$. This transform is easily inverted analytically when the times between renewals have an exponential distribution, but it is difficult to invert otherwise. Even so, we were able to use it to obtain asymptotic formulas for the mean and variance of the number of renewals, (5-12) and (5-15), respectively. We then showed that for large t, $N(t)$ has a normal distribution and grows linearly in t. Now we turn our attention to a more detailed study of the mean number of renewals.

EXERCISES

5-12 Let $G(\,\cdot\,)$ be the distribution function of a nonnegative random variable with $G(0) = 0$. Let $F(\,\cdot\,)$ have a concentration of mass of amount p, $0 < p < 1$, at zero. Suppose that $F(x) = (1 - p)G(x) + p$ for all $x \geq 0$. If the random variable X has distribution function $F(\,\cdot\,)$ and the random variable Y has distribution function $G(\,\cdot\,)$, how are X and Y related? Find the renewal function corresponding to $F(\,\cdot\,)$ in terms of the renewal function corresponding to $G(\,\cdot\,)$.

5-13 Show that for any renewal process

$$E([N(t)]^2) = M(t) + 2 \int_0^t M(t - s) \, dM(s)$$

5-14 Is (5-12) properly interpreted as stating that the graph of $M(t)$ looks like the figure below?

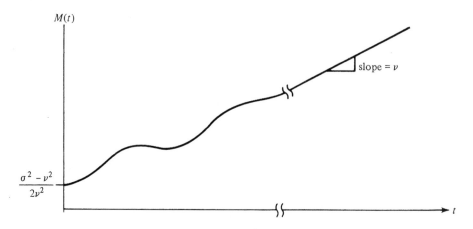

5-15 Find $M(t)$ and $V(t)$ for a renewal process where the times between renewals have a gamma distribution with shape parameter 2 (and arbitrary scale parameter).

5-16 Prove that $F(\infty) = 1$ implies that (w.p.1) $N(t) \to \infty$ as $t \to \infty$, even when $v = \infty$.

5-17 Prove that if $F(\infty) < 1$, then $P\{N(\infty) < \infty\} = 1$ and $M(\infty) < \infty$.

5-18 The times between arrivals of animals at a water hole are i.i.d. and have a gamma distribution with shape parameter 2. For T large, what is an approximate distribution for the number of arrivals in the interval $(0, T]$?

5-19 The distribution function of the time between failures of a machine can be estimated from past failure data by the following procedure. For each fixed $x \geq 0$, define

$$Y_i(x) = \begin{cases} 1 & \text{if } X_i \leq x \\ 0 & \text{otherwise} \end{cases}$$

where X_i is the time between the $(i - 1)$th and ith failures. The machine is new at time 0, and we observe it over $(0, T]$. We estimate $F(x)$ by the proportion of failure times that are no larger than x, that is, by

$$\hat{F}(x; T) = \frac{1}{N(T)} \sum_{i=1}^{N(T)} Y_i(x)$$

where $N(T)$ is the number of failures by time T. Show that $\lim_{T \to \infty} \hat{F}(x; T) = F(x)$. [$\hat{F}(x; T)$ is a *consistent estimator* of $F(x)$.] For T large, give an approximate distribution for $\hat{F}(x; T)$.

5-20 Customers arrive at a telephone booth according to a renewal process with $v = 1$ hour. Each customer pays \$0.25 for the call. What is the annual rate of income for this phone booth? Do you believe that this is a good model for collections at real phone booths? What would you change?

5-4 RENEWAL FUNCTION

The function $M(t) = E[N(t)]$ has a role in renewal theory that is far greater than just a simple expectation. In recognition of its importance, it is called the *renewal function*. In this section we begin to see why it is so important and derive some of its properties.

Insight into the importance of $M(t)$ can be obtained by solving (5-9) for $\tilde{F}(s)$. We obtain

$$\tilde{F}(s) = \frac{\tilde{M}(s)}{1 + \tilde{M}(s)}$$

which shows that $\tilde{M}(\cdot)$ completely determines $\tilde{F}(\cdot)$, hence $M(\cdot)$ uniquely determines $F(\cdot)$. This means that if only $M(t)$, for all $t \geq 0$, were given, then all the statistics of the renewal process could be obtained.

Since $F(\cdot)$ is a distribution function and we always assume $F(0) < 1$, whenever s (the real part when s is complex) is nonnegative,

$$0 < \tilde{F}(s) < 1 \qquad (5\text{-}20)$$

This means (5-9) can be written as

$$\tilde{M}(s) = \sum_{n=1}^{\infty} [\tilde{F}(s)]^n \qquad (5\text{-}21)$$

Inverting term by term yields

$$M(t) = \sum_{n=1}^{\infty} F_n(t) \qquad t \geq 0 \tag{5-22}$$

with $F_n(t)$ as defined in (5-4).† [Equation (5-22) was derived by other methods in Exercise 5-4.] We summarize these results, which were obtained easily from (5-9), in Theorem 5-4.

Theorem 5-4 For any renewal process, a one-to-one correspondence exists between $F(\cdot)$ and $M(\cdot)$, and $M(\cdot)$ has the representation given by (5-22).

Suppose that $F(\cdot)$ has the density $f(\cdot)$, and let $f_n(\cdot)$ be the derivative of $F_n(\cdot)$. Since $F_n(t)$ is a monotone function of t, the monotone convergence theorem (Proposition A-9) and (5-22) yield

$$m(t) \triangleq \frac{d}{dt} M(t) = \sum_{n=1}^{\infty} f_n(t) \tag{5-23}$$

If we exclude $o(\Delta t)$ terms, then $f_n(t)\Delta t = P\{S_n \in (t, \ t + \Delta t)\}$, and so $m(t)\Delta t$ is the probability that the first, or second, or third, etc., renewal occurs during $(t, t + \Delta t)$. In other words, $m(t)\,\Delta t$ is the probability that a renewal occurs during $(t, t + \Delta t)$. For this reason, we call $m(t)$ the *renewal density*. (It is not a true probability density because it does not integrate to 1.)

The Renewal Argument and the Renewal Equation

Many results can be efficiently obtained by the *renewal argument*. This is simply the observation that *at any renewal epoch, another renewal process starts which is a probabilistic replica of the original process*. To see this precisely, consider a renewal process $\{N(t); t > 0\}$ with generating sequence $\{X_i\}_{i=1}^{\infty}$. Fix a positive integer k, and suppose that $S_k = s_k$. Now define a generating sequence $\{X_i'\}_{i=1}^{\infty}$ where for each sample path ω, $X_i'(\omega) = X_{k+i}(\omega)$, and let $\{N'(t - s_k); t > s_k\}$ be its renewal counting process. The situation for $k = 1$ is depicted in Figure 5-2. For any $t' = t - s_k > 0$,

$$P\{N'(t') \geq j\} = P\{X_1' + \cdots + X_j' \leq t'\} = P\{X_{k+1} + \cdots + X_{k+j} \leq t'\}$$
$$= P\{X_1 + \cdots + X_j \leq t'\} \tag{5-24}$$
$$= P\{N(t') \geq j\}$$

since the X's are i.i.d. The renewal argument is a verbal description of (5-24). For $k = 1$, it is clear from Figure 5-2 that

$$N(t) = \begin{cases} 0 & t < X_1 \\ 1 + N'(t - X_1) & t \geq X_1 \end{cases}$$

that is,‡

---

† Some authors use (5-22) to define $M(\cdot)$.

‡ The notation $X \sim Y$ means that the random variables X and Y have the same distribution.

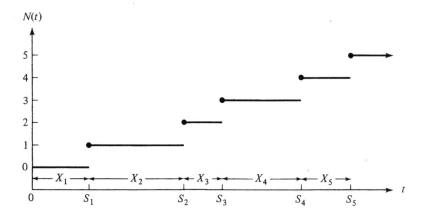

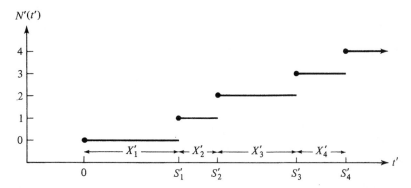

Figure 5-2 $N(t)$ and $N'(t')$ for $k = 1$.

$$N(t) \sim \begin{cases} 0 & t < X_1 \\ 1 + N(t - X_1) & t \geq X_1 \end{cases} \qquad (5\text{-}25)$$

Theorem 5-5 can be proved directly from (5-9) (see Exercise 5-24); we prove it by using the renewal argument.

Theorem 5-5 $M(t)$ satisfies the renewal equation

$$M(t) = F(t) + \int_0^t M(t - x) \, dF(x) \qquad (5\text{-}26)$$

and the only solution of (5-26) that has an LST is given by (5-22).

PROOF From (5-25),

$$E[N(t) | X_1] = \begin{cases} 0 & \text{if } X_1 > t \\ 1 + M(t - X_1) & \text{if } X_1 \leq t \end{cases}$$

Unconditioning on X_1 yields

$$M(t) = \int_0^t [1 + M(t - x)] \, dF(x)$$

which is equivalent to (5-26). Taking LSTs on both sides of (5-26) and re-arranging yields (5-21), which establishes uniqueness in the class of Laplace-Stieltjes-transformable functions.† $\qquad\qquad\square$

One reason for establishing (5-26) is that often it is more convenient to compute $M(t)$ from (5-26) than from (5-22). Computing $M(t)$ from (5-22) involves successive calculations of enough terms $F_n(t) = \int_0^t F_{n-1}(t - x) \, dF(x)$ to approximate the infinite sum accurately. In (5-26) we see that $M(t)$ depends on only $M(t - x)$, $x \in [0, t)$, so recursive computation of $M(\cdot)$ is possible. Renewal equations appear in Section 5-5 and in more general settings in Chapter 6.

Recursive computation is particularly convenient when X is a discrete random variable. In this case, let $f_k = F(k) - F(k - 1) = P\{X = k\}$, $k \in I$. Clearly, $M(t)$ is a step function with jumps possible only where t is a nonnegative integer, so $M(\cdot)$ is completely determined by its values for $t \in I$. Thus, (5-26) yields

$$M(k) = F(k) + \sum_{j=0}^k M(k - j) f_j$$

$$= \frac{F(k) + \sum_{j=1}^k M(k - j) f_j}{1 - f_0} \qquad k \in I$$

and the right-side involves only $M(0), \ldots, M(k - 1)$. Setting $k = 0$ provides $M(0) = f_0/(1 - f_0)$, which enables $M(1), M(2), \ldots, M(k)$ to be computed in turn.

When X is not discrete, the same procedure can be used by dividing t into intervals of length Δt and setting $f_j = F(j\Delta t) - F[(j - 1)\Delta t]$. The values of $M(t)$ within each interval are obtained by interpolation. This procedure is approximate; its accuracy and computational requirements increase as Δt decreases.

Often we have occasion to use a generalization of Theorem 5-5. A *renewal-type equation* (or *generalized renewal equation*) is

$$H(t) = G(t) + \int_0^t H(t - y) \, dF(y) \qquad t \geq 0 \qquad (5\text{-}27)$$

In (5-27), $G(\cdot)$ is a given function with $G(t) = 0$ for $t < 0$, but otherwise $G(\cdot)$ is arbitrary, $F(\cdot)$ has the properties of a distribution function of a nonnegative random variable, and $H(\cdot)$ is the function to be solved for.

---

† Solutions of (5-26) that are unbounded for finite t are possible. The results of this theorem are sharpened in Exercise 5-23.

Theorem 5-6 Let $G(\cdot)$ have an LST. Then the unique Laplace-Stieltjes-transformable solution† of (5-27) is

$$H(t) = G(t) + \int_0^t G(t - y) \, dM(y) \qquad t \geq 0 \qquad (5\text{-}28)$$

PROOF This is left as Exercise 5-25. □

Corollary 5-6 For any renewal process,

$$E(S_{N(t)+1}) = v[M(t) + 1] \qquad (5\text{-}29)$$

PROOF Let $H(t) = E(S_{N(t)+1})$. By the renewal argument,

$$H(t \mid X_1) = \begin{cases} X_1 & \text{if } X_1 > t \\ X_1 + H(t - X_1) & \text{if } X_1 \leq t \end{cases}$$

Hence

$$H(t) = \int_0^\infty H(t \mid X_1 = x) \, dF(x)$$

$$= E(X_1) + \int_0^t H(t - x) \, dF(x)$$

This is (5-27) with the constant $E(X_1) = v$ playing the role of $G(t)$. By using (5-28), the solution is simply

$$H(t) = v + \int_0^t v \, dM(x) = v[M(t) + 1] \qquad \square$$

Example 5-6 illustrates the use of the renewal argument in a different setting.

Example 5-6 A conceptual model of how telephone users behave after getting a busy signal, which is consistent with observed behavior, is the following. At time zero, a call is placed and a busy signal is heard. The caller then flips a coin that has probability p of landing heads. If the coin lands tails, no further calls are made. If the coin lands heads, the caller chooses a random time X_1, with $F(x) = P\{X_1 \leq x\}$, and another call is placed at X_1. This call is successful with probability $1 - r$. If another busy signal is heard, the process repeats itself.

Each call other than the original call placed at time zero is called a *reattempt*. In designing telephone systems, often it is important to know the statistical behavior of the reattempts. Let us find the mean number of reattempts made by time t, denoted by $R(t)$.

---

† Theorems 5-5 and 5-6 were stated under conditions that are sufficient for all the applications we will make of them but more general statements are possible. For example, if $G(\cdot)$ is bounded, then (5-28) is the unique solution that is nonnegative and bounded on every finite interval. For details, see Exercise 5-25.

By using the renewal argument,

$$R(t \mid X_1) = \begin{cases} 0 & \text{if } X_1 > t \\ [1 + rR(t - X_1)]p & \text{if } X_1 \leq t \end{cases}$$

Unconditioning on X_1 yields

$$R(t) = pF(t) + pr \int_0^t R(t - x) \, dF(x)$$

This equation can be solved numerically by recursion or by transform methods; let us do the latter. By taking LSTs on both sides and rearranging,

$$\tilde{R}(s) = \frac{p\tilde{F}(s)}{1 - rp\tilde{F}(s)}$$

When $F(t) = 1 - e^{-\mu t}$,

$$\tilde{R}(s) = \frac{p\mu}{s + \mu(1 - rp)}$$

and inverting yields the explicit solution

$$R(t) = \frac{p}{1 - rp} (1 - e^{-\mu(1 - rp)t}) \qquad t \geq 0 \qquad \square$$

Elementary Renewal Theorem

Now let us return to theorems about $M(t)$. What happens to $M(t)$ as $t \to \infty$? Applying the limit theorem (see Section A-3)

$$\lim_{t \to \infty} \frac{G(t)}{t} = \lim_{s \downarrow 0} s\tilde{G}(s) \tag{5-30}$$

to (5-11) yields

$$\lim_{t \to \infty} \frac{M(t)}{t} = 1/\nu \tag{5-31}$$

Unfortunately, the limit on the right-side of (5-30) can exist when the limit on the left-side does not (see Exercise 5-27), so (5-31) is not proved by this argument.† However, the argument suggests that we try to prove (5-31) true. We start by proving Lemma 5-1.

Lemma 5-1 For any renewal process,

$$M(t) \geq \frac{t}{\nu} - 1 \qquad t \geq 0 \tag{5-32}$$

---

† A stronger transform theorem would do the trick. Theorem 4.3 (p. 192) of D. V. Widder, *The Laplace Transform*, Princeton University Press, Princeton, N.J. (1946) shows that the limits in (5-30) exist simultaneously when $G(\cdot)$ is nondecreasing. Since $M(\cdot)$ is clearly nondecreasing, a transform proof is available.

When $v = \infty$, interpret $1/v = 0$.

PROOF The lemma is trivially true if $v = \infty$, so assume $v < \infty$. By the nature of $S_{N(t)+1}$ (see Figure 5-1)

$$t \leq S_{N(t)+1} \qquad (5\text{-}33)$$

Combining (5-29) and (5-33) gives

$$t \leq v[M(t) + 1]$$

and (5-32) follows immediately. $\qquad\qquad\square$

With this lemma at hand, we can easily establish the validity of (5-31).

Theorem 5-7: Elementary renewal theorem For any renewal process,

$$\lim_{t \to \infty} \frac{M(t)}{t} = \frac{1}{v}$$

with the right-side interpreted as zero when $v = \infty$.

PROOF By virtue of Lemma 5-1, it is sufficient to prove

$$\limsup_{t \to \infty} M(t)/t \leq 1/v$$

We do this for the case when X is (essentially) bounded [i.e., for some $B < \infty$, $F(B) = 1$]; the unbounded case is left to Exercise 5-30.

By their definitions, $t + X_{N(t)} \geq S_{N(t)+1}$ for each t. From (5-29),

$$t + B \geq E(t + X_{N(t)}) \geq E(S_{N(t)+1}) = v(M(t) + 1)$$

Rearranging yields

$$\frac{M(t)}{t} \leq \frac{1}{v} + \frac{B/v - 1}{t}$$

and hence

$$\limsup_{t \to \infty} \frac{M(t)}{t} \triangleq \limsup_{t \to \infty \atop s > t} \frac{M(s)}{s}$$

$$\leq \frac{1}{v} \qquad\qquad\square$$

Example 5-7 One way to replenish an inventory is to order J items at epochs $S_J, S_{2J}, \ldots$, where S_i denotes the epoch of the ith demand. Suppose it costs $C(J)$ dollars to order J items, $J \in I_+$. For example, $C(J)$ could be $K + cJ$, where K represents the fixed cost to process an order and c is the purchase cost per item. Assume that the times between demand epochs are i.i.d. random variables with mean v; hence $E(S_J) = Jv$. We want to find the asymptotic mean rate of incurring ordering costs.

Our assumptions imply that $\{S_{nJ}\}_1^\infty$ generates a renewal process for any

$J \in I_+$. Let $M(t; J)$ be the renewal function for that process and $C_J(t)$ be the expected costs incurred by time t. Since each order costs $C(J)$, we have

$$C_J(t) = C(J)M(t; J)$$

Applying the elementary renewal theorem yields

$$C_J \triangleq \lim_{t \to \infty} \frac{C_J(t)}{t} = \frac{C(J)}{Jv}$$

A quantity discount occurs in many situations. This would be represented by having $C(\cdot)$ be a nondecreasing concave function; that is, $C(J + 1) - C(J)$ is nonincreasing. Let us treat J as a continuous variable† and assume $C(\cdot)$ has a second derivative. Then elementary calculus methods show that C_J achieves its local minima at those values of J satisfying

$$J \frac{d}{dJ} C(J) = C(J) \qquad\qquad \square$$

Fundamental Renewal Theorems

Theorems 5-8 and 5-9 below are very useful in applications and in deriving new results about renewal processes. To state them precisely, we must introduce some special concepts. A random variable Y is called *arithmetic* if the only values it can assume are contained in the set $\{\xi, 2\xi, 3\xi, \dots\}$ for some number ξ. The largest value of ξ with this property is called the *span* of Y. All integer-valued random variables are arithmetic.

Theorem 5-8: Blackwell's renewal theorem If X is not arithmetic, then as $t \to \infty$,

$$M(t) - M(t - h) \to \frac{h}{v} \qquad\qquad (5\text{-}34)$$

for every fixed $h > 0$. If X is arithmetic, then (5-34) holds when h is a multiple of the span of X. In either case, if $v = \infty$, we interpret the right-side of (5-34) as zero.‡

PROOF See Section 5-9. $\qquad\qquad \square$

---

† Formally, what we do is replace $C(\cdot)$ by a function $h(\cdot)$, say, whose domain is all real numbers. We require $h(\cdot)$ to agree with $C(\cdot)$ on the integers; it can be defined in any suitable way for nonintegers. This fruitful method of obtaining a first approximation to an optimal decision is used in the sequel without further comment, but you should always ponder what is being accomplished.

‡ These (apparently) fine points become very important in Chapter 7 where *every* interrenewal time is arithmetic and some of them have an infinite mean.

The expansion (5-12) for $M(t)$ is the intuitive basis for Theorem 5-8. A rigorous proof of the theorem is deferred until Section 5-9. A rough interpretation of (5-34) is that $m(t) \rightarrow 1/v$ as $t \rightarrow \infty$. It is easy to see why the case of arithmetic random variables must be considered separately in Theorem 5-8. If X were degenerate and took on only the value 1, then if $h < 1$, $M(t) - M(t - h)$ is 1 or 0 according to whether $(t - h, t] \cap I$ is nonempty or empty.

The next theorem, 5-9, requires a rather delicate hypothesis that at first glance may appear forbidding; a little bit of study shows that it is actually quite simple. The issue is to find weak conditions to place on a function $H(\cdot)$ so that

$$\lim_{t \to \infty} \int_0^t H(t - x) \, dM(x) = \frac{1}{v} \int_0^\infty H(x) \, dx$$

For the integral on the right to converge, $\lim_{x \to \infty} H(x) = 0$ must hold. This means that for large t, most of the contribution to the integral on the left comes from large values of x. If the limit could be brought inside the integral, an appeal to Blackwell's theorem would finish the proof. This program can be carried out when we restrict the notion of an integrable function.

We define a class of integrable functions based on approximating inscribed and circumscribed step functions. Choose $\Delta > 0$, and for $k\Delta < x < (k + 1)\Delta$ define

$$l_k = \inf H(x) \quad \text{and} \quad u_k = \sup H(x) \quad k = 0, 1, \dots. \tag{5-35}$$

Definition 5-2 A nonnegative function $H(\cdot)$ is *directly Riemann-integrable* if the lower Riemann sum $\Delta \sum_{k=0}^\infty l_k$ and the upper Riemann sum $\Delta \sum_{k=0}^\infty u_k$ are both finite and converge, as $\Delta \rightarrow 0$, to the same finite value. That value is the definition of $\int_0^\infty H(x) \, dx$.

Every bounded function that is continuous except for jumps is directly Riemann-integrable. Included in this class of functions is $F^c(x) = P\{X > x\}$, where X is a nonnegative random variable with finite mean (see Exercise 5-32).

Theorem 5-9: Key renewal theorem† If $H(\cdot)$ is directly Riemann-integrable, then

$$\lim_{t \to \infty} \int_0^t H(t - x) \, dM(x) = \frac{1}{v} \int_0^\infty H(x) \, dx \tag{5-36}$$

If X is arithmetic, then (5-36) holds when t is a multiple of the span of X. In either case, if $v = \infty$, we interpret the right-side of (5-36) as zero.

PROOF Suppose

$$H(u) = \begin{cases} 1 & 0 \leq a \leq u \leq b \leq \infty \\ 0 & \text{otherwise} \end{cases}$$

---

† This theorem was first proved by W. L. Smith, who assumed that $H(\cdot)$ is a monotone function. The version presented here is due to W. Feller.

Then

$$\int_0^t H(t - x) \, dM(x) = \int_{t-b}^{t-a} dM(x) = M(t - a) - M(t - b)$$

$$= M(t) - M(t - b) - [M(t) - M(t - a)]$$

$$\rightarrow \frac{b - a}{v} = \int_0^\infty H(x) \frac{1}{v} \int_0^\infty H(x) \, dx \qquad (5\text{-}37)$$

where Blackwell's theorem was used to obtain the limit. This argument extends directly to situations where $H(\,\cdot\,)$ is a finite step function.† When $H(\,\cdot\,)$ is an arbitrarily selected directly Riemann-integrable function, construct upper and lower approximating step functions $H_u(\,\cdot\,)$ and $H_l(\,\cdot\,)$, respectively, as follows. Fix $\Delta > 0$, and let

$$\delta_k(t) = \begin{cases} 1 & \text{if } k\Delta \le t \le (k + 1)\Delta \\ 0 & \text{otherwise} \end{cases}$$

for $k = 0, 1, \ldots$. Set

$$H_l(t) = \sum_{k=0}^\infty l_k \delta_k(t) \qquad \text{and} \qquad H_u(t) = \sum_{k=0}^\infty u_k \delta_k(t)$$

where l_k and u_k are defined by (5-35). Hence

$$H_l(t) \le H(t) \le H_u(t) \qquad t \ge 0 \qquad (5\text{-}38)$$

Since $M(\,\cdot\,)$ is bounded on all finite intervals, $\int_0^t \delta_k(t - x) \, dM(x)$ is bounded. From (5-37),

$$\lim_{t \to \infty} \int_0^t \delta_k(t - x) \, dM(x) = \frac{\Delta}{v} \qquad k = 0, 1, \ldots$$

Thus,

$$\lim_{t \to \infty} \int_0^t H_l(t - x) \, dM(x) = \frac{\Delta}{v} \sum_{k=0}^\infty l_k$$

and

$$\lim_{t \to \infty} \int_0^t H_u(t - x) \, dM(x) = \frac{\Delta}{v} \sum_{k=0}^\infty u_k$$

Combining the two limits above with (5-38) and the definition of directly Riemann-integrable functions, we see that letting $\Delta \to 0$ completes the proof. $\qquad \square$

The proof above shows that Theorem 5-8 implies Theorem 5-9. The reverse is also true.

---

† Observe that this implies that any discrete function $\{h_n\}$ is directly Riemann-integrable if $\Sigma h_n < \infty$.

Corollary 5-9a Theorem 5-8 $\Leftrightarrow$ Theorem 5-9.

PROOF Choose

$$H(x) = \begin{cases} 1 & 0 \le x \le h \\ 0 & \text{otherwise} \end{cases}$$

Then

$$\int_0^t H(t - x) \, dM(x) = \int_{t-h}^t dM(x) = M(t) - M(t - h)$$

and (5-36) yields (5-34). $\qquad\qquad\qquad\qquad\qquad\qquad\qquad\qquad\qquad\square$

Our care in stating Theorem 5-9 precisely pays off in Corollary 5-9b.

Corollary 5-9b Let $F(\,\cdot\,)$ have a density function $f(\,\cdot\,)$ that is directly Riemann-integrable, and suppose $F(\infty) = 1$. Then

$$\lim_{t \to \infty} m(t) = \frac{1}{\nu} \tag{5-39}$$

PROOF Differentiate both sides† of (5-26) to obtain (see also Exercise 5-35)

$$m(t) = f(t) + \int_0^t m(t - y) f(y) \, dy$$

$$= f(t) + \int_0^t f(t - y) \, dM(y) \tag{5-40}$$

Since $\lim_{t \to \infty} f(t) = 0$ because $F(\infty) = 1$, and since $f(\,\cdot\,)$ is directly Riemann-integrable, we obtain (5-39) by applying Theorem 5-9 to the integral. $\quad\square$

Example 5-8: Poisson process When $f(t) = \lambda e^{-\lambda t}$, the renewal process is a Poisson process, and hence $M(t) = \lambda t \equiv t/\nu$ for all $t \ge 0$. Thus, for all $t \ge 0$,

$$\frac{M(t)}{t} = \nu$$

$$M(t + h) - M(t) = \frac{h}{\nu}$$

$$m(t) = \frac{1}{\nu}$$

and for every integrable function $H(\,\cdot\,)$,

$$\int_0^t H(t - x) \, dM(x) = \frac{1}{\nu} \int_0^t H(x) \, dx$$

---

† Differentiation of definite integrals is explained in Proposition A-13.

and so the asymptotic results of Theorems 5-7 to 5-9 apply at all finite t. Stated another way, these particular properties of the Poisson process hold asymptotically for every renewal process [modulo the restriction on the function $H(\cdot)$ in the last equation]. $\qquad\square$

EXERCISES

5-21 Another way to derive (5-22) is to proceed as follows. Define the indicator functions $1_n(t)$ by

$$1_n(t) = \begin{cases} 1 & \text{if } S_n \le t, \\ 0 & \text{if } S_n > t \end{cases} \qquad n = 1, 2, \ldots$$

Then $N(t) = \sum_{n=1}^{\infty} 1_n(t)$. Now show that $M(t)$ is given by (5-22).

5-22 Prove $M(t) < \infty$ whenever $t < \infty$ and that $\lim_{n \to \infty} F_n(t) = 0$.
Hint: Show

$$F_n(t) \triangleq \int_0^t F_{n-m}(t-s)\, dF_m(s) \le F_{n-m}(t) F_m(t)$$

for $1 \le m \le n - 1$, and then show

$$F_{nr+k}(t) \le [F_r(t)]^n F_k(t)$$

where n, m, r, and k are integers. Now conclude that $F_n(t) \to 0$ geometrically and so $M(t) < \infty$.

5-23 Theorem 5-5 asserts that the only solution of (5-26) which has an LST is (5-22), but even more can be established. Prove that (5-22) is the only solution of (5-26) which is bounded on every finite interval, and show that it vanishes on $(-\infty, 0)$. *Hint*: First use the result of Exercise 5-22 to verify that $M(\cdot)$ is bounded on every finite interval of the form $(0, t]$. Assume there are two such solutions, and let $g(\cdot)$ be their difference. Show by repeated substitution that

$$g(t) = \int_0^t g(t-s)\, dF(s) = \int_0^t g(t-s)\, dF_n(s)$$

Apply the results of Exercise 5-22 to establish that if $g(\cdot)$ is bounded, then $g(t) \equiv 0$.

This argument can also be used to strengthen Theorem 5-6.

5-24 Derive (5-26) directly from (5-9).

5-25 Show that (5-28) is the only solution to (5-27) that has an LST.

5-26 Use the renewal argument to prove $S_{n(t)+1} \sim X_1 + S_{N(t)}$. Give another proof of this result, using only the definition of S_n and the properties of $X_1, X_2, \ldots$.

5-27 Exhibit a function $g(\cdot)$ such that $\lim_{s \to 0} s\tilde{g}(s) = B$ but $\lim_{t \to \infty} g(t)/t$ does not exist. Exhibit another function $h(\cdot)$ such that $\lim_{s \to 0} \tilde{h}(s) = A$ but $\lim_{t \to \infty} h(t)$ does not exist.

5-28 Show that (5-32) holds as an equality for all $t \ge 0$ if, and only if, $\{N(t); t \ge 0\}$ is a Poisson process.

5-29 Find the flaw in the proof of this *false theorem*: For every renewal process, $M(t) \le t/\nu$. Proof: We always have

$$t + X_{N(t)} \ge S_{N(t)+1}$$

Taking expectations on both sides yields

$$t + \nu \ge \nu[M(t) + 1]$$

and a simple rearrangement completes the demonstration.

5-30 Complete the proof of the elementary renewal theorem. *Hint*: Use (5-32) to prove $\lim_{t \to \infty} \inf$

$M(t)/t = 1/v$, and then correct the fallacy described in Exercise 5-29. One way is to fix a positive number A and define a new renewal process where Y_n is the time between the $(n-1)$th and nth renewals, with

$$Y_n = \begin{cases} X_n & \text{if } X_n \leq A \\ A & \text{if } X_n > A \end{cases}$$

Let $v_A = E(Y_n)$, $n = 1, 2, \ldots$, and $M'(t)$ be the renewal function for this process. First show

$$\limsup_{t \to \infty} M'(t)/t \leq 1/v_A$$

and then finish the proof by letting $A \to \infty$. Why is (5-31) not a consequence of the strong law of large numbers for renewal processes (Theorem 5-3)?

5-31 Give an example of a function $g(\cdot)$ such that $\lim_{t \to \infty} g(t) = A$ but $\lim_{t \to \infty} [g(t+h) - g(t)] \neq Ah$ (or, even stronger, that the latter limit does not exist). This shows that Blackwell's theorem is stronger than the elementary renewal theorem.

5-32 Show that $F^c(t) \triangleq P\{X > t\}$ for every nonnegative random variable with finite mean is directly Riemann-integrable.

5-33 The number of cases of claret brought to a large wine store is governed by a renewal process. Let v_a be the mean time between arrivals. The time each case stays in the store is independent of when it, or any other case, was brought in and of when the other cases are picked up by customers. Let v_s be the mean time the case stays in the store and $X(t)$ be the number of cases in the store at time t. Find $\lim_{t \to \infty} E[X(t)]$. This model can be cast as a queueing model. How would you classify this model with the notation in Section 3-3?

5-34 (Continuation)   Let $R(t)$ be the number of cases of wine that have been removed from the store by time t. Find the asymptotic removal rate $\lim_{t \to \infty} R(t)/t$.

5-35 Argue probabilistically that (5-40) is true.

5-36 When X_1 is a discrete random variable, $M(\cdot)$ is not differentiable. Let $f_k = P\{X_1 = k\}$, $k \in I$, and define $m_n = M(n) - M(n-1)$, $n \in I$. Call m_n the *discrete renewal density*. Prove that m_n satisfies

$$m_n = \frac{f_n + \sum_{k=1}^n m_{n-k} f_k}{1 - f_0} \qquad n \in I$$

5-37 Use transform techniques to prove that if $\lim_{t \to \infty} \int_0^t H(t-x)\, dM(x)$ exists, then it must equal $\int_0^\infty H(x)\, dx/v$.

5-5 RECURRENCE TIMES

In this section we characterize the important random variables which describe the local behavior of a renewal process observed at some fixed time t. Refer to Figure 5-3. The random variable A_t, defined by

$$A_t = t - S_{N(t)}$$

represents the time since the last renewal before t. In Example 5-1, A_t is the time since the last circuit pack was installed. The random variable A_t is called the *deficit* (or the *age* or the *backward recurrence time*) at t.

The random variable B_t, defined by

$$B_t = S_{N(t)+1} - t$$

represents the time from t until the next renewal occurs. In terms of Example 5-1,

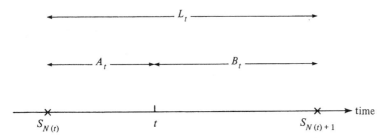

Figure 5-3 Recurrence times.

it is the time until the next circuit pack failure. It is called the *excess* (or the *residual life* or the *forward recurrence time*) at t.

The random variable L_t, defined by

$$L_t = S_{N(t)+1} - S_{N(t)}$$

is the length of the interrenewal time in progress at t. In Example 5-1, it represents the lifetime of the circuit pack found in service at t. We call L_t the *spread*, or the *recurrence time*, at t.

Let

$$A_t(z) = P\{A_t \le z\}$$
$$B_t(z) = P\{B_t \le z\}$$

and

$$L_t(z) = P\{L_t \le z\}$$

Distribution of the Deficit

Let us now find $A_t(z)$ by a direct probabilistic analysis. We use the language of Example 5-1 for definiteness. Since the circuit pack that is working at t could not possibly have been installed more than t time units ago, A_t must be smaller than z for $z > t$. Hence

$$A_t(z) = 1 \qquad z > t \tag{5-41a}$$

For $z \le t$, refer to Figure 5-4.

From Figure 5-4, the only way A_t can be less than or equal to z is if a circuit pack is installed at some time u after $t - z$ and this circuit pack lasts beyond time

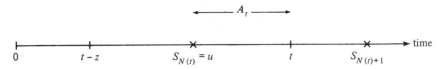

Figure 5-4 Deficit when $z \le t$.

t. A circuit pack installed at time u lasts beyond time t with probability $F^c(t - u)$ and a circuit pack is replaced at time u with density $m(u)$ [see (5-23)], so

$$A_t(z) = \int_{t-z}^{t} F^c(t - u)m(u) \, du \qquad z \le t \qquad (5\text{-}41b)$$

Combining (5-41a) and (5-41b) yields

$$A_t(z) = \begin{cases} \int_{t-z}^{t} F^c(t - u)m(u) \, du & z \le t \\ 1 & z > t \end{cases} \qquad (5\text{-}42)$$

Suppose t is very large and we ask for

$$A(z) \triangleq \lim_{t \to \infty} A_t(z) = \lim_{t \to \infty} P\{A_t \le z\}$$

For every fixed z, (5-42) yields

$$A(z) = \lim_{t \to \infty} \int_{t-z}^{t} F^c(t - u)m(u) \, du$$

Proceeding intuitively, we see that Corollary 5-9b states that $m(u) \, du \to du/v$ as $u \to \infty$. By making the change of variable $y = t - u$ and assuming the limit can be brought inside the integral, we obtain the *equilibrium deficit* distribution:

$$A(z) = \lim_{t \to \infty} \int_{0}^{z} F^c(y)m(t + y) \, dy$$
$$= \frac{1}{v} \int_{0}^{z} F^c(y) \, dy \qquad (5\text{-}43)$$

Consequently,

$$a(z) \triangleq \frac{d}{dz} A(z) = \frac{F^c(z)}{v} \qquad (5\text{-}44)$$

We purposely did not derive (5-43) in a rigorous manner, but instead used an intuitive analysis to show essentially what is going on. Rigorous methods make the above ideas precise (Exercise 5-38).

Distribution of the Excess

Now we use a formal argument to obtain $B_t(z)$. First, observe that $B_t \le z$ if $t < X_1 \le t + z$ and $B_t > z$ if $X_1 > t + z$. If $X_1 = x \le t$, the renewal argument yields (see Figure 5-5)

$$P\{B_t \le z \,|\, X_1 = x \le t\} = P\{B_{t-x} \le z\}$$

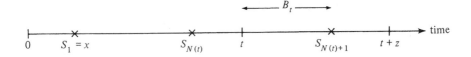

Figure 5-5 Analysis of the excess.

Now unconditioning on the value of X_1 yields

$$B_t(z) = F(t + z) - F(t) + \int_0^t B_{t-x}(z) \, dF(x)$$

This is a renewal-type equation (z is just a parameter), and by Theorem 5-6 its solution is

$$B_t(z) = F(t + z) - F(t) + \int_0^t [F(t + z - x) - F(t - x)] \, dM(x)$$

$$= F(t + z) - \int_0^t F^c(t + z - x) \, dM(x) - \left[F(t) - \int_0^t F^c(t - x) \, dM(x) \right]$$

Since $\int_0^t F^c(t - x) \, dM(x) = M(t) - \int_0^t F(t - x) \, dM(x)$, Equation (5-26) shows that the term in brackets is zero. Thus,

$$B_t(z) = F(t + z) - \int_0^t F^c(t + z - x) \, dM(x) \qquad (5\text{-}45)$$

To obtain the *equilibrium excess* distribution $B(z) \triangleq \lim_{t \to \infty} B_t(z)$, observe that $F^c(\cdot)$ satisfies the conditions of the key renewal theorem (Theorem 5-9). From (5-45),

$$B(z) \triangleq \lim_{t \to \infty} B_t(z) = 1 - \frac{1}{\nu} \int_0^\infty F^c(y + z) \, dy$$

$$= 1 - \frac{1}{\nu} \int_0^\infty F^c(y) \, dy \qquad (5\text{-}46)$$

$$= \frac{1}{\nu} \int_0^\infty F^c(y) \, dy$$

where we used the fact (Exercise 5-3) that $\int_0^\infty F^c(y) \, dy = \nu$.

Comparing (5-43) and (5-46) shows that the equilibrium age and excess distributions are identical. Their common mean, $E(B)$ say, is given by

$$E(B) = \frac{1}{v} \int_0^\infty y F^c(y) \, dy = \frac{v_2}{2v} \qquad (5\text{-}47a)$$

where v_2 is the second moment of the distribution $F(\cdot)$. Often it is convenient to write (5-47a) in terms of the coefficient of variation of $F(\cdot)$, which is its standard deviation divided by its mean and is denoted by c_F. With this notation, (5-47a) becomes

$$E(B) = \frac{c_F^2 + 1}{2v} \qquad (5\text{-}47b)$$

which shows that $E(B)$ is larger or smaller than v according as c_F is larger or smaller than 1.

Density of the Spread Random Variable

Both methods just presented can be used to obtain $L_t(z)$. It is left to you to show that when $F(\cdot)$ has a derivative $f(\cdot)$, the associated density function $l_t(z)$ is given by

$$l_t(z) = f(z)[M(t) - M(t - z)] + \begin{cases} f(z) & \text{if } z > t \\ 0 & \text{if } z \le t \end{cases} \qquad (5\text{-}48a)$$

and

$$l(z) = \lim_{t \to \infty} l_t(z) = \frac{z f(z)}{v} \qquad (5\text{-}48b)$$

The mean of $l(z)$ is $2E(B)$ (why?). It is larger than v unless the times between renewals are a constant. Equation (5-48b) was derived heuristically in Section 5-1.

A summary of these results is given by Theorem 5-10.

Theorem 5-10 For a renewal process with $F(x) = P\{X \le x\}$:
(a) The distribution function of the deficit is given by (5-42), and $a(z) = \lim_{t \to \infty} dA_t(z)/dz$ exists and is given by (5-44).
(b) The distribution function of the excess is $B_t(z)$, given by (5-45), and $b(z) = \lim_{t \to \infty} dB_t(z)/dz$ exists and is also given by (5-44).
(c) $\lim_{t \to \infty} E(A_t) = \lim_{t \to \infty} E(B_t) = E(X^2)/[2E(X)]$
(d) The distribution function of the spread is

$$L_t(z) = \int_{t-z}^t [F(z) - F(t - u)] \, dM(u) + \begin{cases} 0 & \text{if } z \le t \\ F(z) - F(t) & \text{if } z > t \end{cases}$$

and

$$\lim_{t \to \infty} L_t(z) = \frac{1}{E(X)} \int_0^t u \, dF(u)$$

PROOF You are asked to prove part d in Exercise 5-39. $\square$

Examples

Example 5-9: Poisson process Here $F(x) = 1 - e^{-\lambda x}$ and has the density function $f(x) = \lambda e^{-\lambda x}$. Using this density in (5-42) yields

$$A_t(z) = \begin{cases} 1 - e^{-\lambda z} & \text{if } z \le t \\ 1 & \text{if } z > t \end{cases}$$

The age distribution is the same as the original distribution as long as $t \ge z$. That is, $A_t(z)$ is the truncated exponential distribution. From (5-45)

$$B_t(z) = 1 - e^{-\lambda(t+z)} - \lambda \int_0^t e^{-\lambda(t+z-y)} \, dy$$

$$= 1 - e^{-\lambda z} \equiv F(z)$$

Hence the number of events during $(t, t + T]$ has the same distribution as the number of events during $(0, T]$ for any $t \ge 0$ and $T \ge 0$. This is another way of expressing the memoryless property of the exponential distribution. From (5-48a)

$$l_t(z) = \lambda^2 z e^{-\lambda z} + \begin{cases} \lambda e^{-\lambda z} & \text{if } z > t \\ 0 & \text{if } z \le t \end{cases}$$

Thus, when the times between renewals have an exponential distribution, the asymptotic properties obtained for arbitrary renewal processes hold for all finite t [except for the natural truncation in $A_t(z)$ and $l_t(z)$]. Recall, in this connection, Example 5-8. This property is a direct consequence of the memoryless property and indicates why the exponential distribution is often the easiest distribution to use in stochastic models. $\square$

Example 5-10: (s, S) inventory systems The (s, S) inventory policy for replenishing inventories is described in Section 2-7. Periodically, say at the start of each week, the inventory is observed. If the inventory is at least s, then no action is taken; if the inventory is less than s, say x, then the amount $S - x$ is ordered so that the amounts on hand plus on order equal S. (Of course, we choose $s \le S$.) In Chapter 7 of Volume II, it is shown that this policy is optimal in a wide variety of settings. Assume that the weekly de-

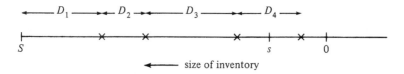

Figure 5-6 Renewal diagram of the inventory levels.

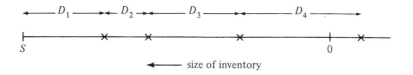

Figure 5-7 Figure 5-6 modified by increasing D_4.

mands on the inventory are i.i.d. random variables with distribution function $F(\cdot)$, orders are filled instantly, and the inventory initially contains S items. We want to determine the distribution of the number of items in inventory when an order is placed.

Let D_i be the demand in the ith period. In Figure 5-6 we diagram the size of the inventory as a function of time. In this sketch, it took four periods for the inventory to drop below s, that is, for the cumulative demand to exceed $S - s$. Let Y_i be the amount on hand when the ith order is placed. Since each order sets the inventory level to S, Y_1, Y_2, ... are i.i.d. From Figure 5-6, it is clear that we can consider D_1, $D_1 + D_2$, ... as renewal epochs and that $P\{Y_1 = j\}$ is the probability that the excess random variable at "time" $S - s$ equals $s - j$. Figure 5-6 also shows that the first replenishment epoch is $N(S - s) + 1$, where $N(\cdot)$ is the renewal process associated with the renewal epochs D_1, $D_1 + D_2$, ... ; hence the expected number of periods that elapse between orders is $M(S - s) + 1$.

In Figure 5-6, Y_1 is explicitly shown to be positive; but that is by no means necessary. Figure 5-7 shows what happens to the inventory when D_4 takes on a much larger value, but D_1, D_2, and D_3 remain the same. In Figure 5-7, j is negative, and we are presented with the modeling issue: What happens to those requests that cannot be filled immediately? There are two common answers to this question, and we present them both.

One answer is to *backlog* these demands, in which case the demand is recorded, the customer waits until the next order is received, and then the demand is satisfied. We interpret negative values of Y_1 as the amount on back order in this case. The other answer is to forfeit sales of these items; i.e., the demands are not filled. Here we interpret $P\{Y_1 \le 0\}$ as the probability that the sale of at least one item is forfeited and $P\{Y_1 = j\}$ for $j < 0$ as the probability that sales of j items are forfeited. ☐

Example 5-11 We can use (5-47a) to establish rigorously the constant in (5-12). Take expectations on both sides of the definition of B_t:

$$E(B_t) = E(S_{N(t)+1}) - t = v[M(t) + 1] - t$$

Hence,

$$M(t) - \frac{t}{v} = \frac{E(B_t)}{v} - 1 \rightarrow \frac{\sigma^2 - v^2}{2v^2} \qquad \square$$

EXERCISES

5-38 Derive (5-42) and (5-43) rigorously.

5-39 Derive (5-48a), (5-48b), and part d of Theorem 5-10.

5-40 For any $F(\,\cdot\,)$, show $E(B^2) = v_3/(3v_1)$, where v_3 is the third moment of $F(\,\cdot\,)$.

5-41 What issues arise in defining A_t and B_t when $P\{X = 0\} > 0$? Find $E(B)$ when $P\{X = j\} = (1 - p)p^j, j \in I$ and $0 < p < 1$.

5-42 Let $G(x, y; t) = P\{L_t \le x, B_t \le y\}$ and $G^c(x, y; t) = 1 - G(x, y; t)$. Use the renewal argument to derive the equations

$$G^c(x, y; t) = F^c(t + y) + F^c(x)[M(t) - M(t + y - x)]$$

$$+ \int_0^{t+y-x} F^c(t + y - u)\, dM(u) \qquad \text{for } y \le x \le t + y$$

and

$$G^c(x, y; t) = F^c(x)[1 + M(t)] \qquad \text{for } x > t + y$$

5-43 (Continuation) Prove that

$$G^c(x, y) = \lim_{t \to \infty} G^c(x, y; t)$$

exists and, moreover,

$$G^c(x, y) = \frac{(x - y)F^c(x)}{v} + \frac{1}{v} \int_x^\infty F^c(u)\, du$$

Show that asymptotically the joint density of spread and excess, $g(x, y)$ say, is given by

$$g(x, y) = \frac{f(x)}{v} \qquad \text{for } y \le x$$

Check this result by obtaining (5-44) and (5-48b) from it.

5-44 Consider a renewal process where the times between renewals have the density function $f(\,\cdot\,)$. Prove that if the equilibrium excess and deficit random variables are independent, then the renewal process is a Poisson process.

5-45 For the (s, S) inventory system with backlogging, (a) derive a formula for the probability that at the epoch an order is placed, more than j demands will be backlogged and (b) provide a specific formula when $P\{D_i = n\} = (1 - p)p^{n-1}$, $n = 1, 2, \ldots$ and $0 < p < 1$. *Hint:* The calculations are trivial when the answer to (a) is ignored and the fact that the geometric distribution is memoryless is used.

5-6 DELAYED, EQUILIBRIUM, AND ALTERNATING RENEWAL PROCESSES

Delayed Renewal Processes

In this section we expand our definition of a renewal process by not requiring that all the times between renewals $X_1, X_2, \ldots$ have the same distribution $F(\,\cdot\,)$. We allow the distribution of X_1 to be different, $G(\,\cdot\,)$ say. We continue to assume that $X_2, X_3, \ldots$ are i.i.d. random variables with distribution function $F(\,\cdot\,)$ and mean μ_F and they are independent of X_1. The counting process of the number of renewals by time t is called a *delayed* (or *generalized*) *renewal process*. This is a very useful generalization, and an application is illustrated in the next example.

Example 5-12 In Example 5-4, the times between renewals represent the number of bits in a computer record. Many computer-based record-keeping systems allow a record to be split between tracks. Consider a group of records that take up many tracks. Let K_i denote the number of records on track i, T be the number of bits on each track, and $F(\,\cdot\,)$ be the distribution function of bits in a record.

From the figure below, it is apparent that K_1 is the number of renewals in "time" T of a renewal process with interevent times having distribution function $F(\,\cdot\,)$. But this is not true of K_2; the number of bits until the first renewal epoch after T is the excess at T. All other record lengths in $(T, 2T]$ do have distribution function $F(\,\cdot\,)$. We can regard the end-of-record points during $(T, 2T]$ as forming a renewal process on $(0, T]$, where the first record length has a different distribution from all the others (in particular, for this example, it has the excess distribution at T).

We can repeat this argument for all the remaining intervals, showing that the number of end-of-record points in the various time intervals of length T generally are not identically distributed. These points generally are not independent either, because if one observed no records ending during the first interval, that could indicate that a record will end soon after the second interval begins. As an example, suppose record lengths are 55 bits with probability $\frac{1}{3}$ and 110 bits with probability $\frac{2}{3}$ and $T = 100$ bits. Then $P\{K_2 = 1 \mid K_1 = 1\} = \frac{4}{9}$, but $P\{K_2 = 1 \mid K_1 = 0\} = \frac{2}{3}$. When $F(\,\cdot\,)$ is the geometric distribution function, then A_T has the same geometric distribution function (and only then), so $K_1, K_2, \ldots$ will be i.i.d. $\qquad\square$

The formal definition of a delayed renewal process is given now.

Definition 5-3 Let X_1, $X_2, \ldots$ be independent random variables with $G(x) = P\{X_1 \le x\}$ and $F(x) = P\{X_i \le x\}$ for $i = 2, 3, \ldots$. Let $S_n = X_1 + \cdots + X_n$ for $x \in I_+$ and $S_0 = 0$, and define $N_G(t)$ by

$$N_G(t) = \inf\{n : S_n \le t\} \qquad t \ge 0$$

The process $\{N_G(t); t \ge 0\}$ is called a *delayed* (or *generalized*) *renewal process*.

We always assume $G(0) < 1$ and $G(\infty) = 1$, so X_1 is finite and not identically zero. Let $v_G = E(X_1)$ and $v_{2G} = E(X_1^2)$. The methods we developed for the ordinary renewal process apply directly to the delayed process.

Let $M_G(t)$ be the mean number of renewals by time t, that is, the renewal function for the delayed renewal process. After the first renewal, the delayed process behaves exactly as an ordinary process. By applying the renewal argument, we obtain

$$M_G(t \mid X_1) = \begin{cases} 0 & \text{if } X_1 > t \\ 1 + M(t - X_1) & \text{if } X_1 \le t \end{cases}$$

Hence,

$$M_G(t) = G(t) + \int_0^t M(t - x)\, dG(x) \tag{5-49}$$

Taking LSTs on both sides of (5-49) and using (5-9) yield

$$\tilde{M}_G(s) = \tilde{G}(s)[1 + \tilde{M}(s)] = \tilde{G}(s)\left[1 + \frac{\tilde{F}(s)}{1 - \tilde{F}(s)} \right]$$

$$= \frac{\tilde{G}(s)}{1 - \tilde{F}(s)} \tag{5-50}$$

Assume $G(\cdot)$ and $F(\cdot)$ have finite third moments, and expand $\tilde{G}(s)$ and $\tilde{F}(s)$ in a Maclaurin's series:

$$\tilde{G}(s) = 1 - v_G s + \tfrac{1}{2} v_{2G} s^2 + o(s^2)$$

and

$$\tilde{F}(s) = 1 - v_F s + \tfrac{1}{2} v_{2F} s^2 + o(s^2)$$

Substituting these expansions in (5-50) yields

$$\tilde{M}_G(s) = \frac{1 - v_G s + \tfrac{1}{2} v_{2G} s^2 + o(s^2)}{v_F s - \tfrac{1}{2} v_{2F} s^2 + o(s^2)}$$

Performing the long division, we obtain

$$\tilde{M}_G(s) = \frac{1}{v_F} \frac{1}{s} + \frac{v_{2F} - 2v_F v_G}{2v_F^2} + o(1)$$

and a formal inversion yields

$$M_G(t) = \frac{t}{v_F} + \frac{v_{2F} - 2v_F v_G}{2v_F^2} + o(1) \tag{5-51}$$

for large values of t. Thus (5-51) suggests that the elementary renewal, key renewal, and Blackwell's renewal theorems all hold for the generalized renewal function because asymptotically $M_G(t)$ is proportional to t. Indeed, all this is true. Let us prove one of these assertions.

Theorem 5-11 Let $H(\cdot)$ be a directly Riemann-integrable function and $M_G(t) = E[N_G(t)]$. Then

$$\lim_{t \to \infty} \int_0^t H(t - x) \, dM_G(x) = \frac{1}{v} \int_0^\infty H(x) \, dx \qquad (5\text{-}52)$$

PROOF Let $Z(t) = \int_0^t H(t - x) \, dG(x)$. From (5-49),†

$$\int_0^t H(t - x) \, dM_G(x) = Z(t) + \int_0^t Z(t - x) \, dM(x) \qquad (5\text{-}53)$$

The following facts are easily established (Exercise 5-47): $Z(\,\cdot\,)$ is directly Riemann-integrable, $\lim_{t \to \infty} Z(t) = 0$, and $\lim_{t \to \infty} \int_0^t Z(x) \, dx = \int_0^\infty H(x) \, dx$. These facts and the key renewal theorem applied to (5-53) yield (5-52). □

Theorem 5-11 and Equation (5-51) indicate that the long-run behavior of $M_G(t)$ does not depend on $G(\,\cdot\,)$. This is quite reasonable because for large t, there will be many renewals whose spacing is governed by $F(\,\cdot\,)$ but only one that is governed by $G(\,\cdot\,)$, so the effect of the former overwhelms the effect of the latter. This is the intuitive reason why all the asymptotic theorems derived for the ordinary renewal process also hold for the delayed renewal process.

Equilibrium Renewal Process

Now we use the delayed renewal process to study what is known as the equilibrium renewal process. In (5-46) the equilibrium excess distribution, which is the distribution of the remaining life of the component in service infinitely long after the process started, was obtained. In terms of Example 5-12, it is the distribution of the length of the partial record at the start of a track n as $n \to \infty$.

Definition 5-4 An *equilibrium renewal process* is a delayed renewal process with

$$G(t) = \frac{1}{v} \int_0^t F^c(y) \, dy$$

The LST of $G(\,\cdot\,)$ is

$$\tilde{G}(s) = \frac{1 - \tilde{f}(s)}{v_F}$$

Substituting this expression into (5-50) yields

$$\tilde{M}_G(s) = \frac{1}{s} \frac{1 - \tilde{f}(s)}{1 - \tilde{f}(s)} \frac{1}{v_F} = \frac{1}{v_F} \frac{1}{s}$$

† One way to establish (5-53) is to take the LST of both sides and use the convolution theorem (Property A-12a).

which can be inverted, by inspection, to obtain

$$M_G(t) = \frac{t}{v_F} \qquad t \geq 0 \tag{5-54}$$

This form of the renewal function holds only for large t when $G(\,\cdot\,)$ is chosen arbitrarily. But with our particular choice of $G(\,\cdot\,)$, the asymptotic form of $M_G(t)$ is valid for all finite values of t.

A more striking example of the consequences of this choice for $G(\,\cdot\,)$ is that the excess distribution is independent of t (Exercise 5-49). Let us investigate the distribution of the spread. It is clear [if you derived (5-48) for yourself] that when $t \geq z$, Equation (5-48a) holds for the delayed renewal process when $M(\,\cdot\,)$ is replaced by $M_G(\,\cdot\,)$. Substituting (5-54) into (5-48a) yields

$$l_t(z) = f(z)\left(\frac{t}{v} - \frac{t-z}{v}\right) = \frac{z f(z)}{v_F} \qquad t \geq z$$

which is (5-48b).

Many other similar results can be obtained; doing so is left to you. The important thing to note is that our descriptions of a renewal process now possess their asymptotic properties at finite times. That is why such a renewal process is called *equilibrium*. The intuitive idea behind "starting observations at random" is that the process has been operating for so long that either the initial conditions are forgotten or their effects have been eliminated. The equilibrium renewal process is a mathematical model of a renewal process that is "started at random."

Alternating Renewal Processes

Another application of the delayed renewal process is the analysis of the *alternating renewal process*. This process is illustrated by Example 5-13.

Example 5-13 In Example 5-1, a replacement model is described which assumes that a failed circuit pack is immediately detected and replaced. Suppose this assumption is not accurate. Assume that a time interval of length Y_i elapses between the epoch of the ith circuit pack failure and the epoch when the $(i + 1)$th circuit pack is placed in position. We can interpret Y_i as the total elapsed time required to detect a failed circuit pack, obtain a replacement, and install it. As with the ordinary renewal structure, we are interested in determining the number of replacements that occur during $(0, t]$. But now another issue arises that was not meaningful before, and that is to determine the probability that a working circuit pack is in place at time t.

A convenient way to diagram the important epochs and random variables is shown in Figure 5-8. In the figure, the crosses represent epochs when a circuit pack fails, and the circles represent epochs when a circuit pack is installed. Let $Z(t) = 0$ when a working circuit pack is in place, and $Z(t) = 1$ otherwise. When $Z(t) = 0$, we say we are in an X-interval; when $Z(t) = 1$, we are in a Y-interval. $\square$

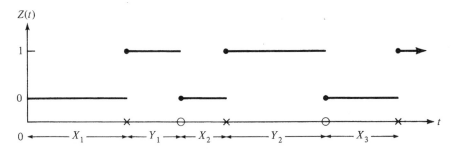

Figure 5-8 Diagram of an alternating renewal process.

The alternating renewal process is most conveniently described in an informal way,† by using the language of Example 5-13. The basic assumptions are that $\{X_i\}_1^\infty$ and $\{Y_i\}_1^\infty$ are sets of i.i.d. random variables and that the X's and Y's are independent of one another. Let $F_1(x) = P\{X_1 \le x\}$ and $F_2(y) = P\{Y_1 \le y\}$. The important random variables to study are $Z(t)$, $N_1(t) \triangleq$ the number of failures by time t, and $N_2(t) \triangleq$ the number of repairs by time t. Also define the renewal functions $M_1(t) = E[N_1(t)]$ and $M_2(t) = E[N_2(t)]$.

The time until the first repair is $X_1 + Y_1$, the time between the first and second repairs is $X_2 + Y_2$, and so forth. It is clear that $\{N_2(t); t \ge 0\}$ is a renewal process where the times between events have the distribution $F(\cdot)$ say, where

$$F(t) = \int_0^t F_1(t - y) \, dF_2(y)$$

Hence, the analysis of $\{N_2(t); t \ge 0\}$ offers no new problems. From (5-9) we obtain

$$\tilde{M}_2(s) = \frac{\tilde{F}_1(s)\tilde{F}_2(s)}{1 - \tilde{F}_1(s)\tilde{F}_2(s)} \tag{5-55}$$

The time to the first failure is X_1. The time between the first and second failures is $Y_1 + X_2$, the time between the second and third failures is $Y_2 + X_3$, and so forth. Since the X's and Y's are i.i.d., $Y_i + X_{i+1}$ has distribution function $F(\cdot)$, X_1 has distribution function $F_1(\cdot)$, and $\{N_1(t); t \ge 0\}$ is a delayed renewal process. From (5-50),

$$\tilde{M}_1(s) = \frac{\tilde{F}_1(s)}{1 - \tilde{F}_1(s)\tilde{F}_2(s)} \tag{5-56}$$

Let $P_0(t) = P\{Z(t) = 0\}$; it is the probability that a working circuit pack is in place at time t. A working pack will be in place at time t if either the first pack lasts beyond t or there is a replacement at some time $u < t$ and that pack lasts

---

† Most treatments of the alternating renewal process emphasize the properties of $\{Z(t); t \ge 0\}$. Counting processes are the heart of renewal theory, so $\{(N_1(t), N_2(t); t \ge 0\}$ is *the* alternating renewal process. A more general treatment of this definitional issue is given in Section 9-2.

beyond t (that is, for at least $t - u$). Hence,[†]

$$P_0(t) = F_1^c(t) + \int_0^t F_1^c(t - u)\,dM_2(u) \tag{5-57}$$

The LST version of (5-57) is obtained from (5-55) and (5-57), namely,

$$\tilde{P}_0(s) = \frac{1 - \tilde{F}_1(s)}{1 - \tilde{F}_1(s)\tilde{F}_2(s)} \tag{5-58}$$

Applying the key renewal theorem to (5-57) yields

$$\begin{aligned} \lim_{t \to \infty} P_0(t) &= \lim_{t \to \infty} \int_0^t F_1^c(t - u)\,dM_2(u) \\ &= \int_0^\infty \frac{F_1^c(u)\,du}{E(X_1) + E(Y_1)} \\ &= \frac{E(X_1)}{E(X_1) + E(Y_1)} \end{aligned} \tag{5-59}$$

which does not depend on the distribution functions $F_1(\,\cdot\,)$ and $F_2(\,\cdot\,)$.

Example 5-14: Computer reliability Return to Example 4-8, except now assume that the lengths of the operating intervals have distribution function $F_1(\,\cdot\,)$ and that the repair times have distribution function $F_2(\,\cdot\,)$. Let $1/\lambda$ and $1/\mu$ be the means of these distribution functions, respectively. Then (5-57) is a generalization of the forward Kolmogorov equations for that example, (5-58) is the generalization of (4-21a), and $\lim_{t \to \infty} P_0(t) = \mu/(\lambda + \mu)$ as before. $\square$

Example 5-15: M/M/1 queue When no customers are in the system, we say that the system is in an idle period; conversely, when at least one customer is present, we say that the system is in a busy period. Let $Z(t)$ denote the number of customers in the system at time t, and assume that $Z(0) = 0$. Let X_i denote the length of the ith idle period and Y_i denote the length of the ith busy period.

Looking at Figure 5-9, we see that X_1 is the time to the first arrival, which has distribution function $1 - e^{-\lambda t}$. The memoryless property of the exponential distribution implies that X_2 has the same distribution function as the interarrival time in progress when the first busy period ends (u_3 in Figure 5-9), which is the distribution function of X_1. The same reasoning applies to the distribution function of X_3, X_4, etc. It is clear that the idle periods are independent, so $X_1, X_2, \ldots$ are i.i.d.

Since all busy periods start when a customer arrives at an empty system, the time until the first service completion after the start of any busy period is

---

[†] You should also derive (5-57) by using the renewal argument.

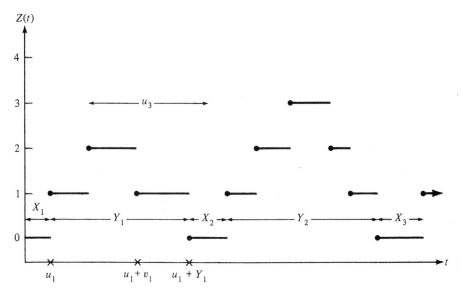

Figure 5-9 A sample path of an M/M/1 queue.

distributed as V_1 and the time until the next arrival epoch is distributed as U_1. Hence, Y_1, Y_2, ... are i.i.d. and independent of the X's, so the busy and idle periods generate an alternating renewal process.†

From (5-59) we obtain

$$p_0 = \lim_{t \to \infty} P\{Z(t) = 0\} = \frac{1/\lambda}{1/\lambda + E(Y_1)}$$

In Exercise 4-22 you showed that

$$p_0 = 1 - \rho \qquad \rho = \lambda/\mu$$

Hence the mean length of a busy period is given by

$$E(Y_1) = \frac{1}{\mu(1 - \rho)} \qquad \qquad \square$$

EXERCISES

5-46 Let $\{N_G(t); t \geq 0\}$ be a delayed renewal process. Prove that (w.p.1) $\lim_{t \to \infty} N_G(t)/t = 1/v_F$.

5-47 Prove the three facts about $Z(\cdot)$ that are used in the proof of Theorem 5-11.

5-48 Show that $M_G(t) = t/v_F$ only if $G(t) = \int_0^t F^c(y)\, dy/v_F$.

† Nowhere in this argument did we use in an essential way the fact that service times are exponential or that there is a single server, so the statement holds for M/G/c queues.

5-49 Show that the excess distribution of an equilibrium renewal process is independent of t. Use this result to prove that an equilibrium renewal process has stationary increments [i.e., that the distribution of $N(t) - N(s)$ depends on only $t - s, t > s$].

5-50 Invert $\tilde{P}_0(s)$ when $F_1(t) = 1 - e^{-\lambda t}$ and $F_2(t) = 1 - e^{-\mu t}$. Where have we obtained this formula before?

5-51 Use LSTs to derive the formula

$$P_0(t) = M_2(t) - M_1(t) + 1$$

Obtain the formula by a direct probabilistic argument.

5-52 In an alternating renewal process, suppose the first interval is an X-interval with probability $v_1/(v_1 + v_2)$. If the first interval is an X-interval, let its distribution function be $\int_0^t F_1^e(y)\, dy/v_1$. Similarly, if it is a Y-interval, its distribution function is $\int_0^t F_2^e(y)\, dy/v_2$, and this occurs with probability $v_2/(v_1 + v_2)$. Find $P_0(t)$ and interpret your answer.

5-53 An infinite-capacity buffer (i.e., one so large that the probability of not having room for an arriving message is negligible) in a teletypewriter is fed by a stream of incoming messages. These messages are made up of characters; the number of characters in the ith message is X_i. The X_i's are i.i.d., and for convenience we treat them as continuous variables with common distribution function $F(\cdot)$. The messages arrive at the buffer on a transmission line that sends one character per unit time. When a full message has been received, it is transmitted out of the buffer in zero time and the buffer is empty until the next message starts to arrive. The time interval that starts when the ith message has been completely received and ends when the $(i + 1)$th message starts to arrive is denoted by Y_i. The Y_i's are i.i.d. and independent of the message lengths; they have distribution function $G(\cdot)$. Let $Z(t)$ be the number of characters in the buffer at time t.

    (a) Find an equation to determine $P_0(t) = P\{Z(t) = 0\}$ when time 0 is the start of the first message.

    (b) Find $\lim_{t \to \infty} P\{Z(t) = 0\}$.

    (c) Find $\lim_{t \to \infty} P\{Z(t) \leq x\}$.

5-54 Refer to Example 5-13. Let $U(t) = 1 - Z(t)$ and $u(t) = E[\int_0^t U(y)\, dy]$; here $u(t)$ represents the expected amount of time, during $(0, t]$, that a working circuit pack is in place.

    (a) Find a functional equation for $u(t)$.

    (b) Solve for $u(t)$ explicitly when $F_1(t) = 1 - e^{-\lambda t}$ and $F_2(t) = 1 - e^{-\mu t}$.

    (c) Give a rigorous proof that, in general, $\lim_{t \to \infty} u(t)/t = v_1/(v_1 + v_2)$.

5-55 The following problem occurs in studying a priority queueing model in Section 11-5. Consider an alternating renewal process that starts with an X-interval as shown in Figure 5-8. Let T be an exponentially distributed random variable with mean $1/\lambda$. If a Y-interval is in progress at T, let the random variable $Y(T)$ be the remaining length of that Y-interval; if a Y-interval is not in progress at T, then $Y(T) \triangleq 0$. Show that the LST of $Y(T)$ is given by

$$1 - \frac{\tilde{M}_1(\lambda)[s - \lambda + \lambda \tilde{F}_2(s) - s\tilde{F}_2(\lambda)]}{s - \lambda}$$

and

$$E[Y(T)] = \frac{\tilde{M}_1(\lambda)[\lambda E(Y_1) + \tilde{F}_2(\lambda) - 1]}{\lambda}$$

5-7* MORE ABOUT THE POISSON PROCESS

In Example 4-7, the Poisson process is defined as a special case of a birth-and-death process. The definition uses the probabilistic structure of the transitions as its axioms. In Example 5-5, we observe that the Poisson process is the renewal counting process when the times between renewals have an exponential distribution. In subsequent sections, the memoryless property of the exponential distribution often provides the basis for simple derivations and conveniently used results, while the corresponding results for other interevent distributions may be hard to obtain and frequently are not in the form of simple formulas. This reason is sufficient to make the Poisson process an important one, and our purpose in this section is to describe some of its other special properties.

Many models in the literature of operations research assume that one or more of the stochastic processes in the model are Poisson processes. The resulting analysis then makes use of the memoryless property of the times between renewals, so the Poisson process assumption is often crucial to both the solution itself and the method of obtaining the solution.†

Axiomatic Construction of a Poisson Process

One might think that a Poisson process is unlikely to be a good model of observed data and that assuming a process to be Poisson is very restrictive. In fact, in a remarkably large number of instances, predictions based on Poisson assumptions perform well. The Palm-Khintchine theorem given in Section 5-8 provides a theoretical justification for using the Poisson process in many settings. The next theorem can be interpreted as a qualitative description of the Poisson process that suggests it will adequately model many phenomena.

> **Theorem 5-12: Axiomatic characterization of the Poisson process** If $\{N(t);$ $t \geq 0\}$ is a counting process that is (w.p.1) finite for finite values of t, $N(0) = 0$, and has the following properties:
> (a) *Stationary increments*: For any $t, s \geq 0$, the distribution of $N(t + s) - N(t)$ depends on s only
> (b) *Independent increments*: For any $t, s \geq 0$, $N(t + s) - N(t)$ is independent of $\{N(u); u \leq t\}$
> (c) *Orderliness*: The jumps of $N(t)$ are (w.p.1) of unit magnitude
> then $\{N(t); t \geq 0\}$ is a Poisson process with some finite rate $\lambda \geq 0$.

Theorem 5-12 asserts that a counting process is a Poisson process if the jumps in all intervals of the same length are identically distributed and independent of past jumps (stationary and independent increments) and the events occur one at a time (orderliness). Properties (a), (b), and (c) are sometimes used as a definition of a Poisson process; the properties are about the jumps of the process, so that this other definition is qualitatively different from our definition, which is that $\{N(t); t \geq 0\}$ is a pure birth process with a constant birth rate.

---

† Queueing models have a repeated emphasis on Poisson arrival processes. Many dynamic inventory models assume demands form a Poisson process, and reliability models frequently assume failures form a Poisson process.

PROOF We begin by showing that for some finite $\lambda \geq 0$,

$$P\{N(t) = 0\} = e^{-\lambda t} \qquad t \geq 0 \tag{5-60}$$

From independent increments,

$$P\{N(t + s) = 0\} = P\{N(t) = 0\}P\{N(t + s) - N(t) = 0\}$$

Stationary increments implies that the last probability equals $P\{N(s) = 0\}$, hence

$$P\{N(t + s) = 0\} = P\{N(t) = 0\}P\{N(s) = 0\}$$

If we let $g(t) = P\{N(t) = 0\}$, the above equation is written as

$$g(t + s) = g(t)g(s) \qquad t, s \geq 0 \tag{5-61}$$

The solution† to the functional equation (5-61) is either $g(t) \equiv 0$ or $g(t) = e^{-\lambda t}$ for some finite $\lambda \geq 0$. Eliminate the former case by observing that if $g(t) \equiv 0$, then, by dividing $(0, t]$ into n equal parts,

$$P\{N(t) \geq n\} \geq \left[1 - g\left(\frac{t}{n}\right)\right]^n$$

which contradicts the assumption that $P\{N(t) < \infty\} = 1$ for $t < \infty$. Thus, (5-60) is established.

The proof is completed by showing that (5-60) implies that $N(t)$ is a pure birth process with a constant birth rate. Stationary, independent increments and (5-60) yield

$$P\{N(t + s) = j \mid N(t) = j\} = P\{N(s) = 0\}$$
$$= e^{-\lambda s} = 1 - \lambda s + o(s)$$

and

$$P\{N(t + s) \geq j + 1 \mid N(t) = j\} = P\{N(s) > 0\}$$
$$= 1 - e^{-\lambda s} = \lambda s + o(s)$$

From orderliness, $P\{N(t + s) > j + 1 \mid N(t) = j\} = o(s)$, so

$$P\{N(t + s) = j + 1 \mid N(t) = j\} = \lambda s + o(s)$$

Thus we have shown that $\{N(t); t \geq 0\}$ is a pure birth process with birth rate λ in each state, which is our definition of a Poisson process. □

Now consider a quantitative model for generating a Poisson counting process. Suppose there is a population of k "customers," which may be, for example, telephone subscribers, users of an inventory, or components of a physical system that are liable to fail. Assume that the probability that a customer will appear by

---

† This follows from the fact that $\lambda e^{-\lambda t}$ is the only density with the memoryless property; it can be proved also by strictly analytical methods.

time t is proportional to t (that is, αt for some constant $\alpha < 1/t$) and that the customers act independently. The probability that n of these customers arrive by time t, say $p_k(n)$, is given by

$$p_k(n) = \binom{k}{n} (\alpha t)^n (1 - \alpha t)^{k-n} \qquad n = 0, 1, \ldots, k \qquad (5\text{-}62)$$

As $k \to \infty$, suppose $\alpha \downarrow 0$, so that $\alpha k = \lambda$, where λ is a positive constant. Then the number of customers that appear in any subinterval of $(0, t]$ is independent of the number that appear in any other subinterval and has the same distribution for all subintervals of the same length. (Why is this not true for finite k?) Thus the counting process has stationary and independent increments. It is also orderly because the probability of two or more customers arriving in an interval of length Δt is $(\alpha \Delta t)^2 + (\alpha \Delta t)^3 + \cdots$, which is certainly $o(\Delta t)$.

To see that this model results in a Poisson distribution, let $k \to \infty$ in (5-62). We obtain

$$
\begin{aligned}
p(n) = \lim_{k \to \infty} p_k(n) &= \lim_{k \to \infty} \frac{k!}{n!(k-n)!} \left(\frac{\lambda t}{k}\right)^n \left(1 - \frac{\lambda t}{k}\right)^k \\
&= \frac{(\lambda t)^n}{n!} \lim_{k \to \infty} \left(1 - \frac{\lambda t}{k}\right)^k \lim_{k \to \infty} \frac{k}{k} \frac{k-1}{k} \times \cdots \times \frac{k-n+1}{k} \\
&= \frac{(\lambda t)^n}{n!} e^{-\lambda t}
\end{aligned}
$$

Thus the Poisson process is a good model for the counting process when a large number of customers are acting independently, the probability that a customer appears by time t is proportional to t, and the proportionality constant is small.

Events of a Poisson Process Occur at Random

The Poisson process sometimes is called the "completely random process," perhaps as a consequence of this next theorem, which states that given the number of events in an interval, the event epochs are independent and uniformly distributed over the interval.

Theorem 5-13 Let $\{N(t); t \geq 0\}$ be a Poisson process and $S_1, S_2, \ldots$ be the event epochs. Then†

$$Pd\{S_1 = s_1, S_2 = s_2, \ldots, S_k = s_k \mid N(T) = k\} = \frac{k!}{T^k}$$

whenever $0 \leq s_1 \leq s_2 \leq \cdots \leq s_k \leq T$ and is zero otherwise.

---

† Recall that $Pd\{A = a\}$ denotes the probability density of the random variable A evaluated at a.

PROOF Let λ be the rate of the process and $S_n = X_1 + \cdots + X_n$, where the X's have their usual interpretation. Then

$$\text{Pd}\{S_1 = s_1, \ldots, S_k = s_k \,|\, N(t) = k\}$$

$$= \frac{\text{Pd}\{X_1 = s_1, X_2 = s_2 - s_1, \ldots, X_k = s_k - s_{k-1} \,|\, N(T) = k\}}{P\{N(T) = k\}}$$

$$= \frac{\lambda^k e^{-\lambda s_1} e^{-\lambda(s_2 - s_1)} \cdots e^{-\lambda(s_k - s_{k-1})} e^{-\lambda(T - s_k)}}{(\lambda T)^k e^{-\lambda T}/k!}$$

$$= \frac{k!}{T^k} \qquad 0 \le s_1 \le s_2 \le \cdots \le s_k \le T \qquad \qquad \square$$

An important application of this theorem is obtained by observing that there are $k!$ ways to arrange $S_1, S_2, \ldots, S_k$. This yields Corollary 5-13.

Corollary 5-13 Let $\{N(t);\ t \ge 0\}$ be a Poisson process, and suppose $N(T) = k \ge 1$. Let the random variable Y_i have the same distribution as S_i, $i = 1, 2, \ldots, k$. Then

$$\text{Pd}\{Y_1 = y_1, \ldots, Y_k = y_k \,|\, N(T) = k\} = T^{-k} \qquad (5\text{-}63)$$

PROOF The random variables $Y_1, \ldots, Y_k$ are not ordered; that is, $Y_1 > Y_2$ is possible even though $S_1 > S_2$ is impossible. Since there are $k!$ ways to arrange k different objects, (5-63) follows immediately from Theorem 5-13.

$$\square$$

This corollary states that when $N(T) = k$ is given and the condition $S_1 \le S_2 \le \ldots \le S_k$ is ignored, the event epochs have the same distribution as k i.i.d. random variables that are uniformly distributed over $[0, T]$. In particular, suppose we are told that an event epoch has occurred during $[0, T]$; we are not told if it is the first, or second, or third, etc. event. Corollary 5-13 asserts that the position of this epoch is uniformly distributed over $[0, T]$ and independent of the position (and the number) of other event epochs during $[0, T]$.

Examples

Example 5-16: Periodic-review inventory policy In a periodic-review inventory system, the inventory level is reviewed periodically, say at times T, $2T, \ldots$, and T is called the *length* of the period (e.g., a month). Suppose we were to backlog all the demands that occur during the period and place an order at the end of the period to satisfy these requests. This policy has the virtue that no items are held in inventory and the fault that customers usually have to wait before their orders are filled. An example of an organization that might use this policy is a library; for reasons of efficiency, new books might be purchased only on the last day of each month. This is an example of the base stockage policy described in Section 2-6; the base stock (S) is zero.

Assume that demands form a Poisson process with rate† λ. Set $a_i = (i - 1)T$, and let $X_i(t)$ be the number of demands that occur during the first t units of period i, that is, during $(a_i, a_i + t]$, $0 \leq t \leq T$. From the memoryless property we deduce that, for each i, $\{X_i(t); 0 \leq t \leq T\}$ is a Poisson process with rate λ. Assume that backlogged demands incur the holding-cost rate h, the cost of purchasing n items is $K + nc$, and at the end of period i (time $a_i + T$), $X_i \triangleq X_i(T)$ items are ordered.

Let us compute the expected holding cost in period i. From (5-63), if $X_i > 0$, the mean holding time for each demand is $T/2$; hence the mean holding cost, given X_i, is $hX_iT/2$ (even when $X_i = 0$). Taking expectations yields $h\lambda T^2/2$.

It is shown in Section 6-2 that in the long run, the cost per unit time equals the expected cost incurred in the first period divided by the expected length of that period. Let $C(T)$ be the asymptotic cost rate when the period is T. Then

$$C(T) = \frac{h\lambda T}{2} + \frac{K}{T} + c\lambda$$

The minimum value of $C(T)$ is achieved by

$$T^* = \sqrt{\frac{2K}{h\lambda}}$$

and

$$C(T^*) = \sqrt{2\lambda Kh} + c\lambda \qquad \square$$

Example 5-17: Queue-scanning problem In a stored-program computer, the central processing unit (CPU) cannot directly "see" if a job has arrived at an empty queue, but must periodically instruct a sensor to check the idle queues. Suppose that the queue is checked at times T, $2T$, $3T$, etc. This reduces the time which the CPU can devote to its other activities. We can model this as a fixed charge R (representing lost work) for each scan. Often it is reasonable to treat the CPU as an infinitely fast server; those jobs found in the queue at a scan epoch will be served in zero time. To model the negative value usually associated with the lost time of a job waiting to be processed by the CPU, apply a holding-cost rate per job h. We want to find that value of T, say T^*, that minimizes the asymptotic cost rate when the jobs arrive at the queue according to a Poisson process.

Observe that the mathematical formulation of this problem is identical to the formulation of Example 5-16 with $c = 0$, so we have already solved it!

$\square$

---

† Choose the units of λ to be commensurate with the unit of T.

Example 5-18: M/G/∞ queue In the M/G/∞ queue, all arrivals enter service at the moment they arrive. The service times of all customers are i.i.d. non-negative random variables with distribution function $G(\cdot)$. This model often is used to approximate an M/G/∞ queue where c is so large that congestion is rare and not a central issue. We want to find the probability that a customer who arrives before time t has not been served yet.

A potential application of this model is the gathering of migration data. Suppose dated tags are placed on animals when the animals are in some specified region R_1 and the tags are removed when the animals reach another region R_2. Assume that animals enter region R_1 according to a Poisson process (these are the arrivals) and that the migration times from R_1 to R_2 are i.i.d. random variables (these are the service times). The number of animals whose tags have not been recovered corresponds to the number of arrivals in an M/G/∞ queue who have not been served yet. The remainder of the example uses queueing terminology.

Let $N = N(t)$ be the number of arrivals in $(0, t]$. If $N > 0$, give each arrival a number chosen at random and without replacement from $\{1, 2, \ldots, N\}$. Let T_k be the arrival epoch of customer k; from (5-63), the T_k's are i.i.d. uniformly on $(0, t]$. Let V_k be the service time of customer k, assume $N = n$, and let $h_k(t) = P\{\text{customer } k \text{ is in service at } t \mid N = n\}$, $k = 1, 2, \ldots, n$. Then, for any $n \in I_+$,

$$h_k(t) = P\{T_k + V_k > t \mid N = n\}$$

$$= \begin{cases} 1 & \text{if } V_k > t \\ P\{T_k > t - x \mid N = n\} & \text{if } V_k = x \le t \\ 0 & \text{otherwise} \end{cases}$$

$$= 1 - G(t) + \frac{1}{t} \int_0^t x \, dG(x) \triangleq h(t)$$

which is independent of k and n. Applications of this result are given in Exercises 5-60 to 5-62. □

Compound Poisson Process

Now let us develop two generalizations of the Poisson process. The first is the *compound Poisson process*, which is the process obtained if orderliness is dropped from Theorem 5-12 and replaced with Property (d):

(d) *Stationary jumps*: Let Z_n be the size of the nth jump, where $\{Z_n; n = 1, 2, \ldots\}$ are i.i.d. random variables.

Some authors restrict the term *compound Poisson process* to processes where the Z_n's are integer-valued and nonnegative, but we do not make this restriction. A compound Poisson process may be used to model group arrivals at a service facility (Z_n is the number of patrons in the nth arriving group). It also may be used to model cumulative purchases (Z_n is either the number of items purchased or the bill of the nth departing customer).

Let $J(t)$ be the number of jumps that occur during $(0, t]$; it is a Poisson process with rate λ. Thus

$$N(t) \triangleq Z_1 + \cdots + Z_{J(t)} \qquad t \geq 0$$

where an empty sum is taken as zero.

When Z_1 is nonnegative and integer-valued, so is $N(t)$. In this case, form the generating function $\hat{N}(z; t) = \sum_{n=0}^{\infty} z^n P\{N(t) = n\}$, and let $\hat{A}(\cdot)$ be the generating function of Z_1. Conditioning on $J(t)$ yields

$$\hat{N}(z; t) = \sum_{j=0}^{\infty} [\hat{A}(z)]^j P\{J(t) = j\}$$

$$= \sum_{j=0}^{\infty} \frac{[\hat{A}(z)]^j e^{-\lambda t}(\lambda t)^j}{j!} = \exp\{-\lambda t[1 - \hat{A}(z)]\}$$

This generating function can be used to obtain the mean and variance of $N(t)$; more direct methods for obtaining these moments are given in Section 6-2.

Nonstationary Poisson Process

The second generalization is the *nonstationary* (or *nonhomogeneous*) *Poisson process*; the use of these terms naturally implies that we should use the adjectives *stationary* or *homogeneous* when referring to a Poisson process characterized by Theorem 5-12 if there is a chance of ambiguity. The nonstationary Poisson process is obtained by dropping the stationary increments property in Theorem 5-12 and replacing it with Property (*e*):

(*e*) *Time-dependent increments*:

$$\lim_{\Delta t \to 0} \frac{P\{N(t + \Delta t) - N(t) > 0\}}{\Delta t} = \lambda(t)$$

Thus, if we neglect terms of size $o(\Delta t)$, then $\lambda(t) \Delta t$ is the probability of a jump during $(t, t + \Delta t]$. In the stationary Poisson process, $\lambda(t)$ is the same for all t.

The nonstationary Poisson process can be employed to model arrival processes in which the arrival rate varies with the time of day. For example, telephone calling rates climb during morning business hours, drop off at lunchtime, climb during the afternoon, and drop off toward evening. Computer and document reproduction centers, fire alarms, and requests for ambulances also exhibit time-varying arrival rates.

A nonstationary Poisson process can be analyzed by converting it to a stationary Poisson process with a different time scale. If $\lambda(t) = 0$ for $0 \leq t < s$ for some $s > 0$ and $\lambda(s) > 0$, the process will have no jumps before time s and will be uninteresting before time s; so we lose no generality by assuming that $\lambda(0) > 0$.

From Property (e), and orderliness, the expected number of events occurring in $(t, t + \Delta t]$ is $\lambda(t) \, \Delta t + o(\Delta t)$. Hence the expected number of events in $(0, t]$ is

$$\Lambda(t) = \int_0^t \lambda(u) \, du$$

In a homogeneous Poisson process with rate $\lambda(0)$, the corresponding expected value is $\lambda(0)t$. This suggests *considering a nonstationary process on the modified time scale,* where τ's replace t's, where τ is defined by

$$\tau = \tau(t) \triangleq \frac{\Lambda(t)}{\lambda(0)} \qquad (5\text{-}64)$$

The purpose of this time-scale transformation is to convert the nonstationary process $\{N(t); t \geq 0\}$ to a stationary process. The transformation does not affect the properties of orderliness and independent increments (why?). Let us see how it makes the increments stationary.

From (5-64),

$$\lambda(0) \frac{d\tau}{dt} = \lambda(t)$$

To avoid unnecessary complications in the argument, assume $\lambda(t) > 0$, so that τ is a strictly increasing, continuous function of t. Then

$$\frac{dt}{d\tau} = \frac{\lambda(0)}{\lambda(t)}$$

Now

$$P\{N(\tau + \Delta\tau) - N(\tau) > 0\} + o(\Delta\tau) = P\{N(t + \Delta t) - N(t) > 0\} + o(\Delta t)$$

$$= \lambda(t) \, \Delta t + o(\Delta t)$$

so

$$\lim_{\Delta\tau \to 0} \frac{P\{N(\tau + \Delta\tau) - N(\tau) > 0\}}{\Delta\tau} = \lambda(t) \lim_{\Delta t \to 0} \frac{\Delta t}{\Delta\tau} + \lim_{\Delta t \to 0} \frac{o(\Delta t)}{\Delta t} \frac{\Delta t}{\Delta\tau}$$

$$= \lambda(t) \frac{dt}{d\tau} = \lambda(0)$$

Hence $\{N(\tau); \tau > 0\}$ has stationary increments and is a Poisson process. This means that

$$P\{N(t) = k\} = P\{N(\tau) = k\} = \frac{[\lambda(0)\tau]^k}{k!} e^{-\lambda(0)\tau}$$

$$= \frac{[\Lambda(t)]^k}{k!} e^{-\Lambda(t)} \qquad (5\text{-}65)$$

so the probability law of the nonstationary process on the original time scale has been obtained. It is the same as the law of the stationary Poisson process, with $\Lambda(t)$ taking the place of λt.

Processes Which Are Not Poisson

In models where assumptions are made about the nature of the fundamental random variables, we may want to establish that a derived stochastic process is Poisson. For example, the fundamental assumptions about the $M/M/1$ queue are about the nature of the arrival and service processes. Let $R(t)$ be the number of departures during $(0, t]$. The process $\{R(t); t \geq 0\}$ is often of interest, and its properties can be deduced from the fundamental assumptions.

The following argument shows that it is possible (in fact, likely) that $\{R(t); t \geq 0\}$ is a Poisson process in the steady state. Let $X(t)$ be the number of customers in the system at time t, λ be the arrival rate, and $1/\mu$ be the mean service time. Recall (4-31) and Exercise 4-22, and assume that

$$P\{X(0) = n\} = (1 - \rho)\rho^n \qquad n \in I, \qquad \rho = \lambda/\mu \qquad (5\text{-}66)$$

i.e., that the system is in the steady state. Let R_1 be the departure epoch of the next customer. If $X(0) > 0$, then R_1 is distributed as a service time (from the memoryless property of the exponential distribution function). Hence the LST of R_1 is

$$E(e^{-sR_1} \mid X(0) > 0) = \frac{\mu}{s + \mu}$$

If $X(0) = 0$, then R_1 is distributed as the sum of an interarrival time (the memoryless property again) and a service time. Hence

$$E(e^{-sR_1} \mid X(0) = 0) = \frac{\lambda}{s + \lambda} \frac{\mu}{s + \mu}$$

Unconditioning with (5-66) yields

$$E(e^{-sR_1}) = \frac{\mu}{s + \mu} \left[\rho + (1 - \rho) \frac{\lambda}{s + \lambda} \right] = \frac{\lambda}{s + \lambda}$$

Hence R_1 has an exponential distribution with mean $1/\lambda$. This is *not sufficient* to prove $\{R(t); t \geq 0\}$ is a Poisson process because it has not been shown that $\{R(t); t \geq 0\}$ is a renewal process. Consider this example.

Example 5-19 Let $X_1, X_2, \ldots$ be interevent times, where

$$X_1 = X_2 = \cdots \qquad \text{and} \qquad P\{X_1 \leq t\} = 1 - e^{-\lambda t}$$

The counting process for these random variables is not a renewal process, yet all the interevent times are exponentially distributed. The reason is that the interevent times are not independent. (They are as dependent as can be.)

$\square$

It is not difficult to prove

$$P\{R(t) = k\} = \frac{e^{-\lambda t}(\lambda t)^k}{k!} \qquad k \in I, t \geq 0$$

Even this is not sufficient to prove that $\{R(t); t \geq 0\}$ is a Poisson process. The next example exhibits a process that is not Poisson, but where the number of events in any interval has the Poisson distribution.

Example 5-20† A process for which (5-63) does not hold cannot be Poisson. We construct a process for which (5-63) fails yet the number of points in any interval has the Poisson distribution. Let the random variable K represent the number of points in the interval. For convenience, place the points in $[0, 1]$, but any interval of finite length would serve. Let $P\{K = k\} = \lambda^k e^{-\lambda}/k!$, $k \in I$, and observe a value of K, say k. If $k \neq 3$, place the k points uniformly and independently in $[0, 1]$, that is, so that (5-63) holds. Write this as

$$F_k(s_1, \dots, s_k) \triangleq P\{S_{i_1} \leq s_1, \dots, S_{i_k} \leq s_k \mid K = k\}$$

$$= s_1 \cdots s_k \qquad k \neq 3 \tag{5-67a}$$

If $k = 3$, place the three points so that

$$F_3(s_1, s_2, s_3) = s_1 s_2 s_3 + \varepsilon (s_1 - s_2)^2(s_1 - s_3)^2(s_2 - s_3)^2$$

$$\cdot s_1 s_2 s_3 (1 - s_1)(1 - s_2)(1 - s_3) \tag{5-67b}$$

For $\epsilon > 0$ and small, $F_3(\,\cdot\,)$ is a distribution function.

Let $Y(t)$ be the number of points in $[0, t]$. Since (5-67) and (5-63) are inconsistent, $\{Y(t); 0 \leq t \leq 1\}$ is not a Poisson process. Define

$$G(a, b; j, k) = P\{\text{exactly } j \text{ of } S_1, \dots, S_k \in (a, b]\}$$

for $0 \leq a < b \leq 1$, $0 \leq j \leq k$. By the symmetry of the construction, all $\binom{k}{j}$ collections of the points are equally likely. Thus

$$G(a, b; j, k) = \binom{k}{j} P\{S_1, \dots, S_j \in (a, b], \quad S_{j+1}, \dots, S_k \notin (a, b]\} \tag{5-68}$$

For $k \neq 3$, the uniform property expressed by (5-67a) immediately yields

$$G(a, b; j, k) = \binom{k}{j} (b - a)^j [1 - (b - a)]^{k-j} \qquad j = 0, 1, \dots, k \tag{5-69}$$

For $k = 3$, observe that $F_3(s_1, s_2, s_3) = s_1 s_2 s_3$ whenever either two or more s_i are equal or at least one $s_i = 1$. Verify for yourself that when $k = 3$, all the joint probabilities required to compute the probability on the right-side of (5-68) have this property, hence (5-69) holds for $k = 3$ as well. Thus

$$P\{Y(b) - Y(a) = j\} = \sum_{k=0}^{\infty} G(a, b; j, k) P\{K = k\}$$

$$= e^{-\lambda} \sum_{k=j}^{\infty} \frac{(b - a)^j [1 - (b - a)]^{k-j} \lambda^k}{j!(k - j)!}$$

$$= \frac{[\lambda(b - a)]^j e^{-\lambda(b-a)}}{j!} \qquad j \in I \qquad \square$$

---

† This example is due to L. Shepp.

The point of the preceding discussion is to show that it is crucial to prove that the interrenewal times are independent in order to establish that a counting process is a Poisson process. A proof that $\{R(t); t \geq 0\}$ is indeed a Poisson process is given in Section 8-8. In order to avoid the difficulty of proving that the times between departures are independent, the proof given there does not use renewal theory at all.

EXERCISES

5-56 Let $\{N(t); t \geq 0\}$ be a homogeneous Poisson process with rate $\lambda > 0$. Prove

$$P\{N(u) = i \mid N(T) = k\} = \binom{k}{i} p^i (1-p)^{k-i}$$

where $u < T$, $0 \leq i \leq k \geq 1$, and $p = u/T$.

5-57 Let $\{N(t); t \geq 0\}$ be a homogeneous Poisson process, and let $S_1, S_2, \ldots$ be the event epochs. Find $E(S_i \mid N(T) = k)$ for each $i = 1, 2, \ldots, k$. Also find $E(T - S_k \mid N(T) = k)$.

5-58 Let $Z_1, Z_2, \ldots, Z_k$ be i.i.d. random variables that are uniformly distributed on $[0, T]$. Let W_1 be the smallest, W_2 the second smallest, etc., up to W_k (W_j is called the jth-*order statistic*). Find the joint density function of $W_1, W_2, \ldots, W_k$. Interpret Theorem 5-13 in light of your answer.

5-59 Consider a queueing system with a Poisson arrival process. Suppose there are $k > 0$ arrivals in the interval $[s, s + T]$. What is the expected time at which the first arrival occurred? The kth? What is the probability that an arrival occurred during $(t, t + \Delta t)$, $s < t < s + T$?

5-60 Let $X(t)$ be the number of customers being served at time t in an $M/G/\infty$ queue. Show that

$$P\{X(t) = j \mid X(0) = 0\} = \frac{e^{-\alpha} \alpha^j}{j!} \qquad j = 0, 1, \ldots$$

where $\alpha = \lambda t h(t)$ and $h(t)$ is defined in Example 5-18.

5-61 (Continuation) Prove that

$$\lim_{t \to \infty} P\{X(t) = j\} = \frac{e^{-\rho} \rho^j}{j!} \qquad j = 0, 1, \ldots$$

where $\rho = \lambda E(V_1)$. Compare this to Exercise 4-24.

5-62 (Continuation) Show that when $X(0) = 0$, the number of departures by time t has a Poisson distribution with mean $\lambda t [1 - h(t)]$. Does this necessarily mean that the departure process is a nonstationary Poisson process? Why?

5-63 Let $\{N(n); n = 0, 1, \ldots\}$ be a renewal process with generating sequence $X_1, X_2, \ldots$, where $P\{X_1 = j\} = (1 - p)p^j$, $j = 0, 1, \ldots$, and $0 < p < 1$. Show that

$$P\{S_1 = i_1, S_2 = i_2, \ldots, S_k = i_k \mid N(n) = k\} = \frac{1}{\binom{n+k}{k}}$$

for $0 \leq i_1 \leq i_2 \leq \cdots \leq i_k \leq n$.

5-64 (Continuation) Show that $E(S_1 \mid N(6) = 2) = 2$ and $E(S_2 \mid N(6) = 2) = 4$. [Can you provide a simple proof that $E(S_i \mid N(n) = k) = ni/(k + 1)$, $i = 1, 2, \ldots, k$? We can prove it only by going through some messy calculations.]

5-65 In Example 5-16 let the holding cost be assessed at the end of each period and assume that $(1 - p)p^j$ is the probability of j demands in a period. What is the expected value of the holding cost incurred during a cycle?

5-66 Change Exercise (5-63) so that $P\{X_1 = j\} = (1 - p)p^{j-1}$, $j = 1, 2, \ldots$. Find $P\{S_1 = i_1, S_2 = i_2, \ldots, S_k = i_k \mid N(n) = k\}$.

5-67 Let $\{N(t); t \geq 0\}$ be a renewal process where the times between renewals have a density function $f(\cdot)$. Show that $\{N(t); t \geq 0\}$ is a Poisson process if, and only if, the equilibrium excess and deficit random variables are independent.

5-68 Let $\{N_i(t); t \geq 0\}$ be Poisson processes with rate λ_i, $i = 1, 2, \lambda_1 \neq \lambda_2$. Flip a fair coin once and let

$$N(t) = \begin{cases} N_1(t) & \text{if the coin lands heads} \\ N_2(t) & \text{if the coin lands tails} \end{cases}$$

Show that $\{N(t); t \geq 0\}$ satisfies (5-63) but is not a Poisson process. Is it a renewal process?

5-69 Let $\{N(t); t \geq 0\}$ be a counting process. Prove that it is a Poisson process if, and only if,

$$P\{N(t_i) = k_i, i = 1, \ldots, n\} = e^{-\lambda t_n} \lambda^{k_n} \prod_{i=1}^{n} \frac{(t_i - t_{i-1})^{k_i - k_{i-1}}}{(k_i - k_{i-1})!}$$

holds for some $\lambda > 0$, any $n \in I_+$, each $k_i \in I$ with $0 = k_0 \leq k_i \leq \ldots \leq k_n$, and any $0 = t_0 \leq t_1 \leq \ldots \leq t_n$.

5-8* SUPERPOSITION OF RENEWAL PROCESSES

Let $\{N_i(t); t \geq 0\}$ be independent renewal processes, $i = 1, 2, \ldots, m$, and define $N(t)$ by

$$N(t) = N_1(t) + N_2(t) + \cdots + N_m(t) \qquad t \geq 0$$

We call $\{N(t); t \geq 0\}$ the *superposition* of $\{N_1(t); t \geq 0\}, \ldots, \{N_m(t); t \geq 0\}$. Exercise 5-9 shows that the superposition of independent Poisson processes is a Poisson process, but Exercise 5-10 shows that the superposition of independent renewal processes need not be a renewal process. In this section, we show that the superposition of a large number of independent equilibrium renewal processes, each with small intensity, is asymptotically a Poisson process.

This theorem, due to C. Palm and A. Y. Khintchine, provides a theoretical justification for using the Poisson process in a variety of models. When many customers, each of whom has a comparatively small calling rate, are attached to a telephone exchange, the theorem predicts that the number of calls attempted is a Poisson process. This prediction typically is confirmed by data. Similarly, demands for a given item in a retail store may be modeled as the superposition of the requests of many independent customers, each of whom shops for that item infrequently. These conditions lead to a Poisson demand process.

Assumptions and Notation

When m equilibrium renewal processes are superposed, let $N_{jm}(\cdot)$ be the counting process for the jth one. In the jth process, let X_{1jm} be the epoch of the first renewal and X_{2jm} be the time between the first and second renewal epochs. Let

$$F_{jm}(t) = P\{X_{2jm} \leq t\} \qquad \text{and} \qquad \lambda_{jm} = \frac{1}{E(X_{2jm})}$$

Then from Definition 5-4,

$$G_{jm}(t) \triangleq P\{X_{1jm} \leq t\} = \lambda_{jm} \int_0^t F_{jm}^c(y) \, dy$$

The theorems in this section use the following assumptions.

Assumption 1 For all m sufficiently large,

$$\lambda_{1m} + \cdots + \lambda_{mm} = \lambda < \infty \tag{5-70}$$

Assumption 2 Given $\epsilon > 0$, for each $t > 0$ and m sufficiently large,

$$F_{jm}(t) \leq \epsilon \qquad j = 1, \ldots, m \tag{5-71}$$

For each m, define

$$N_{0m}(t) = N_{1m}(t) + \cdots + N_{mm}(t) \tag{5-72}$$

Assumption 2 asserts that as m increases, the processes being combined have renewals very infrequently. Assumption 1 shows that $N_{0, m+1}(t)$ is not formed by adding another process to $N_{0m}(t)$. As m increases, the processes being combined are changed so that (at least for large m) the asymptotic rate at which renewals occur is a constant.

Palm's Theorem

Theorem 5-14: Palm's theorem Under Assumptions 1 and 2, for each $t > 0$,

$$\lim_{m \to \infty} P\{N_{0m}(t) = 0\} = e^{-\lambda t} \tag{5-73}$$

PROOF Choose $\epsilon > 0$ and m so large that (5-70) and (5-71) hold. For each $j = 1, \ldots, m$,

$$P\{N_{jm}(t) = 0\} = 1 - G_{jm}(t) = 1 - \lambda_{jm} \int_0^t [1 - F_{jm}(y)] \, dy$$

$$= 1 - \lambda_{jm} t + \lambda_{jm} \int_0^t F_{jm}(y) \, dy$$

According to (5-71), the integral is smaller than ϵt; hence there is some number, α_j say, such that $0 \leq \alpha_j < 1$ and

$$P\{N_{jm}(t) = 0\} = 1 - \lambda_{jm} t + \alpha_j \lambda_{jm} \epsilon t \tag{5-74}$$

Exercise 5-70 asks you to show that (5-74) implies

$$| \ln P\{N_{jm}(t) = 0\} + \lambda_{jm} t | = | \ln(1 + \alpha_j \lambda_{jm} \epsilon t)| < \lambda_{jm} \epsilon c(t) \tag{5-75}$$

for some function $c(t) > 0$ that is independent of j.

Since the superposed streams are independent and (5-70) holds, (5-75) yields

$$| \ln P\{N_{0m}(t) = 0\} + \lambda t | = | \sum_{j=1}^{m} \ln P\{N_{jm}(t) = 0\} + \lambda_{jm} t |$$

$$\leq \sum_{j=1}^{m} | \ln P\{N_{jm}(t) = 0\} + \lambda_{jm} t | \qquad (5\text{-}76)$$

$$< \lambda \epsilon c(t)$$

Since ϵ is arbitrary, (5-76) asserts

$$| \ln P\{N_{0m}(t) = 0\} + \lambda t | \to 0$$

as $m \to \infty$, which implies (5-73). $\qquad \square$

Since Palm's theorem is concerned only with the time to the first renewal of each of the superposed processes, it is not necessary to interpret the theorem in the context of superposition of renewal processes. One can apply it to independent random variables $X_{1m}, \ldots, X_{mm}$, where X_{jm} has distribution function $G_{jm}(\,\cdot\,)$. The postulated form of $G_{jm}(\,\cdot\,)$ requires that it possess a nonincreasing density function. In this setting, Palm's theorem has the following important interpretation.

Consider a device composed of many very reliable components. Suppose the device fails whenever any of its components fails and that the components fail independently. Palm's theorem asserts that the time to the first failure of such a device has an exponential distribution.

Palm-Khintchine Theorem

Palm mistakenly thought Theorem 5-14 proved that $\{N_{0m}(t); t \geq 0\}$ approaches a Poisson process as $m \to \infty$. The nature of this error is shown in Example 5-19; it must be shown that the times between events are independent. The remainder of the proof was provided by A. Y. Khintchine. Khintchine's proof uses these two lemmas.

Lemma 5-2 If Assumptions 1 and 2 hold, then

$$\lim_{m \to \infty} P\{N_{jm}(t) \geq 2 \text{ for some } j \leq m\} = 0 \qquad (5\text{-}77)$$

PROOF Choose $\epsilon > 0$ and m so large that (5-70) and (5-71) hold. Let $\psi_{jm}(t) = P\{N_{jm}(t) \geq 2\}$, $j = 1, \ldots, m$. Since $N_{jm}(t) \geq 2$ if, and only if, $X_{1jm} + X_{2jm} \leq t$,

$$\psi_{jm}(t) = \int_{0}^{t} F_{jm}(t - y) \, dG_{jm}(y) = \lambda_{jm} \int_{0}^{t} F_{jm}(t - y) F_{jm}^{c}(y) \, dy$$

Since distribution functions are nondecreasing functions of their argument and complementary distribution functions are no larger than 1,

$$\psi_{jm}(t) \le \lambda_{jm} F_{jm}(t) \int_0^t dy = \lambda_{jm} t F_{jm}(t) \qquad j = 1, \ldots, m$$

Thus, for m large enough,

$$P\{N_{jm}(t) \ge 2 \text{ for some } j \le m\} \le \sum_{j=1}^m \psi_{jm}(t)$$

$$\le \sum_{j=1}^m \lambda_{jm} t F_{jm}(t) \le \lambda t \epsilon \qquad (5\text{-}78)$$

Since ϵ is arbitrary, (5-78) is equivalent to (5-77). $\qquad\qquad \square$

Lemma 5-3 If Assumptions 1 and 2 hold, for $t \ge 0$,

$$\lim_{m \to \infty} P\{N_{0m}(t) = k\} = \frac{e^{-\lambda t}(\lambda t)^k}{k!} \qquad k = 0, 1, \ldots \qquad (5\text{-}79)$$

PROOF Independence of the superposed streams and (5-72) imply the generating-function relation

$$\hat{P}_m(z) \triangleq \sum_{k=0}^\infty z^k P\{N_{0m}(t) = k\} = \prod_{j=1}^m \sum_{k=0}^\infty z^k P\{N_{jm}(t) = k\} \qquad (5\text{-}80)$$

Choose $\epsilon > 0$ and m so large that (5-70) and (5-71) hold. Exercise 5-71 asks you to prove

$$\lambda_{jm} < \epsilon \qquad j = 1, \ldots, m \qquad (5\text{-}81)$$

Combining (5-74) and (5-81) yields

$$P\{N_{jm}(t) = 0\} = 1 - \lambda_{jm} t + \epsilon q_1 t \qquad j = 1, \ldots, m \qquad (5\text{-}82a)$$

for some number q_1 that is independent of m (it may depend on j). From (5-81) and Lemma 5-2, we conclude that there are numbers q_2 and q_3 which are independent of m such that

$$P\{N_{jm}(t) \ge 2\} = q_3 \epsilon \lambda_{jm} t \qquad j = 1, \ldots, m \qquad (5\text{-}82b)$$

and

$$P\{N_{jm}(t) = 1\} = \lambda_{jm} t + q_2 \epsilon \lambda_{jm} t \qquad j = 1, \ldots, m \qquad (5\text{-}82c)$$

From (5-82a), (5-82b), and (5-82c),

$$\sum_{k=0}^\infty z^k P\{N_{jm}(t) = k\} = 1 - \lambda_{jm} t(1 - z) + q_4(z) \epsilon \lambda_{jm} t \qquad (5\text{-}83)$$

where $q_4(z)$ is independent of m and bounded. Substitute (5-83) into (5-80) and take logarithms. From the Taylor series $ln(1 - x) = -x - x^2/2 - x^3/3 - \cdots$, we obtain

$$ln \hat{P}_m(z) = \sum_{j=1}^m [-\lambda_{jm} t(1 - z) + \alpha_1 \epsilon + \alpha_2 \epsilon^2 + \cdots]$$

for some numbers $\alpha_1, \alpha_2, \ldots$ which are independent of m. Hence

$$\lim_{m \to \infty} \hat{P}_m(z) = e^{-\lambda t(1-z)} \qquad t \geq 0 \tag{5-84}$$

Applying the continuity property of generating functions (i.e., Property A-15a) and inverting (5-84) by inspection yield (5-79). □

Example 5-20 shows that Lemma 5-3 is not sufficient to conclude that asymptotically $\{N_{0m}(t); t \geq 0\}$ is a Poisson process. That conclusion will follow from the characterization of the Poisson process given in Exercise 5-69.

Theorem 5-15: Palm-Khintchine theorem Under Assumptions 1 and 2, as $m \to \infty$, $\{N_{0m}(t); t \geq 0\}$ approaches a Poisson process.

PROOF According to Exercise 5-69, it is sufficient to establish

$$\lim_{m \to \infty} P\{N_{0m}(t_i) = k_i; i = 1, \ldots, n\} = e^{-\lambda t_n} \lambda^{k_n} \prod_{i=1}^{n} \frac{u_i^{l_i}}{l_i!} \tag{5-85}$$

where $n \in I_+$, each $k_i \in I$ with $0 = k_0 \leq k_1 \leq \cdots \leq k_n$, $0 = t_0 \leq t_1 \leq \cdots \leq t_n$,

$$u_i \triangleq t_i - t_{i-1} \qquad \text{and} \qquad l_i \triangleq k_i - k_{i-1} \qquad i = 1, \ldots, n \tag{5-86}$$

From (5-86),

$$t_n = \sum_{i=1}^{n} u_i \triangleq u \qquad \text{and} \qquad k_n = \sum_{i=1}^{n} l_i \triangleq k$$

If we let $\Delta_m(u_i) = N_{0m}(t_i) - N_{0m}(t_{i-1})$, then (5-85) is equivalent to

$$\lim_{m \to \infty} P\{\Delta_m(u_i) = l_i; i = 1, \ldots, n\} = e^{-\lambda u} \lambda^k \prod_{i=1}^{n} \frac{u_i^{l_i}}{l_i!} \tag{5-85'}$$

We now prove that (5-85') is true.

Let C_1 be the condition that $N_{jm} \leq 1$ for each $j \leq m$ and C_2 be the complementary condition. For notational ease, let $\mathcal{E}_m = \{\Delta_m(u_i) = l_i; i = 1, \ldots, n\}$. Thus

$$P\{\mathcal{E}_m\} = P\{\mathcal{E}_m, C_1\} + P\{\mathcal{E}_m, C_2\}$$

Since $P\{\mathcal{E}_m, C_2\} \leq P\{C_2\}$, Equation (5-77) shows

$$\lim_{m \to \infty} P\{\mathcal{E}_m, C_2\} = 0$$

hence

$$\lim_{m \to \infty} P\{\mathcal{E}_m\} = \lim_{m \to \infty} P\{\mathcal{E}_m, C_1\} \tag{5-87}$$

Observe that $\{\mathcal{E}_m\} \Rightarrow \{N_{0m}(u) = k\}$; hence

$$P\{\mathcal{E}_m\} = P\{\mathcal{E}_m \mid N_{0m}(u) = k\} \tag{5-88}$$

Combining (5-87) and (5-88) yields

$$\lim_{m \to \infty} P\{\mathcal{E}_m\} = \lim_{m \to \infty} P\{\mathcal{E}_m, C_1, N_{0m}(u) = k\}$$

$$= \lim_{m \to \infty} P\{\mathcal{E}_m \mid C_1, N_{0m}(u) = k\} \lim_{m \to \infty} P\{C_1, N_{0m}(u) = k\} \qquad (5\text{-}89)$$

Use (5-77) to show $\lim_{m \to \infty} P\{C_2, N_{0m}(u) = k\} = 0$. Thus

$$\lim_{m \to \infty} P\{C_1, N_{0m}(u) = k\} = \frac{e^{-\lambda u}(\lambda u)^k}{k!} \qquad (5\text{-}90)$$

where the right-side is obtained from Lemma 5-3. Now work on the other limit on the right-side of (5-89). Exercise 5-72 asks you to prove that when C_1 and $\{N_{0m}(u) = k\}$ occur, as $m \to \infty$, the probability that a renewal from some process occurs during $(t_{i-1}, t_i]$ approaches u_i/u and is independent of the location of the $k - 1$ other renewals. The formula for multinomial probabilities yields

$$\lim_{m \to \infty} P\{\mathcal{E}_m \mid C_1, N_{0m}(u) = k\} = \frac{k!}{l_1! \cdots l_n!} \left(\frac{u_1}{u}\right)^{l_1} \cdots \left(\frac{u_n}{u}\right)^{l_n} \qquad (5\text{-}91)$$

Substituting (5-90) and (5-91) in (5-89) yields (5-85′). $\qquad \Box$

Notice that only the first two interrenewal times have a part in the proofs of Theorems 5-14 and 5-15. Hence these theorems remain true when the remaining interrenewal times have arbitrary distribution functions and/or depend on the lengths of the first two renewals of the superposed stream.

EXERCISES

5-70 Show that (5-74) implies (5-76).

5-71 Prove that (5-71) implies (5-81).

5-72 Prove that when C_1 and $\{N_{0m}(u) = k\}$ occur, as $m \to \infty$, the probability that a renewal from some process $j\,(j < m)$ occurs during $(t_{i-1}, t_i]$ approaches u_i/u.

5-9** A PROBABILISTIC PROOF OF BLACKWELL'S RENEWAL THEOREM

The original proof of Blackwell's renewal theorem used the analytical properties of the renewal function and is purely formal. In this section we present a proof, due to T. Lindvall,† that has an intuitive basis and requires only probabilistic methods to fill in the details.

The most important version of the theorem to be proved is this one.

† T. Lindvall, "A Probabilistic Proof of Blackwell's Renewal Theorem," *Ann. Prob.* **5**: 482–485 (1977).

Theorem 5-16: Blackwell's renewal theorem If X_1 is not arithmetic, then as $t \to \infty$,

$$M(t) - M(t - h) \to \frac{h}{v} \tag{5-92}$$

for every fixed $h > 0$, where $v = E(X_1) < \infty$.

The proof encompasses the case where X_1 is arithmetic but is not valid when $E(X_1) = \infty$. This is the idea of the proof. In the case of an equilibrium renewal process, (5-54) states that $M(t) - M(t - h) = h/v$ for all $t > h$. Let $\{N(t); t \geq 0\}$ be the given renewal process, and let $F(t) = P\{X_1 \leq t\}$, $v = E(X_1)$, $M(t)$ be the renewal function, and $S_n = X_1 + \cdots + X_n$ be the epoch of the nth renewal. Consider an equilibrium renewal process $\{N'(t); t \geq 0\}$, and let $S_n' = X_0' + \cdots + X_{n-1}'$ be the epoch of the nth event, $P\{X_0' \leq t\} \triangleq g(t) = F^c(t)/v$, $F(t) = P\{X_1' \leq t\}$, and $M'(t)$ be the renewal function. Assume that these two processes are independent, i.e., that $\{X_i\}_{i=0}^{\infty}$ and $\{X_i'\}_{i=0}^{\infty}$ are mutually independent.

If, for some pair of indices m and n, $S_n = S_m' = T$ (that is, at some epoch T both processes experience a renewal), the above hypotheses imply that $S_{n+j} \sim S_{m+j}', j = 1, 2, \ldots$. Hence

$$M(t) = M'(t) + n - m \qquad t > T \tag{5-93}$$

and (5-93) together with (5-54) yields (5-92).

When the times between renewals are not discrete random variables, it is too much to ask for both processes to have a renewal at the same epoch. The details of the proof are concerned with showing that for any fixed $\delta > 0$, there are finite m and n such that

$$P\{0 \leq S_m' - S_n < \delta\} = 1 \tag{5-94}$$

and then that (5-94) can be used to obtain (5-92).

This method of proof is easily extended to cover arithmetic X's, but a different proof is used when $v = \infty$.

Two Required Lemmas

Lemma 5-4: Hewitt-Savage zero-one law Let $X_1, X_2, \ldots$ be i.i.d. random variables and a_i be the realization of X_i. Let $\mathscr{A}$ be a set with the property that if $(a_1, \ldots, a_m, \ldots, a_n, \ldots) \in \mathscr{A}$, then $(a_1, \ldots, a_n, \ldots, a_m, \ldots) \in \mathscr{A}$ also.† Then either $P\{\mathscr{A}\} = 0$ or $P\{\mathscr{A}\} = 1$.

A proof of this lemma may be found in Feller (1971, p. 124). A special case of the use we make of this lemma is seen in the following example.

---

† Thus, if two arbitrary coordinates of an element of $\mathscr{A}$ are interchanged, the resulting point is also an element of $\mathscr{A}$. This condition automatically extends to permutations involving any finite number of coordinates, and we say that $\mathscr{A}$ is *invariant under finite permutations of its coordinates*.

Example 5-21 Let $S_k = X_1 + \cdots + X_k$, $k \in I_+$, r be any interval of real numbers, and $\mathcal{A}$ be the event $\{S_k \in r$ infinitely often$\}$. Write $\mathcal{A}$ as $\{\omega : S_k(\omega) \in r\}$, where ω is a sample path of the process $\{X_k; k \in I_+\}$. If $X_1(\omega) + \cdots + X_m(\omega) + \cdots + X_n(\omega) + \cdots + X_k(\omega) \in r$ for infinitely many k, then $X_1(\omega) + \cdots + X_n(\omega) + \cdots + X_m(\omega) + \cdots + X_k(\omega) \in r$ for infinitely many k, so $\mathcal{A}$ is invariant under finite permutations of its coordinates. By Lemma 5-4, either $P\{\mathcal{A}\} = 0$ or $P\{\mathcal{A}\} = 1$. $\qquad\square$

The next lemma shows that in a renewal process where the times between renewals are not arithmetic, asymptotically every (nonempty) time interval (no matter how small) has positive probability of containing a renewal. The following definition is used to state the lemma precisely.

Definition 5-5 A point x is a *point of increase* of the distribution function $F(\,\cdot\,)$ if $F(b) - F(a) > 0$ for every open interval (a, b) containing x.

Lemma 5-5 Let $F(\,\cdot\,)$ be the distribution function of a nonnegative random variable with $F(0) < 1$ and S be the set formed by the points of increase of $F(\,\cdot\,)$ and all its n-fold convolutions.
(a) If $F(\,\cdot\,)$ is *not* arithmetic, then for any given $\epsilon > 0$, for x sufficiently large, $(x, x + \epsilon)$ contains points of S.
(b) If $F(\,\cdot\,)$ is arithmetic with span ξ, then S contains all points of the form $n\xi$ for n sufficiently large.

A proof of this lemma may be found in Feller (1971, p. 147).

Proof of Theorem 5-16

Call $\{N(t); t \geq 0\}$ the *original* process and $\{N'(t); t \geq 0\}$ the *primed* process, and define

$$\eta(t) = N(t) + 1 \qquad \eta'(t) = N'(t) + 1$$

Let Z_i denote the time between the ith renewal in the original process and the next renewal in the primed process; Z_i is the excess random variable of the primed process at epoch S_i and is given by

$$Z_i \triangleq S'_{\eta'(t)} - S_i \qquad i = 1, 2, \ldots$$

For any fixed $\delta > 0$, define the event $\mathcal{A}_i$ by

$$\mathcal{A}_i = \{Z_j < \delta \text{ for some } j \geq i\} \qquad i = 1, 2, \ldots$$

In particular, $\mathcal{A}_1 = \{Z_j < \delta \text{ for some } j\}$. By proving that $\mathcal{A}_1$ occurs, (5-94) will be established. Figure 5-10 portrays the quantities we have just defined. Clearly

$$\mathcal{A}_1 \supset \mathcal{A}_2 \supset \ldots \supset \bigcap_{i=1}^{\infty} \mathcal{A}_i \triangleq \mathcal{A}_\infty = \{Z_i < \delta \text{ infinitely often}\} \tag{5-95}$$

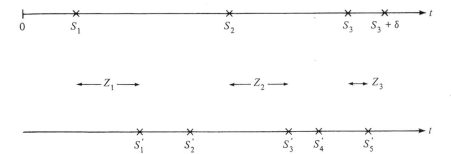

Figure 5-10 Renewal epochs in the original and primed processes.

From Exercise 5-73 the distribution function of Z_i is the same for all i, and hence (Exercise 5-75) the fidi's of $\{Z_{i+n}; n \in I_+\}$ are the same for all i, so $P\{\mathcal{A}_i\}$ is the same for all i. In particular,

$$P\{\mathcal{A}_1\} = P\{\mathcal{A}_\infty\} \tag{5-96}$$

Apply Lemma 5-4 to the sequence of random variables $(X_1, X_1'), (X_2, X_2'), \ldots$ to obtain

$$P\{\mathcal{A}_\infty \mid X_0' = t\} = 0 \text{ or } 1 \qquad t \geq 0 \tag{5-97}$$

(see Exercise 5-76). From (5-95) and the theorem of total probability,

$$\int_0^\infty P\{\mathcal{A}_1 \mid X_0' = t\} g(t) \, dt = P\{\mathcal{A}_1\} = P\{\mathcal{A}_\infty\}$$

$$= \int_0^\infty P\{\mathcal{A}_\infty \mid X_0' = t\} g(t) \, dt \tag{5-98}$$

By assumption, $F(\,\cdot\,)$ is not arithmetic, so Lemma 5-5 asserts

$$P\{\mathcal{A}_\infty \mid X_0' = t\} > 0 \qquad t \geq 0 \tag{5-99}$$

Combining (5-96) through (5-99) yields

$$P\{\mathcal{A}_\infty\} = 1 = P\{\mathcal{A}_1\} \tag{5-100}$$

A consequence of (5-100) is that it is meaningful to talk about the first epoch at which $Z_i < \delta$. Let

$$K = \min \{i : Z_i < \delta\} \qquad J = \min \{j : S_j' \geq S_K\}$$

so that $J = \eta'(S_K)$ and $0 \leq Z_K = S_J' - S_K < \delta$.

Now introduce the following counting functions for intervals:

$$H(a, b] = \begin{cases} N(b) - N(a) & \text{if } a < b \\ 0 & \text{if } a \geq b \end{cases}$$

$$H'(a, b] = \begin{cases} N'(b) - N'(a) & \text{if } a < b \\ 0 & \text{if } a \geq b \end{cases}$$

Choose $h > 0$ and for each t look at the intervals

$$(a, b] = (t, t + h] \cap (0, S_K]$$

and

$$(a', b'] = (t + Z_K, t + Z_K + h] \cap (S_J', \infty)$$

An example of these intervals is depicted in Figure 5-11.

Define $H''(t, t + h]$ by

$$H''(t, t + h] = H(a, b] + H'(a', b']$$

It is apparent that (Exercise 5-77)

$$H''(t, t + h] \sim N(t, t + h] \tag{5-101}$$

The remainder of the proof makes precise this observation: Since K and S_J' are finite, as $t \to \infty$, $(a, b] \to$ the empty set, $(a', b') \to (t + Z_K, t + Z_K + h]$, and (5-101) combined with the stationarity of the primed process yields the desired result.

From (5-101) and visual help from Figure 5-11,

$$M(t + h) - M(t) = E(H''(t, t + h])$$

$$= E(H(a, b]) + E[N'(t + Z_K + h) - N'(t + Z_K)]$$

$$- E(H'(t + Z_K, t + Z_K + h] \cap (0, S_J'])$$

$$\triangleq M_1(t) + M_2(t) - M_3(t) \qquad \text{say} \tag{5-102}$$

Since $Z_K < \delta$ by construction,

$$\frac{h - \delta}{v} = M'(t + h) - M'(t + \delta) \le M_2(t)$$

$$\le M'(t + h + \delta) - M'(t) = \frac{h + \delta}{v} \tag{5-103}$$

which proves that $M_2(t) \to h/v$ uniformly in t as $\delta \downarrow 0$.

Argue that (Exercise 5-79) $P\{N(t+h) - N(t) > n\} \le P\{N(h) + 1 > n\}$, $n \in I_+$, and thus

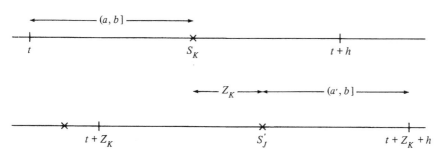

Figure 5-11 $(a, b]$ and $(a', b']$ when $t < S_K < t + h$ and $t + Z_K < S_J' < t + Z_K + h$.

$$M(t + h) - M(t) \leq M(h) + 1 \tag{5-104}$$

Consequently (by using the notation $1\{S_k > t\}$, for the indicator function of the event $\{S_k > t\}$) for any $a > 0$,

$$M_1(t) \leq E([N(t + h) - N(t)] 1\{S_K \geq t\})$$

$$= E([N(t + h) - N(t)] 1\{S_K \geq t\} \mid N(t + h) - N(t) \leq a) P\{N(t + h) - N(t) \leq a\}$$

$$+ E([N(t + h) - N(t)] 1\{S_K \geq t\} \mid N(t + h) - N(t) > a) P\{N(t + h) - N(t) \leq a\}$$

$$\leq aP\{S_K \geq t\} + E[N(h) \mid N(h) \geq a] P\{N(h) > 0\} \tag{5-105}$$

Since $M(t) < \infty$ for all $t < \infty$, the second term on the last line must vanish as $a \to 0$. Hence, $E[N(h) \mid N(h) \geq a] \to 0$ as $a \to \infty$ for any fixed $h > 0$. Since $P\{S_K < \infty\} = 1$, for any a, no matter how large, t may be chosen so large that $aP\{S_K \geq t\}$ is arbitrarily close to zero. Hence, (5-105) yields

$$\lim_{t \to \infty} M_1(t) = 0 \tag{5-106}$$

Similarly (Exercise 5-80),

$$\lim_{t \to \infty} M_3(t) = 0 \tag{5-107}$$

is shown; since δ can be chosen arbitrarily small, (5-102), (5-103), (5-106), and (5-107) complete the proof. $\qquad\square$

To prove Blackwell's theorem when the times between renewals are arithmetic, observe that choosing δ less than the span of $F(\cdot)$ yields $Z_K = 0$ and the method of proof remains valid. The proof for $v = \infty$ requires analytical methods.†

EXERCISES

5-73 Let B_t be the excess random variable of an equilibrium renewal process at epoch t and $b_t(\cdot)$ be its density function. Show that $b_t(\cdot)$ is the same for all $t \geq 0$.

5-74 Prove that an equilibrium renewal process is stationary in the sense that the fidi's of $\{S'_{\eta(t)+n} - t; n \in I_+\}$ are the same as the fidi's of $\{S'_n; n \in I_+\}$ for all $t \geq 0$.

5-75 Prove that the fidi's of $\{Z_{i+n}; n \in I_+\}$ are the same for all $i \in I_+$.

5-76 Prove (5-97).

5-77 Derive (5-101).

5-78 Explain why $M_2(t)$, as defined in (5-102), need not equal h/v for arbitrary values of Z_K. That is, explain why it is necessary to establish (5-102).

5-79 Establish (5-104).

5-80 Prove (5-107).

† See T. Lindvall, "A Probabilistic Proof of Blackwell's Renewal Theorem," *Ann. Prob.* **5**: 482–485 (1977).

BIBLIOGRAPHIC GUIDE

The major theorems of renewal theory were first published between 1941 and 1955. W. Feller proved the elementary renewal theorem via Laplace transform methods in 1941. The proof of that theorem given in Section 5-4 uses the argument in a paper of J. L. Doob published in 1948. In that paper, Doob also obtains the strong law of large numbers for renewal processes, existence and uniqueness of solutions to the renewal equation, and the formula for the equilibrium excess d.f. In 1948, D. Blackwell published his renewal theorem. The asymptotic normality of the number of renewals was obtained by Feller in 1949. W. L. Smith's key renewal theorem was published in 1954. In 1943, C. Palm introduced the concept of regeneration epochs and studied the equilibrium renewal process, which he approached through his theory of event streams; an account of this work is contained in Khintchine (1969). In that book, which first appeared in Russian in 1955, Khintchine presented the axiomatic derivation of the Poisson process and completed Palm's proof about the superposition of stationary renewal processes.

Interest in renewal theory was spurred by the publication in 1950 of the first edition of Feller (1968). Feller used discrete renewal processes, which he called recurrent-event processes, to analyze many interesting problems and simplify the analysis of discrete Markov chains. The expository paper Smith (1958) unified the work to date and was influential in promoting subsequent applications and refinements. For a list of references published before 1958, see Smith (1958).

Most books on applied stochastic processes have a chapter on renewal theory; some are given in the list of references.

References

Çinlar, Erhan, *Introduction to Stochastic Processes*, Prentice-Hall, Englewood Cliffs, N.J. (1975).

Cox, D. R., *Renewal Theory*, Wiley, New York (1962).

Feller, William, *An Introduction to Probability Theory and Its Applications*, vol. 1, 3d ed., Wiley, New York (1968).

Feller, William, *An Introduction to Probability Theory and Its Applications*, vol. 2, 2d ed., Wiley, New York (1971).

Karlin, Samuel, and Howard S. Taylor, *A First Course in Stochastic Processes*, 2d ed., Academic Press, New York (1975).

Khintchine, A. Y., *Mathematical Methods in the Theory of Queueing*, Hafner, New York (1969).

Neuts, Marcel F., *Probability*, Allyn and Bacon, Boston (1973).

Ross, Sheldon M., *Applied Probability Models with Optimization Applications*, Holden-Day, San Francisco (1970).

Smith, Walter L., "Renewal Theory and Its Ramifications," *J. Roy. Stat. Soc.* **B20**: 243–302 (1958).

SIX

RENEWAL-REWARD AND REGENERATIVE PROCESSES

The general idea of a renewal-reward process is to associate a reward with each renewal. In Example 5-2, a cost is associated with each machine failure. The main theorems are similar to the strong law of large numbers for renewal processes and the elementary renewal theorem. They establish the long-run rate at which rewards are earned. A regenerative process is a generalization of a renewal-reward process where, roughly speaking, the rewards are a stochastic process. We show that a wide class of regenerative processes have limiting probabilities. Enhancing regenerative processes with a reward structure extends the concept of the renewal-reward process and leads to powerful convergence theorems. Examples from queueing and inventory models and simulation are given in this chapter. The concepts and theorems in Sections 6-1 through 6-4 are used repeatedly in Chapters 7 to 13.

The renewal argument (page 118) is a powerful means to analyze renewal processes. The connecting link among the topics in this chapter is the exploitation of the renewal argument to study the various processes.

6-1 DEFINITION OF THE RENEWAL-REWARD PROCESS AND EXAMPLES

Start with a renewal process $\{N(t); t \geq 0\}$ having renewal epochs $S_n = X_1 + \cdots + X_n$. Associate with each X_i the random variable C_i, which is interpreted as the *cost of the ith renewal,* and assume that C_i depends on only X_i. Define

and

$$C_l(t) = C_1 + C_2 + \cdots + C_{N(t)} \tag{6-1}$$

$$C_u(t) = C_1 + C_2 + \cdots + C_{N(t)} + C_{N(t)+1} \tag{6-2}$$

with an empty sum taken as zero. Thus, $C_l(t)$ is the sum of the costs associated with those renewals which occurred by time t, and $C_u(t)$ is the sum of the costs of those renewals which have started by time t. Both processes $\{C_l(t);\ t \geq 0\}$ and $\{C_u(t);\ t \geq 0\}$ are called *renewal-reward* processes. The main result does not depend on which representation is chosen.

Example 6-1: Batch demands Suppose that demand epochs for a product occur according to the renewal process $\{N(t);\ t \geq 0\}$ and at the ith demand epoch the random quantity B_i is requested. Interpret B_i as the reward associated with the ith renewal, and assume that $B_1, B_2, \ldots$ are i.i.d. Let $D(t)$ be the number of items demanded by time t. Then

$$D(t) = B_1 + B_2 + \cdots + B_{N(t)}$$

and $\{D(t);\ t \geq 0\}$ is a renewal-reward process. In the special case where B_i is independent of X_i, we call $\{D(t);\ t \geq 0\}$ a *cumulative process.* To specialize even further, when $\{N(t);\ t \geq 0\}$ is a Poisson process, $\{D(t);\ t \geq 0\}$ is a *compound Poisson process.*†   □

Example 6-2: Machine maintenance Let $\{N(t);\ t \geq 0\}$ be a renewal process that governs the failure epochs of a machine. At each failure epoch, another machine is installed at cost R (possibly random). While a machine is running, it incurs maintenance costs (for grease, oil, operational checks, etc.). Suppose that the maintenance cost incurred by a machine which has run for time u is $\gamma \ln(1 + u)$. When the machine fails, the cost R' (possibly random) is incurred; this may represent the cost of repair, a disposal cost, etc. Then $C_i = R_i + \gamma \ln (1 + X_i) + R'_i$, where R_i and R'_i are values of R and R' associated with failure i, assumed to be i.i.d., and

$$C_l(t) = C_1 + C_2 + \cdots + C_{N(t)}$$

is the cost of those failures that have occurred by time t. The cumulative cost incurred by time t is

$$C(t) \triangleq C_l(t) + \gamma \ln (1 + t - S_{N(t)}) \tag{6-3}$$

Suppose that $\gamma = R_1 \equiv 0$, R'_1 is a constant, and the accounting procedure is to record each cost as soon as it is apparent that the cost will be incurred. Then the costs recorded by time t are

$$C_u(t) = R'[N(t) + 1] \qquad\qquad □$$

---

† Some authors restrict this term to processes where the B_i's are integer-valued and nonnegative.

We can generalize (6-3) as follows. When $S_{N(t)} < t < S_{N(t)+1}$, let $\epsilon(t)$ be the reward earned during $(S_{N(t)}, t]$. Define

$$C(t) \triangleq C_l(t) + \epsilon(t) \tag{6-4}$$

The process $\{C(t); t \geq 0\}$ is considered in the sequel as a third type of renewal-reward process.

In (6-1) and (6-2), $C_l(t)$ and $C_u(t)$ depend on both the number of renewals and the costs. The C's are i.i.d. because the X's are, and so (6-1) and (6-2) are sums of i.i.d. random variables containing a random number of terms. Let us digress for a moment and study this sort of sum.

6-2 SUMS OF A RANDOM NUMBER OF RANDOM VARIABLES

Let

$$Z_N = Y_1 + Y_2 + \cdots + Y_N \tag{6-5}$$

where the Y_i's are i.i.d. and N is a random variable which assumes only the values $0, 1, 2, \ldots$. The simplest situation is when N is independent of the Y_i's. Then

$$
\begin{aligned}
E(Z_N) &= \sum_{n=0}^{\infty} E(Z_N \mid N = n) P\{N = n\} \\
&= \sum_{n=0}^{\infty} E(Y_1) n P\{N = n\} \tag{6-6} \\
&= E(Y_1) E(N)
\end{aligned}
$$

From the *conditional variance formula* (Exercise 6-1),

$$
\begin{aligned}
\mathrm{Var}\,(Z_N) &= E[\mathrm{Var}\,(Z_N \mid N)] + \mathrm{Var}\,[E(Z_N \mid N)] \\
&= E[N\,\mathrm{Var}\,(Y_1)] + \mathrm{Var}\,[NE(Y_1)] \tag{6-7} \\
&= \mathrm{Var}\,(Y_1) E(N) + E^2(Y_1)\,\mathrm{Var}\,(N)
\end{aligned}
$$

Suppose that Y_1 is a nonnegative r.v.; then so also is Z_N for each value of N. The LST

$$\tilde{Z}(s) \triangleq \int_0^{\infty} e^{-sz}\, dP\{Z_N \leq z\}$$

can be obtained in the same manner as (6-6). Let $G(\,\cdot\,)$ be the distribution function of Y_1 and $\tilde{G}(\,\cdot\,)$ be its LST. Then

$$
\begin{aligned}
\tilde{Z}(s) &= \sum_{n=0}^{\infty} \tilde{Z}(s \mid N = n) P\{N = n\} \\
&= \sum_{n=0}^{\infty} [\tilde{G}(s)]^n P\{N = n\}
\end{aligned}
$$

which is the generating function of N with transform variable $\tilde{G}(s)$. Letting $\hat{A}(u) = \sum_{n=0}^{\infty} u^n P\{N = n\}$, we get

$$\tilde{Z}(s) = \hat{A}[\tilde{G}(s)] \tag{6-8}$$

We can use (6-8) to obtain the moments of Z_N; in particular, we can utilize it as a lengthier vehicle to derive (6-6) and (6-7) if we forget them.

Wald's Equation

Now consider the more complicated situation where N depends on the Y's and the Y's are not necessarily i.i.d. We do not let N depend on the Y's in an arbitrary fashion. We insist that the event $\{N = n\}$ can depend on $Y_1, Y_2, \ldots, Y_n$ but *not* on $Y_{n+1}, Y_{n+2}, \ldots$. Such an N is called a *stopping time* for $Y_1, Y_2, \ldots$, a name chosen from the original motivation for these results — sequential sampling.

Suppose that you have collected data $Y_1, \ldots, Y_n$ on the failure times of a device in order to estimate the mean time to failure. The devices are tested one at a time, and after each failure you must decide whether to stop, that is, $N = n$, or to continue, that is, $N > n$. Naturally, your decision can depend on only the data you have already collected. It cannot depend on any of the data $Y_{n+1}, Y_{n+2}, \ldots$ that you would not collect if N were equal to n.

The restriction that N is a stopping time is natural in many contexts; in particular, we show it to hold in a renewal-reward process setting. The surprising result is that (6-6) is valid when N is a stopping time; the proof is simple and elegant.

> **Theorem 6-1: Wald's equation** Let $Y_1, Y_2, \ldots$ be random variables with common mean $E(Y_1)$, and let Z_N be given by (6-5). If $E(Y_1) < \infty$ and N is a stopping time for $Y_1, Y_2, \ldots$, with $E(N) < \infty$, then
>
> $$E(Z_N) = E(Y_1)E(N)$$

PROOF Define ξ_n for each $n = 0, 1, \ldots$ by

$$\xi_n = \begin{cases} 1 & \text{if } N \geq n \\ 0 & \text{if } N < n \end{cases}$$

Then

$$Z_N = \sum_{i=1}^{\infty} Y_i \xi_i$$

Thus

$$E(Z_N) = E\left(\sum_{i=1}^{\infty} Y_i \xi_i\right) = \sum_{i=1}^{\infty} E(Y_i \xi_i)$$

because the expectation can be moved inside the summation sign. (Why?) But $\{\xi_n = 1\}$ depends on only $Y_1, \ldots, Y_{n-1}$ because N is a stopping time, and so Y_i and ξ_i are independent. Thus,

$$E(Z_N) = \sum_{i=1}^{\infty} E(Y_1)E(\xi_i)$$

$$= E(Y_1) \sum_{i=1}^{\infty} E(\xi_i) = E(Y_1) \sum_{i=1}^{\infty} P\{N \geq i\}$$

$$= E(Y_1) \sum_{n=0}^{\infty} P\{N > n\} = E(Y_1)E(N)$$

where the last equality follows from Exercise 5-3. □

Corollary 6-1 If $E(N) = \infty$ and $E(Y_1) > 0$, then $E(Z_N) = \infty$.

PROOF Let $N_k = \min\{k, N\}$, $k \in I$. For each k, $E(N_k) < \infty$ and Theorem 6-1 applies. Since $E(N_k)$ is monotone in k,

$$E(Z_N) = \lim_{k \to \infty} E(Y_1)E(N_k) = \infty$$ □

The following result is a generalization of Theorem 6-1.

Theorem 6-2: Wald's identity Let $Y_1, Y_2, \ldots$ be i.i.d. and N be a stopping time. If $\tilde{G}(s) = E(e^{-sY_1}) < \infty$ in some interval (s_0, s_1), then in this interval

$$E([G(s)]^{-N}e^{+sZ_N}) = 1$$

PROOF See Feller (1971, p. 603). □

Applications to Renewal-Reward Processes

Now let us return to renewal-reward processes. If the rewards are independent of the renewal times (i.e., we are dealing with a cumulative process), then $N(t)$ is independent of $C_1, C_2, \ldots, C_{N(t)}$ and (6-6) and (6-7) can be applied.

Theorem 6-3 Let $M(t) = E[N(t)]$ and $V(t) = \text{Var}[N(t)]$. If C_i is independent of X_i for each $i = 1, 2, \ldots$, then

$$E[C_l(t)] = E(C_1)M(t) \tag{6-9}$$

$$E[C_u(t)] = E(C_1)[M(t) + 1] \tag{6-10}$$

$$\text{Var}[C_l(t)] = \text{Var}(C_1)M(t) + E^2(C_1)V(t) \tag{6-11}$$

and

$$\text{Var}[C_u(t)] = \text{Var}(C_1)[M(t) + 1] + E^2(C_1)V(t) \tag{6-12}$$

PROOF Apply (6-6) and (6-7) to (6-1) and (6-2). □

In the more general case, e.g., Example 6-2, C_i depends on X_i. If $N(t)$ were independent of $C_1, C_2, \ldots, C_{N(t)}$, we could apply Theorem 6-1 to obtain $E[C_l(t)]$, but unfortunately that is not the case. To see why, recall that $N(t)$ is defined as the value of n such that

$$X_1 + X_2 + \cdots + X_n \le t \qquad \text{and} \qquad X_1 + X_2 + \cdots + X_{n+1} > t$$

so the event $\{N(t) = n\}$ depends on X_{n+1}, which means that $N(t)$ is not, in general, independent of $C_{N(t)+1}$.

EXERCISES

6-1 Prove that for any pair of random variables X and Y

$$\text{Var}(X) = E[\text{Var}(X \mid Y)] + \text{Var}[E(X \mid Y)]$$

This is called the *conditional variance formula*.

6-2 Produce a counterexample to Theorem 6-1 when N is *not* a stopping time.

6-3 LIMIT THEOREMS FOR RENEWAL-REWARD PROCESSES

The strong law of large numbers for renewal processes and the elementary renewal theorem (Theorems 5-3 and 5-7) are two important results in renewal theory. Theorems 6-4 and 6-5 below are the extensions of these theorems to renewal-reward processes.

Theorem 6-4: Strong law of large numbers for renewal-reward process If $E(|C_1|)$ and $E(X_1)$ are both finite, then (w.p.1)

$$\lim_{t \to \infty} \frac{C_l(t)}{t} = \lim_{t \to \infty} \frac{C_u(t)}{t} = \frac{E(C_1)}{E(X_1)}$$

PROOF This is left as Exercise 6-3. □

The limits in Theorem 6-4 are called *asymptotic reward rates*. Theorem 6-4 has an important and intuitive interpretation. Call each renewal a *cycle*. Then $E(C_1)$ is the expected reward in a cycle, and $E(X_1)$ is the expected duration of a cycle. Theorem 6-4 asserts that the asymptotic reward rate equals the expected reward in the first (or typical, or average, or arbitrary) cycle divided by the expected duration of that cycle. Thus, for any reward structure, the long-run cost per unit time can be found from a very simple calculation.

Now we establish a result analogous to the elementary renewal theorem.

Theorem 6-5 If $E(|C_1|)$ and $E(X_1)$ are both finite, then

$$\lim_{t \to \infty} \frac{E[C_l(t)]}{t} = \lim_{t \to \infty} \frac{E[C_u(t)]}{t} = \frac{E(C_1)}{E(X_1)}$$

The proof is divided into a pair of lemmas.

Lemma 6-1 If $E(C_1)$ and $E(X_1)$ are both finite, then

$$\lim_{t \to \infty} \frac{E[C_u(t)]}{t} = \frac{E(C_1)}{E(X_1)}$$

PROOF Since

$$\{N(t) + 1 = n\} \Leftrightarrow \{X_1 + \cdots + X_{n-1} \leq t, X_1 + \cdots + X_n > t\}$$

the event $\{N(t) + 1 = n\}$ is a stopping time for $C_u(t)$. Applying Wald's theorem to (6-2) yields

$$E[C_u(t)] = E(C_1)[M(t) + 1] \tag{6-13}$$

and an appeal to the elementary renewal theorem finishes the proof. □

From (6-1) and (6-2), $C_u(t) - C_l(t) = C_{N(t)+1}$. The inspection paradox shows that $X_{N(t)+1}$ is not distributed the same as X_1, and (5-48a) shows that $E(X_{N(t)+1}) = \infty$ is possible when $E(X_1) < \infty$. Since $C_{N(t)+1}$ may depend on $X_{N(t)+1}$, there is no a priori guarantee that

$$\lim_{t \to \infty} \frac{E(C_{N(t)+1})}{t} = 0 \tag{6-14}$$

If (6-14) can be established, it and Lemma 6-1 will prove

$$\lim_{t \to \infty} \frac{E[C_l(t)]}{t} = \frac{E(C_1)}{E(X_1)} \tag{6-15}$$

Lemma 6-2 If $E(|C_1|)$ and $E(X_1)$ are both finite, then (6-14) is valid.

PROOF† For notational convenience, let $d(t) = E(C_{N(t)+1})$. Define $r(t) = \int_t^\infty E(C_1 | X_1 = x)\, dF(x)$. Since $r(\cdot)$ never increases and

$$|E(C_1)| \leq E(|C_1|) = \int_0^\infty E(|C_1| : X_1 = x)\, dF(x) < \infty$$

we know that

$$\lim_{t \to \infty} r(t) = 0 \quad \text{and} \quad r(t) \leq E(|C_1|) \tag{6-16}$$

The renewal argument yields

$$d(t | X_1) = \begin{cases} d(t - X_1) & \text{if } X_1 \leq t \\ E(C_1 | X_1 > t) & \text{if } X_1 > t \end{cases}$$

By unconditioning on X_1 we obtain the renewal-type equation

---

† This proof is taken from Ross (1970).

$$d(t) = r(t) + \int_0^t d(t - x) \, dF(x)$$

whose solution is (see Theorem 5-6)

$$d(t) = r(t) + \int_0^t r(t - x) \, dM(x) \tag{6-17}$$

According to (6-16), given any $\epsilon > 0$, we can choose a value of t, say T, which is so large that $|r(t)| < \epsilon$ whenever $t \geq T$. Do so; then for $t > T$,

$$|d(t)| \leq |r(t)| + \int_0^{t-T} |r(t - x)| \, dM(x) + \int_{t-T}^t |r(t - x)| \, dM(x)$$

$$\leq \epsilon + \epsilon M(t - T) + E(|C_1|)[M(t) - M(t - T)]$$

Apply the elementary renewal theorem to obtain

$$\lim_{t \to \infty} \frac{|d(t)|}{t} \leq \epsilon \lim_{t \to \infty} \frac{M(t - T)}{t} = \frac{\epsilon}{E(X_1)}$$

for any $\epsilon > 0$. But this is precisely a statement of (6-14). $\qquad \square$

Example 6-2 describes a model where costs are incurred continuously and one is interested in those costs incurred by time t, which may not be a renewal epoch. That example motivates the following explicit form for the rewards. Let R_i and R_i' be random rewards that are earned at the start and end of the ith renewal interval, respectively. During each renewal, rewards accumulate at rate $c(\cdot)$, and so

$$C_i = R_i + \int_0^{X_i} c(y) \, dy + R_i' \qquad i \in I_+ \tag{6-18}$$

The total reward accumulated by time t is

$$C(t) \triangleq C_l(t) + R_{N(t)+1} + \int_{S_{N(t)}}^t c(y) \, dy \qquad t \geq 0 \tag{6-19}$$

In order for $C_1, C_2, \ldots$ to be i.i.d., assume that $\{R_i + R_i'\}_{i=1}^\infty$ are i.i.d. This setting motivates Theorem 6-6.

Theorem 6-6 Assume that C_i is a monotone function of X_i, $i \in I_+$, and that $C_1, C_2, \ldots$ are i.i.d. random variables. Let $C(t)$ be the reward earned by time t [that is, it is given by (6-4)]. If $E(|C_1|) < \infty$, then

$$\lim_{t \to \infty} \frac{C(t)}{t} = \frac{E(C_1)}{E(X_1)} \quad \text{(w.p.1)} \tag{6-20}$$

and

$$\lim_{t \to \infty} \frac{E[C(t)]}{t} = \frac{E(C_1)}{E(X_1)} \tag{6-21}$$

PROOF Assume each C_i is nonnegative. Then

$$C_l(t) \le C(t) \le C_u(t) \qquad (6\text{-}22)$$

and (6-20) and (6-21) follow from Theorems 6-4 and 6-5, respectively. When each C_i is negative, the inequalities in (6-22) are reversed and (6-20) and (6-21) are obtained as before. When there are no restrictions on the signs, apply the above arguments to the positive and negative parts and combine them. □

Examples

Example 6-3: (s, S) inventory systems In Example 5-10, it was shown that the epochs where an order is placed can be regarded as renewal epochs. According to Theorem 6-6, we can obtain the asymptotic (and expected) cost rate for this policy by finding the expected cost in the first cycle and dividing by $M(S - s) + 1$ (recall that this is the expected length of the first cycle).

The expected cost in the first cycle is obtained as follows. For convenience, let $Q = S - s$. The ordering cost in the first cycle is

$$K + p(X_1 + X_2 + \cdots + X_{n(Q)+1})$$

where X_i is the demand on the ith day, K is the charge for placing an order, p is the cost per unit, and $N(\cdot)$ is the renewal process defined in Example 5-10. Since $N(Q) + 1$ is a stopping time, Wald's theorem applies, and the expected ordering cost in a cycle is

$$K + p[M(Q) + 1]E(X_1)$$

Let $G(y)$ be the expected total inventory-related (e.g., holding and shortage) cost during a cycle that starts with y items on hand, and let $g(y)$ be the expected cost on the first day of the cycle. Of course, each cycle starts with S items on hand, and so we want to find $G(S)$ in particular. Using the renewal argument, we obtain

$$G(S \mid X_1 = x) = \begin{cases} g(S \mid X_1 = x) & \text{if } x > Q \\ g(S \mid X_1 = x) + G(S - x) & \text{if } x \le Q \end{cases}$$

and so

$$G(S) = g(S) + \int_0^Q G(S - x)\, dF(x)$$

$$= g(S) + \int_0^Q g(S - x)\, dM(x)$$

Hence, the asymptotic cost rate is

$$pE(X_1) + \frac{K + g(S) + \int_0^Q g(S - x)\, dM(x)}{1 + M(Q)}$$

It is easy to obtain $g(\cdot)$ when a holding-cost rate of h is charged for each item on hand at the close of the business day and a backlog penalty of b is charged for each item backlogged. Then

$$g(y) = E[h(y - X_1)^+ + b(X_1 - y)^+]$$ □

Example 6-4: Queue with nonstationary Poisson arrivals Suppose the arrivals to a single server form a nonstationary Poisson process. (See page 151 for the definition of this process.) Suppose, also, that there is no waiting room; those customers who arrive when the server is free are served, and all other customers are lost. Let $\lambda(t)$ be the *intensity function* of the nonstationary Poisson process; that is, $\lambda(t)\,\Delta t + o(\Delta t)$ is the probability that a customer arrives during $(t, t + \Delta t]$. Assume that

$$\lim_{T \to \infty} \frac{1}{T} \int_0^T \lambda(t)\,dt = \lambda$$

with $0 < \lambda < \infty$; that is, the asymptotic arrival rate is λ.

Intuitively, a nonstationary Poisson process is "more variable" than a stationary Poisson process with the same arrival rate. One might conjecture† that this increased variability would lead to an increase in the proportion of lost customers. This example shows that this conjecture is not valid in general.

Choose δ between 0 and 1, and choose

$$\lambda(t) = \begin{cases} 1/\delta & \text{if } i \le t \le i + \delta, i \in I \\ 0 & \text{otherwise} \end{cases} \tag{6-23}$$

Hence, the asymptotic arrival rate is $\lambda = 1$. Choose the service times to be the constant v, with

$$\delta < v < 1 - \delta \tag{6-24}$$

From (6-23), the number of arrivals during $[i, i + \delta]$ is a batch of size B_i, say, which has a Poisson distribution with mean 1. The independent increment property of a Poisson process (see page 145) implies that $B_0, B_1, \ldots$ are i.i.d. Assume the server is free at time zero.

To obtain the asymptotic proportion of lost customers, let C_i be the number of customers lost during $[i, i + 1)$, and set each $X_i \equiv 1$, $i \in I$. From (6-24), $C_i = (B_i - 1)^+$, and so $C_0, C_1, \ldots$ are i.i.d. with

$$E(C_0) = E[(B_0 - 1)^+] = \sum_{n=1}^{\infty} (n - 1)P\{B_1 = n\}$$

$$= E(B_1) - P\{B_1 \ge 1\} = e^{-1} \tag{6-25}$$

Let $A(t)$ be the number of customers who arrive by time t and $C(t)$ be the

† See S. Fond and S. M. Ross, "A Heterogeneous Arrival and Service Loss Model," *Nav. Res. Logist. Quart.* **25**: 483–488 (1978), for a statement of this conjecture.

number of customers who are lost by time t. It is easy to verify that $C(t)$ satisfies the hypothesis of Theorem 6-4, so

$$\lim_{t \to \infty} \frac{C(t)}{t} = \frac{E(C_0)}{E(X_0)} = e^{-1} \qquad \text{(w.p.1)} \tag{6-26}$$

Theorem 6-6 can also be used (how?) to show

$$\lim_{t \to \infty} \frac{A(t)}{t} = \lambda = 1 \qquad \text{(w.p.1)} \tag{6-27}$$

From (6-26) and (6-27), the asymptotic proportion of customers who are lost, π say, is

$$\pi \triangleq \lim_{t \to \infty} \frac{C(t)}{A(t)} = \frac{\lim\limits_{t \to \infty} C(t)/t}{\lim\limits_{t \to \infty} A(t)/t} \tag{6-28}$$

$$= e^{-1} \qquad \text{(w.p.1)}$$

Now suppose that the arrivals form a stationary Poisson process with arrival rate 1. Let X_1 be the epoch when the first customer to enter service completes service, and let $S_n\ (=X_1 + \cdots + X_n)$ be the epoch when the nth customer to enter service completes service, $n \in I_+$. Set $S_0 = 0$, as usual. Let C_i be the number of customers who are lost during $(S_{i-1}, S_i]$, $i \in I_+$. The independent increment property of the Poisson arrival process implies that $X_1, X_2 \ldots$ and $C_1, C_2 \ldots$ are i.i.d. Since X_i is the sum of an interarrival time and a service time,

$$E(X_1) = 1 + v$$

When a customer enters service, those customers who arrive during the next v units of time are lost, so

$$E(C_1) = v$$

Again let $C(t)$ be the number of customers lost by time t. Since (6-27) is valid for stationary Poisson arrivals, the asymptotic proportion of customers lost, p say, is

$$p \triangleq \lim_{t \to \infty} \frac{C(t)}{A(t)} = \lim_{t \to \infty} \frac{C(t)}{t}$$

$$= \frac{E(C_1)}{E(X_1)} \qquad \text{(w.p.1)} \tag{6-29}$$

$$= \frac{v}{1 + v}$$

When $\delta < \frac{1}{10}$, $v = \frac{9}{10}$ satisfies (6-24), and (6-28) and (6-29) yield

$$p = \tfrac{9}{19} > e^{-1} = \pi \qquad\qquad \square$$

EXERCISES

6-3 Prove that with the hypotheses of Theorem 6-4, both $C_l(t)/t$ and $C_u(t)/t$ converge to $E(C_1)/E(X_1)$ as $t \to \infty$.

6-4 Represent the alternating renewal process as a renewal-reward process.

6-5 Show that when $c(y) = c$ in (6-18), Theorem 6-6 is an immediate consequence of the strong law of large numbers for renewal processes and the elementary renewal theorem.

6-6 A machine can be in two states, up or down. It is up at time zero and alternates between being up and down. The lengths of successive up times $X_1, X_2, \ldots$ are i.i.d. with distribution function $F(\,\cdot\,)$ and the lengths of successive down times $Y_1, Y_2, \ldots$ are i.i.d. with distribution function $G(\,\cdot\,)$. Up and down times are mutually independent. Let $u(t)$ be the expected amount of time that the machine is up during $(0, t]$.

    (a) Find a functional equation for $u(t)$.
    (b) Solve for $u(t)$ explicitly when $f(t) = \lambda e^{-\lambda t}$ and $g(t) = \mu e^{-\mu t}$.
    (c) Give a rigorous proof that, in general,

$$\lim_{t \to \infty} \frac{u(t)}{t} = \frac{v_f}{v_f + v_g}$$

6-7 State and prove the analog of Blackwell's renewal theorem for a renewal-reward process.

6-8: The periodic replacement policy Items are replaced at times $0, Y, 2Y, \ldots$ at a cost of C_p each; these are called planned replacements and are done even if the item in place is working. Items also are replaced when they fail at a cost of C_f each, with $C_f > C_p$. The potential lifetimes are i.i.d. with distribution function $F(\,\cdot\,)$. Find the asymptotic cost rate of this policy. When is this policy better than having no planned replacements? Why is the answer trivial when the times between failures have an exponential distribution?

6-9: The age replacement policy Whenever an item has been in service for time t_c, it is replaced at a cost of C_a. Items that fail are replaced immediately for a cost of C_f. What is the asymptotic cost rate of this policy, and when is it better than having no age replacements?

6-10 Describe settings in which the age replacement policy would be appropriate and settings in which the periodic replacement policy would be appropriate. Discuss the computational aspects of obtaining the optimal values of Y and t_c.

6-4 REGENERATIVE PROCESSES

A process that is related to the renewal-reward process is the regenerative process.

> **Definition 6-1: Regenerative process** A stochastic process $\{Y(t); \, t > 0\}$ is a *regenerative process* if there exists an epoch, S_1 say, such that the continuation of the process beyond S_1 is a probabilistic replica of the process beginning at time zero.

The existence of S_1 implies the existence of epochs $S_2, S_3, \ldots$ with the same regenerative property, and S_i is called the ith *regeneration epoch*. Similarly, we say that the process regenerates itself at a regeneration epoch. By defining $S_0 = 0$, $X_i = S_i - S_{i-1}$, $i = 1, 2, \ldots$, we have, as usual, $S_i = X_1 + X_2 + \cdots + X_i + S_0$. The definition of the regeneration epochs implies that the X_i's are i.i.d. and nonnegative, so S_i can be interpreted as the ith renewal epoch of the renewal process generated by $\{X_i\}_{i=1}^{\infty}$.

Many models in operations research have the regenerative property. The (s, S) inventory model in Example 6-3 regenerates when an order is placed, so the stochastic process representing the amount of inventory is a regenerative process. In Example 6-7 below, theorems about regenerative processes are used to compute easily the steady-state distribution of the amount in inventory. Other applications that are mentioned include the $GI/G/c$ queue and the analysis of simulation experiments.

Definition 6-1 is verbal and intuitive; it is conveniently applied when the regeneration epochs are easily identified. Here is a formal definition.

Definition 6-1a The stochastic process $\{Y(t); t \geq 0\}$ is *regenerative*, with regeneration epochs $\{S_i\}_{i=1}^{\infty}$, $0 = S_0 < S_1 < \cdots$, if for any positive integer k and epochs $0 \leq t_1 < t_2 < \cdots < t_k$, the random vectors $[Z_0(t_1), \ldots, Z_0(t_k), S_1 - S_0], [Z_1(t), \ldots, Z_1(t_k), S_2 - S_1], \ldots$ are mutually independent and identically distributed, where

$$Z_n(t) = \begin{cases} Y(t - S_n) & \text{if } 0 \leq t < S_{n+1} - S_n \\ 0 & \text{otherwise} \end{cases}$$

Observe that $Y(t) = \sum_{n=0}^{\infty} Z_n(t - S_n)$. The essence of Definition 6-1a is that there are epochs $S_1, S_2, \ldots$ such that the given process can be broken into a segment $Z_0(\,\cdot\,)$ on the interval $[0, S_1)$, a segment $Z_1(\,\cdot\,)$ on the interval $[S_1, S_2)$, $\ldots$, such that the segments are independent and stochastically identical.

Existence of Limits

This is the most important theorem about regenerative processes.

Theorem 6-7 Let $\{Y(t); t \geq 0\}$ be a regenerative process with state space $S \subset I$. Let $P_k(t) \triangleq P\{Y(t) = k\}$, and assume $v = E(X_1) < \infty$. If either
(a) the sample paths $Y(\,\cdot\,, \omega)$ are in the set† $D[0, \infty)$ with probability 1
or
(b) $P\{X_1 \leq t\}$ has a density on some interval
then

$$\lim_{t \to \infty} P_k(t) = p_k \qquad k \in S \qquad (6\text{-}30)$$

exists, with $p_k \geq 0$ and $\Sigma_{k \in S}\, p_k = 1$. Moreover, let T_k be the amount of time

---

† The set of functions $D[0, \infty)$ is the set of all real-valued functions defined on $[0, \infty)$ that are right-continuous and have limits from the left. The processes that arise in operations research models typically have sample paths in $D[0, \infty)$.

$Y(\,\cdot\,) = k$ during $[S_0, S_1]$. Then

$$p_k = \frac{E(T_k)}{v} \tag{6-31}$$

If S_1 is arithmetic, (6-30) and (6-31) hold when t is a multiple of the span.

PROOF Let $Q_k(t) = P\{Y(t) = k, S_1 > t\}$ and $F(t) = P\{X_1 \le t\}$. The renewal argument yields

$$P_k(t) = Q_k(t) + \int_0^t P_k(t - x)\, dF(x)$$

This renewal-type equation has the solution

$$P_k(t) = \int_0^t Q_k(t - x)\, dM(x) + Q_k(t) \tag{6-32}$$

where $M(\,\cdot\,)$ is the renewal function of the renewal process generated by $\{X_i\}_{i=1}^{\infty}$. It has been shown† that either (a) or (b) implies that $Q_k(\,\cdot\,)$ is directly Riemann-integrable. Applying the key renewal theorem to (6-32) yields

$$p_k \triangleq \lim_{t \to \infty} P_k(t) = \frac{1}{v} \int_0^{\infty} Q_k(x)\, dx \tag{6-33}$$

It is clear that $p_k \ge 0$; thus

$$\sum_{k=0}^{\infty} Q_k(t) = P\{S_1 > t\} = 1 - F(t)$$

Hence

$$\sum_{k \in S} p_k = \frac{1}{v} \int_0^{\infty} [1 - F(x)]\, dx = 1$$

which proves (6-30). Now define

$$I_k(t) = \begin{cases} 1 & \text{if } Y(t) = k \text{ and } S_1 > t \\ 0 & \text{otherwise} \end{cases}$$

Then $T_k = \int_0^{\infty} I_k(t)\, dt$, and

$$\int_0^{\infty} Q_k(t)\, dt = \int_0^{\infty} E[I_k(t)]\, dt = E \int_0^{\infty} I_k(t)\, dt = E(T_k) \tag{6-34}$$

so (6-33) and (6-34) establish (6-31). □

Theorem 6-7 is used frequently to prove that limiting probabilities exist. The difficult step in applying it usually lies in showing that $v < \infty$ holds. Equation

† See Douglas R. Miller, "Existence of Limits in Regenerative Processes," *Ann. Math. Stat.* **43**: 1275–1282 (1972).

(6-31) asserts that the limiting probabilities can be obtained by averaging over the first (or any arbitrarily chosen) cycle. It is similar to the type of result obtained in Theorems 6-4 and 6-5.

It is notationally convenient to introduce a random variable Y^* such that $p_k = P\{Y^* = k\}$. Then (6-30) asserts that $Y(t)$ *converges* to Y^* in the sense that $\lim_{t \to \infty} P\{Y(t) = k\} = P\{Y^* = k\}$, and we write

$$\operatorname*{dlim}_{t \to \infty} Y(t) = Y^* \tag{6-30'}$$

Equation (6-31) gives a representation of the distribution of Y^*. This mode of convergence for random variables is called *convergence in distribution,* or *weak convergence.* Unless all the random variables $\{Y(t); t \geq 0\}$ and Y^* are integer-valued, or the distribution function of Y^* is continuous [this could happen in (6-35) below], (6-30') expresses a stronger mode of convergence than convergence in distribution. The difference between these concepts is slight and technical and is not pursued here.

Condition (*b*) is easy to check if the distribution function of X_1 is given explicitly. In Example 6-5 below, this distribution function is not known, but condition (*a*) is easy to verify. It follows directly from the definition that $D[0, \infty)$ includes all step functions and excludes functions having infinitely many jumps in a finite time (e.g., exploding birth-and-death processes).

If S_1 is integer-valued, it is natural to require that $t \to \infty$ on integer values in (6-30). If, for some integer k larger than 1, S_1 takes values in $\{k, 2k, 3k, \ldots\}$ only, then S_1 is *periodic* with span k, and (6-30) holds only when t is a multiple of k. If S_1 is integer-valued, it is not periodic (it is *aperiodic*) if $P\{S_1 = 1\} > 0$; see Exercise 6-11.

In Theorem 6-7, the requirement that $S \subset I$ is not necessary, as seen below.

Corollary 6-7a If $S \subset (-\infty, \infty)$, $v < \infty$, and either (*b*) from Theorem 6-7 is valid or
(*a'*) for any $x \in S$, $Y(t; \omega)$ crosses x at most finitely many times in any finite interval of time (w.p.1)
then

$$H(x) \triangleq \lim_{t \to \infty} P\{Y(t) \leq x\} \qquad x \in S \tag{6-35}$$

exists for every real number x, with $H(\infty) = 1$. Furthermore,

$$H(x) = \frac{E(T_x)}{v} \tag{6-36}$$

where T_x is the amount of time $Y(\cdot) \leq x$ during $[0, S_1]$.

PROOF This is left as Exercise 6-12. □

As with delayed renewal processes, the epoch of the first regeneration need not have the same distribution function as the lengths of remaining regeneration intervals. If $S_0 > 0$, then $\{Y(t); t \geq 0\}$ is a *delayed* (or *generalized*) *regenerative process*. Here is the formal definition.

Definition 6-2 Let S_0 be a nonnegative random variable. The process $\{Y_d(t); t \geq 0\}$ is a *delayed regenerative process* if there is a regenerative process $\{Y(t); t \geq 0\}$ such that whenever $S_0 = s$, $Y_d(t) = Y(t - s)$ for all $t \geq s$.

Corollary 6-7b In a delayed regenerative process, if $P\{S_0 < \infty\} = 1$, then (6-30) and (6-31) are valid.

PROOF This is left as Exercise 6-13. $\qquad\square$

Examples

Example 6-5: *GI/G/c* queue Let $X(t)$ be the number of customers present at time t. Assume $X(0) = 0$. The epochs when a customer arrives to an empty system, $\tau_0, \tau_1, \ldots$ say, are regeneration points for $\{X(t); t \geq 0\}$ because previous arrival and service times have no effect on the future of the process. (See Example 5-15 for further elaboration.) The sample paths of this process are step functions; hence they are in $D[0, \infty)$. Let $B_i = \tau_i - \tau_{i-1}$ be the length of the ith busy period. When $E(B_1) < \infty$ and B_1 is not arithmetic, Theorem 6-7 asserts that the limiting probabilities

$$p_i = \lim_{t \to \infty} P\{X(t) = i\} \qquad i \in I \qquad (6\text{-}37)$$

exist.

A sufficient condition for $E(B_1) < \infty$ is easy to state and difficult to prove.† Let C_i be the number of customers served during the ith busy period, and let U_i and V_i be the interarrival and service times respectively, $i \in I_+$. Then

$$E(V_1) < cE(U_1) \qquad (6\text{-}38a)$$

and

$$P\{U_1 > V_1\} > 0 \qquad (6\text{-}38b)$$

together‡ imply $E(C_1) < \infty$. Since $B_1 = U_1 + \cdots + U_{C_1}$, Equation (6-38) implies $E(B_1) < \infty$, by Wald's equation. If U_1 is not arithmetic, neither is B_1.

If $X(0) \neq 0$, then (6-38b) implies $P\{C_1 < \infty\} = 1$ (how?); hence Corollary 6-7b implies (6-37) holds for any initial distribution for $X(0)$. $\qquad\square$

---

† See Ward Whitt, "Embedded Renewal Processes in the *GI/G/s* Queue," *J. Appl. Prob.* **9**: 650–658 (1972).

‡ For $c = 1$, (6-38b) is not required to prove $E(C_1) < \infty$. Equation (6-38b) is superfluous for (6-37) to hold, but it is essential for the given proof (see Exercise 6-14).

Example 6-6: Birth-and-death processes The memoryless property of the exponential distribution implies that the entry into *any* fixed state in the state space of the process is a regeneration epoch. In particular, choose state zero; then (4-108) establishes that $E(X_1) = E(T_{00}) < \infty$ if, and only if, $\sum_0^\infty b_j < \infty$. When the initial state is $i > 0$, let $S_0 = S_0(i)$ be the epoch of the first entry to state zero. Exercise 7-41 establishes

$$P\{S_0 < \infty\} = 1 \Leftrightarrow \sum_0^\infty (\lambda_j b_j)^{-1} = \infty \tag{6-39}$$

This proves the "if" part of Corollary 4-4a.

Theorem 4-4 asserts that the limiting probabilities exist if the sum in (6-39) diverges and $\sum_0^\infty b_j = \infty$; so Theorem 6-7 does not imply that limiting probabilities fail to exist if $E(X_1) = \infty$. ☐

Connection with Renewal-Reward Processes

The renewal-reward process can be generalized to the regenerative-reward process. Let $r(k)$ be the *reward rate* earned while the regenerative process $\{Y(t); t \ge 0\}$ is in state k. That is, if $Y(t) = k$ for all $t \in [t_0, t_1]$, then the reward earned during that interval is $r(k)(t_1 - t_2)$.

We do not treat delayed regenerative processes explicitly because doing so may obscure the essential ideas with additional notation. It should become clear that the limit theorems we obtain are valid for delayed processes whenever the expected reward earned in $[0, S_0]$ is finite.

Theorem 6-7 gives conditions for $\{Y(t); t \ge 0\}$ to have a limiting distribution; the next theorem shows how to express the asymptotic reward rate in terms of the first cycle.

Theorem 6-8 Let $\{Y(t); t \ge 0\}$ be a regenerative process with $E(S_1) < \infty$ and state space $S \subset I$. Let $r(\cdot)$ be such that

$$E\left(\int_0^{S_1} |r[Y(u)]|\, du\right) < \infty$$

Then

$$\lim_{t \to \infty} \frac{1}{t} \int_0^t r[Y(u)]\, du = \frac{E(\int_0^{S_1} r[Y(u)]\, du)}{E(S_1)} \quad \text{(w.p.1)}$$

$$= \lim_{t \to \infty} \frac{1}{t} E\left(\int_0^t r[Y(u)]\, du\right) \tag{6-40}$$

PROOF Let T_k be the amount of time $Y(\cdot) = k$ during $[0, S_1]$; then

$$\int_0^{S_1} r[Y(u)]\, du = \sum_{k \in S} T_k r(k)$$

Since $\{Y(t); t \geq 0\}$ regenerates at $S_1, S_2, \ldots$, the random variables $\{C_i\}$ defined by

$$C_i = \int_{S_i}^{S_{i+1}} r[Y(u)] \, du \qquad i \in I_+$$

are i.i.d., and we can consider them as rewards imposed on the renewal process that has renewal epochs $\{S_i\}_{i=0}^{\infty}$. If $r(\cdot) \geq 0$, applying Theorem 6-6 to this renewal-reward process yields (6-40). If $r(\cdot) < 0$, apply Theorem 6-6 to the negative of this process. Otherwise, apply Theorem 6-6 to the positive and negative parts (as in the proof of Theorem 6-6) and recombine. $\square$

The asymptotic reward rate also can be expressed in terms of the limiting probabilities.

Corollary 6-8a If $\{Y(t); t \geq 0\}$ satisfies the hypotheses of Theorem 6-7, then

$$\lim_{t \to \infty} \frac{1}{t} \int_0^t r[Y(u)]du = \sum_{k \in S} p_k r(k) \qquad (\text{w.p.1})$$

$$= \lim_{t \to \infty} \frac{1}{t} E\left(\int_0^t r[Y(u)]du \right) \qquad (6\text{-}41)$$

If S_1 is arithmetic, then (6-41) holds when t is a multiple of the span.

PROOF Observe

$$\frac{E\left(\int_0^{S_1} r[Y(u)] \, du \right)}{E(S_1)} = \sum_{k \in S} r(k) \frac{E(T_k)}{E(S_1)}$$

$$= \sum_{k \in S} r(k) p_k$$

where the second equality follows from (6-31). $\square$

When S is uncountable, Corollary 6-7a shows that the middle term in (6-41) is replaced by $\int_{-\infty}^{\infty} r(x) \, dH(x)$, where $H(\cdot)$ is defined by (6-35). Choosing $r(x) = x^{\alpha}$ yields this result.

Corollary 6-8b If for some α, $E(\int_0^{S_1} |[Y(u)]^{\alpha}| \, du) < \infty$, then

$$\lim_{t \to \infty} \frac{1}{t} \int_0^t [Y(u)]^{\alpha} \, du = E[(Y^*)^{\alpha}] \qquad (\text{w.p.1}) \qquad (6\text{-}42)$$

where $Y^* = \text{dlim}_{t \to \infty} Y(t)$.

The following corollary of Theorem 6-8 is so useful that it is granted the title of theorem.

Theorem 6-9: Ergodic theorem for regenerative processes Let $\{Y(t); t \geq 0\}$ satisfy the hypotheses of Theorem 6-7 and $T_k(t)$ be the amount of time $Y(\cdot) = k$ during $[0, t]$. Then (w.p.1)

$$\lim_{t \to \infty} \frac{T_k(t)}{t} = p_k \qquad k \in \mathcal{S}$$

PROOF Fix $k \in I_+$, and choose

$$r[Y(t)] = \begin{cases} 1 & \text{if } X(t) = k \\ 0 & \text{otherwise} \end{cases}$$

Then $\int_0^t r[Y(u)]\, du = T_k(t)$. Now apply Theorem 6-8. □

This theorem states that for a regenerative process, the *time-average occupancy of a state over an entire sample path equals the asymptotic probability that the process is in that state* (w.p.1).

In a $GI/G/1$ queue satisfying (6-38), let D_n denote the delay of the nth arriving customer. The process $\{D_n; n \in I_+\}$ is regenerative (Exercise 6-15), so a discrete-parameter version of Corollary 6-8*b* asserts $\sum_{n=1}^N E(D_n)/N \to E(D^*)$, where $D^* = \text{dlim}_{n \to \infty} D_n$. A particular consequence of the next theorem is that the stronger statement, $\lim_{n \to \infty} E(D_n) = E(D^*)$, holds true.

Theorem 6-10 Let $\{Y_n; n \in I_+\}$ be a regenerative process such that
(a) $E(S_1) < \infty$
and
(b) $Y_n \geq 0$ and $E(Y_n) < \infty$ for all $n \in I_+$
hold. Let $Y^* = \text{dlim}_{n \to \infty} Y_n$. Then

$$\lim_{n \to \infty} E(Y_n) = E(Y^*) \qquad (6\text{-}43)$$

If S_1 is periodic, then (6-43) holds when n is a multiple of the period.

PROOF Let $Q_k(\cdot)$ be as in the proof of Theorem 6-7, and

$$h_n \triangleq \sum_{k \in \mathcal{S}} k Q_k(n) \qquad n \in I_+$$

Multiply both sides of (6-32) by k, and sum over $k \in \mathcal{S}$ to obtain

$$E(Y_n) = h_n + \sum_{i=1}^n h_{n-i} m_i \qquad (6\text{-}44)$$

where $\{m_i\}$ is the discrete renewal density of the renewal process composed of the regeneration epochs. The proof of the key renewal theorem shows that $\{h_i\}$ is directly Riemann-integrable if

$$\sum_{i=1}^{\infty} h_i < \infty \qquad (6\text{-}45)$$

holds. Assume (6-45) holds, and apply the key renewal theorem to (6-44), obtaining [note that (6-45) implies $h_n \to 0$]

$$\lim_{n \to \infty} E(Y_n) = \frac{1}{E(S_1)} \sum_{i=1}^{\infty} h_i \qquad (6\text{-}46)$$

To verify that (6-45) holds and to obtain an explicit formula for the sum, observe that

$$
\begin{aligned}
\sum_{i \in S} h_i &= \sum_{i=1}^{\infty} E(Y_i \mid S_1 > i) P\{S_1 > i\} \\
&= \sum_{i=1}^{\infty} \sum_{n=i+1}^{\infty} E(Y_i \mid S_1 = n) P\{S_1 = n\} \\
&= \sum_{n=1}^{\infty} P\{S_1 = n\} \sum_{i=1}^{n} E\{Y_i \mid S_1 = n\} \\
&= E(Y_1 + \cdots + Y_{S_1})
\end{aligned}
\qquad (6\text{-}47)
$$

Hypotheses (a) and (b) and Wald's equation imply that the right-side of (6-47) is finite; hence (6-45) holds. Substituting (6-47) into (6-46) and using Corollary 6-8b yield (6-43). ☐

The restriction to nonnegative random variables is not required for (6-43) to hold.

Corollary 6-10 If (b) of Theorem 6-10 is replaced by

(b′) $E(\mid Y_1 \mid + \cdots + \mid Y_{S_1} \mid) < \infty$

then (6-43) is valid.

PROOF The proof of Theorem 6-10 can be applied to the positive and negative parts of each Y_n, with (b′) asserting that the right-side of (6-47) is finite. In order to combine the expectations of the positive and negative parts of Y_n to form $E(Y_n)$, we require $E(\mid Y_n \mid) < \infty$. This is not an obvious consequence of (b′) (why?), but it does follow from (b′); the proof by contradiction given below is due to S. Halfin.

Suppose $E(\mid Y_n \mid) = \infty$ for some $n \in I_+$; let k be the smallest index for which that is so. From the regenerative property, we can assume without loss of generality that $P\{S_1 \geq k\} > 0$ (why?). From (b′),

$$E(\mid Y_1 \mid + \cdots + \mid Y_{S_1} \mid \text{ given } S_1 < k) P\{S_1 < k\}$$
$$+ E(\mid Y_1 \mid + \cdots + \mid Y_{S_1} \mid \text{ given } S_1 \geq k) P\{S_1 \geq k\} < \infty$$

which implies $E(\mid Y_k \mid \text{ given } S_1 \geq k) < \infty$. This, in turn, implies

$$E(\mid Y_k \mid \text{ given } S_1 < k) = \infty \qquad (6\text{-}48)$$

When $S_1 = j < k$, the process $\{Y_n; \ n = j + 1, \ j + 2, \ldots\}$ is stochastically identical to $\{Y_n; n \in I_+\}$ by the regenerative property. However, in the former process, (6-48) implies that the $(k - j)$th random variable has infinite expectation, which contradicts the assumption that k is the smallest index for which $E(|Y_n|) = \infty$. $\qquad\square$

The following problem in the analysis of queues motivated the next theorem. For an M/G/1 queue that has reached the steady state, we would like to find the remaining service time of the customer in service, given that there is one. This is equivalent to the time until the next departure epoch, given that the system is not empty. The standard way to compute this is to argue thus. If we consider only those times when the server is working, then the service times $V_1, V_2, \ldots$ generate a renewal process so that the remaining service is the equilibrium-excess distribution. This argument is suspect, because of the inspection paradox, even though it provides the correct answer.

A generalization of the above setting is to consider that each renewal interval is divided into a beginning (A interval), a middle (B interval), and an end (C interval), as shown below. We allow up to two of these subintervals always to have zero length, and we assume that $\{A_i\}_{i=1}^{\infty}$, $\{B_i\}_{i=1}^{\infty}$, and $\{C_i\}_{i=1}^{\infty}$ consist of i.i.d. random variables. For each i, A_i, B_i, and C_i are *not* independent of one another because $A_i + B_i + C_i \equiv X_i$.

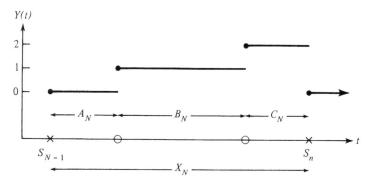

To make matters precise, define the following processes. Let $\{Y(t); t \geq 0\}$ be defined by

$$Y(t) = \begin{cases} 0 & \text{if } t \in \text{some } A \text{ interval} \\ 1 & \text{if } t \in \text{some } B \text{ interval} \\ 2 & \text{if } t \in \text{some } C \text{ interval} \end{cases} \qquad (6\text{-}49)$$

Let $\{N(t); t \geq 0\}$ be the renewal process generated by $X_1, X_2, \ldots$, and let $\{R(t); t \geq 0\}$ be defined by

$$R(t) = \begin{cases} S_{N-1} + A_N - t & \text{if } Y(t) = 0 \\ S_{N-1} + A_N + B_N - t & \text{if } Y(t) = 1 \\ S_N - t & \text{if } Y(t) = 2 \end{cases} \qquad (6\text{-}50)$$

where $N = N(t)$. Thus $\lim_{t \to \infty} P\{R(t) \le x \mid Y(t) = 1\}$ is the asymptotic distribution function of the remainder of a B interval when attention is restricted only to B intervals.

Theorem 6-11 Let $Y(t)$ and $R(t)$ be given by (6-49) and (6-50), $G(y) = P\{B_1 \le y\}$, $v_G = E(B_1)$, and $v_F = E(X_1) < \infty$. Then

$$\lim_{t \to \infty} P\{R(t) \le x \mid Y(t) = 1\} = \frac{1}{v_G} \int_0^x G^c(y) \, dy$$

When X_1 is arithmetic, t is restricted to be a multiple of the span.

PROOF† Define the indicator function

$$I(t; x) = \begin{cases} 1 & \text{if } R(t) > x \text{ and } Y(t) = 1 \\ 0 & \text{otherwise} \end{cases}$$

Thus,

$$P\{R(t) > x, \ Y(t) = 1\} = P\{I(t; x) = 1\} \tag{6-51}$$

Manifestly, the Y and I processes both regenerate at renewal epochs X_1, $X_1 + X_2$, etc. The sample paths of these processes are step functions, so they are in $D[0, \infty)$, and so Theorem 6-7 establishes the existence of the limits that appear below.

From (6-31) and (6-51),

$$\lim_{t \to \infty} P\{R(t) > x, \ Y(t) = 1\} = \frac{E(\int_0^{X_1} I(t; x) \, dt)}{v_F} \tag{6-52}$$

By referring to the figure below, it is apparent that

$$\int_0^{X_1} I(t; x) \, dt = \begin{cases} B_1 - x & \text{if } B_1 > x \\ 0 & \text{otherwise} \end{cases}$$

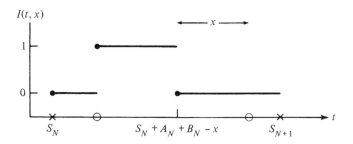

† This theorem and its proof are taken from Linda Green, "A Limit Theorem on Subintervals of Interrenewal Times," *Oper. Res.* **30** (1982).

Hence, by using integration by parts,

$$E\left(\int_0^{X_1} I(t; x) \, dt\right) = \int_x^\infty (y - x) \, dG(y)$$

$$= \int_x^\infty G^c(y) \, dy \tag{6-53}$$

Another consequence of (6-31) is that

$$\lim_{t \to \infty} P\{Y(t) = 1\} = \frac{v_G}{v_F} \tag{6-54}$$

Equation (6-54) implies that

$$\lim_{t \to \infty} P\{R(t) > x \mid Y(t) = 1\} = \frac{\lim_{t \to \infty} P\{R(t) > x, Y(t) = 1\}}{\lim_{t \to \infty} P\{Y(t) = 1\}} \tag{6-55}$$

is meaningful; substituting (6-52), (6-53), and (6-54) in (6-55) yields the desired result. $\qquad\square$

More Examples

Example 6-7 Theorem 2-6 asserts that for each sample path of a queue, ω say, the proportion of time the facility is empty is given by

$$p_0(\omega) = 1 - \frac{\lambda(\omega)}{\mu(\omega)}$$

where $\lambda(\omega)$ is the arrival rate and $\mu(\omega)$ is the departure rate while the server is busy. For the $GI/G/1$ queue with arrival rate λ and mean service time $1/\mu$, the strong law of large numbers for renewal processes asserts that (with probability 1) $\lambda(\omega) = \lambda$ and $\mu(\omega) = \mu$. Hence Theorem 6-9 implies that when $\lambda < \mu$,

$$\lim_{t \to \infty} P\{\text{system empty at } t\} = 1 - \frac{\lambda}{\mu}$$

In other words, Theorem 2-6 for a deterministic process holds as a theorem for regenerative processes. $\qquad\square$

Example 6-8: Example 6-3 continued In Example 6-3, it was shown that for an arbitrary cycle in an (s, S) inventory system, the expected number of consecutive days before the inventory drops below q is $M(S - q) + 1$, for $S \geq q \geq s$. Let I_n be the inventory at the *start* of day n. Then Theorem 6-9 implies that

$$\lim_{n \to \infty} P\{I_n \geq j\} = \begin{cases} 0 & \text{if } j > S \\ \dfrac{M(S - j) + 1}{M(S - s) + 1} & \text{if } s \leq j \leq S \\ 1 & \text{if } j < s \end{cases}$$

Hence

$$p_j \triangleq \lim_{n \to \infty} P\{I_n = j\} = \begin{cases} \dfrac{M(0) + 1}{M(S - s) + 1} & \text{if } j = S \\[2ex] \dfrac{M(S - j) - M(S - j - 1)}{M(S - s) + 1} & \text{if } s \leq j \leq S - 1 \\[2ex] 0 & \text{otherwise} \end{cases}$$

The inventory costs used in Example 6-3 are imposed on the end-of-day inventory. (What modeling problems occur if these costs are imposed at the start of each day?) Let J_n be the inventory level at the *end* of day n. Then

$$J_n = (I_n - D_n)^+$$

where D_n is the demand on day n. Clearly $\{J_n; n \geq 0\}$ regenerates at the same epochs that $\{I_n; n \geq 0\}$ does, so Corollary 6-8a yields

$$\pi_j \triangleq \lim_{n \to \infty} P\{J_n = j\} = \begin{cases} 0 & \text{if } j > S \\[1ex] \displaystyle\sum_{i=s}^{S} p_i f_{j-1} & \text{if } S \geq j > 0 \\[2ex] \displaystyle\sum_{i=s}^{S} p_i F_{i-1}^c & \text{if } j = 0 \end{cases}$$

where $f_i = P\{D_1 = i\}$ and $F_i^c = P\{D_i > i\}$.

Let K_n be the number of items back-ordered on day n, so

$$K_n = (D_n - I_n)^+$$

Reasoning as above yields

$$g_j \triangleq \lim_{n \to \infty} P\{K_n = j\} = \begin{cases} \displaystyle\sum_{i=s}^{S} p_i f_{i+j} & j > 0 \\[2ex] \displaystyle\sum_{i=0}^{S} \pi_i & j = 0 \end{cases}$$

If a holding cost of h_j is incurred when $J_n = j$ and a back-order charge of b_j is incurred when $K_n = j$, then the asymptotic rate at which these costs are charged is, by Corollary 6-8a, $\sum_{j=1}^{S} h_j \pi_j + \sum_{j=1}^{\infty} b_j g_j$. In Example 6-3, $h_j = h \cdot j$ and $b_j = b \cdot j$. $\qquad \square$

Example 6-9: Proof of Theorem 4-8 Let $\{X(t); t \geq 0\}$ be a birth-and-death process that has an honest limiting distribution $\{p_j\}$. Corollary 4-4a and Theorem 4-11 show that this implies that the first-passage time from any initial state to state zero is finite. Since times of entries into state zero are regeneration epochs, $\{X(t); t \geq 0\}$ is a (possibly delayed) regenerative process.

Fix $j \in S$ and set

$$r[X(t)] = I_j(t) = \begin{cases} 1 & \text{if } X(t) = j \\ 0 & \text{otherwise} \end{cases}$$

Let S_0 be the first epoch when $X(t) = 0$. Then $\int_0^{S_0} I_j(t)\, dt < \infty$, and so Corollary 6-8a implies

$$\lim_{T \to \infty} \frac{1}{T} \int_0^T I_j(t)\, dt = p_j \qquad \text{(w.p.1)}$$

for each $j \in I$, and (4-75) is proved. The proof of (4-76) is similar (Exercise 6-17).

Application to the Analysis of Simulation Experiments

In a discrete-event simulation, the process being simulated changes only at a discrete set of points. Let $\{Y(t); t \geq 0\}$ be the process being simulated, and suppose the purpose of the simulation is to estimate some "steady-state" characteristic which we denote by $E[r(Y^*)]$. For example, $Y(t)$ could be the number of customers present at t in a $GI/G/c$ queue. To obtain the mean number of customers present as $t \to \infty$, choose $r(y) = y$; to obtain the limiting probability that 10 or fewer customers are present, choose $r(y) = 1$ if $y \leq 10$ and $r(y) = 0$ otherwise.

Suppose $\{Y(t); t \geq 0\}$ is a regenerative process with finite mean cycle times. Let

$$C_i \triangleq \int_{S_i}^{S_{i+1}} r[Y(u)]\, du \qquad i \in I_+$$

Theorem 6-8 and its corollaries imply

$$\theta \triangleq E[r(Y^*)] = \frac{E(C_1)}{E(S_1)}$$

Since the pairs $(C_1, S_1), (C_2, S_2), \ldots$ are i.i.d.,

$$\bar{C}_n \triangleq \frac{1}{n} \sum_{i=1}^{n} C_i \qquad \text{and} \qquad \bar{S}_n \triangleq \frac{1}{n} \sum_{i=1}^{n} S_i$$

are unbiased estimates of $E(C_1)$ and $E(S_1)$, respectively, and a reasonable (but biased) estimate of θ based on n cycles is

$$\hat{\theta}_n \triangleq \frac{\bar{C}_n}{\bar{S}_n} \tag{6-56}$$

The bias in $\hat{\theta}_n$ is due to the (possible) correlation between C_i and S_i. As $n \to \infty$, $\hat{\theta}_n \to \theta$.

An estimator should be accompanied by a measure of its variability. Toward this end, Crane and Iglehart (1975) established the following proposition.

Proposition 6-1 In a regenerative simulation, for large n, the probability is *approximately* α that

$$\frac{\bar{C}_n \bar{S}_n - k_n a_n - D_n^{1/2}}{\bar{S}_n^2 - k_n s_n} \leq \theta \leq \frac{\bar{C}_n \bar{S}_n - k_n a_n + D_n^{1/2}}{\bar{S}_n^2 - k_n s_n}$$

where

$$a_n = \frac{1}{n-1} \sum_{i=1}^{n} (C_i - \bar{C}_n)(S_i - \bar{S}_n)$$

$$s_n = \frac{1}{n-1} \sum_{i=1}^{n} (S_i - \bar{S}_n)^2$$

$$k_n = \left[\Phi^{-1}\left(1 - \frac{\gamma}{2}\right) \right]^2$$

$$c_n = \frac{1}{n-1} \sum_{i=1}^{n} (C_i - \bar{C}_n)^2$$

$$D_n = (\bar{C}_n \bar{S}_n - k_n a_n)^2 - (\bar{S}_n^2 - k_n s_n)(\bar{C}_n^2 - k_n c_n)$$

and $\Phi(\,\cdot\,)$ is the distribution function of a normal random variable with mean 0 and variance 1.

PROOF See Crane and Iglehart (1975) and references therein. □

The regenerative method for estimating θ is to run the simulation for n cycles and form the estimator given by (6-56). An approximate $100(1 - \alpha)$ percent confidence interval is given by Proposition 6-1. The precision of an estimator typically improves with increasing sample size. If there is a choice in specifying regeneration points, choosing the state that is likely to yield the shortest mean cycle time tends to produce the greatest number of cycles for a given cost of a simulation run.

The regenerative method of estimation does not work well when $E(S_1)$ is large because a very long run is required to obtain sufficiently many cycles for good precision. A heuristic procedure based on "almost regeneration epochs" has been proposed for handling this situation.†

EXERCISES

6-11 Let X be a discrete random variable. Show that X is aperiodic if $P\{X = 1\} > 0$.

6-12 Prove Corollary 6-7a.

6-13 Prove Corollary 6-7b.

6-14 In Example 6-5, assume (6-38b) is false. Exhibit a $GI/G/2$ queue for which $\{X(t); t \ge 0\}$ never regenerates.

6-15 In Example 6-5, let D_n be the delay (wait in queue) of the nth arriving customer. Show that $\{D_n; n \in I_+\}$ regenerates at arrival epochs where all servers are free. Prove that $\mathrm{dlim}_{n \to \infty} D_n = D^*$ for some random variable D^*.

6-16 Let W_n be the waiting time of the nth arriving customer in a $GI/G/c$ queue in which customers are served FIFO. Assume that at time zero a customer arrives to an empty system and that (6-38) is

---

† F. L. Gunther and R. W. Wolff, "The Almost Regenerative Method for Stochastic System Simulations," *Oper. Res.* **28**: 375–386 (1980).

satisfied. Let C be the number of customers served in the first busy period and $I_n(t)$ be defined by

$$I_n(t) = \begin{cases} 1 & \text{if } W_n \le t \\ 0 & \text{if } W_n > t \end{cases}$$

Show that $\lim P\{W_n \le t\} = E[\sum_{n=1}^{C} I_n(t)]/E(C)$, and discuss the role of this result in the design and analysis of a simulation of a $GI/G/c$ queue.

6-17 Formulate Theorems 6-7, 6-8, and 6-9 for a discrete-parameter process. Prove (4-76).

6-18 What are the regeneration epochs in a $GI/G/1/1$ queue? Let $M(t)$ be the mean number of arrivals during $(0, t]$, $t > 0$, and $G(\cdot)$ be the service time distribution function; also let p_n be the probability that the nth arriving customer is denied access to the server (i.e., is blocked). Prove

$$p \triangleq \lim_{n \to \infty} p_n = \frac{\int_0^\infty M(t)\, dG(t)}{1 + \int_0^\infty M(t)\, dG(t)}$$

Let $F(\cdot)$ be the interarrival time distribution function, and assume $G(t) = 1 - e^{-\mu t}$. Show that $p = \tilde{F}(\mu)$, where $\tilde{F}(\cdot)$ is the LST of $F(\cdot)$.

6-19 State and prove versions of Theorems 6-8 and 6-9 for continuous random variables.

6-20 Extend Theorem 6-10 to a regenerative-reward process. In particular, prove that if $E(|Y_1^\alpha| + \cdots + |Y_{S_1}^\alpha|) < \infty$ for some given α, then $E(Y_n^\alpha) \to E[(Y^*)^\alpha]$, where $Y^* = \text{dlim}_{n \to \infty} Y_n$.

6-21 Let B_n be defined as in Theorem 6-11, L_t be the spread, and D_t be excess at epoch t for the renewal process generated by $\{B_n\}_{n=1}^\infty$. State precisely, and then prove, that when attention is focused on epochs only in some B interval, the asymptotic joint distribution of spread and excess of a B interval is $\lim_{t \to \infty} P\{L_t \le x, D_t \le y\}$, as given in Exercise 5-43.

6-22 In Example 6-7, suppose L_i days elapse between the day when the ith order is placed and the day it is received. Assume $L_1, L_2, \ldots$ are i.i.d. random variables. What are the regeneration points for $\{I_n; n \in I_+\}$? Derive conditions for the mean time between regenerations to be finite.

6-5* BRANCHING PROCESSES

Branching processes are analyzed by recursive arguments similar to those used in examining regenerative process. An important result obtained in this section is the distribution function of the busy period of an M/G/1 queue.

Underlying Chance Mechanism

Various processes are defined with respect to the following underlying chance mechanism.

We consider particles that live for one unit of time and have the ability to produce new particles of like kind. Just before expiring, each particle produces k new particles with probability p_k, $k \in I$. Each particle acts independently of all the other particles. At time zero, a single particle, called the *progenitor*, forms the zeroth generation. The particles spawned by the particles of the nth generation form the $(n + 1)$th generation.

Feller (1968) provides examples of this mechanism in nuclear chain reactions, survival of family names, and gene mutations. In the analysis of the busy period of an M/G/1 queue, the zeroth generation is a customer who arrives when the server is free, and the $(n + 1)$th generation is formed by those customers who arrive while the nth generation is in service.

Extinction Probabilities

Let S_n be the size of the nth generation. We seek the probability that the particle system is extinct by time n, which is $P\{S_n = 0\} \triangleq d_n$. Define the generating functions

$$\hat{P}_n(z) = \sum_{k=0}^{\infty} z^k P\{S_n = k\} \qquad n \in I$$

In particular,

$$\hat{P}_1(z) \triangleq P(z) = \sum_{k=0}^{\infty} z^k p_k.$$

Suppose the progenitor spawns k offspring. The nth generation of the progenitor is formed by adding the members of the $(n-1)$th generation of each of the offspring. Hence $S_n = 0$ if $k = 0$, and

$$S_n = S_{n-1}^{(1)} + \cdots + S_{n-1}^{(k)} \qquad k \in I_+ \tag{6-57}$$

where $S_{n-1}^{(i)}$ is the number in the $(n-1)$th generation of offspring i. The summands in (6-57) are i.i.d. and distributed as S_{n-1}; taking generating functions on both sides of (6-57) yields

$$\hat{P}_n(z) = P[\hat{P}_{n-1}(z)] \qquad n \in I_+ \tag{6-58}$$

In principle, one can calculate $\hat{P}_n(z)$ from (6-58); only in special cases can $\hat{P}_n(z)$ be obtained analytically.

From (6-58) we obtain

$$d_n = \hat{P}_n(0) = P(d_{n-1}) \qquad n \in I_+ \tag{6-59}$$

where $d_0 = 1$. Trivially,

$$p_0 = 0 \Rightarrow d_n = 0 \qquad \text{and} \qquad p_0 = 1 \Rightarrow d_n = 1 \qquad n \in I_+$$

so suppose $0 < p_0 < 1$. It is intuitively clear that

$$d_1 < d_2 < d_3 < \cdots \tag{6-60}$$

and (6-60) can be obtained formally from (6-59) by noting that $P(z)$ is an increasing function of z for $0 \le z \le 1$. Hence, there exists

$$d \triangleq \lim_{n \to \infty} d_n \le 1 \tag{6-61}$$

Since $P(z)$ is a continuous function of z, at least for $0 \le z \le 1$, (6-59) implies that d satisfies

$$d = P(d) \qquad 0 \le d \le 1 \tag{6-62}$$

Since

$$d = \lim_{n \to \infty} P\{S_n = 0\} = \lim_{n \to \infty} P\{\text{extinction by } n\}$$

it is of considerable interest to characterize the solution of (6-62). The content of Theorem 6-12 below is intuitive; the method of proof is used in a variety of settings.

Theorem 6-12 Let $v = \sum_{k=0}^{\infty} k p_k$ and assume $0 < p_0 < 1$. If $v \leq 1$, the solution of (6-62) is $d = 1$; if $v > 1$, d is the unique solution of (6-62) with $0 < d < 1$.

PROOF For $0 \leq z \leq 1$, the first and second derivatives of $P(z)$ are nonnegative, so the graph of $P(z)$ starts at $p_0 = P(0)$ and increases, at an increasing rate, to $1 = P(1)$. The slope at $z = 1$ is v. A graphical solution of (6-62) is shown in Figure 6-1.

It is always true that $d = 1$ solves (6-62). When $v \leq 1$, this is the only solution because as z decreases from 1, $P(z)$ decreases more slowly than z. When $v > 1$, there is exactly one solution of (6-62) with $0 < d < 1$ (why?), d' say. This smaller solution must be the required value because the monotonicity of $P(\cdot)$ and (6-59) imply $d_1 = p_0 < P(d') = d'$, and by induction, $d_{n+1} = P(d_n) < P(d') = d'$. □

Size of the Total Progeny

The number of particles spawned in generations 0 through n is

$$C_n \triangleq 1 + S_1 + \cdots + S_n \qquad n \in I_+ \tag{6-63}$$

We are interested in the distribution of C_n. In particular, the limit as $n \to \infty$ is the size of the total progeny, including the progenitor.

Let

$$\hat{H}_n(z) = \sum_{k=0}^{\infty} z^k P\{C_n = k\} \qquad n \in I_+$$

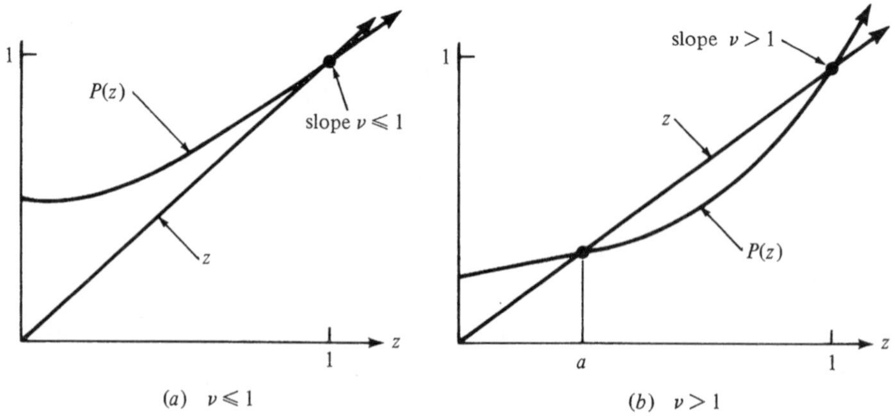

(a) $v \leq 1$ (b) $v > 1$

Figure 6-1 Graphical solution of (6-62).

Conditioning on the number spawned in the first generation, as in the derivation of (6-58), we obtain

$$\hat{H}_n(z) = z \sum_{k=0}^{\infty} [\hat{H}_{n-1}(z)]^k p_k = zP[\hat{H}_{n-1}(z)] \qquad n \in I_+ \qquad (6\text{-}64)$$

In particular, $\hat{H}_1(z) = zP(z)$.

Since $p_0 = 1$ leads to trivialities, assume $p_0 < 1$. For $0 \le z < 1$,

$$\hat{H}_2(z) = zP[zP(z)] < zP(z) = H_1(z)$$

and by induction

$$\hat{H}_1(z) > \hat{H}_2(z) > \hat{H}_3(z) > \cdots \qquad (6\text{-}65)$$

Since each $\hat{H}_n(\cdot) \ge 0$, the sequence in (6-65) has a limit, $\hat{H}(z)$ say. By the continuity theorem for generating functions (Property A-15a),

$$\hat{H}(z) = \sum_{k=0}^{\infty} z^k \lim_{n \to \infty} P\{C_n = k\}$$

i.e., it is the generating function of the size of the total progeny, which is the random variable $C^* \triangleq \text{dlim}_{n \to \infty} C_n$. The continuity theorem and (6-64) imply that $\hat{H}(\cdot)$ satisfies

$$\hat{H}(z) = zP[\hat{H}(z)] \qquad 0 \le z < 1 \qquad (6\text{-}66)$$

That is, for each fixed z, $\hat{H}(z)$ solves

$$x = zP(x) \qquad 0 \le x, z < 1 \qquad (6\text{-}66')$$

Theorem 6-13 Let $v = \sum_{k=0}^{\infty} kp_k$ and assume $0 < p_0 < 1$. Then $\hat{H}(z)$ is the unique positive solution of (6-66') and $\hat{H}(1) = d$. In particular, $P\{C^* < \infty\} = 1$ if, and only if, $v < 1$.

PROOF Recall from the proof of Theorem 6-12 that $P(z)$ increases with z at an increasing rate, for $0 \le z < 1$. Observe that $zp_0 = zP(0) > 0$, $zP(1) < 1$, and from (6-62), $zP(d) = zd < d$. A graphical solution of (6-66') is shown in Figure 6-2. Since $P(1) = 1$, the monotonicity properties of $P(\cdot)$ show that (6-66') has a unique positive solution. (If $p_0 = 0$, $x = 0$ is the desired solution.) Observe that $\hat{H}(z) < d$ for $0 \le z < 1$ and $\lim_{z \uparrow 1} \hat{H}(z) = d$. From Theorem 6-12, $d = 1$ if, and only if, $v < 1$, which proves the last assertion. $\square$

Useful information may be obtained from (6-66).

Corollary 6-13a When $v < 1$,

$$E(C^*) = \frac{1}{1-v} < \infty \qquad (6\text{-}67)$$

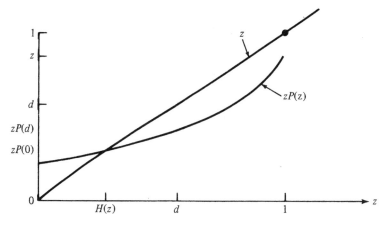

Figure 6-2 Graphical solution of (6-61a).

and

$$\text{Var}\,(C^*) = \frac{E(S_1^2 - S_1)}{(1 - v)^3} + \frac{2v - 1}{(1 - v)^2} + \frac{1}{1 - v} \tag{6-68}$$

PROOF Apply the formulas

$$E(C^*) = \frac{d}{dz}\,\hat{H}(z)\big|_{z=1}$$

and

$$E[C^*(C^* - 1)] = \frac{d^2}{dz^2}\,\hat{H}(z)\big|_{z=1}$$

to (6-66). $\qquad\Box$

Applications to Busy Periods of an M/G/1 Queue

Let λ be the arrival rate, $G(\,\cdot\,)$ be the service time distribution function with LST $\tilde{G}(\,\cdot\,)$, V_i be the service time of the ith arrival, and $\rho = \lambda E(V_1)$. Let C be a generic r.v. representing the number of customers served in a busy period, and let $\hat{H}(\,\cdot\,)$ be the generating function of C.

Corollary 6-13b The generating function of the number of customers served in a busy period of an M/G/1 queue satisfies (6-66) with $P(z) = \tilde{G}(\lambda - \lambda z)$. So $P\{C < \infty\} = 1$ if, and only if, $\rho < 1$. If $\rho < 1$, then

$$E(C) = \frac{1}{1 - \rho} < \infty \tag{6-69}$$

and

$$\text{Var}\,(C) = \frac{\lambda^2\,\text{Var}\,(V_1) + 1}{(1 - \rho)^3} \qquad (6\text{-}70)$$

PROOF Let the customer who initiates a busy period be the progenitor of a branching process where the members of the $(n + 1)$th generation are the customers who arrive while the nth generation is in service. The number of customers served in a busy period is the total progeny of this branching process. By conditioning on the value of V_1,

$$p_k = P\{S_1 = k\} = \int_0^\infty e^{-\lambda t} \frac{(\lambda t)^k}{k!}\,dG(t)$$

$$P(z) = \sum_{k=0}^\infty z^k p_k = \int_0^\infty e^{-\lambda t} \sum_{k=0}^\infty \frac{(\lambda tz)^k}{k!}\,dG(t)$$

$$= \tilde{G}(\lambda - \lambda z)$$

and

$$v = E(S_1) = \int_0^\infty \lambda t\,dG(t) = \rho$$

The desired conclusions follow from Theorem 6-13 and Corollary 6-13a.

□

The LST of the busy period distribution function is obtained by similar arguments.

Proposition 6-2 In an M/G/1 queue, let the generic random variable B represent the length of a busy period, $B(\,\cdot\,)$ be the distribution function of B, and $\tilde{B}(\,\cdot\,)$ be the LST of $B(\,\cdot\,)$. Then (a) $\tilde{B}(\,\cdot\,)$ satisfies

$$\tilde{B}(s) = \tilde{G}[s + \lambda - \lambda\tilde{B}(s)] \qquad (6\text{-}71)$$

and is the root of (6-71) with smallest absolute value; (b) $P\{B < \infty\} = 1$ if, and only if, $\rho < 1$; and (c) if $\rho < 1$,

$$E(B) = \frac{E(V_1)}{1 - \rho} < \infty \qquad (6\text{-}72)$$

and

$$\text{Var}\,(B) = \frac{\text{Var}\,(V_1) + \rho[E(B)]^2}{(1 - \rho)^3} \qquad (6\text{-}73)$$

PROOF Consider the customer who initiates a busy period to be the progenitor of a branching process, as in the proof of Corollary 6-13b. The length of the busy period is the sum of the service times of each member of the total progeny. Let V_1 be the service time of the progenitor and S_1 be the size of the first generation; number these offspring in order of arrival. Suppose that after the progenitor completes service, the customers are served in this order: the first offspring and its total progeny, the second offspring and its entire progeny, etc. Since the service times are i.i.d. and the arrival process is Poisson and independent of the service times, this order of service does not affect $B(\cdot)$ (why?). The time required to serve an offspring and its entire progeny has distribution function $B(\cdot)$, and these times are i.i.d. Hence the LST of $B(\cdot)$, conditional on $V_1 = x$ and $S_1 = k$, is

$$\tilde{B}(s \mid V_1 = x, S_1 = k) = e^{-sx}[\tilde{B}(s)]^k$$

Since

$$P\{S_1 = k \mid V_1 = x\} = \frac{e^{-\lambda x}(\lambda x)^k}{k!}$$

the LST of $B(\cdot)$ conditional on $V_1 = x$ is

$$\tilde{B}(s \mid V_1 = x) = e^{-(s+\lambda)x} \sum_{k=0}^{\infty} \frac{[\lambda x \tilde{B}(s)]^k}{k!}$$

$$= \exp\{-[s + \lambda - \lambda\tilde{B}(s)]x\}$$

and (6-71) follows by unconditioning on V_1.

The remainder of the proof is left as Exercise 6-23. ☐

The distribution function $B(\cdot)$ may be obtained by inverting (6-71),

$$B(t) = \sum_{n=1}^{\infty} \int_0^t e^{-\lambda z} \frac{(\lambda x)^{n-1}}{n!} \, dG_n(x)$$

where $G_n(\cdot)$ is the n-fold convolution of $G(\cdot)$ with itself. For details, see Theorem I.3.3 in Takács (1962). It is more insightful to derive the density function by using a combinatorial argument. We will need this lemma.

Lemma 6-3 Let $V_1, \ldots, V_n$ be i.i.d. nonnegative random variables. Let $T_1, \ldots, T_n$ be ordered observations of n i.i.d. random variables that are uniformly distributed on $[0, t]$. Then

$$P\{V_1 + \cdots + V_k \leq T_k, k = 1, \ldots, n \mid \sum_{i=1}^{n} V_i = y\}$$

$$= \begin{cases} \dfrac{t - y}{t} & \text{if } y \leq t \\ 0 & \text{if } y \geq t \end{cases} \tag{6-74}$$

PROOF See the proof of Equation (41) in Section 5.6 of Cox and Smith (1961). ☐

In the $M/G/1$ queue, given that n arrivals appeared during $[0, t]$, the service times $V_1, V_2, \ldots, V_n$ and the arrival epochs $T_1, T_2, \ldots, T_n$ satisfy the hypothesis of Lemma 6-3 by assumption and by Theorem 5-13, respectively.

Proposition 6-3 In an $M/G/1$ queue, assume the service times have density function $g(\cdot)$. Then the density function for a busy period is given by

$$b(t) = e^{-\lambda t}\left[g(t) + \sum_{n=1}^{\infty} \frac{(\lambda t)^n}{n!} \int_0^t \frac{x}{t} g_n(t - x)g(x) \, dx \right] \qquad (6\text{-}75)$$

where $g_n(\cdot)$ is the n-fold convolution of $g(\cdot)$ with itself.

PROOF Number the customers served in a busy period in order of arrival, and let time zero be the start of the busy period. One way $\{B = t\}$ can occur is for $V_1 = t$ and no customers to arrive during $(0, t^+]$. The density of this pair of events is $g(t)e^{-\lambda t}$, which is the first term of (6-75). Suppose $V_1 = x < t$. Then $B = t$ if, and only if, $A(t) = n$ (say) customers arrive during $(0, t]$ with $V_2 + \cdots + V_{n+1} = t - x$ *and* the arrival epochs of these customers are such that the busy period ends when customer $n + 1$ completes service. Let $A(t)$ be the number of customers who arrive during $(0, t]$ (this does not include the customer who initiates the busy period), and let T_j be the arrival epoch of the jth such customer; if $A(t) = n$, set $\tau_j = t - T_{n+1-j}$, $j = 1, 2, \ldots, n$. From Figure 6-3, it is apparent that we require

$$T_1 \leq V_1, \; T_2 \leq V_1 + V_2, \ldots, \; T_n \leq V_1 + \cdots + V_n \qquad (6\text{-}76)$$

Since the T_j are i.i.d. and uniformly distributed over $(0, t]$, so are the τ_j, and the probability that the events in (6-76) all occur is identical to the probability that all the events

$$\tau_1 \geq V_{n+1}, \; \tau_2 \geq V_{n+1} + V_n, \ldots, \; \tau_n \geq V_{n+1} + \cdots + V_2 \qquad (6\text{-}76')$$

occur. Since the service times are i.i.d., the probability that all the events in (6-76') occur, given $V_2 + \cdots + V_{n+1} = t - x$, is (by Lemma 6-3)

$$P\{V_2 + \cdots + V_{k+1} \leq \tau_k, \, k = 1, \ldots, n \, \Big| \sum_{j=2}^{n+1} V_j = t - x\} = \frac{x}{t}$$

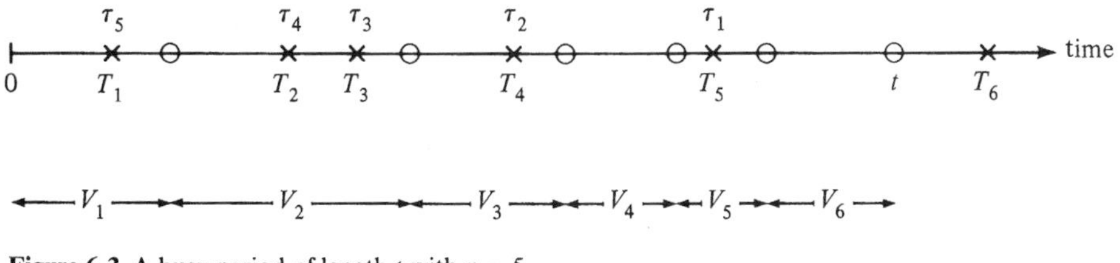

Figure 6-3 A busy period of length t with $n = 5$.

for $x < t$. Thus,

$$b(t \mid V_1 = x, A(t) = n) = \frac{x}{t} g_n(t - x) \qquad n \in I_+$$

Unconditioning first on $A(t)$ and then on V_1, followed by bringing the integral inside the sum, yields the second term in (6-75). □

Busy periods are used extensively in studying queues where one class of customers has priority over another (Section 11-5). They also can be utilized to obtain the LST of the delay distribution.†

Example 6-10: M/G/1 queue with customers served LIFO Suppose customers are served in reverse order of arrival in a single-server queueing system with Poisson arrivals with rate λ and service-time distribution function $G(\,\cdot\,)$. Let D_n be the delay of the nth arriving customer, $D^* = \mathrm{dlim}_{n \to \infty} D_n$, and $\tilde{D}(\,\cdot\,)$ be the LST of D^*. Assume $\rho < 1$.

From Theorem 6-7 and (6-69), the limiting probability that a customer arrives when the server is free is $1 - \rho$. Hence

$$P\{D^* = 0\} = 1 - \rho \qquad (6\text{-}77)$$

Suppose a customer named C_0 arrives when the server is busy. Let the remaining service time of the customer in service at C_0's arrival epoch be R, and let N (a random variable) more customers named $C_1, \ldots, C_N$ arrive before service is completed. Observe that the effect of C_i on the delay of C_0 is C_i's service time plus the sum of the service times of its progeny, i.e., the length of a busy period. Let $\tilde{D}_b(\,\cdot\,)$ be the LST of D^* given that an arriving customer finds the server busy. Then

$$\tilde{D}_b(s \mid R = y, N = n) = e^{-sy} [\tilde{B}(s)]^n \qquad n \in I$$

Hence

$$\tilde{D}_b(s \mid R = y) = e^{-sy} \sum_{n=0}^{\infty} \frac{e^{-\lambda y} [\lambda y \tilde{B}(s)]^n}{n!}$$

$$= \exp\{ -[s + \lambda - \lambda \tilde{B}(s)] y \}$$

and thus

$$\tilde{D}_b(s) = \tilde{R}[s + \lambda - \lambda \tilde{B}(s)] \qquad (6\text{-}78)$$

where $R(y) \triangleq P\{R \le y\}$. Combining (6-77) and (6-78) yields

---

† When customers are served FIFO, the LST of the steady-state delay distribution function is obtained via busy-period analysis in Section 5.10 of Leonard Kleinrock, *Queueing Systems*, vol. 1, Wiley, New York (1975).

$$\tilde{D}(s) = 1 - \rho + \rho\tilde{R}[s + \lambda - \lambda\tilde{B}(s)] \tag{6-79}$$

It remains to find $R(\cdot)$. Theorem 6-11 asserts that as $t \to \infty$, the distribution function of the remaining service time of the customer in service at t (given that there is one) approaches the equilibrium-excess distribution function of $G(\cdot)$. Since the probability that a Poisson arrival occurs during $(t, t + s)$ is proportional to s and independent of t, it is natural to suppose that is what occurs for the nth customer as $n \to \infty$. This is a correct supposition that is proved in Section 11-2. Hence

$$\tilde{R}(s) = \frac{1 - \tilde{G}(s)}{sv} \tag{6-80}$$

where v is the mean service time.

From (6-79) and (6-80), we can obtain the moments (Exercise 6-29)

$$E(D^*) = \frac{\lambda v_{2G}}{2(1 - \rho)} \tag{6-81a}$$

and

$$\text{Var}(D^*) = \frac{\lambda v_{3G}}{3(1 - \rho)^2} + [E(D^*)]^2 \frac{1 + \rho}{1 - \rho} \tag{6-81b}$$

EXERCISES

6-23 Complete the proof of Proposition 6-2.

6-24 Give a direct argument for (6-76').

6-25 Show that the density function of the busy period of an M/M/1 queue, obtained in Theorem 4-10, satisfies Proposition 6-3.

6-26 Suppose the service time of the customer who initiates a busy period has a different distribution function from the other customers served in the busy period. This may arise because a setup time for the server is required at the start of each busy period. Find the analogs of Corollary 6-13b and Propositions 6-2 and 6-3 for this model. What is $\lim_{n \to \infty} P\{V_n \le x\}$?

6-27 Suppose, as in the start-up and shut-down models in Section 4-6, that at the start of a busy period the server commences service when n customers are present and serves continually until no customers are present. Let $p(t)$ be the probability the server is idle at time t. Show that when $\rho < 1$, no matter how many customers are present at time zero, $\lim_{t \to \infty} p(t) = 1 - \rho$. For any value of ρ, derive the Laplace transform of $p(\cdot)$.

6-28 In a GI/G/1 queue, suppose customer 1 arrives at time zero and initiates a busy period. Let U_n be the interarrival time between the nth and $(n + 1)$th arrivals and V_n be the service time of the nth arrival; i.e., the second arrival occurs at epoch U_1 and has service time V_2. Let $X_n = U_n - V_n$ and $S_n = X_1 + \cdots + X_n$, $n \in I_+$. Let C be the number of customers served in this period. Show

$$C = \inf\{n : S_n > 0\}$$

6-29 Derive (6-81). *Hint:* Expand $\tilde{R}(s)$ in a Taylor's series up to terms of order s^2 and then differentiate $\tilde{D}(s)$.

BIBLIOGRAPHIC GUIDE

Renewal-reward processes were introduced in Smith (1955). Discrete-time regenerative processes were introduced by W. Feller, who called them *recurrent-event* processes; the full scope of Feller's theory may be found in Feller (1968). The continuous-time version of Feller's theory first appeared in Smith (1955). One of the goals of Smith's investigations was a theory to cope with the $GI/G/c$ queue. An account of the early work on these topics may be found in Smith (1958). The treatment given here follows Cox (1962), Çinlar (1975), and Ross (1970). We do not include the analogs of the Blackwell and key renewal theorems; these may be found in Brown and Ross (1972).

The proposal to exploit the regenerative structure when queueing simulations are analyzed is due to Cox and Smith (1961). For the implementation of this proposal, see Crane and Iglehart (1975), Fishman (1978), and Crane and Lemoine (1977).

The material on branching processes in Section 6-5 is derived from Feller (1968, chap. 12). A richer discussion is given in chap. 8 of Karlin and Taylor (1975). The classical reference work is the monograph by Harris (1963).

The material on the busy period of an $M/G/1$ queue follows Cox and Smith (1961); this is the standard analysis given in texts on queueing theory. Takács (1962) presents many detailed results. The power of ballot theorems to analyze queueing processes is described in Takács (1967).

References

Brown, Mark, and Sheldon M. Ross: "Asymptotic Properties of Cumulative Processes," *SIAM J. Appl. Math.* **22**: 93–105 (1972).
Çinlar, Erhan: *Introduction to Stochastic Processes*, Prentice-Hall, Englewood Cliffs, N.J. (1975).
Cox, D. R.: *Renewal Theory*, Wiley, New York (1962).
—— and Walter L. Smith: *Queues*, Wiley, New York (1961).
Crane, M. A., and D. L. Iglehart: "Simulating Stable Stochastic Systems, III: Regenerative Processes and Discrete Event Simulations," *Oper. Res.* **23**: 33–45 (1975).
—— and A. J. Lemoine: *An Introduction to the Regenerative Method for Simulation Analysis*, Springer-Verlag, New York (1977).
Feller, William: *An Introduction to Probability Theory and Its Applications*, vol. 1, 3d ed., Wiley, New York (1968).
——: *An Introduction to Probability Theory and Its Applications*, vol. 2, 2d ed., Wiley, New York (1971).
Fishman, George S.: *Principles of Discrete Event Simulation*, Wiley, New York (1978).
Harris, T.: *The Theory of Branching Processes*, Springer-Verlag, New York (1963).
Karlin, Samuel, and Howard S. Taylor: *A First Course in Stochastic Processes*, 2d ed., Academic, New York (1975).
Ross, Sheldon M.: *Applied Probability Models with Optimization Applications*, Holden-Day, San Francisco (1970).
Smith, W. L.: "Regenerative Stochastic Processes," *Proc. Roy. Soc.* **A232**: 6–31 (1955).
——: "Renewal Theory and Its Ramifications," *J. Roy. Stat. Soc.* **B20**: 243–302 (1958).
Takács, Lajos: *Introduction to the Theory of Queues*, Oxford, New York (1962).
——: *Combinatorial Methods in the Theory of Stochastic Processes*, Wiley, New York (1967).

SEVEN

MARKOV CHAINS

7-1 DEFINITION AND EXAMPLES

Markov chains are widely used to model phenomena in engineering, physical and social science, and business. Besides being important in their own right, Markov chains introduce more complicated processes.

In Section 4-5 it is shown that, with mild restrictions, birth-and-death processes possess limiting probabilities which are easy to calculate. One reason Markov chain models are used in practice is that in many cases, they also have limiting probabilities. These limiting probabilities satisfy a system of linear equations that is often amenable to analytical or numerical solution methods. Sections 7-3, 7-4, and 7-5 set up machinery to prescribe which Markov chains have limiting probabilities; these probabilities are obtained in Section 7-6.

Definition and Elementary Properties

A Markov chain, abbreviated MC, is a discrete-parameter stochastic process with (at most) a countably infinite state space and the property that the past history of the process is completely summarized by the present state. It is conventional (but not necessary) to label the states by consecutive nonnegative integers. This is the formal definition.

Definition 7-1 The stochastic process $\{Y_n; n \in I\}$ is a *Markov chain* with state space $S \subseteq I$ if each Y_n assumes values only in S and

$$P\{Y_{n+1} = j \mid Y_0 = i_0, \ldots, Y_n = i_n\} = P\{Y_{n+1} = j \mid Y_n = i_n\} \qquad (7\text{-}1)$$

holds for all $n \in I$ and $i_0, i_1, \ldots, i_n, j \in \mathcal{S}$. A Markov chain is called *finite* (*infinite*) when $\mathcal{S}$ is *finite* (*infinite*). We refer to (7-1) as the *Markov property*.

Equation (7-1) is a special case of the general Markov property given in (3-4), so a Markov chain is a discrete-parameter Markov process with a discrete state space. Let

$$a_j = P\{Y_0 = j\} \qquad j \in \mathcal{S}, \qquad \sum_{j \in \mathcal{S}} a_j = 1 \tag{7-2a}$$

and

$$p_{ij}^{(n)} = P\{Y_{n+1} = j \mid Y_n = i\} \qquad i, j \in \mathcal{S}, \qquad n \in I \tag{7-2b}$$

and define the row vector $\mathbf{a} = (a_j; j \in \mathcal{S})$ and the matrices $P^{(n)} = (p_{ij}^{(n)})$. We call $\mathbf{a}$ the *vector of initial probabilities* and $P^{(n)}$ the *matrix of transition probabilities at stage n*. Manifestly,

$$p_{ij}^{(n)} \geq 0 \tag{7-3a}$$

and

$$1 = P\{Y_{n+1} \in \mathcal{S} \mid Y_n = i\} = \sum_{j \in \mathcal{S}} p_{ij}^{(n)} \qquad i \in \mathcal{S}, \qquad n \in I \tag{7-3b}$$

so $P^{(n)}$ has nonnegative elements and all the row sums are 1. Such matrices are called *stochastic*.

For any discrete-parameter stochastic process,

$$P\{Y_0 = i_0, Y_1 = i_1, \ldots, Y_n = i_n\} = P\{Y_n = i_n \mid Y_0 = i_0, \ldots, Y_{n-1} = i_{n-1}\}$$
$$\times P\{Y_0 = i_0, \ldots, Y_{n-1} = i_{n-1}\} \qquad \square$$

For a Markov chain, (7-1) asserts that first term on the right is $P\{Y_n = i_n \mid Y_{n-1} = i_{n-1}\}$, hence

$$P\{Y_0 = i_0, \ldots, Y_n = i_n\} = P\{Y_n = i_n \mid Y_{n-1} = i_{n-1}\}P\{Y_0 = i_0, \ldots, Y_{n-1} = i_{n-1}\}$$

Proceeding by induction yields

$$P\{Y_0 = i_0, \ldots, Y_n = i_n\} = P\{Y_0 = i_0\}P\{Y_1 = i_1 \mid Y_0 = i_0\} \cdots P\{Y_n = i_n \mid Y_{n-1} = i_{n-1}\}$$

$$\tag{7-4}$$

Since a discrete-parameter stochastic process is characterized by its fidi's (see Section 3-1), it follows from (7-2) and (7-4) that *a Markov chain is determined by the vector of initial probabilities and the matrices of transition probabilities.*

Birth-and-death processes are continuous-time Markov processes, so it should not be surprising that Markov chains have some properties similar to properties of birth-and-death processes. Define the *k-step transition functions* by

$$p_{ij}(n; k) = P\{Y_{n+k} = j \mid Y_n = i\} \qquad i, j \in \mathcal{S}, n, k \in I \tag{7-5a}$$

For $0 \le m \le k$, invoking the Markov property yields

$$
\begin{aligned}
p_{ij}(n; k) &= \sum_{l \in S} P\{Y_{n+k} = j \mid Y_n = i, \ Y_{n+m} = l\} P\{Y_{n+m} = l \mid Y_n = i\} \\
&= \sum_{l \in S} P\{Y_{n+k} = j \mid Y_{n+m} = l\} P\{Y_{n+m} = l \mid Y_n = i\} \\
&= \sum_{l \in S} p_{lj}(n + m; k - m) p_{il}(n; m) \\
&= \sum_{l \in S} p_{il}(n; m) p_{lj}(n + m; k - m)
\end{aligned}
\tag{7-6}
$$

which is the *Chapman-Kolmogorov equation* for Markov chains [cf. (4-13)].
From the calculus of conditional probabilities,

$$
\begin{aligned}
P\{Y_1 = j\} &= \sum_{i \in S} P\{Y_0 = i\} P\{Y_1 = j \mid Y_0 = i\} \\
&= \sum_{i \in S} a_i p_{ij}^{(0)} \qquad j \in S
\end{aligned}
$$

Let $\mathbf{P}_n = (P\{Y_n = j\}; j \in S)$, which is called the *vector of state probabilities* at epoch n. Thus, the equation above can be written

$$
\mathbf{P}_1 = \mathbf{a} P^{(0)}
$$

Use (7-6) with $m = 1$ to obtain

$$
\begin{aligned}
P\{Y_2 = j\} &= \sum_{i \in S} P\{Y_0 = i\} P\{Y_2 = j \mid Y_0 = i\} \\
&= \sum_{i \in S} a_i \sum_{l \in S} p_{il}^{(0)} p_{lj}^{(1)} \qquad j \in S
\end{aligned}
$$

and a simple induction shows

$$
\mathbf{P}_n = \mathbf{a} P^{(0)} P^{(1)} \cdots P^{(n-1)} \qquad n \in I_+
\tag{7-7a}
$$

An MC has *time homogeneous* transition probabilities, and the MC is called *homogeneous*, when

$$
P\{Y_{n+k} = j \mid Y_n = i\} = P\{Y_k = j \mid Y_0 = i\} = p_{ij}(k, 0) \triangleq p_{ij}(k)
\tag{7-5b}
$$

holds for every $n \in I_+$. By setting $k = 1$ and choosing $n = 0, 1, \ldots$, in turn, (7-5b) yields $P^{(0)} = P^{(1)} = \cdots$. Let P denote the common $P^{(k)}$, so (7-7a) reduces to

$$
\mathbf{P}_n = \mathbf{a} P^n \qquad n \in I
\tag{7-7b}
$$

Unless otherwise noted, only homogeneous chains are considered from now on,† and the term *Markov chain* must be interpreted as homogeneous Markov chain.
From (7-4) and (7-7b) it is evident that for any stochastic matrix P there is a Markov chain having P as its transition matrix (Exercise 7-17). It is usual to refer

---

† Nonhomogeneous chains can be described as homogeneous chains, so no loss of generality is involved (see Exercise 7-6).

to "the states of P" as an abbreviation for "the states of the Markov chain with transition matrix P."

Digraph Representation

Another similarity with birth-and-death processes is the use of digraphs to portray the transition matrix. The nodes of the digraph represent the states; if $p_{ij} > 0$, a directed arc from node i to node j with label p_{ij} is drawn. For example, the digraph for

$$P = \begin{pmatrix} 1 & 0 & 0 & 0 & 0 \\ \frac{1}{2} & 0 & \frac{1}{2} & 0 & 0 \\ 0 & \frac{1}{2} & 0 & \frac{1}{2} & 0 \\ 0 & 0 & \frac{1}{2} & 0 & \frac{1}{2} \\ 0 & 0 & 0 & 0 & 1 \end{pmatrix}$$

is shown in Figure 7-1.

Digraphs are used when Markov chains are classified in Section 7-5.

Examples

Example 7-1: Gambler's ruin model Two players, A and B, start a coin-matching game with two fair coins each. A trial of the game consists of the players simultaneously flipping a coin. Player A wins if both coins match; otherwise, player B wins. All the trials are independent of one another. The game ends when one player has all four coins.

Let Y_n be the number of coins held by A after the nth trial and define, for $n = 1, 2, \ldots,$

$$X_n = \begin{cases} 1 & \text{if A wins on trial } n \\ 0 & \text{if game ends before trial } n \\ -1 & \text{if B wins on trial } n \end{cases}$$

Then $Y_0 = 2$,

$$Y_n = Y_{n-1} + X_n = X_1 + \cdots + X_n \qquad n = 1, 2, \ldots$$

and since the trials are independent,

$$P\{Y_{n+1} = j \mid Y_1 = i_1, \ldots, Y_n = i_n\}$$
$$= P\{X_{n+1} = j - i_n \mid X_1 = i_1, X_2 = i_2 - i_1, \ldots, X_n = i_n - i_{n-1}\}$$
$$= P\{X_{n+1} = j - i_n \mid Y_n = i_n\}$$
$$= P\{Y_{n+1} = j \mid Y_n = i_n\} \qquad n = 1, 2, \ldots$$

which verifies that the Markov property holds. For this process, the state

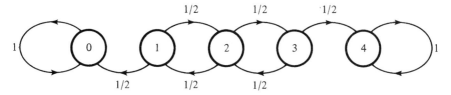

Figure 7-1 The digraph of P.

space is $\{0, 1, 2, 3, 4\}$ and $\mathbf{a} = (0, 0, 1, 0, 0)$. Player A wins a coin on a trial with probability $\frac{1}{2}$, and the game ends when Y_n equals 0 or 4, in which case $Y_{n+k} = Y_n$ for $k = 1, 2, \ldots$. Thus,

$$P\{Y_{n+1} = i + 1 \mid Y_n = i\} = P\{Y_{n+1} = i - 1 \mid Y_n = i\} = \frac{1}{2} \qquad i = 1, 2, 3$$

$$P\{Y_{n+1} = i \mid Y_n = i\} = 1 \qquad i = 0, 4$$

and all other transition probabilities are zero. Hence, the transition matrix is

$$P = \begin{pmatrix} 1 & 0 & 0 & 0 & 0 \\ \frac{1}{2} & 0 & \frac{1}{2} & 0 & 0 \\ 0 & \frac{1}{2} & 0 & \frac{1}{2} & 0 \\ 0 & 0 & \frac{1}{2} & 0 & \frac{1}{2} \\ 0 & 0 & 0 & 0 & 1 \end{pmatrix}$$

The digraph of this transition matrix is given in Figure 7-1.

Some obvious quantities one might want to compute are the probability that A wins all the coins (that is, B is ruined) and the distribution of the number of trials until the game ends.

The development of probability theory was strongly influenced by the analysis of gambling. This model is an example of a *gambler's ruin* process, and we study it in more detail in Section 7-8. A different setting for this process is provided by reexamining the model of data storage in Example 5-4. Assume that all record lengths contain the same number of characters. This might occur because a fixed number of bytes are reserved for each record or as an approximation when the mean record length is large compared to the fluctuations around the mean. Suppose a track contains room for four records and initially it contains two records. From time to time, transactions occur that either add or delete one record from the track. When the track becomes full or empty, it is removed from service, and no further changes in the content of the track occur. Adding a record corresponds to player A's winning a trial. We do not require that this occur with probability $\frac{1}{2}$; that is, we allow the coins to be unfair. Let Y_n denote the number of records on the track after the nth transaction. When successive transactions are mutually independent, then $\{Y_n; n \in I\}$ is described by a gambler's ruin process. □

Example 7-2: Machine maintenance In the examples of machine maintenance (Section 2-2, Examples 4-1 and 4-4) described in previous chapters, the machine has been characterized as working or failed. Many machines, such as an automobile or a television set, can be in a condition where they are working, but not well; a binary classification scheme may conceal useful information.

Suppose a machine is examined once a day and classified as working perfectly, has a minor defect, or has failed. Associate these classifications with states 0, 1, and 2, respectively; and let Y_n be the state of the machine on day n. Let w be the probability that a new machine works perfectly and $1 - w$ be the probability it has a minor defect; hence $\mathbf{a} = (w, 1 - w, 0)$. Assume that the number of consecutive days for which the machine works perfectly has the geometric distribution $u^{k-1}(1 - u)$, $k = 1, 2, \ldots$, and on a day it ceases to work perfectly it fails with probability g. Similarly, assume that the number of consecutive days for which a minor defect persists has the distribution $d^{k-1}(1 - d)$, $k = 1, 2, \ldots$, and that a minor defect is always followed by a failure. Last, assume that repairing the machine takes a random number of days, with distribution $\rho^{k-1}(1 - \rho)$, $k = 1, 2, \ldots$, and that the repair process produces a perfectly operating machine with probability ϕ and a machine with a minor defect with probability $1 - \phi$. Postulate that at each epoch when a new state is entered, the future behavior of the machine does not depend on the past behavior; e.g., repair times are independent of running times.

Since $\{Y_n; n \in I\}$ is a discrete-parameter process with state space $(0, 1, 2)$, it is a Markov chain if (7-1) can be verified. The key fact used in establishing (7-1) is the memoryless property of the geometric distribution. An example of the calculations is the following.

The time that a minor defect persists is independent of the states of the machine before that minor defect occurs. This means that when $Y_n = 1$, no matter how long the current minor defect has been present, the probability that it causes a failure at time $n + 1$ is $1 - d$. Thus

$$P\{Y_{n+1} = 2 \mid Y_n = 1, Y_{n-1} = i_{n-1}, \ldots, Y_0 = i_0\} = (1 - d)P\{Y_n = 1\}$$

which verifies (7-1) for $j = 2$ and $i_n = 1$ and shows $p_{12} = 1 - d$. Since transitions from state 1 to state 0 cannot occur, $p_{10} = 0$ and $p_{11} = 1 - p_{12} = d$. The remaining calculations are done similarly (see Exercise 7-2), and

$$P = \begin{pmatrix} u & (1-u)(1-g) & (1-u)g \\ 0 & d & 1-d \\ (1-\rho)\phi & (1-\rho)(1-\phi) & \rho \end{pmatrix}$$

The digraph for this transition matrix is

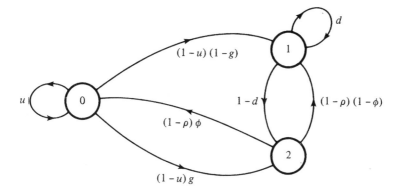

The types of results one might seek from this model are the probability that a perfectly operating machine is available on day 12, the long-run proportion of time the machine is working, and the distribution of the times between visits to the repair shop. □

Example 7-3: M/D/1 queue In the $M/D/1$ queue, customers arrive at a single server according to a Poisson process, and the service time for all customers is the same constant. Let λ be the arrival rate and v be the service time, and assume that the customers are served FIFO. Let $X(t)$ be the number of customers in the system at time t. The process $\{X(t); t \geq 0\}$ cannot be a Markov chain because it does not have a discrete-parameter set; more important, it is not a Markov process at all. Since the service times are constant, some departure epochs [downward jumps in $X(\cdot)$] after time t are completely determined by some arrival epochs that occur before time t. For example, by referring to Figure 7-2, it is clear that

$$P\{X(t_1 + v) = 0 \,|\, X(t) = 1, t_1 \leq t < t_1 + v; X(t) = 0, t_1 - \Delta \leq t < t_1\} = 1$$

and

$$P\{X(t_1 + v) = 0 \,|\, X(t) = 1, t_1 + \delta \leq t < t_1 + v; X(t) = 0, t < t_1 + \delta\} = 0$$

which violates the Markov property.

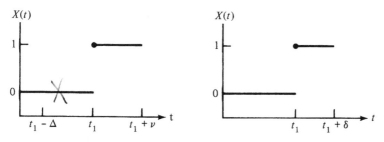

Figure 7-2 Explanation of why $X(t)$ is not Markov.

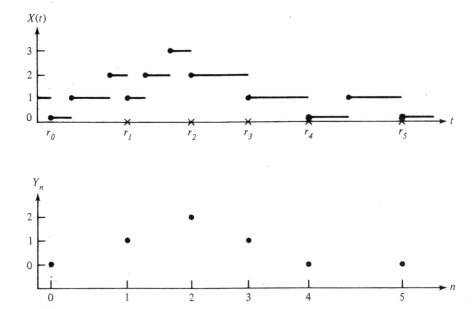

Figure 7-3 The relation between $X(t)$ and Y_n.

There is a way in which $\{X(t); \, t \geq 0\}$ can be partially described by a Markov chain, namely via an *embedded* chain. We choose a countable subset of the original parameter set, $[0, \infty)$ in this case, and consider only those random variables with indices in the subset. In particular, we choose those epochs when a service completion occurs. Let R_n be the epoch when the nth arriving customer departs (having finished service), and set $Y_n = X(R_n + 0)$; that is, Y_n is the number in the system just after customer n departs. We use the convention $R_0 \equiv 0$. The relation between $X(t)$ and Y_n is illustrated in Figure 7-3; observe that the index n refers to a customer, and not to an epoch, and that $nv \neq R_n$. Also note that the epochs R_n are random variables which are determined as the process evolves. The fact that a probabilistic mechanism is employed to determine the embedding points is of no consequence.

We claim that $\{Y_n \, ; \, n \in I\}$ is a Markov chain. It is a discrete-parameter process with a countably infinite state space, so to support the MC claim, (7-1) has to be confirmed. Let the random variable A_n denote the number of arrivals when customer $n + 1$ is in service, $n \in I$. In Figure 7-3, $A_0 = A_1 = 2$, $A_2 = A_3 = 0$, and $A_4 = 1$. If $Y_n > 0$, the number left behind by the $(n + 1)$th departure is the number left behind by the nth departure less 1 (since customer $n + 1$ was one of the customers left behind by customer n) plus the number of arrivals during the service time of customer $n + 1$. In symbols, $Y_{n+1} = Y_n - 1 + A_n$. If $Y_n = 0$, the customers left behind by customer $n + 1$ are precisely those customers who arrived while customer $n + 1$ was in service [look at $(R_0, R_1]$ and $(R_4, R_5]$ in Figure 7-3]; hence, $Y_{n+1} = A_n$. Thus,

$$Y_{n+1} = \max \{Y_n - 1, 0\} + A_n \qquad n \in I \qquad (7\text{-}8)$$

Note that given the value of Y_n, knowledge of $Y_{n-1}, Y_{n-2}, \ldots, Y_0$ would not alter our knowledge of Y_{n+1}. The memoryless property of the exponential interarrival times allows the conclusion that $A_1, A_2, \ldots$ are i.i.d. (see Example 5-9) and A_n is independent of $Y_n, Y_{n-1}, \ldots, Y_0$, so (7-8) verifies (7-1). Moreover,

$$c_k \triangleq P\{A_0 = k\} = \frac{e^{-\lambda v}(\lambda v)^k}{k!} \qquad k \in I$$

so the transition matrix for $\{Y_n ; n \in I\}$ is given by

$$P = \begin{pmatrix} c_0 & c_1 & c_2 & c_3 & \cdots \\ c_0 & c_1 & c_2 & c_3 & \cdots \\ 0 & c_0 & c_1 & c_2 & \cdots \\ 0 & 0 & c_0 & c_1 & \cdots \\ \cdots & \cdots & \cdots & \cdots & \cdots \end{pmatrix}$$

In Figure 7-3, $Y_0 \equiv 0$; this would be represented by $\mathbf{a} = (1, 0, 0, \ldots)$. This is not necessary, and any initial distribution could be assigned that reflects our knowledge about the number of customers in the system at $t = 0$ (which must correspond to a departure epoch).

From the embedded chain, $P\{Y_n = j\} = \mathbf{a}P^n$ by invoking (7-7b). In Section 7-5 it is shown that $\pi_j = \lim_{n \to \infty} P\{Y_n = j\}$ exists, and in Section 7-6 it is shown that π_j can be computed easily. Our original interest was in $X(t)$, the number in the system at t. In Exercise 9-12 (and for more general models with Poisson arrivals in Section 11-2), it is shown that

$$\lim_{n \to \infty} P\{Y_n = j\} = \lim_{t \to \infty} P\{X(t) = j\} \qquad j \in I$$

so the asymptotic probabilities for the process $\{X(t); t \geq 0\}$, which is not Markovian and does not have a discrete-parameter set, can be determined from the asymptotic probabilities of the embedded Markov chain $\{Y_n ; n \in I\}$.

The embedded chain also can be used to obtain waiting-time distributions. When customers are served FIFO, those customers in the system when customer n departs must be precisely those who arrived while customer n was in the system. Let the random variable W_n denote the waiting time (i.e., the time from arrival to departure) of customer n. Then

$$P\{Y_n = j \mid W_n = t\} = \frac{e^{-\lambda t}(\lambda t)^j}{j!}$$

and

$$P\{Y_n = j\} = \int_0^\infty P\{Y_n = j \mid W_n = t\} \, dP\{W_n \leq t\}$$

$$= \int_0^\infty \frac{e^{-\lambda t}(\lambda t)^j}{j!} \, dP\{W_n \leq t\} \qquad j \in I$$

Form the probability generating function (PGF) of $P\{Y_n = j\}$ by multiplying both sides by z^j and summing over $j \in I$:

$$\sum_{j=0}^{\infty} z^j P\{Y_n = j\} = \sum_{j=0}^{\infty} z^j \int_0^{\infty} \frac{e^{-\lambda t}(\lambda t)^j}{j!} \, dP\{W_n \le t\}$$

$$= \int_0^{\infty} e^{-\lambda t} \sum_{j=0}^{\infty} \frac{(z\lambda t)^j}{j!} \, dP\{W_n \le t\} \qquad (7\text{-}9)$$

$$= \int_0^{\infty} e^{-t\lambda(1-z)} \, dP\{W_n \le t\}$$

where Fubini's theorem justifies interchanging the integral and the summation. Thus, the LST of the waiting time W_n is expressed† in terms of the PGF of Y_n. $\qquad\square$

Example 7-4: (s, S) inventory model Section 2-6 and Examples 5-10, 6-3, and 6-8 describe the (s, S) inventory model and focus on those operating characteristics which are easily obtained from the renewal structure of the model. Other operating characteristics are obtained more easily by considering a certain Markov chain.

Let D_n be the (random) demand on day n, $n \in I$, and Y_n be the inventory on hand at the start of day n, $n \in I$. Assume orders are placed at the end of any day where $Y_n - D_n < s$ and are delivered by the start of day $n + 1$, resulting in $Y_{n+1} = S$. Assume that $\{D_n\}_0^{\infty}$ are i.i.d. with distribution function $F(\,\cdot\,)$, and let $f_j = P\{D_0 = j\}$, $j \in I$. The evolution of the process $\{Y_n\,;\ n \in I\}$ is governed by

$$Y_{n+1} = \begin{cases} Y_n - D_n & \text{if } Y_n - D_n \ge s \\ S & \text{if } Y_n - D_n < s \end{cases}$$

for each $n \in I$, and the state space is $\{s, s + 1, \ldots, S\}$; thus, $\{Y_n\,;\ n \in I\}$ is a Markov chain. Let $Q = S - s$; the transition matrix is given by

	S	$S-1$	$S-2$	$\ldots$	$s+1$	s
S	$f_0 + F^c(Q)$	f_1	f_2	$\ldots$	$f_{Q}-1$	f_Q
$S-1$	$F^c(Q-1)$	f_0	f_1	$\ldots$	f_{Q-2}	f_{Q-1}
$S-2$	$F^c(Q-2)$	0	f_0	$\ldots$	f_{Q-3}	f_{Q-2}
$\cdot$	$\cdot$	$\cdot$	$\cdot$		$\cdot$	$\cdot$
$\cdot$	$\cdot$	$\cdot$	$\cdot$		$\cdot$	$\cdot$
$s+1$	$F^c(1)$	0	0	$\ldots$	f_0	f_1
s	$F^c(0)$	0	0	$\ldots$	0	f_0

---

† Contrast this analysis with the analysis of the M/M/1 queue in Example 4-12, where W_n is expressed in terms of the number of customers in the system at the epoch of customer n's arrival.

where the labels of the row and column indices are displayed on the borders of the matrix for clarity.

This formulation of the (s, S) inventory model is convenient for obtaining such quantities as the distribution of the inventory level on day n; the distribution of the inventory level one week from today, given that there are i items on hand today; and the probability that no orders are placed during the next week, given that i items are currently on hand. □

Example 7-5: Personnel flows in a graded social system The concepts and theorems in the theory of Markov chains may be applicable in studying processes that are not stochastic processes at all. Consider a population of fixed size N whose members are divided into J distinct strata. For example, the population could be the faculty of a college that is neither expanding nor contracting, and the strata could be their academic rank (e.g., instructor $= 1$, assistant professor $= 2$, associate professor $= 3$, professor $= 4$). Suppose the population is observed from time to time (e.g., at the start of each academic year), and let $Y_n(j)$ be the number of people in stratum j at observation epoch $n, j \in \{1, 2, ..., J\}$ and $n \in I$. Assume that $\mathbf{a}^{(0)} \triangleq (Y_0(1), ..., Y_0(J))$ is specified and that at each observation epoch the proportion of the population that moves from stratum i to stratum j is p_{ij}. Set $P = (p_{ij})$ and $\mathbf{Y}_n = (Y_n(1), ..., Y_n(J))$; clearly P is a stochastic matrix. In the academic setting, we can interpret p_{ij} as the proportion of faculty of rank i who are promoted to rank j when $j > i$ and p_{ii} as the proportion in grade i who remain there. It seems appropriate to assume that demotions do not occur; when $j < i$, then p_{ij} is the proportion of those in grade i who leave the college and are replaced by someone placed in grade j (replacements can be included in the other two cases as well).

The dynamics of the movements among strata are given by

$$\mathbf{Y}_{n+1} = \mathbf{Y}_n P, \qquad n \in I$$

A simple induction shows

$$\mathbf{Y}_n = \mathbf{a}^{(0)} P^n \qquad n \in I \tag{7-10}$$

Set $p_n(j) = Y_n(j)/N$ and $a(j) = Y_0(j)/N$. The former is the proportion of the population in stratum j at epoch n, and the latter is the corresponding quantity at epoch 0. Form the vectors $\mathbf{P}_n$ and $\mathbf{a}$ in the obvious way; dividing both sides of (7-10) by N yields

$$\mathbf{P}_n = \mathbf{a} P^n \qquad n \in I$$

which is formally identical to (7-7b).

It is possible, but not mandatory, to interpret this model stochastically. One can assume that each member of the population moves from stratum to stratum according to a Markov chain with transition matrix P independently of the movements of the rest of the population. When the population size is large, the strong law of large numbers asserts that the proportion of people moving between strata i and j will be close to p_{ij}. A deterministic interpretation of the model is that a policy has been established which results in $p_{ij}N$ people moving from stratum i to stratum j at the start of each year; no chance events are involved.

A model such as this could be used to predict the distribution of the population among the strata. If changes in personnel practices can be described by changes in the transition matrix, then the effects of different personnel practices on the occupancy of each stratum can be analyzed. ☐

A discrete-parameter process that is not a Markov chain frequently may be redefined as a Markov chain by choosing another state space. The idea is to *expand the state space* so that enough information is conveyed by the "expanded states" to achieve the Markov property. This is illustrated below.

Example 7-6: Example 7-2 revisited Instead of geometrically distributed repair times, assume that all repair times require exactly two days. In the notation of Example 7-2, $\{Y_n ; n \in I\}$ is not a Markov chain because

$$P\{Y_5 = 2 \,|\, Y_4 = 2, \; Y_3 \neq 2\} = 1 \neq 0 = P\{Y_5 = 2 \,|\, Y_4 = 2, \; Y_3 = 2\}$$

Define the process $\{Y'_n; n \in I\}$ with state space $\{0, 1, 2, 3\}$, where states 0 and 1 are as before, $Y'_n = 2$ means a machine is in the first day of repair on day n, and $Y'_n = 3$ means the machine is in the second day of repair on day n. It is left for you to verify that $\{Y'_n; n \in I\}$ is a Markov chain with transition matrix

$$\begin{pmatrix} u & (1-u)(1-g) & (1-u)g & 0 \\ 0 & d & 1-d & 0 \\ 0 & 0 & 0 & 1 \\ \pi & 1-\pi & 0 & 0 \end{pmatrix}$$

☐

In Example 7-6, the addition of one state permitted a Markov chain representation of the underlying phenomenon. Things are not always that simple. If repair takes k days, $k - 1$ additional states are needed. When Y_n depends on $Y_{n-k}, \ldots, Y_{n-1}, \{Y'_n; n \in I\}$ with $Y'_n = (Y_{n-k}, \ldots, Y_n)$ is a Markov chain (prove this for yourself). If the state space for the Y-process has $m + 1$ elements, the state space for the Y'-process can have as many as $(m + 1)^k$ elements, and valid formulas such as (7-7b) may be intractable.

EXERCISES

7-1 Generalize Example 7-1 in these ways. Players A and B start with n_A and n_B coins, respectively, and player A has probability p of winning a trial. Write out the transition matrix.

7-2 Complete the demonstration that the process $\{Y_n ; n \in I\}$ in Example 7-2 is a Markov chain.

7-3 A machine has two key components, the power supply and the steering mechanism. The power supply is made of semiconductors and either works perfectly or does not work at all; the probability that it fails after k days of operation is $(1 - b)b^{k-1}$, $k \geq 1$. The steering mechanism is mechanical and goes out of adjustment after k days of operation with probability $(1 - a)a^{k-1}$, $k \geq 1$. Once the steering mechanism is out of adjustment, it will operate imperfectly for $k \geq 1$ days, with probability $(1 - d)d^{k-1}$, after which it fails. The machine functions if, and only if, the power supply and steering mechanism function. Make the same assumptions about the repair processes as in Example 7-2, and use the notation of that example. Show that the mechanism we have just described is represented by the MC given in Example 7-2 with $u = ab$ and $g = (1 - b)/(1 - ab)$.

7-4 The analysis in Example 7-3 is applicable to the M/G/1 queue. Show that for the M/G/1 queue with service-time distribution function $G(\ \cdot\ )$, (7-4) holds, $P\{A_0 = k\} = \int_0^\infty e^{-\lambda t}(\lambda t)^k/(k!) \, dG(t)$, and the transition matrix of the chain embedded at departure epochs has the same form as in Example 7-3. Display in detail the transition matrix for the M/M/1 queue. Explain why this technique does not apply to the M/G/c queue.

7-5 Let $\{X(t); t \geq 0\}$ be a birth-and-death process with birth rate λ_i and death rate μ_i in state i. Let S_n denote the epoch of the nth transition, and define $Y_n = X(S_n)$, $n \in I$, with $S_0 = 0$. Show that $\{Y_n ; n \in I\}$ is a Markov chain with state space I and transition matrix

$$\begin{pmatrix} 0 & 1 & 0 & 0 & 0 & \cdots \\ q_1 & 0 & p_1 & 0 & 0 & \cdots \\ 0 & q_2 & 0 & p_2 & 0 & \cdots \\ 0 & 0 & q_3 & 0 & p_3 & \cdots \\ \multicolumn{6}{c}{\dotfill} \end{pmatrix}$$

where $p_i = \lambda_i/(\lambda_i + \mu_i)$ and $q_i = \mu_i/(\lambda_i + \mu_i)$, $i \in I_+$.

7-6 Let $\{Y_n ; n \in I\}$ be a nonhomogeneous MC. Expand the state space to form $Y'_n = (Y_n, n)$. Show that $\{Y_n ; n \in I\}$ is a homogeneous MC.

7-7 Consider a renewal process where the times between renewals are discrete random variables with $f_i = P\{X_1 = i\}$, $i = 1, 2, \ldots, k$. Let S_n be the epoch of the nth renewal. Show that $\{S_n, n \in I\}$ is a Markov chain and display the transition matrix. Use (7-7b) to compute $P\{S_2 = j\}$.

7-8 Suppose the times between events in an alternating renewal process have geometric distributions. Formulate this as a two-state Markov chain. (The wording of this exercise is purposefully vague; part of your task is to describe precisely what process is a two-state Markov chain.) For small values of n, demonstrate that (7-7b) provides the discrete analog of (5-57).

7-9 Find an embedded Markov chain for the GI/M/1 queue which will enable you to obtain the waiting-time distribution of the nth arriving customer when the order of service is FIFO. Do the same for the GI/M/c queue.

7-10 Do Exercises 7-4 and 7-9 when no more than K customers can be held in the queue; customers who arrive when all K queueing positions are filled depart without affecting the system in any way.

7-11 This exercise introduces the concept of a *partially observable* Markov chain. A company owns a machine that is described by the model in Example 7-2. An executive of the company receives a daily report on the status of the machine, but the report states only whether the machine is working or being repaired. Let X_n be the information received by the executive on day n, where

$$X_n = \begin{cases} 0 & \text{if } Y_n = 0 \text{ or } 1 \\ 1 & \text{if } Y_n = 2 \end{cases}$$

Let P be the transition matrix of $\{Y_n ; n \in I\}$. Show that $\{X_n ; n \in I\}$ is a Markov chain when

$$P = \begin{pmatrix} p_{00} & p_{01} & \alpha \\ p_{10} & p_{11} & \alpha \\ p_{20} & p_{21} & p_{22} \end{pmatrix}$$

and is *not* a Markov chain when, e.g.,

$$P = \begin{pmatrix} 0 & 1 & 0 \\ 0 & 0 & 1 \\ 1 & 0 & 0 \end{pmatrix}$$

For the latter case, show how to expand the state space (only one more state is needed) so that the X-process is a Markov chain. Do not make the trivial choice of $Y_n \equiv X_n$.

7-12 In New Jersey, a voter is allowed to change party affiliation for primary elections only by abstaining from primary elections for one year. Suppose experience shows that a Democrat will abstain one-half the time and a Republican one-fourth the time. A voter who abstained for exactly one year joins the other party 75 percent of the time, rejoins the original party 10 percent of the time, and abstains again 15 percent of the time. A voter who has abstained two years in a row is equally likely to join either party. Formulate this process as a Markov chain. Do you believe that the Markov property is valid in this setting?

7-13 In California, voters can change parties every year regardless of their past behavior. Suppose that Democrats switch to being Republicans 25 percent of the time and turn Independent 10 percent of the time and that Republicans switch to being Democrats 20 percent of the time and turn Independent 20 percent of the time. Each year, half of the Independents join a party, and they are as likely to join the Democrats as the Republicans. If someone voted in the Republican primary in 1970, explain how to find, but *do not calculate*, the probability they will be an Independent in 1984.

7-14 Suppose a communications network transmits numbers in a binary system, that is, 0s and 1s. At each stage of the network, there is probability $q > 0$ that the number will be passed incorrectly to the next stage. Let Y_0 be the number entered into the system and Y_n be the number recorded at stage n. What conditions are needed for $\{Y_n; n \in I\}$ to be a Markov chain? What is $P\{Y_{10} = 1 \mid Y_0 = 1, Y_8 = 0\}$ when the Y-process is a Markov chain?

7-15 Let $D_1, D_2, \ldots$ be i.i.d. random variables representing demands on an inventory system. If a replenishment is ordered on day n, it will arrive on day $n + m$ with probability $q^{m-1}(1 - q)$, $m \in I_+$; all replenishment orders are for the amount $Q > 0$. Let J_n be the number of items on hand at the *end* of day n and O_n be the number of orders outstanding at the end of day n. The ordering policy is this: Place an order at the end of day n if $Y_n \triangleq J_n + QO_n \leq R$, where R is given. Formulate a Markov chain that is suitable for analyzing this process. Prove that the process you formulate is a Markov chain. How would your formulation change if the lead time were not geometrically distributed?

7-16 The results of a market survey of 1,000 consumers faced with a choice of three brands (A, B, and C) is summarized in this matrix:

$$\begin{array}{c} \quad \\ A \\ B \\ C \end{array} \begin{pmatrix} A & B & C \\ 125 & 75 & 50 \\ 75 & 100 & 75 \\ 50 & 100 & 350 \end{pmatrix}$$

The row labels indicate the brand purchased last period, and the column labels represent the brand purchased during the current period. Using these data, formulate a Markov chain model. What assumptions about brand switching do you need to make? Do you think the data presented are sufficient to formulate a good model? If not, what additional data would you want and how would you reformulate the model?

7-17 Let P be a stochastic matrix. Specify a Markov chain that has P as its transition matrix. (What does it take to specify a stochastic process? You might want to review Section 3-1.)

7-18 The $GI/G/1$ queue in discrete time is a rough model of a node in a data communications network. Messages may arrive at epochs $\Delta, 2\Delta, \ldots$; the times between arrivals are i.i.d. random variables and let a_k be the probability that successive messages arrive k units apart. Each message consists of a random number of "packets"; these random variables are i.i.d., and we let b_k be the probability that a message contains k packets. The service time of a packet is the constant Δ.

Let Y_n be the number of messages at the node at epoch $n\Delta$. Show that $\{Y_n\,;\;n\in I\}$ is a Markov chain when $a_k = p(1-p)^{k-1}$ and $b_k = q(1-q)^{k-1}$; $k\in I_+$, $0 < p,\ q < 1$. Since arrivals and service completions may occur simultaneously, the order of these events must be specified. Construct the transition matrix when (*a*) service completions occur first and (*b*) arrivals occur first.

7-19 (Continuation) Do Exercise 7-18 when $a_k = p(1-p)^k$ and $b_k = q(1-q)^k$, $k\in I$. What is the qualitative difference in the two functional forms? Which one is more appealing for the intended application?

7-20 (Continuation) Explain why $\{Y_n\,;\;n\in I\}$ is *not* a Markov chain when $\{a_k\}$ and $\{b_k\}$ are not geometric distributions. Explain, in detail, how to construct a Markov chain for this situation by expanding the state space to include the time since the last arrival and elapsed service time of the customer in service (if any).

7-2 ALGEBRAIC SOLUTION OF FINITE MARKOV CHAINS

Equation (7-7*b*) shows that the transient state probabilities of a Markov chain are dependent on P^n. This section describes an algebraic method for calculating P^n when the state space is finite.

Sylvester's Formula

Theorem 7-1: Sylvester's formula Let M be a square matrix of order $s + 1$ $(s < \infty)$. Let $\lambda_0, \lambda_1, \ldots, \lambda_s$ be the eigenvalues of M, and assume they are distinct. Then

$$M^n = \sum_{k=0}^{s} \lambda_k^n Z_k(M) \tag{7-11}$$

where the matrices $Z_k(M)$ are given by

$$Z_k(M) = \frac{\prod_{r\neq k}(M - \lambda_r I)}{\prod_{r\neq k}(\lambda_k - \lambda_r)} \qquad k = 0, 1, \ldots, s \tag{7-12}$$

and I is the identity matrix.

A proof of this form of Sylvester's formula can be found in Hildebrand (1965, section 1.22), and the modifications needed when the eigenvalues are not distinct may be found in Gantmacher (1959, vol. II, sec. 5). The hard work in applying Theorem 7-1 is to obtain the eigenvalues.

Examples

The following examples illustrate some of the different types of behavior that finite Markov chains exhibit.

Example 7-7 Take

$$P = \begin{pmatrix} \frac{1}{2} & \frac{1}{2} \\ \frac{1}{3} & \frac{2}{3} \end{pmatrix}$$

The eigenvalues are $\lambda_0 = 1$ and $\lambda_1 = \frac{1}{6}$. From (7-12),

$$Z_0(P) = \begin{pmatrix} \frac{2}{5} & \frac{3}{5} \\ \frac{2}{5} & \frac{3}{5} \end{pmatrix} \qquad \text{and} \qquad Z_1(P) = \begin{pmatrix} \frac{3}{5} & -\frac{3}{5} \\ -\frac{2}{5} & \frac{2}{5} \end{pmatrix}$$

Substitution into (7-11) yields

$$P^n = \begin{pmatrix} \frac{2}{5} & \frac{3}{5} \\ \frac{2}{5} & \frac{3}{5} \end{pmatrix} + (\tfrac{1}{6})^n \begin{pmatrix} \frac{3}{5} & -\frac{3}{5} \\ -\frac{2}{5} & \frac{2}{5} \end{pmatrix}$$

In particular, to five decimal places,

$$P^6 = \begin{pmatrix} 0.40013 & 0.59987 \\ 0.39991 & 0.60009 \end{pmatrix}$$

which is very close to $Z_0(P)$.

Observe that one eigenvalue is unity and the Z-matrix associated with it is a stochastic matrix with identical rows. The other eigenvalue is less than 1, and the associated Z-matrix has rows which sum to zero. This implies that† $\lim_{n \to \infty} P^n$ exists and is a stochastic matrix, $\lim_{n \to \infty} P\{Y_n = j \mid Y_0 = i\}$ is the same for $i = 0$ and $i = 1$, and the limit is approached geometrically (why?). □

Example 7-8 Take

$$P = \begin{pmatrix} 0 & \frac{1}{2} & \frac{1}{2} \\ \frac{1}{2} & 0 & \frac{1}{2} \\ 0 & 0 & 1 \end{pmatrix}$$

The eigenvalues are $\lambda_0 = 1$, $\lambda_1 = \frac{1}{2}$, $\lambda_2 = -\frac{1}{2}$. From (7-12),

$$Z_0 = \begin{pmatrix} 0 & 0 & 1 \\ 0 & 0 & 1 \\ 0 & 0 & 1 \end{pmatrix} \quad Z_1 = \begin{pmatrix} \frac{1}{2} & \frac{1}{2} & -1 \\ \frac{1}{2} & \frac{1}{2} & -1 \\ 0 & 0 & 0 \end{pmatrix} \quad Z_2 = \begin{pmatrix} \frac{1}{2} & -\frac{1}{2} & 0 \\ -\frac{1}{2} & \frac{1}{2} & 0 \\ 0 & 0 & 0 \end{pmatrix}$$

and substituting into (7-11) yields

$$P^n = \begin{pmatrix} (\tfrac{1}{2})^{n+1}[1 + (-1)^n] & (\tfrac{1}{2})^{n+1}[1 - (-1)^n] & 1 - (\tfrac{1}{2})^n \\ (\tfrac{1}{2})^{n+1}[1 - (-1)^n] & (\tfrac{1}{2})^{n+1}[1 - (-1)^n] & 1 - (\tfrac{1}{2})^n \\ 0 & 0 & 1 \end{pmatrix}$$

The properties observed in Example 7-7 are present in this example. Since $p_{22} = 1$ and $P\{Y_n = 2 \mid Y_0 = i\} > 0$ for some n, $P\{Y_n = 2 \mid Y_0 = i\} \to 1$ as $n \to \infty$ for $i = 0, 1, 2$. □

Example 7-9 Take

$$P = \begin{pmatrix} 0 & 1 \\ 1 & 0 \end{pmatrix}$$

† Let $A(1)$, $A(2)$, ... be $n \times m$ matrices. We define $\lim_{k \to \infty} A(k)$ as the matrix $A = (a_{ij})$ with $a_{ij} = \lim_{k \to \infty} a_{ij}(k)$, where $a_{ij}(k)$ is the (i, j)th element of $A(k)$.

so Y_n alternately takes on the values 0 and 1. The eigenvalues are ± 1, and Theorem 7-1 yields

$$P^n = \frac{1}{2}\begin{pmatrix} 1 + (-1)^n & 1 - (-1)^n \\ 1 - (-1)^n & 1 + (-1)^n \end{pmatrix}$$

The significant feature of this example is that P^n does not have a limit as $n \to \infty$, but $\lim_{n\to\infty} P^n/n$ does exist. $\qquad\square$

Example 7-10 Let P_1 and P_2 denote the transition matrices in Examples 7-7 and 7-8, respectively, and take

$$P = \left(\begin{array}{c:c} P_1 & 0 \\ \hdashline 0 & P_2 \end{array}\right)$$

The eigenvalues of P are the union of the eigenvalues of P_1 and P_2, namely, $\{1, \frac{1}{6}, 1, \frac{1}{2}, -\frac{1}{2}\}$. Since 1 appears twice, Theorem 7-1 does not apply. By direct multiplication, it is apparent that

$$P^n = \left(\begin{array}{c:c} P_1^n & 0 \\ \hdashline 0 & P_2^n \end{array}\right)$$

Observe that $\lim_{n\to\infty} P^n$ exists, but $\lim_{n\to\infty} P\{X_n = j \mid X_0 = i\}$ depends on i. $\qquad\square$

Analytical Results

Even with modern computers frequently it is difficult to obtain the eigenvalues of a stochastic matrix, so Theorem 7-1 cannot always be used to obtain $P\{X_n = j \mid X_0 = i\}$. The typical analytical response to this situation is to study limiting probabilities, and much of this chapter is devoted to asymptotic behavior. The limiting probabilities will serve as a useful surrogate for the probability law of Y_n when they are approached rapidly. The next theorem establishes the general validity of some of the observations in Examples 7-7 and 7-8.

Theorem 7-2 Let P be a stochastic matrix of finite order $s + 1$, with distinct eigenvalues. Then

(a) $\lambda_0 = 1$ is always an eigenvalue
(b) all the rows of $Z_0(P)$ sum to 1
(c) $|\lambda_k| \le 1$ for $k = 1, \ldots, s$
(d) the rows of $Z_k(P)$ sum to zero,† $k = 1, \ldots, s$
(e) if $|\lambda_k| < 1$ for $k = 1, \ldots, s$, then $P^\infty \triangleq \lim_{n\to\infty} P^n$ exists and $P^\infty = Z_0(P)$, which is stochastic
(f) and

$$|(P^n - P^\infty)_{ij}| \le |\lambda'|^n \sum_{k=1}^s |[Z_k(P)]_{ij}| \qquad (7\text{-}13)$$

where $\lambda' \triangleq \max\{\lambda_k : k = 1, \ldots, s\}$.

† Such matrices are called *differential*.

PROOF Since the rows of P sum to 1, $\det(P - I) = 0$, establishing (a). The largest possible magnitude for an eigenvalue of a finite nonnegative matrix is the maximum of the row sums [see, e.g., Gantmacher (1959, vol. II, p. 83)]. All the row sums of a stochastic matrix are 1, so (c) is established. Let $\mathbf{1}$ denote an $s + 1$-column vector consisting only of ones. Since P is stochastic, $P\mathbf{1} = \mathbf{1}$, or equivalently, $(P - \lambda_0 I)\mathbf{1} = \mathbf{0}$ (= the $s + 1$ vector of zeros). Since $\prod_{r \neq k} (P - \lambda_r I)$ is the same no matter how the terms are ordered, for $k \neq 0$ we can always choose $P - \lambda_0 I$ as the last term. The row sums of $Z_k(P)$ are

$$\left[\prod_{r \neq 1,k} (P - \lambda_r I) \right](P - \lambda_0 I)\mathbf{1} = \mathbf{0}$$

which yields (d). Since P^n is a stochastic matrix, $(d) \Rightarrow (b)$. From (7-11), (a), and the hypothesis of (e), $P^\infty \rightarrow Z_0(P)$. Since $(P^\infty)_{ij} = P\{\lim_{n \to \infty} X_n = j \mid X_0 = i\} \geq 0$, $(b) \Rightarrow Z_0(P)$ is stochastic. From (7-11), the definition of λ_1, and the (triangle) inequality $|a + b| \leq |a| + |b|$,

$$|(P^n - P^\infty)_{ij}| = \left| \left[\sum_{k=1}^{s} \lambda_k^n Z_k(P) \right]_{ij} \right|$$

$$\leq \sum_{k=1}^{s} |\lambda_k|^n |[Z_k(P)]_{ij}|$$

$$\leq |\lambda_1|^n \sum_{k=1}^{s} |[Z_k(P)]_{ij}| \qquad \square$$

An immediate consequence of (e) is that with the hypothesis that P has distinct eigenvalues, $Z_0(P)$ is a stochastic matrix when there is a unique eigenvalue of magnitude 1. Example 7-9 shows that the latter condition is not necessary. A proof without the condition can be constructed by using the algebraic representation of P^n given in Feller (1968, chap. 16) or Isaacson and Madsen (1976, chap. 4).

Generating Function of the State Vector

Theorem 7-1 is useful for obtaining $\mathbf{P}_n$ via (7-7). Markov chains with rewards are examined in Chapter 4 of Volume II. For those models it is necessary to know the generating function of $\mathbf{P}_n$, namely, $\hat{\mathbf{P}}(z) \triangleq \sum_{n=0}^{\infty} z^n \mathbf{P}_n$. The reason is this. Suppose a reward r_j is earned each time the Markov chain $\{X_n; n \in I\}$ enters state j, and assume rewards are discounted, with discount factor z. The expected discounted reward at stage n is $z^n \mathbf{P}_n \mathbf{r}$, where $\mathbf{r}$ is the transpose of $(r_0, \ldots, r_s)$. Hence, the sum of the expected discounted rewards for stages $0, 1, \ldots$ is $\hat{\mathbf{P}}(z)\mathbf{r}$.

To obtain $\hat{\mathbf{P}}(z)$, start with (7-7b), which implies

$$\mathbf{P}_n P = \mathbf{P}_{n+1} \qquad n \in I$$

Multiply both sides by z^n, $|z| \leq 1$, and sum over all $n \in I$, to obtain

$$\hat{\mathbf{P}}(z)P = \sum_{n=0}^{\infty} z^n \mathbf{P}_{n+1} = \frac{\hat{\mathbf{P}}(z) - \mathbf{P}_0}{z}$$

Hence

$$\hat{\mathbf{P}}(z)(I - zP) = \mathbf{P}_0 \tag{7-14}$$

It is routine to solve (7-14) when $(I - zP)^{-1}$ exists, a fact which follows from this theorem.

Theorem 7-3 Let D be a finite square matrix with the property that $\lim_{n \to \infty} D^n = 0$. Then $(I - D)^{-1}$ exists and is given by

$$(I - D)^{-1} = \sum_{n=0}^{\infty} D^n \tag{7-15}$$

PROOF By direct calculation,

$$(I - D)(I + D + \cdots + D^{n-1}) = I - D^n \tag{7-16}$$

and the right-side approaches I as $n \to \infty$ because $D^n \to 0$. By choosing n sufficiently large, $\det(I - D^n)$ can be made as close to $\det I = 1$ as desired; in particular, choose n large enough that $\det(I - D^n) \neq 0$. Taking determinants on both sides of (7-16) yields

$$\det(I - D) \det(I + D + \cdots + D^{n-1}) = \det(I - D^n) \neq 0$$

Hence $\det(I - D) \neq 0$, which implies $(I - D)^{-1}$ exists.

Multiplying both sides of (7-16) by $(I - D)^{-1}$ on the left gives

$$(I + D + \cdots + D^{n-1}) = (I - D)^{-1}(I - D^n)$$

Letting $n \to \infty$ on both sides yields (7-15). □

Choose $D = zP$ with $|z| < 1$. Then $D^n = z^n P^n \to 0$; hence $(I - zP)^{-1}$ exists, and the solution of (7-14) is

$$\hat{\mathbf{P}}(z) = \mathbf{P}_0 (I - zP)^{-1} = \mathbf{P}_0 \sum_{n=0}^{\infty} z^n P^n \tag{7-17}$$

EXERCISES

7-21 In Example 7-2, find P^n with these data: $f = \frac{1}{3}$, $\phi = \frac{4}{5}$, $u = \frac{7}{10}$, $d = \frac{2}{5}$, and $\rho = \frac{1}{2}$.

7-22 (Continuation) Show that $|\lambda_1| = |\lambda_2| = \sqrt{13}/100$. Calculate a few values of $|\lambda_1|^n$ to get a feeling for how fast the bound in (7-13) approaches zero.

7-23 (Continuation) Let $\pi_3 = \lim_{n \to \infty} p_{23}(n)$. Show that (to two significant figures) $p_{23}(2) - \pi_3 = 0.20$ and that the bound given by (7-13) is 0.24.

7-24 (Continuation) Suppose each period is a day and $r_0 = \$10/\text{day}$, $r_1 = \$5/\text{day}$, and $r_2 = $

−\$2/day. Using an annual interest rate of 7.3 percent, calculate the expected discounted reward over an infinite planning horizon.

7-25 Find the formula for P^n when

$$P = \begin{pmatrix} a & 1-a \\ 1-b & b \end{pmatrix}$$

for $0 \le a, b \le 1$.

7-3 FIRST-PASSAGE TIMES

In the context of Example 7-2, an interesting random variable is the number of days that a perfectly working machine lasts before requiring repair. From the following definition you will see that this can be described as a first-passage time between states 0 and 2. Another reason for studying first-passage times is their strong connection with the n-step transition probabilities and the existence and computation of limiting probabilities.

> **Definition 7-2** Let i and j be two states (possibly $i = j$) of the Markov chain $\{Y_n ; n \in I\}$. Suppose that for a given n, $Y_n = i$. Then
>
> $$T_{ij} \triangleq \inf \{k : k \in I_+, \, Y_{n+k} = j\}$$
>
> is a *first-passage time* from state i to state j. And T_{ii} is called a *first return* to state i.

Some subtleties are involved when $i = j$ because the process does not have to change values during a first return. Figure 7-4 illustrates various first passages for a partial realization of a chain with three states. In the figure, the T_{ij} labels are associated with intervals; for example, $[2, 4]$ is called T_{10}. According to the definition, T_{10} is a (random) number, but often it is convenient to think of that

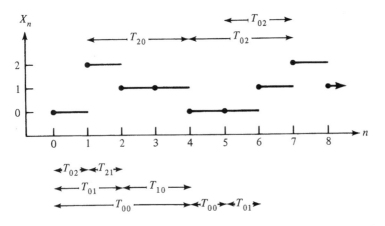

Figure 7-4 Some first-passage times.

number as the length of an interval during which a first passage from state 1 to state 0 is "taking place." Notice that various intervals are given the same label; for example, $[0, 1]$ and $[4, 7]$ are both labeled T_{02}. For this reason, T_{ij} should be called a *generic* first-passage time; it is a symbol we attach to any (and all) first-passage times from state i to state j. Since the definition does not depend on the sample path before state i is entered, both $[4, 7]$ and $[5, 7]$ deserve the label T_{02}. Since inf $\{k : k > 0, Y_{4+k} = 0\} = 1$, the first return to state 0 after epoch 4 occurs at epoch 5, and $[4, 6]$ is *not* a first-passage interval from state 0 to itself.

During the evolution of a Markov chain there may be many first passages from state i to state j; let $T_{ij}^{(k)}$ denote the length of the kth one to start. Since the Markov property implies that given the current state, the future evolution of a Markov chain is independent of its past, for each pair (i, j), $T_{i,j}^{(1)}$, $T_{i,j}^{(2)}$, ... are identically distributed random variables. Furthermore, for each state j, $T_{jj}^{(1)}$, $T_{jj}^{(2)}$, ... are mutually independent, and when $Y_0 = i \neq j$, $T_{ij}^{(1)}$, $T_{jj}^{(1)}$, $T_{jj}^{(2)}$, ... are mutually independent. This observation justifies using T_{ij} as a generic first-passage time and allows us to define properly for any $k \in I_+$,

$$g_{ij}(n) \triangleq P\{T_{ij} = n\} \qquad i, j \in S, \quad n \in I_+$$

We call $g_{ij}(\cdot)$ the *first-passage-time probability function*. Notice that $\{T_{jj}^{(k)}\}$ are the kind of random variables that induce renewal processes; this observation will be of more than passing interest.

Proposition 7-1 For all pairs of states,

$$g_{ij}(1) = p_{ij}, \qquad g_{ij}(n) = \sum_{k \neq j} p_{ik} g_{kj}(n-1) \qquad n = 2, 3, \ldots \qquad (7\text{-}18)$$

By *convention*, set $g_{ij}(0) = 0$.

PROOF Notice $\{T_{ij} = 1\}$ if, and only if, a transition from i to j is made in one step. For $n \geq 2$, $\{T_{ij} = n\}$ if, and only if, after state i is entered, the first transition is to some state $k \neq j$ and then a first passage from k to j is executed in $n - 1$ steps.   □

In Example 7-1 (the gambler's ruin model), a transition from state 0 to state 1 in n steps is impossible for any n. In such situations, set $T_{ij} = \infty$.

Let $\mu_{ij} = E(T_{ij})$. Suppose $Y_0 = i$; conditioning on the first transition gives

$$E(T_{ij} \mid Y_1 = k) = \begin{cases} 1 & \text{if } k = j \\ 1 + \mu_{kj} & \text{if } k \neq j \end{cases}$$

from which we get

$$\mu_{ij} = 1 + \sum_{k \neq j} p_{ik} \mu_{kj} \qquad (7\text{-}19)$$

for all $i, j \in S$. In (7-19) it is understood that the sum includes only those k for which $p_{ik} > 0$, and if there are no such indices, the sum is assigned the value 0.

An important feature of (7-19) is that it demonstrates that typically, in order to obtain a particular mean first-passage time, many have to be computed simultaneously. In Section 7-7, sufficient conditions for (7-19) to have a unique finite solution and matrix methods for solving (7-19) when the number of states is finite are described.

Example 7-11: Example 7-1 continued The special structure of the transition matrix of the gambler's ruin model in Example 7-1 allows (7-18) to be solved easily. Clearly

$$g_{00}(1) = g_{44}(1) = 1 \qquad g_{00}(n) = g_{44}(n) = 0 \qquad \text{for } n \geq 2$$

Suppose we are particularly interested in T_{24}. Since $p_{24} = 0$, $g_{24}(1) = 0$. From (7-18),

$$g_{24}(n) = p_{23}g_{34}(n-1) + p_{21}g_{14}(n-1)$$
$$= \tfrac{1}{2}[g_{34}(n-1) + g_{14}(n-1)] \qquad n \geq 2$$

$$g_{34}(1) = p_{34} = \tfrac{1}{2}$$

$$g_{34}(n) = \tfrac{1}{2}g_{24}(n-1) \qquad\qquad n \geq 2$$

$$g_{14}(n) = 0 \qquad\qquad n \leq 2$$

$$g_{14}(n) = \tfrac{1}{2}g_{24}(n-1) \qquad\qquad n \geq 3$$

Substituting the expressions for $g_{34}(\cdot)$ and $g_{14}(\cdot)$ into the representation of $g_{24}(\cdot)$ yields

$$g_{24}(2) = \tfrac{1}{4}$$

and

$$g_{24}(2+n) = \tfrac{1}{2}\left[\tfrac{1}{2}g_{24}(n) + \tfrac{1}{2}g_{24}(n)\right]$$
$$= \frac{g_{24}(n)}{2} \qquad n = 1, 2, \ldots$$

By iteration, it is readily established that

$$g_{24}(2n+1) = 0 \qquad g_{24}(2n) = \frac{1}{2^{n+1}} \quad n = 1, 2, \ldots$$

Observe that

$$P\{T_{24} < \infty\} = \sum_{n=1}^{\infty} g_{24}(n) = \tfrac{1}{2}$$

which implies $\mu_{24} = \infty$. This result is reasonable because the symmetry of P indicates that with probability $\tfrac{1}{2}$, state 0 will be reached (gambler's ruin) before state 4, so $P\{T_{24} = \infty\} = \tfrac{1}{2}$. $\qquad\square$

EXERCISES

7-26 In Example 7-1, find $g_{11}(\,\cdot\,)$, $g_{22}(\,\cdot\,)$, and $g_{33}(\,\cdot\,)$.

7-27 For any Markov chain, prove that if $g_{ii}(n) > 0$ implies $n \in \{\xi, 2\xi, \dots\}$ for some integer ξ, then $p_{ii}(n) > 0$ implies $n \in \{\xi, 2\xi, \dots\}$.

7-28 For the embedded MC of the M/D/1 queue (Exercise 7-4), argue that $\mu_{i0} = i\mu_{10}$. Show that when $\rho < 1$, $\mu_{10} = 1/(1 - \rho)$, where

$$\rho \triangleq \sum_{n=0}^{\infty} n \int_{0}^{\infty} \frac{e^{-\lambda t}(\lambda t)^n}{n!} \, dG(t) = \lambda v_G$$

7-4 EMBEDDED RENEWAL PROCESSES IN MARKOV CHAINS

We want to show how certain transitions of a Markov chain can be isolated to form a renewal process. Let $Y_0 = i$ be the initial state of the Markov chain $\{Y_n;\ n \in I\}$, and pick some state $j \in S$ (possibly $j = i$). Let X_k be the time between the kth and $(k + 1)$th visits to state j, $k \in I_+$, and $N_{ij}(n)$ be the number of visits to state j up to and including the nth transition, $n \in I_+$. Then if $i = j$,

$$X_k = T_{jj}^{(k)} \qquad k \in I_+$$

and $\{N_{jj}(n);\ n \in I_+\}$ is an ordinary renewal counting process where the time between renewals has probability function $g_{jj}(\,\cdot\,)$. If $j \neq i$, then

$$X_1 = T_{ij}^{(1)} \qquad \text{and} \qquad X_{k+1} = T_{jj}^{(k)} \qquad k \in I_+$$

Thus, $\{N_{ij}(n);\ n \in I_+\}$ is a delayed renewal counting process in which the time until the first renewal has the probability function $g_{ij}(\,\cdot\,)$ and the time between the succeeding renewals has probability function $g_{jj}(\,\cdot\,)$.

We call a visit to state j a *recurrent event* because the embedded renewal process statistically restarts itself each time state j is entered; i.e., each entry into state j is a regeneration epoch. The results and insights of renewal theory can be used in the analysis of Markov chains via this embedded renewal process.

Renewal Functions

In renewal theory, a central role is played by the mean number of renewals. In the context of a Markov chain, the renewal function is $M_{ij}(n) \triangleq E[N_{ij}(n)]$.

Define, for $Y_0 = i$, $k \in I_+$, and some fixed state j,

$$Z_{ij}(k) = \begin{cases} 1 & \text{if } Y_k = j \\ 0 & \text{if } Y_k \neq j \end{cases} \tag{7-20}$$

Then

$$M_{ij}(n) = \sum_{k=1}^{n} E[Z_{ij}(k)] = \sum_{k=1}^{n} p_{ij}(k) \tag{7-21}$$

Define $m_{ij}(0) = 0$ and

$$m_{ij}(n) = M_{ij}(n) - M_{ij}(n-1) \qquad n \in I_+$$

This is the discrete renewal density (see Exercise 5-36). From (7-21),

$$m_{ij}(n) = p_{ij}(n) \qquad n \in I_+ \tag{7-22}$$

so $m_{ij}(n)$ is the probability of a renewal at epoch n.

Relationship between Transition Probabilities and First-Passage Times

The following result relates n-step transition probabilities to first-passage times. It leads to limit theorems and stresses the importance of renewal theoretic methods in analyzing Markov chains.

Proposition 7-2 For any pair of states i and j (possibly $i = j$),

$$p_{ij}(n) = g_{ij}(n) + \sum_{k=1}^{n} g_{ij}(n-k)p_{jj}(k) \qquad n \in I \tag{7-23}$$

PROOF Use the renewal argument to write

$$m_{ij}(n \mid T_{ij}^{(1)} = k) = \begin{cases} 0 & \text{if } k > n \\ 1 & \text{if } k = n \\ m_{jj}(n-k) & \text{if } k < n \end{cases}$$

Unconditioning on $T_{ij}^{(1)}$ and using (7-22) yield (7-23). $\qquad \qquad \square$

What (7-23) asserts is that the only way the MC can be in state j at stage n is by first entering state j at stage $k \leq n$ and then returning to state j (not necessarily for the first time) in the remaining $n - k$ stages.

Since the random variables $\{T_{jj}^{(k)}\}_{k=1}^{\infty}$ are integer-valued, they are arithmetic. Their span is the largest integer ξ such that $\{n: g_{jj}(n) > 0\} \subset \{\xi, 2\xi, \ldots\}$.

Proposition 7-3 For any pair of states i and j (possibly $i = j$), if the span of T_{jj} is 1, then

$$\lim_{n \to \infty} p_{ij}(n) = \frac{P\{T_{ij} < \infty\}}{\mu_{jj}} \tag{7-24a}$$

If the span of T_{jj} is $\xi > 1$, then

$$\lim_{n \to \infty} p_{ij}(n_1 + n\xi) = \frac{P\{T_{ij} < \infty\}}{\mu_{jj}} \tag{7-24b}$$

where n_1 is any element of $\{n: g_{ij}(n) > 0\}$.

PROOF Assume the span is 1. For $i \neq j$, write (7-23) as

$$p_{ij}(n) = \sum_{k=0}^{n} g_{ij}(n - k) p_{jj}(k)$$

$$= \sum_{k=0}^{n} g_{ij}(n - k) m_{jj}(k) \qquad (7\text{-}25)$$

Since $g_{ij}(\cdot)$ is a step function, it is automatically directly Riemann-integrable, and applying the key renewal theorem to (7-25) yields

$$\lim_{n \to \infty} p_{ij}(n) = \frac{1}{\mu_{jj}} \sum_{k=0}^{\infty} g_{ij}(k)$$

which is (7-24a). For $i = j$, use (7-22) and Blackwell's renewal theorem to write

$$\lim_{n \to \infty} p_{ii}(n) = \lim_{n \to \infty} m_{ii}(n) = \frac{1}{\mu_{ii}} \qquad (7\text{-}26)$$

Since $\mu_{ii} < \infty \Rightarrow P\{T_{ii} < \infty\} = 1$ and $P\{T_{ii} < \infty\} < 1 \Rightarrow u_{ii} = \infty$, (7-26) $\Rightarrow$ (7-24a) when $i = j$. Modifications when the span is greater than 1 are straightforward (see Exercise 7-30). $\qquad \square$

EXERCISES

7-29 For $\mu_{jj} < \infty$, derive (7-24a) from Theorem 6-7. What additional statement is obtained by this method?

7-30 Complete the proof of Proposition 7-3 by making the necessary modifications when the span is greater than 1.

7-31 Let $\{Y_n; n \in I\}$ be a Markov chain with $Y_0 = k$. Define the random variables $X_n, n \in I_+$, by

$$X_n = \begin{cases} 1 & \text{if } Y_n = k \\ 0 & \text{if } Y_n \neq k \end{cases}$$

and assume that μ_{kk} and $\sigma_{kk}^2 = \text{Var}\,(T_{kk})$ are finite. Show that after a large number of transitions N, $[\sum_{n=0}^{N} (X_n - N/\mu_{kk})]/\sigma_{kk}$ is approximately normally distributed with mean 0 and variance 1.

7-5 CLASSIFICATION OF MARKOV CHAINS

Proposition 7-3 shows that $\lim_{n \to \infty} p_{ij}(n)$ exists, with an obvious modification when the span of T_{jj} is larger than 1. Examples 7-7 to 7-10 show that for some transition matrices, the limiting probabilities are independent of the initial state, and for others they are not. In this section, Markov chains are classified in a manner that describes their sample-path behavior and limiting probabilities. As you might conjecture from Proposition 7-3, first-passage time properties are important in the classification scheme.

Classification of States

Each state of the Markov chain $\{Y_n; n \in I\}$ is classified by its first-return-time behavior.

Definition 7-3 State j is called *persistent* if $P\{T_{jj} < \infty\} = 1$; otherwise it is called *transient*. It is called *null* if $E(T_{jj}) = \infty$ and *nonnull* if $E(T_{jj}) < \infty$. A persistent nonnull state is called *ergodic*.† State j is called *periodic* if the span of T_{jj} is greater than 1; otherwise, it is called *aperiodic*.

Clearly, all transient states are null.

Proposition 7-4 below shows that the bifurcation into periodic and aperiodic states is not fundamental, but is a way of describing a technical detail. Statements about aperiodic states have an obvious modification for periodic states.

From the definition and Proposition 7-3, we obtain (Exercise 7-33) the following result.

Proposition 7-4 If state j is null, then for all $i \in \mathcal{S}$,

$$\lim_{n \to \infty} p_{ij}(n) = 0 \qquad (7\text{-}27a)$$

If state j is ergodic, then

$$\lim_{n \to \infty} p_{jj}(n) = \frac{1}{E(T_{jj})} > 0 \qquad (7\text{-}27b)$$

If state j is periodic, (7-27b) holds only when n is a multiple of the span of T_{jj}.

Example 7-12: Example 7-1 continued The transition matrix for this example is

$$P = \begin{pmatrix} 1 & 0 & 0 & 0 & 0 \\ \frac{1}{2} & 0 & \frac{1}{2} & 0 & 0 \\ 0 & \frac{1}{2} & 0 & \frac{1}{2} & 0 \\ 0 & 0 & \frac{1}{2} & 0 & \frac{1}{2} \\ 0 & 0 & 0 & 0 & 1 \end{pmatrix}$$

Since $p_{00} = p_{44} = 1$, $E(T_{00}) = E(T_{44}) = 1$, so states 0 and 4 are aperiodic, persistent, and nonnull. Transitions out of states 0 and 4 are impossible, so $P\{T_{0j} < \infty\} = 0$ for $j \neq 0$ and $P\{T_{4j} < \infty\} = 0$ for $j \neq 4$. By conditioning on the first transition,

$$P\{T_{11} < \infty\} = p_{10} P\{T_{01} < \infty\} + p_{12} P\{T_{21} < \infty\}$$

$$= p_{12} P\{T_{21} < \infty\} \leq p_{12} = \tfrac{1}{2}$$

$$P\{T_{33} < \infty\} = p_{34} P\{T_{43} < \infty\} + p_{32} P\{T_{23} < \infty\}$$

$$= p_{32} P\{T_{23} < \infty\} \leq p_{23} = \tfrac{1}{2}$$

† Some authors restrict the use of the term *ergodic* to aperiodic, persistent, nonnull states.

so states 1 and 3 are transient. Similar reasoning (Exercise 7-34) establishes also that state 2 is transient.

From Exercise 7-26, states 1, 2, and 3 are periodic with period 2. □

Let $Z_{ij}(k)$ be defined by (7-20), and set

$$N_j = \sum_{k=1}^{\infty} Z_{jj}(k)$$

which denotes the number of returns made to state j. From the Markov property, when $Y_0 \neq j$, N_j is distributed as the number of returns to state j subsequent to the first entry into state j. The following theorem shows that persistent and transient states are fundamentally different; the latter are visited only finitely many times (hence there is a last visit), while the former are visited repeatedly.

Theorem 7-4 (a) State j is transient if, and only if, $P\{N_j < \infty\} = 1$. (b) Let $\theta_j = P\{T_{jj} < \infty\}$; then

$$E(N_j) = \begin{cases} \dfrac{\theta_j}{1 - \theta_j} & \text{if } \theta_j < 1 \\ \infty & \text{if } \theta_j = 1 \end{cases}$$

PROOF At each visit to state j, θ_j is the probability that there will be another visit; hence

$$P\{N_j = k\} = \theta_j^k (1 - \theta_j) \qquad k \in I \tag{7-28}$$

By definition, state j is transient if $\theta_j < 1$, and (7-28) implies (a). Part (b) is an immediate consequence of (7-28). □

Corollary 7-4 In a Markov chain with a finite number of states, at least one state is ergodic.

PROOF Number the states $0, 1, \ldots, s$, and assume they are all transient. By Theorem 7-4, $P\{\sum_{j=0}^{s} N_j < \infty\} = 1$. After some finite number of transitions, n^* say, $P\{Y_n \in S\} = 0$ for all $n > n^*$, which is absurd. Thus at least one state is persistent. If all the persistent states were null, Proposition 7-4 would yield $\lim_{n \to \infty} P\{Y_n \in S\} = 0$, which is false, so at least one of the persistent states is not null. □

Communicating Classes

In a chain with many states, it would be a very time-consuming task to classify all the states if they had to be classified one at a time. The following concepts lead to theorems that allow sets of states to be classified.

Definition 7-4 State j is *reachable* from state i, written $i \to j$, if $P\{T_{ij} < \infty\} > 0$. If $i \to j$ and $j \to i$, we write $i \leftrightarrow j$ and say that states i and j *communicate*.

Thus $i \rightarrow j$ when there are states $i = i_0, i_1, \ldots, i_{n-1}, i_n = j$ $(n < \infty)$ such that $p_{i_0 i_1} p_{i_1 i_2} \cdots p_{i_{n-1} j} > 0$. In terms of the digraph representation of P, $i \rightarrow j$ if, and only if, there is a finite path from i to j and $i \leftrightarrow j$ if, and only if, there is a finite path from i to j and a finite path from j to i.

Definition 7-5 A nonempty set of states C is called a *communicating class* if (a) $i \leftrightarrow j$ for all $i, j \in C$ and (b) if $i \in C$ and $i \rightarrow j$, then $j \in C$.

A communicating class is a set of states that all communicate with one another and is maximal in the sense that no state which communicates with any member of the set is excluded. In terms of the digraph of P, a communicating class is a set of nodes which has no arcs leaving it (there may be arcs entering it).

Two types of communicating classes deserve distinction.

Definition 7-6 If a communicating class consists of a single state, that state is called *absorbing*. If the only communicating class of the stochastic matrix P is the set of all states, P is *irreducible*. The negation of irreducible is *reducible*.

State i is absorbing if and only if $p_{ii} = 1$; once an absorbing state is entered, it is never left. The set of all states satisfies Definition 7-5b, so P is irreducible if, and only if, every state communicates with every other state.

Let C be a communicating class and choose any $i \in C$. If $p_{ij} > 0$, then $i \rightarrow j$ and $j \in C$ by part (b) of Definition 7-5. Hence

$$\sum_{j \in C} p_{ij} = 1 \qquad i \in C \tag{7-29}$$

so once a chain enters a communicating class, it cannot leave it. We express this property by saying that C is *closed*. Equation (7-29) has the further implication that the matrix P_C formed from P by deleting all rows and columns not in C is a stochastic matrix, so it describes the transitions of some irreducible Markov chain.

Example 7-13 (a) In Example 7-1, states 0 and 4 are absorbing since $P_{00} = P_{44} = 1$. The remaining states do not form a communicating class because for $i \in \{1, 2, 3\}$, $i \rightarrow 0$ and $i \rightarrow 4$, but i cannot be reached from state 0 or 4. Observe that $\{1, 2, 3\}$ satisfies the first condition of Definition 7-6, but not the second.

(b) In Example 7-2,

$$P = \begin{pmatrix} u & (1-u)(1-g) & (1-u)g \\ 0 & d & 1-d \\ (1-\rho)\phi & (1-\rho)(1-\phi) & \rho \end{pmatrix}$$

Assume $u, g, d, \rho < 1$ and $\phi > 0$. Then $p_{01} > 0$, $p_{12} > 0$, and $p_{31} > 0$, so P is easily seen to be irreducible.

(c) In Example 7-3, $p_{0j} = c_j > 0$ for all j, and $p_{i,\,i-1} = c_0 > 0$. For any i, $p_{i0} \geq p_{i,\,i-1} p_{i-1,\,i-2} \cdots p_{10} = (c_0)^i > 0$; hence $p_{ij} \geq c_j (c_0)^i > 0$, and P is irreducible.

(d) In Example 7-10,

$$P = \begin{pmatrix} \frac{1}{2} & \frac{1}{2} & 0 & 0 & 0 \\ \frac{1}{3} & \frac{2}{3} & 0 & 0 & 0 \\ 0 & 0 & 0 & \frac{1}{2} & \frac{1}{2} \\ 0 & 0 & \frac{1}{2} & 0 & \frac{1}{2} \\ 0 & 0 & 0 & 0 & 1 \end{pmatrix}$$

There are two communicating classes, $\{0, 1\}$ and $\{2, 3, 4\}$. □

Class Properties

The most important feature of a communicating class is that *all its members have the same classification*. This is established in the next three theorems.

Theorem 7-5 Let C be a communicating class. Either (a) all states in C are aperiodic or (b) all states in C are periodic with the same period.

PROOF This theorem is trivial for an absorbing state, so assume C contains at least two states. Choose $i \in C$, and let ξ be the period of i (possibly $\xi = 1$). Now choose $j \in C, j \neq i$. Since C is a communicating class, $i \leftrightarrow j$, so there are positive integers n_1 and n_2 such that

$$g_{ij}(n_1) > 0 \quad \text{and} \quad g_{ji}(n_2) > 0 \tag{7-30}$$

Let n' and n'' be the smallest values of n_1 and n_2, respectively, that satisfy (7-30), and let $\alpha = g_{ij}(n')g_{ji}(n'')$. By construction, $\alpha > 0$. Since a sequence of transitions going from i back to itself can pass through state j, but need not do so, we have

$$p_{ii}(n + n' + n'') \geq g_{ij}(n')p_{jj}(n)g_{ji}(n'') = \alpha p_{jj}(n) \tag{7-31}$$

Similarly,

$$p_{ii}(n' + n'') \geq g_{ij}(n')g_{ji}(n'') = \alpha > 0 \tag{7-32}$$

From the result of Exercise 7-27, $P_{ii}(n) > 0$ only when n is an integral multiple of ξ; (7-32) shows that $n' + n''$ is an integral multiple of ξ. Using this fact in conjunction with (7-31) shows that $p_{jj}(n) = 0$ unless n is an integral multiple of ξ. By reversing the roles of i and j in (7-31) and (7-32), it can be seen that $p_{jj}(n) > 0$ for some integral multiple of ξ. Appealing once more to Exercise 7-27 completes the proof. □

The next theorem shows that only persistent states can be reached from a persistent state.

Theorem 7-6 If state i is persistent and $i \to j$, then

(a) $j \to i$

and

(b) j is persistent.

PROOF Observe that $i \to j$ implies the existence of a positive integer n' such that $g_{ij}(n') > 0$. A first return from i back to i may go through state j since $i \to j$, but it may not. Define $\theta_{ji} = P\{T_{ji} < \infty\}$; then

$$P\{T_{ii} = \infty\} = 1 - \theta_{ii} \ge g_{ij}(n')(1 - \theta_{ji}) \ge 0$$

Since i is persistent, $\theta_{ii} = 1$, and by the definition of n', $g_{ij}(n') > 0$. Thus, $\theta_{ji} = 1$, which implies $j \to i$.

To prove (b), observe that (a) implies the existence of positive integers n_1 and n_2 such that

$$p_{ij}(n_1) > 0 \quad \text{and} \quad p_{ji}(n_2) > 0 \tag{7-33}$$

A path of transitions from j back to itself may go through state i, but it need not. Hence, for any integer $n \ge 0$,

$$p_{jj}(n + n_1 + n_2) \ge p_{ji}(n_2)p_{ii}(n)p_{ij}(n_1) \tag{7-34}$$

Summing over n, we obtain

$$\sum_{n=0}^{\infty} p_{jj}(n) \ge \sum_{n=0}^{\infty} p_{jj}(n + n_1 + n_2) \ge p_{ij}(n_1)p_{ji}(n_2) \sum_{n=0}^{\infty} p_{ii}(n)$$

Equations (7-32) and (7-33) imply $\sum_{n=0}^{\infty} p_{jj}(n) = \infty$. $E(N_j) = \sum_{n=0}^{\infty} p_{jj}(n)$, hence an appeal to Theorem 7-4 completes the proof. $\square$

This is the most important classification result.

Theorem 7-7 Let C be a communicating class. Either (a) all states in C are transient, or (b) all states in C are null-persistent, or (c) all states in C are ergodic.

PROOF If C is an absorbing state, (c) holds trivially. Assume C has at least two states; for any distinct pair of states $i, j \in C$, $i \to j$ and $j \to i$. This means that there exist positive integers n_1 and n_2 such that

$$p_{ij}(n_1) > 0 \quad \text{and} \quad p_{ji}(n_2) > 0 \tag{7-35}$$

Let n' and n'' be the smallest values of n_1 and n_2, respectively, that satisfy (7-35), and let

$$\alpha = p_{ij}(n')p_{ji}(n'')$$

By virtue of these definitions, $\alpha > 0$.

If i is persistent, then Theorem 7-6 asserts that j is persistent, and vice versa; this is logically equivalent to the statement that if i is transient, then j is transient, and vice versa, which proves (a).

To prove (b), observe that if j is null-persistent, then Theorem 7-6 implies that i is persistent. Since $i \leftrightarrow i$ and $j \leftrightarrow j$ (Exercise 7-36), $P\{T_{ii} < \infty\} = P\{T_{jj} < \infty\} = 1$. Since j is a null-persistent state, $E(T_{jj}) \triangleq \mu_{jj} = \infty$ and (7-24a) asserts that $\lim_{n \to \infty} p_{jj}(n) = 1/\mu_{jj} = 0$. From (7-34), for any state $i \neq j$,

$$0 = \lim_{n \to \infty} p_{jj}(n) = \lim_{n \to \infty} p(n + n_1 + n_2) \geq p_{ji}(n_2)p_{ij}(n_1) \lim_{n \to \infty} p_{ii}(n)$$

so $\lim_{n \to \infty} p_{ii}(n) = 0$. By (7-24a), this means that i is null, so i is null-persistent, which proves (b).

Part (c) is proved by the law of the excluded middle: If i is ergodic, then j cannot be transient or null-persistent else (a) or (b) would claim i is not ergodic, so j must be ergodic. $\qquad \square$

Combining Theorems 7-5 and 7-7 yields an important result: *All states in a communicating class are of the same type.* Thus it is sufficient to classify one state of a communicating class. When P is irreducible, once the classification of any state is made, the classification of all states is known.

A Test for Transient States

Suppose a communicating class contains infinitely many states. How can we tell if they are transient or persistent? Sometimes the structure of the chain allows a direct computation of $P\{T_{ii} < \infty\}$ for some distinguished state i. At other times the next theorem may help.

Theorem 7-8 Let C be a communicating class with states $0, 1, \ldots$. All the states of C are transient if, and only if, there is a bounded solution, *which is not identically zero*, to the equations

$$y_i = \sum_{j=1}^{\infty} p_{ij} y_j \qquad i \in I_+ \tag{7-36}$$

PROOF Let $u_i(n)$ be the probability of starting in state i and not reaching state 0 on any of the first n transitions, $i \in I_+$. Then

$$u_i(1) = \sum_{j=1}^{\infty} p_{ij} \tag{7-37a}$$

and

$$u_i(n + 1) = \sum_{j=1}^{\infty} p_{ij} u_j(n) \qquad n \in I_+ \tag{7-37b}$$

for all $i \in I_+$. It is probabilistically clear that $u_i(n + 1) \leq u_i(n)$ for each i.

Furthermore, $u_i(n) \geq 0$, so $u_i = \lim_{n \to \infty} u_i(n)$ exists and satisfies

$$u_i = \sum_{j=1}^{\infty} p_{ij} u_j \qquad i \in I_+ \tag{7-38}$$

If the only bounded solution of (7-36) is identically zero, then $u_i \equiv 0$ is the only solution of (7-38). This means that state 0 is entered infinitely many times, so by Theorem 7-4, state 0 is persistent. From Theorem 7-7 this implies that all the states of C are persistent, proving the "only if" part. If (7-36) has a bounded nonzero solution, it can be normalized so that each $|y_i| \leq 1$. From (7-36), (7-37b), and (7-37a) we obtain

$$u_i(1) = \sum_{j=1}^{\infty} p_{ij} \geq \sum_{j=1}^{\infty} p_{ij} |y_j| \geq |y_i|$$

$$u_i(2) = \sum_{j=1}^{\infty} p_{ij} u_i(1) \geq \sum_{j=1}^{\infty} p_{ij} |y_j| \geq |y_i|$$

and by induction,

$$u_i(n) \geq |y_i| \qquad i \in I_+ \tag{7-39}$$

for each n. Thus if a bounded solution to (7-36) is not identically zero, $u_i > |y_i| > 0$ for some i, which makes state 0 (and thereby all states) transient, completing the proof. $\qquad\square$

Example 7-14: Example 7-3, the M/D/1 queue, continued The transition matrix of the embedded chain for the M/D/1 queue is

$$P = \begin{pmatrix} c_0 & c_1 & c_2 & c_3 \cdots \\ c_0 & c_1 & c_2 & c_3 \cdots \\ 0 & c_0 & c_1 & c_2 \cdots \\ 0 & 0 & c_0 & c_1 \cdots \\ & & \cdots \cdots \cdots \cdots \end{pmatrix}$$

where $c_k = e^{-\lambda v}(\lambda v)^k/(k!)$, $k \in I$, λ is the customer arrival rate, and v is the service time for each customer. The purpose of the following analysis is to show the states are transient when $\rho = \lambda v > 1$.

Part (c) of Example 7-13 shows that the MC is irreducible. Thus, all states are of the same type. Since c_k is the probability that k new customers arrive during the service time of a customer, $\rho = \sum_{k=0}^{\infty} k c_k = \lambda v$ is the mean number of customers who arrive during a service time. If $\rho > 1$, customers arrive faster, on the average, than they leave; intuitively, the number of customers present at departure epochs grows without bound. That is, we suspect that all states are transient when $\rho > 1$. Similarly, it seems reasonable that all states are ergodic when $\rho < 1$. (This intuitive argument leaves the classification when $\rho = 1$ ambiguous; in fact, all states are null-persistent when $\rho = 1$.)

Consider the case $\rho > 1$. Let $u_j = P\{T_{j0} < \infty\}$ and set $u_1 = \alpha$. By conditioning on the first event, we obtain

$$u_1 = c_0 + c_1 u_1 + c_2 u_2 + \cdots$$

$$u_2 = c_0 u_1 + c_1 u_2 + \cdots$$

and so forth, i.e.,

$$u_j = \sum_{i=0}^{\infty} c_i u_{j-1+i} \qquad (7\text{-}38')$$

Since $p_{ij} = 0$ for $j < i - 1$, a sequence of transitions from i to 0 must pass through states $i - 1$, $i - 2$, ..., 1 in turn. The rows of P have the same elements except that the position of c_0 is constantly shifted to the right by one place. This observation suggests that $u_j = (u_1)^j$; it is simple to check that $u_j = \alpha^j$ is a solution of $(7\text{-}38')$.

According to Theorem 7-8, if a nonzero bounded solution of

$$y_i = \sum_{j=1}^{\infty} p_{ij} y_j \qquad i \in I_+ \qquad (7\text{-}36)$$

can be displayed, then all states are transient. An interpretation of y_i is $y_i = 1 - P\{T_{i0} < \infty\}$, which suggests we try

$$y_i = 1 - \alpha^i \qquad 0 \leq \alpha \leq 1 \qquad (7\text{-}40)$$

as a solution to $(7\text{-}36)$. Substituting $(7\text{-}40)$ and the particular form of P into $(7\text{-}36)$ yields

$$1 - \alpha^i = \sum_{j=0}^{\infty} c_j (1 - \alpha^{i-1+j})$$

hence

$$\alpha = \sum_{j=0}^{\infty} \alpha^j c_j = e^{-\rho(1-\alpha)} \qquad (7\text{-}41)$$

If a solution to $(7\text{-}41)$ exists with $0 < \alpha < 1$, then $(7\text{-}40)$ provides the desired solution of $(7\text{-}36)$.

Since $(7\text{-}41)$ is satisfied by $\alpha = 1$, $e^{-\rho} > 0$, and both sides of $(7\text{-}41)$ are continuous functions of α, $(7\text{-}41)$ has a solution in $(0, 1)$ if the slope of the right-side at $\alpha = 1$ is larger than 1; see Figure 7-5. This occurs when $\rho > 1$, so *all states are transient when $\rho > 1$.* ◻

The above method is not the most convenient way to show that all states are transient when $\rho > 1$. Some theorems in the next section require considerably less calculation to prove this result; moreover, the conclusion that all states are ergodic when $\rho < 1$ and the value of $\lim_{n \to \infty} P\{Y_n = j \mid Y_0 = i\}$ are obtained.

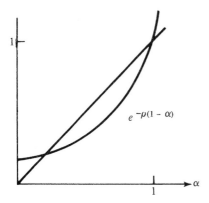

Figure 7-5 Graphical solution of Equation (7-41).

EXERCISES

7-32 Classify the states of the chains with the following transition matrices.

$$(a) \quad \begin{pmatrix} 0 & \frac{1}{3} & \frac{1}{2} \\ \frac{1}{2} & 0 & \frac{1}{2} \\ \frac{2}{3} & \frac{1}{3} & 0 \end{pmatrix}$$

$$(b) \quad \begin{pmatrix} 0 & 0 & 0 & 1 \\ 0 & 0 & 0 & 1 \\ \frac{1}{2} & \frac{1}{2} & 0 & 0 \\ 0 & 0 & 1 & 0 \end{pmatrix}$$

$$(c) \quad \begin{pmatrix} \frac{1}{2} & 0 & \frac{1}{2} & 0 & 0 \\ \frac{1}{4} & \frac{1}{2} & \frac{1}{4} & 0 & 0 \\ \frac{1}{2} & 0 & \frac{1}{2} & 0 & 0 \\ 0 & 0 & 0 & \frac{1}{2} & \frac{1}{2} \\ 0 & 0 & 0 & \frac{1}{2} & \frac{1}{2} \end{pmatrix}$$

7-33 Prove Proposition 7-4.

7-34 In Example 7-12, show that state 2 is transient.

7-35 A transition matrix P is in *canonical form* if the states are renumbered (if necessary) so that those belonging to a communicating class are numbered consecutively and the transient states, if any, are assigned the largest numbers. For example, in a chain with two communicating sets of states and some number of transient states, we write P as the partitioned matrix

$$P = \begin{pmatrix} P_1 & 0 & 0 \\ 0 & P_2 & 0 \\ A_1 & B_1 & T \end{pmatrix}$$

where P_1, P_2, and T are square matrices. Matrices P_1 and P_2 are the transition matrices corresponding to the two communicating classes.

(a) Interpret A_1, B_1, and T.

(b) Show that, for P as above,

$$P^n = \begin{pmatrix} P_1^n & 0 & 0 \\ 0 & P_2^n & 0 \\ A_n & B_n & T^n \end{pmatrix}$$

where A_n and B_n are complicated matrices which you need not specify.

(c) Interpret A_n, B_n, and T_n in terms of n-step transition probabilities.

(d) Put the matrices in Exercise 7-32 in canonical form.

7-36 Prove that $\leftrightarrow$ is an equivalence relation, i.e., that (a) $i \leftrightarrow i$ for all i, (b) $i \leftrightarrow j \Leftrightarrow j \leftrightarrow i$, and (c) $i \leftrightarrow j$ and $j \leftrightarrow k \Rightarrow i \leftrightarrow k$.

7-37 For P, a stochastic matrix, show that P is reducible if, and only if, it can be put in the form

$$P = \begin{pmatrix} A & 0 \\ B & D \end{pmatrix}$$

7-38 Prove that an MC with an irreducible transition matrix P cannot be periodic if at least one diagonal element of P is positive.

7-39 Let the transition matrix of an MC be given by $p_{i0} = q_i$ and $p_{i,i+1} = 1 - q_i$, $i \in I$. Show that the states are transient if, and only if, $\sum_{i=0}^{\infty} q_i < \infty$. Under what conditions are the states null?

7-40 Show that the infinite chain where P is given by

$$\begin{pmatrix} 1 & 0 & 0 & 0 & 0 & \cdots \\ q & 0 & p & 0 & 0 & \cdots \\ 0 & q & 0 & p & 0 & \cdots \\ 0 & 0 & q & 0 & p & \cdots \\ & & \cdots & & & \cdots \end{pmatrix}$$

is ergodic when $p < q$.

7-41 In Exercise 7-40 replace the q and p in row i by q_i and p_i, respectively. Prove that the states are persistent if, and only if,

$$\sum_{i=1}^{\infty} \frac{q_1 q_2 \cdots q_i}{p_1 p_2 \cdots p_i} = \infty$$

Use this result to establish (6-39), i.e., that a birth-and-death process will reach state 0 from any initial state if, and only if, $\sum_{j=0}^{\infty} (\lambda_j b_j)^{-1} = \infty$, where b_j is given by (4-20). Under what conditions are the states ergodic?

7-42 In an MC with a finite number of states, prove that state j is transient if, and only if, there is some state i such that $j \to i$ but not $i \to j$. Is this statement true for chains with an infinite number of states? *Hint*: Do Exercise 7-39.

7-43 Let $\{Y_n; n \in I\}$ be an MC with transition matrix P as shown in Exercise 7-37, with D containing a finite number of elements. Let i and j be transient states, $Y_0 = i$, and N_{ij} be the number of times state j is visited. Show $[E(N_{ij})] = I + \sum_{k=1}^{\infty} D^k = (I - D)^{-1}$.

7-6 LIMITING AND STATIONARY PROBABILITIES

In this section we show how to determine whether a steady state exists and to describe it when it does.

Definition and Existence Conditions
for Limiting Probabilities

The usual intuitive interpretation of a steady state is that the effects of the initial conditions have vanished (cf. Definition 4-2). This motivates the definition of a steady-state distribution of a Markov chain.

> **Definition 7-7** Let $\{Y_n; n \in I\}$ be a Markov chain with state space S and initial probability vector $\mathbf{a}$. The vector $\mathbf{p} = (p_j; j \in S)$ is the *limiting* (or *steady-state*) *distribution* (or *vector*) of the MC if
>
> $$\lim_{n \to \infty} P\{Y_n = j\} = p_j \qquad j \in S \tag{7-42}$$
>
> for any choice of $\mathbf{a}$.

According to Definition 7-7, the existence of $\mathbf{p}$ means that $P^\infty \triangleq \lim_{n \to \infty} P^n$ exists and, moreover, $\mathbf{a}P^\infty$ does not depend on $\mathbf{a}$. Therefore, $\lim_{n \to \infty} P\{Y_n = j \mid Y_0 = i\}$ does not depend on i. Examples 7-9 and 7-10 exhibit MCs which lack these properties. The definition does not require

$$\sum_{j \in S} p_j = 1 \tag{7-43}$$

There are two reasons. First, (7-43) is not true for some important Markov chains, e.g., the symmetric simple random walk described in Section 7-7. Second, when (7-43) is true, it can be deduced. When S is finite, then

$$1 = \lim_{n \to \infty} P\{Y_n \in S\} = \sum_{j \in S} \lim_{n \to \infty} P\{Y_n = j\} = \sum_{j \in S} p_j$$

and (7-43) holds. When S is infinite, bringing the limit inside the sum is not necessarily valid. Theorem 7-10 below provides necessary and sufficient conditions for (7-43) to hold when S is infinite.

Suppose $S = \{0, 1, \dots \}$; for any integer $m \in S$,

$$1 \geq \lim_{n \to \infty} P\{Y_n \leq m\} = \sum_{j=0}^{m} p_j$$

Since $p_j \geq 0$, the sequence of partial sums is monotone and bounded, hence it has a limit and

$$\sum_{j=0}^{\infty} p_j \leq 1 \tag{7-43'}$$

The following result is an immediate consequence of Definition 7-7.

> **Proposition 7-5** If a Markov chain with transition matrix P has the limiting distribution $\mathbf{p}$, then each row of P^∞ is $\mathbf{p}$.

PROOF See Exercise 7-44. □

The classification machinery in Section 7-5 is designed to provide concisely stated conditions under which a limiting distribution exists. Recall that $\mu_{jj} \triangleq E(T_{jj})$ is the mean number of transitions between visits to state j.

Theorem 7-9: Existence of positive limiting probabilities For a Markov chain $\{Y_n; n \in I\}$ with state space S, a limiting distribution **p** with

$$p_j \triangleq \lim_{n \to \infty} P\{Y_n = j\} = \frac{1}{\mu_{jj}} > 0 \qquad j \in S \qquad (7\text{-}44)$$

exists if, and only if,
    (a) the transition matrix is irreducible
    (b) the states are ergodic
If the states are periodic, (7-44) holds only when n is a multiple of the period.

PROOF From the definitions of an irreducible transition matrix and of an ergodic state, $P\{T_{ij} < \infty\} = 1$ for all $i, j \in S$. Thus, Proposition 7-3 yields

$$\lim_{n \to \infty} p_{ij}(n) = \frac{1}{\mu_{jj}} > 0$$

that is, (a) and (b) imply (7-44) for all $j \in S$. If there were a nonergodic state, state k say, then $\mu_{kk} = \infty$ and Proposition 7-3 would yield $\lim_{n \to \infty} p_{kk}(n) = 0$. If the transition matrix were not irreducible, then there would exist $i, j \in S$ such that state j could not be reached from state i; hence $P\{T_{ij} < \infty\} = 0$ and Proposition 7-3 would yield $\lim_{n \to \infty} p_{ij}(n) = 0$. Thus, (a) and (b) are necessary for $p_j > 0$ to exist for all $j \in S$.   □

It should be apparent from the proof of Theorem 7-9 that if the states are persistent-null instead of ergodic, then **p** exists with $p_j = 0$ for all $j \in S$. This is the situation when (7-43) is invalid.

Suppose S consists of a single communicating class C and a set of transient states T. Limiting probabilities exist whenever a transition out of the transient states occurs with probability 1. If T consists of a finite number of states, eventual entrance into C is ensured. If T consists of infinitely many states, Theorem 7-16 provides a way of checking the above condition.

Corollary 7-9 Let $\{Y_n; n \in I\}$ be a Markov chain with state space S, with $S = C \cup T$, where C is a communicating class of ergodic states and T is a set of transient states. If, for each $i \in T$, $\lim_{n \to \infty} P\{Y_n \in C \mid Y_0 = i\} = 1$, then a limiting probability vector **p** exists. Furthermore, $p_j > 0$ for $j \in C$ and $p_j = 0$ for $j \in T$.

PROOF See Exercise 7-45.   □

Stationary Probabilities

Equation (7-19) indicates that μ_{jj} may be difficult to obtain when there are many states, so (7-44) may not provide a convenient way to calculate p_j. Indeed, p_j is not frequently evaluated via (7-44). The usual method of calculating $\mathbf{p}$ is to obtain a vector that is conceptually distinct from $\mathbf{p}$ but is identical to $\mathbf{p}$ when $\mathbf{p}$ exists.

> **Definition 7-8** Let $\{Y_n; n \in I\}$ be a Markov chain with state space S and transition matrix P. The vector $\boldsymbol{\pi} = (\pi_i; i \in S)$ is a *stationary* (or *invariant*) probability vector of P if
>
> $$\boldsymbol{\pi} = \boldsymbol{\pi}P \qquad (7\text{-}45a)$$
>
> $$\sum_{i \in S} \pi_i = 1 \qquad (7\text{-}45b)$$
>
> $$\pi_i \geq 0 \qquad i \in S \qquad (7\text{-}45c)$$

From $\mathbf{P}_n = \mathbf{P}_{n-1}P$ and (7-45), if $\mathbf{a} = \boldsymbol{\pi}$, then

$$\mathbf{P}_1 = \boldsymbol{\pi}P = \boldsymbol{\pi}$$

$$\mathbf{P}_2 = \mathbf{P}_1 P = \boldsymbol{\pi}P = \boldsymbol{\pi}$$

and by induction,

$$\mathbf{P}_n = \boldsymbol{\pi} \qquad n \in I \qquad (7\text{-}46)$$

Once the state probability vector is given by $\boldsymbol{\pi}$, it remains as $\boldsymbol{\pi}$, so if both $\boldsymbol{\pi}$ and $\mathbf{p}$ exist, they must coincide. It remains to show $\boldsymbol{\pi}$ and $\mathbf{p}$ exist together.

> **Theorem 7-10** If a Markov chain has a limiting probability vector $\mathbf{p}$, with some $p_j > 0$, then $\mathbf{p}$ is the unique solution of (7-45).

PROOF It is automatic that $\mathbf{p}$ satisfies (7-45c). Assume that the ergodic states are aperiodic. For each state j, the Chapman-Kolmogorov equation yields

$$p_{ij}(n+1) = \sum_{k \in S} p_{ik}(n)p_{kj} \qquad (7\text{-}47)$$

If $S = \{0, 1, \ldots, s\}$, taking limits as $n \to \infty$ on both sides of (7-47) yields

$$p_j = \sum_{k=0}^{s} p_k p_{kj} \qquad j = 0, 1, \ldots, s$$

This and (7-43), that is, $\sum_{j=0}^{s} p_j = 1$, show that $\mathbf{p}$ is a solution of (7-45). If S is infinite, choose $m < \infty$; from (7-47),

$$p_{ij}(n+1) \geq \sum_{k=0}^{m} p_{ik}(n)p_{kj} \qquad j \in I$$

Take limits as $n \to \infty$ on both sides, as before, to obtain

$$p_j \geq \sum_{k=0}^{m} p_k p_{kj} \qquad j \in I \qquad (7\text{-}48)$$

For each j, the right-side of (7-48) is nondecreasing in m and bounded, so it has a limit as $m \to \infty$. Hence

$$p_j \geq \sum_{k \in S} p_k p_{kj} \qquad j \in I \qquad (7\text{-}49)$$

We want to show that (7-49) is, in fact, an equality. Assume the opposite, and obtain a contradiction as follows. Sum both sides of (7-49) over all $j \in S$, obtaining, with the aid of $\sum_{j \in S} p_j \leq 1$,

$$1 \geq \sum_{j=0}^{\infty} p_j > \sum_{j=0}^{\infty} \sum_{k=0}^{\infty} p_k p_{kj}$$

$$= \sum_{k=0}^{\infty} p_k \sum_{j=0}^{\infty} p_{kj} \qquad \text{(Fubini's theorem)}$$

$$= \sum_{k=0}^{\infty} p_k$$

This is absurd! Hence, (7-49) is an equality for all j and $\mathbf{p}$ satisfies (7-45a), i.e.,

$$\mathbf{p} = \mathbf{p}P \qquad (7\text{-}50)$$

It remains to show that $\sum_{j \in S} p_j = 1$ holds. From (7-50) and the argument leading to (7-46),

$$\mathbf{p} = \mathbf{p}P^{\infty} \qquad (7\text{-}51)$$

By hypothesis, $p_j > 0$ for some j. Without loss of generality (renumber the states if necessary), suppose $p_0 > 0$. From Proposition 7-5, each row of P^{∞} is $\mathbf{p}$; write (7-51) as

$$p_i = p_i \sum_{j=0}^{\infty} p_j \qquad i \in I$$

Since $\sum_{j=0}^{\infty} p_j > 0$, the equality above holds for $i = 0$ if and only if (7-43) is valid; hence $\mathbf{p}$ satisfies all the requirements of (7-45).

Since limits are unique, (7-46) and the statement following it prove that (7-45) has a unique solution whenever $\mathbf{p}$ exists.

The proof when the ergodic states are periodic is similar (see Exercise 7-49). □

The typical procedure for obtaining $\mathbf{p}$ is to establish first that the hypotheses of either Theorem 7-9 or Corollary 7-9 hold, so that $\mathbf{p}$ exists, and then solve (7-45). The value of $\mu_{jj} (= 1/\pi_j)$ is obtained with no additional effort.

Equation (7-45a) does not have a unique solution, so at least one of the equations $\pi_i = \sum_j \pi_j p_{ji}$ is redundant. When S is finite and $\boldsymbol{\pi}$ is unique, it can be shown that exactly one equation is redundant, and it may be chosen at will (see Exercise 7-50).

In many applications, a convenient way to solve (7-45) is to ignore (7-45b) and find a solution of (7-45a) and (7-45c) with $0 < \sum \pi_i < \infty$. Then $\pi_i' = \pi_i / \sum \pi_j$ solves (7-45).

Examples

Example 7-15: Example 7-7 continued By inspection it is clear that

$$P = \begin{pmatrix} \frac{1}{2} & \frac{1}{2} \\ \frac{1}{3} & \frac{2}{3} \end{pmatrix}$$

is irreducible. Since the state space is finite, both states are ergodic. Also it is obvious that the states are aperiodic. Hence, $\mathbf{p}$ exists and is the unique solution of

$$p_0 = \frac{p_0}{2} + \frac{p_1}{3}$$

$$p_1 = \frac{p_0}{2} + \frac{2p_1}{3}$$

$$1 = p_0 + p_1$$

which is $p_0 = \frac{2}{5}$, $p_1 = \frac{3}{5}$. The calculation of P^∞ by Sylvester's formula in Section 7-2 agrees with Proposition 7-5. $\qquad\square$

Example 7-16: Example 7-5, personnel flows, continued Suppose the transition matrix for Example 7-5 is given by

$$P = \begin{pmatrix} q_0 + h_0 & p_0 & 0 & 0 \\ h_1 & q_1 & p_1 & 0 \\ h_2 & 0 & q_2 & p_2 \\ h_3 & 0 & 0 & q_3 \end{pmatrix}$$

This represents an organization with four strata; p_i is the proportion of those members of stratum i who are promoted one level, and q_i is the proportion who remain in grade i. In an organization that "promotes from within," all hiring is done at the lowest stratum. The matrix P above represents such an organization, and h_i is the proportion of people in stratum i who leave the organization; they are replaced by people hired into stratum 0.

If $q_3 < 1$, P is irreducible. If $q_3 = 1$, state 3 is absorbing and all other states are transient. In either case, (7-45) determines the steady-state proportions. The case $q_3 = 1$ is trivial, so assume $q_3 < 1$. From (7-45),

$$\pi_0 = (q_0 + h_0)\pi_0 + h_1\pi_1 + h_2\pi_2 + h_3\pi_3$$

$$\pi_1 = p_0\pi_0 + q_1\pi_1$$

$$\pi_2 = p_1\pi_1 + q_2\pi_2$$

$$\pi_3 = p_2\pi_2 + q_3\pi_3$$

$$1 = \pi_0 + \pi_1 + \pi_2 + \pi_3$$

The top equation is redundant (why?), so discard it because it is the messiest one! The next three equations can be solved recursively in terms of π_0:

$$\pi_1 = \frac{p_0}{1 - q_1} \pi_0 \quad \pi_2 = \frac{p_0 p_1}{(1 - q_1)(1 - q_2)} \pi_0 \quad \pi_3 = \frac{p_0 p_1 p_2}{(1 - q_1)(1 - q_2)(1 - q_3)} \pi_0$$

The bottom equation determines π_0:

$$\pi_0 = \left[1 + \frac{p_0}{1 - q_1} + \frac{p_0 p_1}{(1 - q_1)(1 - q_2)} + \frac{p_0 p_1 p_2}{(1 - q_1)(1 - q_2)(1 - q_3)} \right]^{-1} \qquad \square$$

Example 7-17: Example 7-4, the (s, S) inventory model, continued The transition matrix for this example is

$$P = \begin{array}{c} S \\ S-1 \\ S-2 \\ \\ s+1 \\ s \end{array} \begin{pmatrix} f_0 + F^c(Q) & f_1 & f_2 & \cdots & f_{Q-1} & f_Q \\ F^c(Q-1) & f_0 & f_1 & \cdots & f_{Q-2} & f_{Q-1} \\ F^c(Q-2) & 0 & f_0 & \cdots & f_{Q-3} & f_{Q-2} \\ \cdots\cdots\cdots\cdots\cdots\cdots\cdots\cdots\cdots\cdots \\ F^c(2) & 0 & 0 & \cdots & f_0 & f_1 \\ F^c(1) & 0 & 0 & \cdots & 0 & f_0 \end{pmatrix}$$

where $Q = S - s$. If $f_1 > 0$, then P is irreducible, and P is aperiodic if $f_0 > 0$. The model is trivial if $f_0 = 1$, so that situation is prohibited. Since $i \to S$ for all i in the state space, it is impossible for there to be more than one communicating class (why?). Since the state space is finite, there exists at least one ergodic state (Corollary 7-4), Theorem 7-10 applies, and (7-45) has a unique solution.

Equation (7-45a) yields

$$(1 - f_0)\pi_S = \sum_{i=0}^{Q} F^c(Q - i)\pi_{S-i}$$

$$(1 - f_0)\pi_{S-1} = f_1 \pi_S$$

$$(1 - f_0)\pi_{S-2} = f_2 \pi_S + f_1 \pi_{S-1}$$

$$\cdots\cdots\cdots\cdots\cdots\cdots\cdots\cdots\cdots\cdots\cdots$$

$$(1 - f_0)\pi_s = f_Q \pi_S + f_{Q-1}\pi_{S-1} + \cdots + f_1 \pi_{s+1}$$

Take the top equation as the redundant one; the remaining equations can be solved recursively in terms of π_S. The recursive solution clearly provides $\pi_{S-i} \geq 0$, and π_S is obtained from $\sum_{i=s}^{S} \pi_i = 1$.

When the daily demands form a sequence of independent Bernoulli random variables with mean p, $f_0 = 1 - p$ and $f_1 = p$. An explicit solution to the equations above is

$$\pi_S = \pi_{S-1} = \ldots = \pi_s = \frac{1}{S + 1}$$

In Example 6-8, π_{S-i} is given in terms of the discrete renewal function associated with the demand distribution $F(\cdot)$. It is easy to show that that representation satisfies the recursive equations derived above; it is not so easy to obtain the renewal representation directly from the recursive equations (Exercise 7-52). $\qquad \square$

A Converse of Theorem 7-10

In order to apply Theorem 7-10, you have to know that at least one state is ergodic. For finite chains this is automatically true (cf. Corollary 7-4).

The following converse of Theorem 7-10 is especially important for infinite chains.

Theorem 7-11 Let $\{Y_n; n \in I\}$ be a Markov chain with transition matrix P and state space S. If P is irreducible and (7-45) has a solution, then all states are ergodic and $\pi_j = p_j = 1/\mu_{jj} > 0$ for all $j \in S$.

PROOF Let π denote a solution of (7-45). The derivation of (7-46) showed

$$\pi = \pi P^n \qquad n \in I_+ \tag{7-52}$$

Since P is irreducible, all the states are of the same type. If the states were null, then $P^n \to 0$ and (7-52) would imply $\pi_j = 0$ for all $j \in S$. But this contradicts the assumption that $\sum_{j \in S} \pi_j = 1$, and the only remaining possibility is that all states are ergodic.

Since all states are ergodic and the chain is irreducible, Theorems 7-9 and 7-10 apply and $\pi_j = p_j = 1/\mu_{jj} > 0$ for all $j \in S$. ☐

The important feature of Theorem 7-11 is that *if P is irreducible* and *some solution of (7-45) is obtained*, then that solution is the unique solution and it coincides with **p**, *which must exist*.

Another useful condition for an MC to have ergodic states is given in A. G. Pakes, "Some Conditions for Ergodicity and Recurrence of Markov Chains", *Oper. Res.* **17**: 1058–1061 (1969).

Ergodic Properties of Markov Chains

The next theorem shows that the limiting probabilities can be interpreted as the asymptotic proportion of time that the chain spends in each state. Theorems of this type are called *ergodic* theorems. Recall that in Section 7-4, $N_{ij}(n)$ is defined as the number of visits to state j up to and including the nth transition when the initial state is i. Since each visit is of length 1, $N_{ij}(n)$ can be interpreted as the total time that the process spends in state j during $[0, n + 1)$.

Theorem 7-12: Ergodic theorem for Markov chains In a Markov chain with state space S and limiting distribution **p**,

$$\lim_{n \to \infty} \frac{N_{ij}(n)}{n} = \frac{1}{\mu_{jj}} = \lim_{n \to \infty} \frac{E[N_{ij}(n)]}{n} = p_j \tag{7-53}$$

for all $i, j \in S$, the first equality holding with probability 1.

PROOF Since $\{N_{ij}(n); n \in I\}$ is a delayed renewal process (see page 136), the strong law of large numbers for renewal processes (Theorem 5-3) and the elementary renewal theorem (Theorem 5-7) apply (see Section 5-4) and assert that the limits in (7-53) exist and equal $1/\mu_{jj}$. The last equality follows from Theorem 7-9. $\qquad\qquad\qquad\qquad\qquad\qquad\qquad\qquad\qquad\qquad\square$

Observe that (7-53) does not need to be modified for periodic states (why?). Let

$$D_{ij}(n) = \begin{cases} 1 & \text{if } Y_{n-1} = i \text{ and } Y_n = j \\ 0 & \text{otherwise} \end{cases}$$

for $n \in I_+$. It follows from Theorem 7-12 and the strong law of large numbers, or by a direct renewal argument, that when $\mathbf{p}$ exists,

$$d_{ij} \triangleq \lim_{n \to \infty} \frac{1}{n} \sum_{k=1}^{n} D_{ij}(k) = p_i p_{ij} \qquad i, j \in \mathcal{S} \tag{7-54}$$

Exercise 7-54 asks you to prove (7-54). The quantity d_{ij} is the asymptotic proportion of transitions that go from state i to state j.

Alternate Characterization of the Steady-State Distribution

In Section 4-5, the equations for the steady-state probabilities of a birth-and-death process are simplified by observing that in the steady state, the net rate of flow of probability between adjacent states is zero. Let us now develop the analogous result for Markov chains.

Partition $\mathcal{S}$ into disjoint subsets A and B. By using (7-54) and Fubini's theorem, the asymptotic proportion of transitions that go from A to B is

$$D_{AB} \triangleq \sum_{\substack{i \in A \\ j \in B}} d_{ij} = \sum_{i \in A} p_i \sum_{j \in B} p_{ij}$$

and the asymptotic proportion of transitions that go from B to A is

$$D_{BA} = \sum_{\substack{j \in B \\ i \in A}} d_{ji} = \sum_{j \in B} p_j \sum_{i \in A} p_{ji}$$

In the steady state, it should be true that $D_{AB} = D_{BA}$. This is trivially true when $\mathbf{p} = \mathbf{0}$, so we dismiss that case from further consideration. If $\mathbf{p}$ exists, then $\boldsymbol{\pi}$ exists and $\mathbf{p} = \boldsymbol{\pi}$ (Theorem 7-10), but $\boldsymbol{\pi}$ may exist when $\mathbf{p}$ does not. A more general statement than $D_{AB} = D_{BA}$ is

$$\sum_{i \in A} \pi_i \sum_{j \in B} p_{ij} = \sum_{j \in B} \pi_j \sum_{i \in A} p_{ji} \tag{7-55}$$

The next theorem establishes (7-55) and more.

Theorem 7-13: Alternate characterization of the stationary distribution Consider a Markov chain with state space $\mathcal{S}$ and transition matrix P. A vector $\boldsymbol{\pi}$ satisfying (7-45b) and (7-45c) solves (7-45a) if, and only if, $\boldsymbol{\pi}$ solves (7-55) for any A and B such that $A \cup B = \mathcal{S}$ and $A \cap B = \emptyset$.

PROOF Assume (7-55) holds and choose $A = \{i\}$. Then (7-55) is

$$\pi_i(1 - p_{ii}) = \sum_{j \neq i} \pi_j p_{ji}$$

for any $i \in S$. Hence π satisfies $\pi = \pi P$ and (7-45) holds true. Now assume that (7-45a) holds and write it as

$$\sum_{j \in S} \pi_i p_{ij} = \pi_i = \sum_{j \in S} \pi_j p_{ji} \qquad i \in S$$

Sum both sides over all $i \in A$ to obtain

$$\sum_{i \in A} \sum_{j \in A} \pi_i p_{ij} + \sum_{i \in A} \sum_{j \in B} \pi_i p_{ij} = \sum_{i \in A} \sum_{j \in A} \pi_j p_{ji} + \sum_{i \in A} \sum_{j \in B} \pi_j p_{ji}$$

Each double sum is no larger than unity, so the first terms on each side cancel and leave

$$\sum_{i \in A} \pi_i \sum_{j \in B} p_{ij} = \sum_{i \in A} \sum_{j \in B} \pi_j p_{ji} = \sum_{j \in B} \pi_j \sum_{i \in A} p_{ji}$$

The last equality is justified by Fubini's theorem. ☐

Example 7-18: M/G/1 queue In Exercise 7-4 you showed that the embedded Markov chain of the $M/G/1$ queue has transition martix

$$P = \begin{pmatrix} c_0 & c_1 & c_2 & c_3 & \cdots \\ c_0 & c_1 & c_2 & c_3 & \cdots \\ 0 & c_0 & c_1 & c_2 & \cdots \\ 0 & 0 & c_0 & c_1 & \cdots \\ \cdots\cdots\cdots\cdots\cdots\cdots \end{pmatrix}$$

where

$$c_k = P\{A_0 = k\} = \int_0^\infty \frac{e^{-\lambda t}(\lambda t)^k}{k!} \, dG(t) \tag{7-56}$$

By inspection, P is irreducible and aperiodic. A limiting probability vector with positive entries exists if, and only if, (7-45) has a solution (Theorems 7-9 and 7-11). You are invited to write out (7-45a) for this example and attempt to show that (7-45) has a solution. It can be done, but the following analysis, based on (7-55) and Theorem 7-13, should be easier.

The structure of P shows that one-step transitions into $B = \{0, 1, \ldots, k\}$ occur only from state $k + 1$, so (7-55) might be easy to solve if we choose $A = \{k + 1, k + 2, \ldots\}$. Do so for $k = 0, 1, \ldots$ in turn; defining $\bar{c}_i = \sum_{k=i+1}^\infty c_k = P\{A_0 > i\}$, from (7-55) we obtain

$$\pi_1 c_0 = \pi_0 \bar{c}_0$$

$$\pi_2 c_0 = \pi_0 \bar{c}_1 + \pi_1 \bar{c}_1$$

$$\pi_3 c_0 = \pi_0 \bar{c}_2 + \pi_1 \bar{c}_2 + \pi_2 \bar{c}_1 \tag{7-57}$$

etc. Observe that

$$\sum_{i=0}^{\infty} \bar{c}_i = \sum_{i=0}^{\infty} P\{A_0 > i\} = E(A_0)$$

$$= \lambda v \triangleq \rho$$

It is apparent from (7-57) that if $\pi_0 > 0$ is known, (7-57) can be solved recursively for π_j, and necessarily $\pi_j > 0$, $j \in I_+$. Also, for any $\alpha > 0$, it is clear that $x_j = \alpha\pi_j$ solves (7-57), and if $\sum_{j=0}^{\infty} \pi_j = 1$, then $\sum_{j=0}^{\infty} x_j = \alpha$. Thus, (7-57) has a solution with $\pi_j > 0$, $j \in I$, and $\sum_{j=0}^{\infty} \pi_j = 1$ if, and only if, it has a solution with $x_0 = 1$ and $\sum_{j=0}^{\infty} x_j = \alpha < \infty$; moreover, $\pi_0 = 1/\alpha$ (note that $x_0 = 1 \Rightarrow \alpha > 1$).

Formally, replace π_i by x_i in (7-57) and set $x_0 = 1$. Summing column by column yields

$$c_0 \sum_{j=1}^{\infty} x_j = \sum_{i=0}^{\infty} \bar{c}_i + \sum_{j=1}^{\infty} x_j \sum_{i=1}^{\infty} \bar{c}_i$$

$$= \rho + (\rho - \bar{c}_0) \sum_{j=1}^{\infty} x_j$$

which can be rewritten as

$$c_0 \alpha = (\rho - 1 + c_0)\alpha + 1 \tag{7-58}$$

A positive finite value of α can solve (7-58) if, and only if, $\rho < 1$, in which case the solution is $\alpha = 1/(1 - \rho)$. Thus, if $\rho < 1$, setting

$$\pi_0 = 1 - \rho \tag{7-59}$$

and iterating (7-57) produce the solution of (7-45), and all states are ergodic. If $\rho \geq 1$, (7-45) has no solution and all states are null.

In digital computation, subtractions are a major source of roundoff errors, particularly when the minuend and subtrahend are nearly equal. The recursive solution of (7-57) involves only additions, so it is a fast and accurate procedure for solving (7-45). However, the coefficients $\bar{c}_i$ might not be evaluated accurately. If $\bar{c}_i$ is computed from $\bar{c}_i = \sum_{k=i+1}^{\infty} c_k$, a truncation error will be introduced because only a finite number of terms can be used (unless, of course, a closed-form evaluation of the sum can be made). If $\bar{c}_i$ is evaluated from $\bar{c}_i = 1 - \sum_{k=0}^{i} c_k$, roundoff errors will appear when $\bar{c}_i$ gets close to zero. Until this occurs, accurate values are obtained. $\qquad\square$

Example 7-19: Example 7-18, M/G/1 queue, continued An alternative analysis of the stationary distribution leads to a closed-form expression for the probability generating function of $\{\pi_i\}$ and the LST of the asymptotic waiting-time d.f. When $\rho < 1$, the analysis in Example 7-18 shows that (7-45) has a unique solution.

Write (7-45a) equation by equation:

$$\pi_0 = c_0 \pi_0 + c_0 \pi_1$$

$$\pi_1 = c_1 \pi_0 + c_1 \pi_1 + c_0 \pi_2 \tag{7-60}$$

$$\cdots\cdots\cdots\cdots\cdots$$

$$\pi_i = c_i \pi_0 + c_i \pi_1 + c_{i-1} \pi_2 + \cdots + c_0 \pi_{i+1},$$

etc. Let $\hat{\pi}(z) = \sum_{i=0}^{\infty} z^i \pi_i$ and $\hat{c}(z) = \sum_{i=0}^{\infty} z^i c_i$. Multiply the ith equation in (7-60) by z^i and sum each column; this yields

$$\hat{\pi}(z) = \hat{c}(z)[\pi_0 + \pi_1 + z\pi_2 + \cdots + z^i \pi_{i+1} + \cdots]$$

Multiplying each term within the brackets by z produces

$$z\pi_0 + \sum_{i=1}^{\infty} z^i \pi_i = \hat{\pi}(z) - \pi_0 + z\pi_0$$

Hence

$$\hat{\pi}(z) = \frac{\hat{c}(z)[\hat{\pi}(z) + (z-1)\pi_0]}{z}$$

from which we get

$$\hat{\pi}(z) = \frac{\pi_0(z-1)\hat{c}(z)}{z - \hat{c}(z)} \tag{7-61}$$

From (7-56),

$$\hat{c}(z) = \sum_{n=0}^{\infty} z^n \int_0^{\infty} \frac{e^{-\lambda t}(\lambda t)^n}{n!} \, dG(t)$$

$$= \int_0^{\infty} e^{-\lambda t(1-z)} \, dG(t) = \tilde{G}(\lambda - \lambda z) \tag{7-62}$$

and

$$\lim_{z \uparrow 1} \frac{d}{dz} \hat{c}(z) = -\lambda \lim_{s \downarrow 0} \frac{d}{ds} \tilde{G}(s) = \lambda v \triangleq \rho \tag{7-63}$$

where v is the mean service time. Substituting (7-59) and (7-62) into (7-61) yields

$$\hat{\pi}(z) = \frac{(1-\rho)(z-1)\tilde{G}(\lambda - \lambda z)}{z - \tilde{G}(\lambda - \lambda z)} \tag{7-64}$$

Equation (7-64) is called the† *Pollaczek-Khintchine formula*. In Section 11-2 it is shown that (7-64) is the PGF of the number of customers present at an arrival epoch and at an arbitrary epoch in the steady state. Let L be the

---

† Several equations pertaining to the asymptotic operating characteristics of the M/G/1 queue are called the Pollaczek-Khintchine formula.

mean of the distribution $\{\pi_i\}$; then

$$L \triangleq \sum_{i=0}^{\infty} i\pi_i = \lim_{z \uparrow 1} \frac{d}{dz} \hat{\pi}(z) = \rho + \frac{\rho^2(\sigma^2/v^2 + 1)}{2(1 - \rho)} \tag{7-65}$$

where σ^2 is the variance of a generic service time. Equation (7-65) also is called the Pollaczek-Khintchine formula. Thus, for fixed values of the arrival rate λ and mean service time v, that is, fixed ρ, the mean number of customers varies linearly with the variance of the service time.

Clearly L is smallest when $\sigma^2 = 0$, which corresponds to constant service times, i.e., the $M/D/1$ queue. For exponentially distributed service times, $\sigma^2 = v^2$ and the second term in (7-65) is twice the value it assumes in the $M/D/1$ queue.

Assume customers are served FIFO, and let W_n be the waiting time of the nth arriving customer and $W_n(t) = P\{W_n \le t\}$. In Example 6-5 it is shown that $W(t) = \lim_{n \to \infty} W_n(t)$ exists. From (7-9), it must be true that

$$\tilde{W}(\lambda - \lambda z) \triangleq \int_0^{\infty} e^{-t\lambda(1-z)} \, dW(t) = \hat{\pi}(z)$$

Making the change of variables $s = \lambda - \lambda z$ and using (7-64) yield

$$\tilde{W}(s) = \hat{\pi}\left(\frac{\lambda - s}{\lambda}\right) = \frac{(1 - \rho)s\tilde{G}(s)}{s - \lambda + \lambda\tilde{G}(s)} \tag{7-66}$$

This, too, is called the Pollaczek-Khintchine formula.

Let $E(W)$ be the mean of $W(\cdot)$. From (7-66),

$$E(W) = \lim_{s \downarrow 0} \frac{d}{ds} \tilde{W}(s) = \frac{v[1 + \rho(\sigma^2/v^2 + 1)]}{2(1 - \rho)} \tag{7-67}$$

which is yet another equation called the Pollaczek-Khintchine formula. Observe from (7-65) and (7-67) that

$$L = \lambda E(W) \tag{7-68}$$

which generalizes (4-89). $\square$

EXERCISES

7-44 Prove Proposition 7-5.

7-45 Prove Corollary 7-9.

7-46 Prove that the MC of Example 7-2 has a limiting distribution, and find it.

7-47 Give an example of a stochastic matrix P such that (7-45) has infinitely many solutions. *Hint*: Only a 2×2 matrix is needed.

7-48 Give an example of an MC that has a stationary distribution but no limiting distribution.

7-49 Prove Theorem 7-10 for periodic states.

7-50 Prove that when the number of states is finite, if (7-45) has a unique solution, then exactly one of the equations in (7-45a) is redundant and it can be chosen at will.

7-51 Show that the parameters in Example 7-16 can be selected to achieve any desired limiting distribution.

7-52 Show that the solutions of Examples 7-17 and 6-8 are identical.

7-53 Let P be an irreducible stochastic matrix. Show that if the only solution of $\mathbf{x} = \mathbf{x}P$, $\sum_{i=0}^{\infty} x_i < \infty$, $\mathbf{x} \geq \mathbf{0}$ is $\mathbf{x} = \mathbf{0}$, then the states of P are not ergodic.

7-54 Give a rigorous proof of (7-54).

7-55 Using the notation in (7-54), prove that $E\left[\sum_{k=1}^{n} D_{ij}(k)\right]/n \to p_i p_{ij}$.

7-56 A transition matrix $P = (P_{ij})$ is called *doubly stochastic* if the column sums are 1:

$$\sum_i p_{ij} = \sum_j p_{ij} = 1$$

For an irreducible and aperiodic doubly stochastic chain with a finite number of states $s + 1$, find the limiting probabilities.

7-57 Let k be a positive integer. Suppose a die (possibly unfair) is rolled repeatedly with the rolls being mutually independent. Let X_n be the number of pips showing on roll n and $S_n = X_1 + \cdots + X_n$. Find $\lim_{n \to \infty} P\{S_n/k \text{ is an integer}\}$.

7-58 Argue probabilistically that (7-57) is true; i.e., derive (7-57) without calculations. Similarly, obtain (7-63) without calculation.

7-59 For the M/G/1 queue in steady state, derive (a) the PGF of the number of customers in *queue* (do not count a customer in service), (b) the mean number of customers in queue, (c) the LST of the *delay* in queue (do not include time spent in service), and (d) the mean delay. All these quantities are called a Pollaczek-Khintchine formula. The Pollaczek-Khintchine formulas in Example 7-19 enable you to do the above without computation. Use your answer to (c) to find the variance of the delay.

7-60 (Continuation) Let $D(t)$ be the distribution function of the steady-state delay random variable and $B(t) = \int_0^t [1 - G(x)] \, dx/v$ [see Section 5-5 for an interpretation of $B(\cdot)$]. Use your solution to Exercise 7-59c to show

$$D(t) = 1 - \rho + \rho \int_0^t D(t - x) \, dB(x) \tag{7-69}$$

and

$$D(t) = 1 - \rho + (1 - \rho) \sum_{n=1}^{\infty} \rho^n B_n(t) \tag{7-70}$$

where $B_n(\cdot)$ is the n-fold convolution of $B(\cdot)$ with itself. Discuss the use of these equations in numerical computation of $D(\cdot)$.

7-61 (Continuation) Let Y_n be the number of customers in the system just before the arrival epoch of customer n. Prove that

$$\lim_{n \to \infty} P\{Y_n = i\} = \pi_i = \lim_{N \to \infty} \frac{1}{N} \sum_{n=1}^{N} 1\{Y_n = i\}$$

where $1\{Y_n = i\}$ is 1 if $Y_n = i$ and 0 otherwise, and the last equality holds with probability 1. (*Hint*: Refer to Exercise 4-50.)

7-62 (Continuation) Suppose that arriving customers are turned away whenever $N - 1$ customers are already in the queue. Show that the embedded MC has

$$P = \begin{pmatrix} c_0 & c_1 & c_2 & \cdots & c_{N-2} & \bar{c}_{N-1} \\ c_0 & c_1 & c_2 & \cdots & c_{N-2} & \bar{c}_{N-1} \\ 0 & c_1 & c_2 & \cdots & c_{N-3} & \bar{c}_{N-2} \\ \hdotsfor{6} \\ 0 & 0 & 0 & \cdots & c_1 & \bar{c}_2 \\ 0 & 0 & 0 & \cdots & c_0 & \bar{c}_1 \end{pmatrix}$$

Show that (7-57) holds for $i = 1, 2, \ldots, N - 1$. Explain why $\pi_0 \neq 1 - \rho$. How would you obtain π for this P from the solution of (7-57)? Is the conclusion of Exercise 7-61 valid for this model?

7-63 (Continuation) Consider an $M/G/1$ queue in which the first customer in each busy period receives exceptional service. That is, customers who arrive when the server is free have service-time distribution function $G_0(\,\cdot\,)$, and all other customers have service-time distribution function $G(\,\cdot\,)$. Let v_0 be the mean of $G_0(\quad)$ and $\rho_0 = \lambda v_0$.

(a) Display the transition matrix for the embedded MC, and prove that a limiting distribution exists if, and only if, $\rho < 1$.

(b) Let π be the limiting distribution and $\hat{\pi}(\,\cdot\,)$ be its generating function. Show

$$\pi_0 = \frac{1 - \rho}{1 - \rho + \rho_0}$$

and

$$\hat{\pi}(z) = \frac{\pi_0[\tilde{G}(\lambda - \lambda z) - z\tilde{G}_0(\lambda - \lambda z)]}{\tilde{G}(\lambda - \lambda z) - z}$$

(c) In the steady state, find the expected number of customers present, the LST of the waiting-time distribution function, and the generating function of the number of customers in the queue.

7-64: The $M/G/1$ queue with batch arrivals Arrivals occur according to a compound Poisson process (i.e., arrival epochs occur according to a Poisson process, and at the ith arrival epoch, a batch of B_i customers arrives, $B_1, B_2, \ldots$ are i.i.d. random variables). The customers are served one at a time; the service times are i.i.d. with distribution function $G(\,\cdot\,)$. Let $b_i = P\{B_1 = i\}$. Find the LST of the steady-state delay. *Hint*: Argue that there is no loss of generality in assuming that if $B_n = i$, the customers in batch n are assigned the numbers $1, 2, \ldots, i$ at random and enter service lowest number first. Then complete the following steps.

(a) Obtain the LST of the delay of customer number 1 in the nth batch as $n \to \infty$.

(b) Obtain the probability that an arriving customer is in a batch of size i.

(c) Find the probability that a customer is kth in a batch, i.e., the long-run proportion of customers that receive number k.

(d) Obtain the LST of the delay of a customer due to members of its batch that precede it into service.

(e) Combine the solutions to (a) and (d) to obtain the solution.

7-65 Show that as $\rho \uparrow 1$ in an $M/G/1$ queue, $P\{(1 - \rho)D^* \le t\}$ approaches the exponential distribution function with mean $[(1 - \rho^2)/(2\lambda) + \lambda v_{2G}/2]/(1 - \rho)$, where v_{2G} is the second moment of the service times.

7-66 Obtain the limiting distributions of the two MCs in Exercise 7-18. What conditions on p and q are required for the limiting distribution to exist?

7-67 Obtain the limiting distribution of the two MCs in Exercise 7-19.

7-7* FURTHER RESULTS ABOUT LIMITING DISTRIBUTIONS

This section extends the results in Section 7-6 on limiting probabilities. Rewards are superimposed on transitions, Theorem 7-12 is strengthened for finite chains, the probability that a chain which starts in a transient state is ultimately absorbed in a communicating class is obtained, and uniqueness of finite solutions of (7-19) is established.

Markov Chains with Rewards

In Chapter 6, renewal processes are enhanced by associating a reward with each renewal to form renewal-reward processes. In a similar fashion, associating a reward with each transition of a Markov chain produces a useful class of processes.†

Start with a Markov chain $\{Y_n; n \in I\}$ having state space S. Suppose for some n that $Y_n = i$ and $Y_{n+1} = j$ and that this is the kth occurrence of a transition from state i to state j. Associate with this transition the *random reward* $r_{ij}^{(k)}$, which is earned at epoch $n + 1$. *Assume* that $r_{ij}^{(1)}$, $r_{ij}^{(2)}$, ... are i.i.d. for all $i, j \in S$, and let r_{ij} denote the generic r.v. Let $R_i(n)$ be the sum of the first n rewards when $Y_0 = i$; if for sample path ω the states of the chain are $Y_1(\omega) = i_1$, $Y_2(\omega) = i_2, \ldots$, then $R_i(0; \omega) = 0$ and

$$R_i(n; \omega) = r_{ii_1}(\omega) + r_{i_1 i_2}(\omega) + \cdots + r_{i_{n-1},i_n}(\omega) \qquad n \in I_+ \qquad (7\text{-}71)$$

Let $E(r_i)$ denote the expected reward earned on a transition out of state i:

$$E(r_i) = \sum_{j \in S} E(r_{ij})p_{ij} \qquad i \in S \qquad (7\text{-}72)$$

By analogy with Corollary 6-8a, it should be true that

$$\lim_{n \to \infty} \frac{R_i(n)}{n} = \sum_{j \in S} E(r_j)p_j \qquad i \in S$$

A proof based on Corollary 6-8a is not hard (Exercise 7-69); an even easier one is available.

Theorem 7-14 Let $\{Y_n; n \in I\}$ have the limiting distribution $\mathbf{p}$ with $p_k > 0$ for some $k \in S$. Assume $E(r_i) < \infty$ for all $i \in S$. Then

$$\lim_{n \to \infty} \frac{R_i(n)}{n} = \sum_{m \in S} E(r_m)p_m = \lim_{n \to \infty} \frac{E[R_i(n)]}{n} \qquad (7\text{-}73)$$

for all $i \in S$, the first equality holding with probability 1.

PROOF As in Section 7-6, let $D_{ij}(n)$ be the number of transitions from state i to state j which occur by the nth transition epoch. Then (7-71) may be written as

$$R_i(n) = \sum_{m \in S} \sum_{j \in S} \sum_{k=1}^{D_{mj}(n)} r_{mj}^{(k)}$$

where the explicit dependence on $Y_0 = i$ is suppressed for notational ease.

---

† Much of Volume II, particularly Chapter 4, concerns the optimization of Markov chains with rewards.

Thus,

$$\frac{R_i(n)}{n} = \sum_{m \in S} \sum_{j \in S} \frac{D_{mj}(n)}{n} \frac{1}{D_{mj}(n)} \sum_{k=1}^{D_{mj}(n)} r_{mj}^{(k)}$$

$$\rightarrow \sum_{m \in S} \sum_{j \in S} p_{mj} p_m E(r_{mj}) \qquad \text{(w.p.1)}$$

$$= \sum_{m \in S} p_m E(r_m)$$

where the middle line is obtained by using (7-54) and the strong law of large numbers. This establishes the first equality. The proof of the second equality is left as Exercise 7-68. □

Example 7-20: Example 7-2, machine maintenance, continued In Example 7-2, let $g = \frac{1}{3}$, $\phi = \frac{4}{5}$, $u = \frac{7}{10}$, $d = \frac{2}{5}$, and $\rho = \frac{1}{2}$. From Exercise 7-21 or 7-46, $p_0 = \frac{24}{53}$, $p_1 = \frac{11}{53}$, and $p_2 = \frac{18}{53}$. Suppose the expected output per period from the machine is 10 units when it is operating perfectly and 5 units when it is operating with a minor defect. Set $E(r_0) = 10$ and $E(r_1) = 5$; then (7-73) asserts that the long-run production rate is $10 \times \frac{24}{53} + 5 \times \frac{11}{53} = \frac{295}{53} = 5.57$ units per period *no matter what the initial state of the machine is.* □

Example 7-21: Example 7-4, the (s, S) inventory model, continued To obtain the frequency of placing orders, set

$$r_{ij} = \begin{cases} 1 & i \neq S, j = S \\ 0 & i \neq S, j \neq S \end{cases}$$

Then $E(r_i) = F^c(Q - i)$ for $i \neq S$. For $i = S$, an order is placed if the demand during the next period exceeds Q; in the same manner as above, $E(r_S) = F^c(Q)$ is obtained. From (7-73), the asymptotic rate at which orders are placed is $\sum_{i=s}^{S} \pi_i F^c(Q - i)$, where $\{\pi_i\}_{i=s}^{S}$ are given in Example 7-17. □

A Sharpening of Theorem 7-12

Exercise 7-22 shows that for Example 7-2, $P_{ij}(16)$ is accurately approximated by p_j, which does not depend on i. However, it seems reasonable that the proportion of the first 16 days that the machine is working perfectly would be larger if the machine worked perfectly when it was purchased (it was in state 0) than if it were broken when purchased (it was in state 2). Similarly, in Example 7-20 it seems reasonable that for every finite n, $E[R_0(n)] > E[R_1(n)] > E[R_2(n)]$. The next theorem addresses this issue.

In Section 7-4, $M_{ij}(n)$ is defined as the expected number of visits to state j after n transitions, when the initial state is i; $M_{ii}(n)$ does *not* include the visit to state i at epoch 0. In the current context, it is important to include the visit to state i at epoch 0. Let $M_{ij}^{\#}(n + 1)$ correspond to $M_{ij}(n)$ in this situation; manifestly

$$M_{ij}^{\#}(n + 1) \triangleq M_{ij}(n) + \begin{cases} 1 & \text{if } i = j \\ 0 & \text{if } i \neq j \end{cases}$$

[The argument of $M_{ij}^{\#}(\ \cdot\ )$ is larger by 1 than the argument of $M_{ij}(\ \cdot\ )$ because, in effect, a transition "at zero" has been introduced.] In matrix notation, use (7-21) to write

$$M^{\#}(n+1) = M(n) + I = I + \sum_{k=1}^{n} P^k \qquad n = 0, 1, \ldots \qquad (7\text{-}74)$$

with the sum interpreted as zero when $n = 0$.

For ease of notation, let $Q = P^{\infty}$. From Proposition 7-5 and Theorem 7-12, $M_{ij}^{\#}(n) - nQ_{ij} = o(n)$ for large n. The object of the ensuing analysis is to obtain an explicit expression for the dominant term in the function $o(n)$. Except for Lemma 7-1, the analysis is restricted to MCs with a *finite* number of states.

Lemma 7-1 Let P be an irreducible stochastic matrix with $P^{\infty} = Q \neq 0$. Then

(a) $QP = PQ = Q$
(b) $Q^n = Q$
(c) $Q^m P^n = P^n Q^m = Q$
(d) $(P - Q)^n = P^n - Q, \quad n \in I_+$

PROOF From Proposition 7-5, Q has identical rows, and the common entry in the jth column is p_j. Since $Q \neq 0$, $p_j = \pi_j$ and satisfies (7-45). Hence, the (i, j)th entry of PQ is $\sum_j p_{ij}\pi_j$, which equals π_i by (7-45). The (i, j)th entry of QP is $\sum_j \pi_j p_{ij} = \pi_i$, establishing (a).

The (i, j)th element of Q^2 is $\sum_i \pi_i \pi_j = \pi_j$, so $Q^2 = Q$ and (b) follows by induction. Part (c) is obtained by combining (a) and (b):

$$Q^m P^n = Q P^n = Q P P^{n-1} = Q P^{n-1} = \cdots = QP = Q$$

Part (d) is trivially true when $n = 1$. For $n = 2$ use (b) and (c) to obtain

$$(P - Q)^2 = P^2 + Q^2 - PQ - QP = P^2 - Q$$

A simple induction completes the proof. ☐

Lemma 7-2 Let P be a *finite* and *irreducible* stochastic matrix, with $P^{\infty} = Q$. Then

$$F \triangleq I + \sum_{n=1}^{\infty} (P^n - Q)$$

exists and has the representation

$$F = [I - (P - Q)]^{-1} \qquad (7\text{-}75)$$

PROOF From part (d) of Lemma 7-1, $P^n - Q = (P - Q)^n$. Since $P^n \to Q$, $(P - Q)^n \to 0$ and Theorem 7-3 asserts that F exists and is given by (7-75). ☐

The matrix F is called the *fundamental matrix* for a Markov chain (with transition matrix P) because of its role in the following theorem and its corollary.

Theorem 7-15 For P as in Lemma 7-2 and $M^{\#}(n)$ defined by (7-74),

$$\lim_{n \to \infty} [M^{\#}(n) - nQ] = F - Q$$

where F is given by (7-75).

PROOF Use $M^{\#}(n) = I + \sum_{k=1}^{n-1} P^k$ to obtain

$$M^{\#}(n) - nQ = I + \sum_{k=1}^{n-1} P^k - nQ$$

$$= I + \sum_{k=1}^{n-1} (P^k - Q) - Q \qquad n = 1, 2, \ldots$$

According to part (d) of Lemma 7-1, $P^k - Q = (P - Q)^k$, hence

$$M^{\#}(n) - nQ = I + \sum_{k=1}^{n-1} (P - Q)^k - Q$$

$$\to F - Q$$

by Lemma 7-2. $\qquad\qquad\square$

A corresponding result can be obtained for Markov chains with rewards. Let $\mathcal{S} = \{0, 1, \ldots, s\}$, $\mathbf{R}(n) = (R_0(n), \ldots, R_s(n))'$, $\rho_i = E(r_i)$, and $\boldsymbol{\rho} = (\rho_0, \ldots, \rho_s)$. Then (7-73) can be written as

$$\frac{E[\mathbf{R}(n)]}{n} \to Q\boldsymbol{\rho}$$

Corollary 7-15 For P as in Lemma 7-2 and $\rho_i < \infty$, $i = 0, 1, \ldots, s$,

$$\lim_{n \to \infty} E[\mathbf{R}(n + 1)] - nQ\boldsymbol{\rho} = (F - Q)\boldsymbol{\rho} \qquad (7\text{-}76)$$

PROOF Recall that rewards are earned at the end of a transition, so that after $n + 1$ transitions occur, n rewards have accumulated. From (7-71) and Wald's equation (Theorem 6-1),

$$E[R_i(n + 1)] = \rho_i + \sum_{j=1}^{s} M_{ij}(n)\rho_j$$

Using matrix notation and (7-74), we obtain

$$E[\mathbf{R}(n)] = [I + M(n)]\boldsymbol{\rho} = M^{\#}(n)\boldsymbol{\rho}$$

Hence,

$$E[\mathbf{R}(n)] - nQ\boldsymbol{\rho} = [M^{\#}(n) - nQ]\boldsymbol{\rho}$$

$$\to (F - Q)\boldsymbol{\rho}$$

by Theorem 7-15. $\qquad\qquad\square$

Example 7-22: Example 7-20 continued The transition matrix for this example is

$$P = \begin{pmatrix} 0.7 & 0.2 & 0.1 \\ 0 & 0.4 & 0.6 \\ 0.4 & 0.1 & 0.5 \end{pmatrix}$$

and the rows of Q are $\pi = (\frac{24}{53}, \frac{11}{53}, \frac{18}{53})$. Straightforward matrix calculations provide, to three decimal places,

$$F = (I - P + Q)^{-1} = \begin{pmatrix} 1.325 & 0.037 & 0.367 \\ -0.742 & 1.166 & 0.164 \\ 0.011 & -0.152 & 1.134 \end{pmatrix}$$

From Theorem 7-15, the expected number of the first 16 periods that the machine is in state 0, if it was purchased in state 0, is approximately

$$M_{00}^{\#}(16) = 16Q_{00} + F_{00} - Q_{00}$$
$$= 15\pi_0 + 1.325 = 8.117$$

If the machine was purchased in state 2, the expected number of the first 16 periods that it is in state 0 is approximately

$$M_{20}^{\#}(16) = 15\pi_0 + F_{20} = 6.803$$

Observe that for large values of n,

$$M_{00}^{\#}(n) - M_{20}^{\#}(n) \approx F_{00} - F_{20} = 1.314 \qquad \square$$

Example 7-23: Example 7-21 continued The expected gain in starting with a perfectly working machine rather than a failed one is $E[R_0(n) - R_2(n)]$. Corollary 7-15 yields

$$E[R_i(n) - R_k(n)] \rightarrow \sum_{j=0}^{s} (F_{ij} - Q_{ij})\rho_j - \sum_{j=0}^{s} (F_{kj} - Q_{kj})\rho_j$$
$$= \sum_{j=0}^{s} (F_{ij} - F_{kj})\rho_j \qquad (7\text{-}77)$$

for any two initial states i and k. In this example, $\rho = (10, 5, 0)$, and F is given in Example 7-22. Thus,

$$E[R_0(n) - R_2(n)] \rightarrow (1.325 - 0.011)10 + (0.037 + 0.152)5 = 14.09$$

A somewhat surprising result is obtained when initial states 1 and 2 are compared. From (7-77),

$$E[R_2(n) - R_1(n)] \rightarrow (0.011 + 0.742)10 - (0.152 + 1.166)5 = 0.94$$

When n is large, starting with a failed machine provides larger expected rewards than starting with a functioning (but partially defective) machine. Can you explain this anomaly? $\qquad \square$

Lemmas 7-1 and 7-2 are not valid for periodic or reducible MCs because the limit P^∞ does not exist. When all the states are ergodic, Exercise 7-79 asks you to prove that the asymptotic proportion of time the MC is in state j when the initial state is i exists for all pairs of states (i, j). These probabilities are the elements of the matrix

$$P^* \triangleq \lim_{N \to \infty} \frac{1}{N} \sum_{n=0}^{N-1} P^n$$

In Section 4-6 of Volume II, it is important to know that P^∞ can be replaced by P^* in Lemmas 7-1 and 7-2.

Proposition 7-6 Let P be a stochastic matrix consisting entirely of ergodic states. Then Lemmas 7-1 and 7-2 are valid when P^∞ is replaced by

$$P^* \triangleq \lim_{N \to \infty} \frac{1}{N} \sum_{n=0}^{N-1} P^n$$

PROOF This is left as Exercise 7-80. □

Here is another fact that is important in Section 4-6 of Volume II.

Proposition 7-7 Let P be the transition matrix of a finite MC consisting of s ergodic states. Then

$$\text{rank}(I - P) + \text{rank}(P^*) = s$$

PROOF This is left as Exercise 7-81. □

Probability of Leaving the Set of Transient States

In Corollary 7-9, the state space is partitioned into a single communicating class and a set of transient states, and it is necessary to be able to determine whether the chain enters the communicating class, w.p.1, when it starts in a transient state. The following analysis provides a way to check this condition.

Let C be the communicating class and T be the set of transient states; both C and T may be countably infinite. Let $\{Y_n; n \in I\}$ be the Markov chain and

$$y_i(n) = P\{Y_n \in C \mid Y_0 = i\} \qquad n \in I, \quad i \in T$$

Clearly, $y_i(0) = 0$. Condition on the first transition to obtain

$$y_i(n + 1 \mid Y_1 = j) = \begin{cases} 1 & \text{if } j \in C \\ y_j(n) & \text{if } j \in T \end{cases}$$

Hence

$$y_i(n + 1) = \sum_{j \in T} p_{ij} y_j(n) + \sum_{j \in C} p_{ij} \qquad i \in T, \quad n \in I \qquad (7\text{-}78)$$

By its nature, $y_i(n)$ is nondecreasing in n and bounded by 1, and so

$y_i \triangleq \lim_{n \to \infty} y_i(n)$ exists. Taking limits as $n \to \infty$ on both sides of (7-78) yields

$$y_i = \sum_{j \in T} p_{ij} y_j + \sum_{j \in C} p_{ij} \qquad i \in T \tag{7-79}$$

If T consists of a finite number of states, $y_i = 1$ for all $i \in T$, as shown in the proof of Corollary 7-4. Indeed, when T is a finite set, the unique solution of (7-79) is $y_i \equiv 1$ (Exercise 7-74). Suppose T consists of infinitely many states; $y_i \equiv 1$ is always a solution of (7-79), but it may not be the only solution such that $y_i \leq 1$ for all $i \in T$.

Example 7-24 Let P be given by

$$P = \begin{pmatrix} 1 & 0 & 0 & 0 & \cdots \\ q & 0 & p & 0 & \cdots \\ 0 & q & 0 & p & \cdots \\ \end{pmatrix}$$

where $q = 1 - p$. State 0 is absorbing, so let $C = \{0\}$. Since P is the transition matrix of the embedded chain of a birth-and-death process with $\lambda_0 = 0$, $\lambda_i = \lambda > 0$ for $i \in I_+$, $\mu_i = \mu$ for $i \in I_+$, and $p = \lambda/(\lambda + \mu)$ (see Exercise 7-5), we suspect that I_+ is a set of transient states. Theorem 7-8 can be used to verify this conjecture, so let $T = I_+$.

Equation (7-79) yields

$$y_1 = py_2 + q$$

$$y_i = py_{i+1} + qy_{i-1} \qquad i = 2, 3, \ldots$$

There are two solutions to this system of equations:

$$y_i = 1 \quad \text{and} \quad y_i = \left(\frac{q}{p}\right)^i \qquad i \in I_+$$

When $p \leq \frac{1}{2}$, $y_i \equiv 1$ is the only probabilistically meaningful solution. When $p > q$, the second solution gives $y_i < 1$, and Corollary 4-10 suggests the conjecture that the smaller value is $\lim_{n \to \infty} P\{Y_n = 0 \mid Y_0 = i > 0\}$. $\square$

If you argue from a specific example to the general situation, then Example 7-24 makes the following theorem plausible.

Theorem 7-16 Let $\{Y_n; \; n \in I\}$ be a Markov chain with state space $S = C \cup T$, with $C \cap T = \varnothing$, where C is a communicating set of states and T is a set of transient states. Then $\lim_{n \to \infty} P\{Y_n \in C \mid Y_0 = i \in T\}$ is given by the smallest nonnegative solution of (7-79).

PROOF Let z_i, $i \in I_+$, be *any* nonnegative solution of (7-79). From (7-78) and (7-79),

$$z_i \geq \sum_{j \in C} p_{ij} = y_i(1) \qquad i \in T$$

Assume $z_i \geq y_i(n)$ for some n; (7-79) yields

$$z_i = \sum_{j \in T} p_{ij} z_j + \sum_{j \in C} p_{ij} \geq \sum_{j \in T} p_{ij} y_j(n) + \sum_{j \in C} p_{ij}$$
$$= y_i(n + 1)$$

Hence $\lim_{n \to \infty} y_i(n)$ is the smallest nonnegative solution of (7-79). □

Finite Solutions of Equation (7-19)

The mean first-passage times between states i and j, denoted μ_{ij}, have been used in various parts of this chapter. Now we discuss how to compute them.

Equation (7-19), repeated as (7-80), relates all the $\{\mu_{ij}\}$:

$$\mu_{ij} = 1 + \sum_{k \neq j} p_{ik} \mu_{kj} \qquad i, j \in S \tag{7-80}$$

If C is a communicating class, $\mu_{ij} = \infty$ when $i \in C$ and $j \notin C$, so it makes sense to seek a finite solution of (7-80) in two situations: when $i, j \in C$ and when i is a transient state and $j \in C$. You are asked to analyze the latter situation in Exercise 7-76; we examine the former now.

Recall that a communicating class can be construed as an irreducible chain. Typically, first one would determine whether C consists of nonnull states via Theorem 7-11. If the states of C are null, there is little point in solving (7-80) [observe that $\mu_{ij} \equiv \infty$ does formally solve (7-80)]. We shall obtain the important result that (7-80) has a unique finite solution when C consists of ergodic states. A theorem to that effect is established with the help of this lemma.

Lemma 7-3 Let P be an irreducible stochastic matrix that represents the transition matrix of a Markov chain with ergodic states. Let $\mathbf{y}$ be a column vector that satisfies

$$P\mathbf{y} = \mathbf{y} \tag{7-81}$$

Then $\mathbf{y} = (c, c, \dots)'$ for some constant c.

PROOF From (7-81), $P^2\mathbf{y} = P(P\mathbf{y}) = P\mathbf{y} = \mathbf{y}$, and, in general, $P^n\mathbf{y} = \mathbf{y}$. The hypotheses imply that $Q = P^\infty$ exists, hence $Q\mathbf{y} = \mathbf{y}$. Since the rows of Q are identical, so are the elements of $\mathbf{y}$. □

Theorem 7-17 If P satisfies the hypothesis of Lemma 7-3, then (7-80) has a unique finite solution.

PROOF First we show that (7-80) has a finite solution. From (7-44), set $\mu_{jj} = 1/\pi_j < \infty$ for each j. The theorem is trivially true if P consists of a single element, so assume there are more. Let the generic random variable T_{jj} be the length of a first return interval to state j, and let the generic random variable V_{ij} be the number of visits to state i between successive entries into state j.

Fix $i \neq j$; since a first return to state j may visit state i, but need not, we have

$$\mu_{jj} \geq E(T_{jj} \mid V_{ij} > 0)P\{V_{ij} > 0\}$$

The first term on the right is at least as large as μ_{ij}, hence

$$\frac{1}{\pi_j} \geq \mu_{ij} P\{V_{ij} > 0\} \tag{7-82}$$

If $P\{V_{ij} > 0\} > 0$ is established, (7-82) implies $\mu_{ij} < \infty$.

Observe that $E(V_{ij}) = P\{V_{ij} > 0\}E(V_{ij} \mid V_{ij} > 0)$. Exercise 7-78 asks you to prove $E(V_{ij}) = \pi_i/\pi_j$. Consequently,

$$P\{V_{ij} > 0\} = \frac{\pi_i}{\pi_j E(V_{ij} \mid V_{ij} > 0)} > 0 \tag{7-83}$$

Combining (7-82) and (7-83) yields

$$\mu_{ij} \leq \frac{1}{\pi_j P\{V_{ij} > 0\}} < \infty$$

for all pairs of states i and j.

To prove uniqueness, write (7-80) in matrix form as

$$M = \Theta + P(M - D) \tag{7-84}$$

where $M = (\mu_{ij})$, Θ is a matrix consisting entirely of 1's, and D is a matrix whose off-diagonal elements are all zero and whose diagonal elements are given by $D_{ii} = \mu_{ii}$; all these matrices are square and have the same order as P. Multiply (7-84) by π:

$$\pi M = \pi \Theta + \pi M - \pi D$$

Hence

$$\pi D = \pi \Theta = 1 \tag{7-85}$$

Suppose M and M^* are two distinct solutions of (7-84). Set $\Delta = M - M^*$ and let D^* be the diagonal matrix with $D_{ii}^* = \mu_{ii}^*$. By (7-85), $\pi D^* = 1$; that is, $D_{ii}^* = 1/\pi_i = D_{ii}$, so $D^* = D$. Thus, Δ satisfies $\Delta = P\Delta$, and so each column of Δ satisfies (7-81). Lemma 7-3 asserts that each column of Δ consists of identical constants; since $\Delta_{ii} = \mu_{ii} - \mu_{ii}^* = 0$, $\Delta = 0$ and so (7-84) has a unique solution. $\qquad\square$

EXERCISES

7-68 Complete the proof of Theorem 7-14.

7-69 Give an alternative proof of Theorem 7-14 that uses Corollary 6-8a.

7-70 In Example 7-17, let $h(i)$ be the holding cost incurred in a period that ends with i items in inventory. Find the asymptotic holding-cost rate.

7-71 (Continuation) In Example 7-17, find the asymptotic proportion of demands that are back-logged and the rate at which orders are placed. Use the methods of this section.

7-72 Show that the fundamental matrix F, defined by (7-75), has the following properties: (a) $PF = FP$, (b) $F\mathbf{1} = \mathbf{1}$, (c) $\pi F = \pi$, where $\pi = \pi P$, and (d) $I - F = Q - PF$.

7-73 Exhibit two different transition matrices which have the same fundamental matrix.

7-74 Show that when T is a finite set, (7-79) has a unique solution.

7-75 Suppose a transition matrix can be put in the form

$$P = \begin{matrix} C_1 \\ C_2 \\ T \end{matrix} \begin{pmatrix} S_1 & 0 & 0 \\ 0 & S_2 & 0 \\ R_1 & R_2 & U \end{pmatrix}$$

Let $y_{ki}(n) = P\{Y_n \in C_k \mid Y_0 = i \in T\}$, $y_{ki} = \lim_{n\to\infty} y_{ki}(n)$, and $\mathbf{y}_k = (y_{k1}, y_{k2}, \dots)$. Derive a system of equations satisfied by $\mathbf{y}_k$. When T is a finite set, show that $\mathbf{y}_k = (I - U)^{-1}R_k\mathbf{1}$. Give a probabilistic interpretation of $(I - U)^{-1}R_k$.

7-76 Find μ_{ij} when i is a transient state and j is in some communicating class. *Hint:* Use Exercise 7-75.

7-77 Consider (7-82) for a finite chain. Let F be the fundamental matrix and F_{dg} be the diagonal matrix formed by setting the off-diagonal elements of F to zero. Show that $M = (I - F + \Theta F_{dg})D$ solves (7-82).

7-78 With the notation and hypotheses used in the proof of Theorem 7-17, prove $E(V_{ij}) = \pi_i/\pi_j$ for any pair of states i and j. *Hint:* Use Theorem 6-10.

7-79 Let P be the transition matrix of the Markov chain $\{Y_n; \; n \in I\}$. Let $N_{ij}(n) \triangleq \# \{k : Y_k = j, \; k = 0, 1, \dots, n; \; Y_0 = i\}$ and $\eta_{ij} \triangleq \lim_{n\to\infty} N_{ij}(n)/n$. Show that η_{ij} exists and is the (i, j)th element of the matrix

$$P^* \triangleq \lim_{N\to\infty} \sum_{n=0}^{N-1} \frac{P^n}{N}$$

What is the form of P^* when there are c communicating classes?

7-80 Prove Proposition 7-6. *Hint:* Review the proof of Lemmas 7-1 and 7-2.

7-81 Prove Proposition 7-7. *Hint:* Establish $s = \text{rank}(I - P + P^*) \le \text{rank}(I - P) + \text{rank}(P^*) \le s$

7-8* SIMPLE-RANDOM-WALK AND GAMBLER'S RUIN PROCESSES

This section describes some of the properties of two special Markov chains.

Simple-Random-Walk Process

The simple random walk usually is associated with the following situation. An inebriated celebrant starts wending his way home. At each step, he walks to the northeast with probability p and to the southeast with probability $q = 1 - p$. His next step is independent of where he is and how he got there. Let S_n be his position after n steps, with $S_0 = 0$. Figure 7-6a shows a realization of the S_n-process; in Figure 7-6b the points in Figure 7-6a are connected by line segments. The latter conforms to the typical path walked by a drunkard in cartoons.

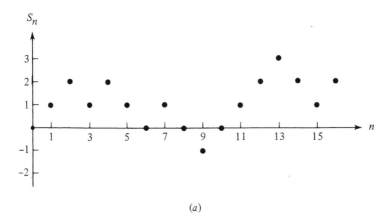

(a)

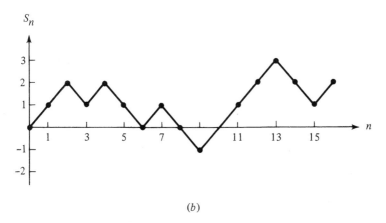

(b)

Figure 7-6 Illustration of a simple random walk.

The simple random walk is an important model in several fields, including finance, ecology, and physics. Its main use in operations research is as a building block for more complicated models, some of which are illustrated in this section. The formal definition of the process is as follows.

Definition 7-9 Let Y_1, Y_2, ... be i.i.d. r.v.'s with $P\{Y_1 = 1\} = p$ and $P\{Y_1 = -1\} = q \triangleq 1 - p$. Fix an integer i, set $S_0 = i$, and $S_n = S_{n-1} + Y_n$ for $n \in I_+$. The process $\{S_n; n \in I\}$ is a *simple random walk*. When $p = q$, the process is called *symmetric*.

The adjective *simple* pertains to the two-point distribution for Y_1. Random walks where Y_1 has an arbitrary distribution are described in Section 10-3.

It follows immediately from Definition 7-9 that $\{S_n; n \in I\}$ is a Markov chain with state space $\{0, \pm 1, \pm 2, \dots\}$ and transition matrix

$$
P =
\begin{array}{c}
\\
\\
2 \\
1 \\
0 \\
-1 \\
-2 \\
\\
\end{array}
\begin{array}{cccccc}
\cdots & -2 & -1 & 0 & 1 & 2 & \cdots \\
\end{array}
\left(
\begin{array}{ccccccc}
& & & & & & \\
\cdots & 0 & 0 & 0 & q & 0 & \cdots \\
\cdots & 0 & 0 & q & 0 & p & \cdots \\
\cdots & 0 & q & 0 & p & 0 & \cdots \\
\cdots & q & 0 & p & 0 & 0 & \cdots \\
\cdots & 0 & p & 0 & 0 & 0 & \cdots \\
& & & & & & \\
\end{array}
\right)
\qquad (7\text{-}86)
$$

From (7-86) it is evident that P is irreducible (so all states have the same classification) and the states are periodic, with period 2. An interesting feature of the symmetric process is given by this theorem.

Theorem 7-18 The symmetric simple random walk is a null-persistent Markov chain.

PROOF First show persistence. Let y_i be the probability that state 0 will be entered within a finite number of steps when the initial state is i. If it can be shown that $y_i = 1$ for all integers i, persistence will be established.

Clearly $y_0 = 1$. Conditioning on the value of Y_1 yields

$$
y_i = \frac{y_{i+1} + y_{i-1}}{2}
$$

hence

$$
y_{i+1} - y_i = y_i - y_{i-1} \triangleq \Delta
$$

If $\Delta > 0$, y_i increases without bound, while if $\Delta < 0$, y_i decreases without bound. Both are absurd since $0 \le y_i \le 1$; hence $\Delta = 0$ and $y_i = y_0 = 1$ for all i.

To show that all states are null, observe that the structure of P implies $\mu_{i0} = |i| \mu_{10}$ (cf. Exercise 7-82 and the proof of Lemma 4-1). Conditioning on the first event [or, formally, substituting in (7-80)] yields

$$
\mu_{00} = 1 + \mu_{10}
$$

and

$$
\mu_{10} = 1 + \frac{\mu_{20}}{2} = 1 + \mu_{10}
$$

The latter equation has no finite solution; according to Theorem 7-17, this means $\mu_{10} = \infty$, hence $\mu_{00} = \infty$. $\qquad \square$

The symmetry in P and its similarity to the embedded chain of a birth-and-death process (Exercise 7-5) lead to a simple proof of the following fact.

Proposition 7-8 When $p \neq q$, the simple random walk is a transient Markov chain.

PROOF This is left as Exercise 7-83. ☐

An explicit expression for the n-step transition probabilities is easy to obtain. Observe that $S_n = S_0 + Y_1 + \cdots + Y_n$, hence

$$P\{S_n = j \mid S_0 = i\} = P\{Y_1 + \cdots + Y_n = j - i\}$$
$$= P\{S_n = j - i \mid S_0 = 0\}$$

so it suffices to determine

$$P_j(n) = P\{S_n = j \mid S_0 = 0\} \qquad j \in S, n \in I \qquad (7\text{-}87)$$

Let the random variable U_n be the number of Y_i's that equal $+1$ and the random variable D_n be the number of Y_i's that equal -1, $1 \leq i \leq n$. Clearly,

$$U_n + D_n = n \qquad n \in I \qquad (7\text{-}88)$$

Thus

$$S_n = S_0 + \sum_{i=1}^{n} Y_i = S_0 + U_n - D_n$$

and when $S_0 = 0$,

$$S_n = j \Leftrightarrow U_n - D_n = j \qquad (7\text{-}89)$$

Solve (7-88) and (7-89) simultaneously to obtain

$$S_n = j \Leftrightarrow U_n = \frac{n+j}{2} \qquad D_n = \frac{n-j}{2}$$

Since U_n must be an integer,

$$P\{S_n = j \mid S_0 = 0\} = 0 \qquad j \text{ odd (even) and } n \text{ even (odd)} \qquad (7\text{-}90a)$$

Consider the complementary case. Since the Y_i's are i.i.d. and have a range of exactly two numbers, U_n is a binomial r.v. and

$$P\{S_n = j \mid S_0 = 0\} = \binom{n}{(n+j)/2} p^{(n+j)/2} q^{(n-j)/2} \qquad (7\text{-}90b)$$

for both j and n even (odd).

Gambler's Ruin Process

The gambler's ruin process is described in Example 7-1 and Exercise 7-1 as a Markov chain. A different view of the process is to construct it as a simple random walk with *barriers*. Let $S_0 = i > 0$ be the initial stake of the gambler and $b - i > 0$ be the stake of the opponent. Let Y_n be the return on the nth trial and S_n be the gambler's fortune after n trials. Also Y_n and S_n are as in Definition 7-9 *except* that there are no further trials once $S_n = 0$ (the gambler is ruined) or $S_n = b$ (the opponent is ruined). Thus, $\{S_n; n \in I\}$ behaves as a simple random walk when $0 < S_n < b$, and 0 and b represent *absorbing* barriers, in the sense that once they are "touched," the process is absorbed there. See Figure 7-7.

The natural object to study is the probability that the gambler is ruined at the nth trial, $R(n; i, b)$, say. Since the gambler's ruin process is a finite Markov chain with two absorbing states 0 and b, $R(n; i, b)$ can be obtained from the general theory (how?). The point of the following analysis is to exploit the special structure of this important process.

Conditioning on Y_1 yields

$$R(n; i, b) = pR(n - 1; i + 1, b) + qR(n - 1; i - 1, b) \qquad (7\text{-}91)$$

for $0 < i < b$ and $n \in I_+$. Clearly,

$$R(n; 0, b) = 0 = R(n; b, b) \qquad n \in I_+ \qquad (7\text{-}92a)$$

and

$$R(0; 0, b) = 1 \qquad R(0; i, b) = 0 \qquad 0 < i \le b \qquad (7\text{-}92b)$$

We will solve (7-91), with boundary conditions (7-92a) and (7-92b), by the method of generating functions. For notational convenience, suppress the dependence on b and let $R_i(n) \triangleq R(n; i, b)$. Define

$$\hat{R}_i(z) = \sum_{n=0}^{\infty} z^n R_i(n)$$

Multiply both sides of (7-91) by z^n, $|z| < 1$, and sum over I_+ to obtain

$$\hat{R}_i(z) = pz\hat{R}_{i+1}(z) + qz\hat{R}_{i-1}(z) \qquad 0 < i < b \qquad (7\text{-}93)$$

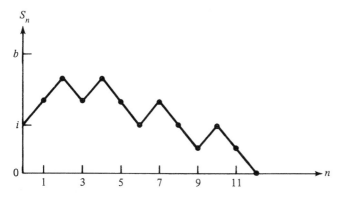

Figure 7-7 A possible sample path of a gambler's ruin process.

Do the same on (7-92a) and (7-92b) to obtain

$$\hat{R}_0(z) = 1 \qquad \hat{R}_b(z) = 0 \qquad (7\text{-}94)$$

For each value of z, (7-93) is a difference equation with boundary conditions (7-94). It can be solved by applying the general theory of linear difference equations [see, e.g., Section 1.7 of Hildebrand (1968)]. We use probabilistic intuition to construct the solution. Exercise 7-91 asks you to show that (7-93) and (7-94) have a unique solution.

Temporarily ignore the boundary conditions (7-94). By analogy with the simple random walk, the number of steps required to reach state 0 from state i is the sum of i i.i.d. random variables with distribution $R_1(\cdot)$; hence $\hat{R}_i(z) = [R_1(z)]^i$. Thus, a solution of (7-93) ought to be $[x(z)]^i$ for some function $x(z)$ taking values in $[0, 1]$. Substituting $\hat{R}_i(z) = [x(z)]^i$ into (7-93) shows that $x(z)$ must satisfy

$$pzx^2(z) - x(z) + qz = 0$$

There are two distinct solutions of this equation:

$$x_1(z) = \frac{1 + \sqrt{1 - 4pqz^2}}{2pz} \quad \text{and} \quad x_2(z) = \frac{1 - \sqrt{1 - 4pqz^2}}{2pz} \qquad (7\text{-}95)$$

It is easily shown (Exercise 7-92) that

$$\hat{R}_i(z) = c_1(z)[x_1(z)]^i + c_2(z)[x_2(z)]^i \qquad (7\text{-}96)$$

solves (7-93), where $c_1(z)$ and $c_2(z)$ are arbitrary functions of z. Recall the boundary equations (7-94), and attempt to choose $c_1(z)$ and $c_2(z)$ to satisfy (7-94). This requires that

$$1 = R_0(z) = c_1(z) + c_2(z)$$

and

$$0 = R_b(z) = c_1(z)x_1^b(z) + c_2(z)x_2^b(z)$$

Solving these two equations simultaneously and substituting the result into (7-96) yield

$$\hat{R}_i(z) = \frac{x_1^b(z)x_2^i(z) - x_1^i(z)x_2^b(z)}{x_1^b(z) - x_2^b(z)}$$

This formula can be simplified by observing that $x_1(z)x_2(z) = q/p$, hence

$$\hat{R}_i(z) = \left(\frac{q}{p}\right)^i \frac{x_1^{b-i}(z) - x_2^{b-i}(z)}{x_1^b(z) - x_2^b(z)} \qquad (7\text{-}97)$$

The probability of ultimate ruin is given by

$$R_i \triangleq \sum_{n=0}^{\infty} R_i(n) = \lim_{z \uparrow 1} \hat{R}_i(z)$$

$$= \left(\frac{q}{p}\right)^i \frac{(q/p)^{b-i} - 1}{(q/p)^b - 1} \qquad 0 \le i < b \qquad (7\text{-}98a)$$

when $p \neq q$. From L'Hospital's rule,

$$R_i = \frac{b - i}{b} \qquad 0 \leq i \leq b \qquad p = q = \tfrac{1}{2} \qquad (7\text{-}98b)$$

Let $W_i(n)$ be the probability that the game ends at the nth trial with the gambler winning. Since the fortune of the gambler's opponent is a gambler's ruin process with p interchanged with q and i interchanged with $b - i$ (why?), $W_i(n)$ is the probability that the opponent is ruined at the nth trial. From (7-97),

$$\hat{W}_i(z) \triangleq \sum_{n=0}^{\infty} z^n W_i(n) = \left(\frac{p}{q}\right)^{b-i} \frac{x_1^i(z) - x_2^i(z)}{x_1^{b-i}(z) - x_2^{b-i}(z)} \qquad (7\text{-}99)$$

Let the *duration* of the game be the number of trials required for *some* player to be ruined. In Markov chain terminology, the duration of the game is the number of transitions until absorption. Its generating function is $\hat{R}_i(z) + \hat{W}_i(z)$ (why?).

Reflecting Barriers

In Example 7-1, the gambler's ruin process is used to describe the occupancy of a file. The file is removed from service when it becomes empty or full. A variation of that model is to remove the file only when it becomes full. In terms of the simple random walk, a *reflecting barrier* is placed at zero. Specifically, take the transition matrix of a Markov chain to be

$$P = \begin{array}{c} \\ \\ 0 \\ 1 \\ 2 \\ \vdots \\ b-2 \\ b-1 \\ b \end{array} \begin{pmatrix} q_0 & p_0 & 0 & \cdots & 0 & 0 & 0 \\ q & 0 & p & \cdots & 0 & 0 & 0 \\ 0 & q & 0 & \cdots & 0 & 0 & 0 \\ & & & \cdots\cdots\cdots\cdots\cdots & & & \\ 0 & 0 & 0 & \cdots & 0 & p & 0 \\ 0 & 0 & 0 & \cdots & q & 0 & p \\ 0 & 0 & 0 & \cdots & 0 & 0 & 1 \end{pmatrix} \qquad (7\text{-}100)$$

The Markov chain corresponding to this transition matrix differs from the gambler's ruin process only in that whenever state 0 is reached, the next transition moves the process to state 1 with probability p_0 or keeps the process at state 0 with probability $q_0 = 1 - p_0$. Thus, each visit to state 0 lasts for k transitions with probability $(1 - p_0)^{k-1} p_0$.

From (7-100) it is apparent that all states can reach state b and that state b is absorbing. This means that states $0, 1, \ldots, b - 1$ are transient (why?). The interesting problem is to determine the distribution of the number of transitions until absorption at b from any initial state. In terms of the file example, this is interpreted as the distribution of the number of transactions until the file is filled.

Let $W_i(n) \triangleq W(n; i, b)$ be the probability of absorption at b at epoch n when the initial state is i. Conditioning on the first event yields

$$W_i(n) = p W_{i+1}(n - 1) + q W_{i-1}(n - 1) \qquad 0 < i < b \qquad (7\text{-}101)$$

The boundary conditions for (7-101) are

$$W_b(0) = 1 \quad \text{and} \quad W_b(n) = 0 \quad n \in I_+ \tag{7-102a}$$

for the absorbing barrier at b, and

$$W_0(n) = p_0 W_1(n-1) + q_0 W_0(n-1) \tag{7-102b}$$

for the reflecting barrier at zero.

Define $\hat{W}_i(z) = \sum_{n=0}^{\infty} z^n W_i(n)$; it is obtained from (7-101), (7-102a) and (7-102b) in the same manner $\hat{R}_i(z)$ is obtained from (7-93) and (7-94). The details are left to Exercise 7-99. The result is

$$\hat{W}_i(z) = \frac{(pz)^{b-i} F_i(z)}{F_b(z)} \quad i = 0, 1, \dots, b \tag{7-103}$$

where

$$F_i(z) = (1 - q_0 z)p(x_1^i - x_2^i) = p_0 qz(x_1^{i-1} - x_2^{i-1})$$

with $x_1 = x_1(z)$ and $x_2 = x_2(z)$ given by (7-95).

A useful quantity that can be extracted from (7-103) is the expected time until absorption, μ_{ib} say. Straightforward, but extensive, calculations yield†

$$\mu_{ib} = \frac{d}{dz} W_i(z)\Big|_{z=1}$$

$$= \frac{b-i}{p-q} + \frac{(p-q-p_0)p}{(p-q)^2 p_0}\left[\left(\frac{q}{p}\right)^i - \left(\frac{q}{p}\right)^b\right] \quad p \neq q \tag{7-104a}$$

and

$$\mu_{ib} = (b-i)\left(\frac{1}{p_0} + b + i - 1\right) \quad p = q \tag{7-104b}$$

Equations (7-104a) and (7-104b) are valid when $p + q < 1$, that is, when $p_{ii} = 1 - p - q > 0$ for $1 \leq i \leq b - 1$. The variance of the time to absorption is also obtained from (7-103) by Hardin and Sweet. For $p > q$, the first term in (7-104a) is the mean first-passage time from i to b of a simple-random-walk process (Exercise 7-97), so the second term represents the effect of the reflecting barrier at zero.

Example 7-25 Firms that rent equipment have inventories which exhibit upward fluctuations because of returned items as well as the usual downward fluctuations caused by demands. As a particular example, consider a nationwide firm that rents trailers. These trailers may be picked up in one city and dropped off in another. In a growing city, the rate at which trailers are

---

† See J. C. Hardin and A. L. Sweet, "Moments of the Time to Absorption in the Random Walk between a Reflecting and an Absorbing Barrier," *SIAM Rev.* **12**: 140–142 (1970).

dropped off typically exceeds the rate at which they are picked up, so periodically trailers are transported to a city with a net outflow from a city with a net inflow.

Suppose the demand and return processes are Poisson with rates λ and μ, respectively, and assume they are independent. Suppose, further, that the following policy is used: When the inventory reaches b, one item is sent to another location. If the current inventory is j, what is the expected time until the inventory is zero?

Let an event denote a demand epoch or a return epoch, and let $\rho = \lambda/\mu$. The times between events are i.i.d. exponential r.v.'s with mean $1/(\lambda + \mu)$, and the probability that the next event is a demand, p' say, is $p' = \lambda/(\lambda + \mu) = \rho/(1 + \rho)$; $q' = 1/(1 + \rho)$ is the probability that the next event is a return.

Let $X(t)$ be the inventory level at time t, let t_n be the epoch of the nth event, and define $S_n = X(t_n^+)$. Manifestly, $\{S_n; n \in I\}$ is a simple random walk with a reflecting barrier at b. If an absorbing barrier is placed at zero and we set $S_0 = j$, then the number of transitions until absorption of $\{S_n; n \in I\}$, say N_j, is the number of events that have occurred by the time the inventory reaches zero.

In the process $\{S_n; n \in I\}$, the initial position is j above the reflecting barrier and $b - j$ below the absorbing barrier, and a transition moves toward the absorbing barrier with probability p'. A moment's reflection should convince you that $\mu_j = E(N_j)$ is given by (7-104), with $i = b - j$, $p = p_0 = p'$, and $q = q_0 = q'$. Hence

$$\mu_j = (1 + \rho)[-j(1 - \rho) + \rho^{-b}(1 - p^j)]/(1 - \rho)^2 \qquad \rho \neq 1 \quad (7\text{-}105a)$$

and

$$\mu_j = j(2b + 1 - j) \qquad\qquad\qquad \rho = 1 \quad (7\text{-}105b)$$

Let T_j be the time required for the inventory to reach zero. Then T_j has the representation $T_j = Z_1 + \cdots + Z_{N_j}$, where $Z_1, Z_2, \ldots$ are i.i.d. exponential r.v.'s with mean $1/(\lambda + \mu)$. Applying Wald's equation (Theorem 6-1) yields

$$E(T_j) = \frac{\mu_j}{\lambda + \mu} \qquad\qquad (7\text{-}106)$$

where μ_j is given by (7-105). $\qquad\qquad\qquad\qquad\qquad\qquad\qquad\square$

Example 7-26: Example 7-25 continued Expand the model of Example 7-25 by introducing a cost structure. Let r denote the expected rental fee paid by a customer, s denote the expectation of the net value of shipping an item to a high-demand location, and h denote the holding-cost rate for items in inventory. The objective of our investigation is to obtain that value of the "ship level" b that maximizes the asymptotic net reward rate.

Figure 7-8 shows a part of a typical realization of the inventory process for the ship levels b and $b - 1$. During $[t_1, t_2)$ the value of the ship level is

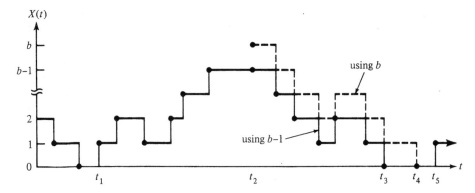

Figure 7-8 A typical realization of the inventroy process.

irrelevant. At epoch t_2, a returned item is shipped using ship level $b - 1$, but the item would be kept if level b were used. Consequently, if ship level b were used instead of level $b - 1$, the inventory would be larger by 1 during $[t_2, t_4)$, there would be one less item shipped (at epoch t_2), and one more rental fee would be earned (at epoch t_4). Observe that epochs t_1 and t_5 are the same when ship levels $b - 1$ and b are used.

Since the ship level acts as a reflecting barrier for the inventory process, $t_4 - t_2$ is the first-passage time denoted by T_b in Example 7-25, and $E(T_b)$ is given by (7-106) with $j = b$. For the cycle $[t_1, t_5]$, ship level b is better than ship level $b - 1$ if, and only if,

$$hE(T_b) < f - r \qquad (7\text{-}107)$$

Exercise 7-101 asks you to show that Theorem 6-5 asserts that (7-107) provides a valid way of comparing the asymptotic reward, and expected reward, per unit time. By its nature, $E(T_b)$ increases with b [this can be shown algebraically by using (7-105)], so the optimal value of b is the largest b that satisfies (7-107). □

EXERCISES

7-82 For the simple-random-walk process, prove that $\mu_{i0} = |i| \mu_{10}$.

7-83 Prove Proposition 7-8.

7-84 Explain where the proof of Theorem 7-18 fails when $p \neq q$.

7-85 Let $\{S_n^* ; n \in I\}$ be a Markov chain formed by modifying the simple-random-walk process and introducing $P\{S_{n+1}^* = j \mid S_n^* = j\} = r = 1 - (p + q) > 0$. Prove that Theorem 7-18 is valid for this process. Find $P\{S_n^* = j \mid S_0^* = 0\}$.

7-86 An alternative method of proving a symmetric simple random walk is a persistent MC is to show $\sum_{n=1}^{\infty} P\{S_{2n} = 0 \mid S_0 = 0\} = \infty$. From (7-90b), $P_{2n} \triangleq P\{S_{2n} = 0 \mid S_0 = 0\} = \binom{2n}{n} (\frac{1}{4})^n$. Stirling's formula is $n! \approx \sqrt{2\pi} \, n^{n+1/2} e^{-n}$, where π has its usual trigonometric meaning, and is remarkably accurate even for $n = 1$ (it approximates 1! by 0.9221). Use Stirling's formula to show $\sum_{n=1}^{\infty} P_{2n}$ diverges.

7-87 (Continuation) A simple random walk in two dimensions makes one-step transitions from (x, y) to exactly one of the neighboring points $(x + 1, y)$, $(x - 1, y)$, $(x, y + 1)$, $(x, y - 1)$; the probability of each of these transitions is $\frac{1}{4}$. Let $\mathbf{0} = (0, 0)$, S_n be the state (an ordered pair) after n transitions, and $P_{2n} = P\{S_{2n} = \mathbf{0} \mid S_n = \mathbf{0}\}$. Show

$$P_{2n} = \left(\frac{1}{4}\right)^{2n} \binom{2n}{n} \sum_{k=0}^{n} \binom{n}{k}^2 = \left(\frac{1}{4}\right)^{2n} \binom{2n}{n}^2$$

Hint: Use the combinatorial identities

$$\sum_{k=0}^{n} \binom{n}{k}^2 = \binom{2n}{n} \quad \text{and} \quad \binom{2n}{n}^2 = \sum_{k=0}^{n} \frac{2n!}{k!k!(2n-k)!(2n-k)!} \tag{7-108}$$

Now show that all states are persistent.

7-88 (Continuation) A simple random walk in three dimensions makes one-step transitions from (x, y, z) to one of the six neighboring points of the form $(x \pm 1, y \pm 1, z \pm 1)$, each with probability $\frac{1}{6}$. Extend the notation of Exercise 7-87 to this case in the natural way, and show

$$P_{2n} = \left(\frac{1}{2}\right)^{2n} \binom{2n}{n} \sum_{j=0}^{n} \sum_{k=0}^{n-j} \left[\frac{1}{3^n} \frac{n!}{j!k!(n-j-k)!}\right]^2$$

Explain why the sum of squares is smaller than the maximum term within brackets, which occurs when j and k are about $n/3$. Use this result to show $P_{2n} \approx 1/\sqrt{n^3}$. Conclude that *all states are transient*.

7-89 For k a positive integer, the Maclaurin series of $(1 + z)^k$ is

$$(1 + z)^k = 1 + \binom{k}{1} z + \cdots + \binom{k}{k} z^k$$

Compare the coefficients of z^n on both sides of $(1 + z)^i (1 + z)^j = (1 + z)^{i+j}$ and derive the first identity in (7-108). Use it to derive the second identity in (7-108).

7-90 In a simple random walk starting at zero, find the probability that state $j > 0$ will be reached before state $k < 0$. (No calculations are required.)

7-91 Argue that iterating (7-93) for $i = 1, 2, \ldots, b - 1$ proves that (7-93) and (7-94) have a unique solution.

7-92 Show, by direct substitution, that (7-96) is a solution of (7-93).

7-93 The generating function $\hat{R}_i(z)$ given by (7-97) can be simplified. Introduce the variable ϕ defined by $\cos \phi = 1/(2z\sqrt{pq})$. Replace i by j in (7-97), and use i as $\sqrt{-1}$. Use the identity

$$e^{i\theta} = \cos \theta + i \sin \theta$$

to establish $x_1(z) = \sqrt{q/p}(\cos \phi + i \sin \phi) = \sqrt{q/p}\, e^{i\phi}$. Find $x_2(z)$ and show

$$\hat{R}_j(z) = \left(\frac{q}{p}\right)^j \frac{\sin(b-j)\phi}{\sin b\phi} \tag{7-109}$$

7-94 Let $D_j = D_j(b)$ be the expected duration of a gambler's ruin game when the gambler starts with j and the opponent starts with $b - j$. Explain why D_j satisfies

$$D_j = 1 + pD_{j+1} + qD_{j-1} \qquad 1 \le j \le b - 1$$

with boundary conditions $D_0 = D_b = 0$. Solve this difference equation directly, or use (7-109) to obtain $D_j = \lim_{z \uparrow 1} (d/dz) [\hat{R}_j(z) + \hat{W}_j(z)]$. (The latter method may involve considerable algebra.) The solution is

$$D_j = \frac{j}{q - p} - \frac{b}{q - p} \frac{1 - (q/p)^j}{1 - (q/p)^b} \qquad p \ne q$$

and

$$D_j = j(b - j) \qquad p = q = \tfrac{1}{2}$$

7-95 Evaluate the limits of (7-97) and (7-98) as $b \to \infty$. What is the interpretation of $b = \infty$ in the gambler's ruin process?

7-96 In a simple random walk, find the probability that the process reaches zero when it starts at $i > 0$. Use your answer to prove Proposition 7-8.

7-97 Consider the process $\{S_n^* ; n \in I\}$ of Exercise 7-85. When $p > q$ and $b > i$, show that the mean first-passage time from i to j is $(b - i)/(p - q)$. What is it when $p \le q$?

7-98 Let $\{S_n; n \in I\}$ be a simple random walk with $S_0 = 0$. Let $\{G_n; n \in I\}$ be a gambler's ruin process where the gambler has initial state k, *wins* with probability q, *loses* with probability p, and is playing an infinitely rich opponent. Let $M_n = \max \{S_0, S_1, \ldots, S_n\}$, and $g_n(k) = P\{G_n > 0 \,|\, G_0 = k\}$. Show $P\{M_n < k\} = g_n(k)$. Obtain $\lim_{n \to \infty} P\{M_n = k\}$, using the answer to Exercise 7-95.

7-99 Derive (7-103).

7-100 The calculations in obtaining (7-104) from (7-103) are simplified by using the variable ϕ defined in Exercise 7-93. Show that $F_j(z)$ can be written as

$$F_j(z) = 2i(pq)^{(j+1)/2} z^{j+1} \left[\sin(j+1)\phi + \left(\frac{1 - p_0}{p} \right) \sin(j-1)\phi - \frac{q_0 \sin j\phi}{\sqrt{pq}} \right]$$

where $i = \sqrt{-1}$. Use this representation to obtain (7-104b).

7-101 Explain why Theorem 6-5 justifies (7-107); no calculations are required.

7-102 In Example 7-26, show that no items should be kept on hand if $h/\lambda > f - r$. Give an intuitive explanation of this formula.

7-103 Formulate the model of Example 7-26 in terms of an M/M/1 queue with a limited waiting room.

7-104 Modify Example 7-26 so that rewards are discounted continuously. What is the analog of (7-107) in this situation?

7-9* ANALYSIS OF GENERATING FUNCTIONS

In the analysis of the M/G/1 queue in Example 7-19 and the gambler's ruin process in Section 7-8, generating functions are an important tool. In this section we present techniques for dealing with generating functions.†

Expansion of a Polynomial

Let z be a complex number, $A(z)$ be a polynomial of degree m, and $B(z)$ be a polynomial of degree n, with $0 < n \le m$:

$$A(z) = \sum_{i=0}^{m} a_i z^i \qquad \text{and} \qquad B(z) = \sum_{i=0}^{n} b_i z^i$$

The coefficients of $A(\,\cdot\,)$ and $B(\,\cdot\,)$ may be complex.

---

† A nodding acquaintance with the rudiments of functions of a complex variable is necessary to understand this section.

By long division, we can obtain a polynomial of degree $m - n$, say $Q(\cdot)$, called the *quotient*, such that

$$A(z) = B(z)Q(z) + R(z) \qquad (7\text{-}110)$$

where $R(\cdot)$ is the *remainder* and has degree smaller than n (Exercise 7-105). Choose $B(z) = z - b$; b can be any complex number. Then $R(z)$ must have degree zero; i.e., it is a constant c say, and (7-110) becomes

$$A(z) = (z - b)Q(z) + c \qquad (7\text{-}111)$$

When $z = b$, $A(b) = c$ and $c = 0 \Leftrightarrow A(b) = 0$; that is, b is a root of $A(\cdot)$. By the fundamental theorem of algebra, $A(\cdot)$ has a root, α_1 say; choosing $b = \alpha_1$ results† in $c = 0$, and (7-111) can be written

$$A(z) = (z - \alpha_1)Q(z) \qquad (7\text{-}112)$$

where $Q(\cdot)$ is a polynomial of degree $m - 1$.

Repeat this procedure on $Q(\cdot)$, on its quotient, etc. After m steps, the quotient is a nonzero constant, q say, and

$$A(z) = q(z - \alpha_1) \cdots (z - \alpha_m) \qquad (7\text{-}113)$$

Since $A(0) = a_0$, and we can assume $a_0 \neq 0$ without loss of generality, (7-113) yields

$$a_0 = A(0) = q(-\alpha_1) \cdots (-\alpha_m)$$

Hence

$$q = \frac{a_0(-1)^m}{\displaystyle\prod_{j=1}^{m} \alpha_j} \qquad (7\text{-}114)$$

Substituting (7-114) into (7-113) yields the expansion

$$A(z) = a_0 \left(\frac{1-z}{\alpha_1} \right) \cdots \left(\frac{1-z}{\alpha_m} \right) \qquad (7\text{-}115)$$

Partial-Fraction Expansions

We say $P(z)$ is a *rational function* of the complex variable z if

$$P(z) = \frac{N(z)}{D(z)} \qquad (7\text{-}116)$$

where $N(\cdot)$ and $D(\cdot)$ are polynomials. A rational function is a *proper fraction* if

---

† Since α_1 may be complex even when all the coefficients of $A(\cdot)$ are real, we must allow $B(\cdot)$ to have complex coefficients.

the degree of $D(\cdot)$ exceeds the degree of $N(\cdot)$. This is an important fact about rational functions.†

Proposition 7-9 A rational function that is a proper fraction may be uniquely written as

$$P(z) = \sum_{j=1}^{m} \sum_{k=1}^{r_j} \frac{c_{jk}}{(\delta_j - z)^k} \tag{7-117}$$

where $\delta_1, \ldots, \delta_m$ are the distinct roots of the denominator, r_j is the multiplicity of δ_j, and $\{c_{jk}\}$ is a set of constants.

When the roots of $D(\cdot)$ are all simple [as is often the case when $P(\cdot)$ is the generating function of a unique stationary distribution of a Markov chain], (7-117) simplifies to

$$P(z) = \frac{c_1}{\delta_1 - z} + \cdots + \frac{c_m}{\delta_m - z} \tag{7-118}$$

where m is the degree of $D(\cdot)$. The c's are obtained via this proposition.

Proposition 7-10 Let $P(\cdot)$ be given by (7-116) where the degree of $D(\cdot)$ is m and the degree of $N(\cdot)$ is less than m. If $D(\cdot)$ has m distinct roots $\delta_1, \delta_2, \ldots, \delta_m$, then $P(\cdot)$ has the representation (7-118), where

$$c_j = \frac{-N(\delta_j)}{D'(\delta_j)} \qquad j = 1, 2, \ldots, m \tag{7-119}$$

PROOF Assume that $N(\cdot)$ and $D(\cdot)$ have no roots in common. Use (7-113) to obtain

$$P(z) = \frac{N(z)}{q(z - \delta_1) \ldots (z - \delta_m)} \tag{7-120}$$

Let $\Delta_j(z) = q \prod_{i \neq j} (z - \delta_j)$. From (7-120), for any $j = 1, 2, \ldots, m$,

$$(\delta_j - z)P(z) = \frac{-N(z)}{\Delta_j(z)} \to \frac{-N(\delta_j)}{\Delta_j(\delta_j)}$$

as $z \to \delta_j$. Since $D(z) = (z - \delta_j)\Delta_j(z)$,

$$D'(\delta_j) = \Delta_j(\delta_j) \tag{7-121}$$

hence

$$\lim_{z \to \delta_j} (\delta_j - z)P(z) = \frac{-N(\delta_j)}{D'(\delta_j)} \qquad j = 1, 2, \ldots, m \tag{7-122}$$

---

† For a proof, see, e.g., sec. 115 of Phillip Franklin, *A Treatise on Advanced Calculus*, Dover, New York (1964).

Multiply both sides of (7-118) by $\delta_j - z$, and then let $z \to \delta_j$; this yields

$$\lim_{z \to \delta_j} (\delta_j - z)P(z) = c_j \qquad j = 1, 2, \ldots, m \qquad (7\text{-}123)$$

Combining (7-122) and (7-123) yields (7-119) when $N(\,\cdot\,)$ and $D(\,\cdot\,)$ have no roots in common. If δ_j is a root of both $N(\,\cdot\,)$ and $D(\,\cdot\,)$, then (7-119) yields $c_j = 0$. Dividing the numerator and the denominator of $P(\,\cdot\,)$ by $z - \delta_j$ eliminates this root from consideration. Repeating this procedure for all common roots reduces $P(\,\cdot\,)$ to the situation where its numerator and the denominator have no roots in common. $\qquad\square$

The next proposition gives a criterion for $D(\,\cdot\,)$ to have only simple roots.

Proposition 7-11 The polynomial $D(z) = d_n z^n + \cdots + d_1 z + d_0$ has multiple roots if, and only if, the determinant

$$
\left.
\begin{aligned}
n-1 &\left\{
\begin{vmatrix}
d_n d_{n-1} \ldots d_1 \; d_0 & & & \\
& d_n \; d_{n-1} \ldots \; d_1 \; d_0 & & \\
& & d_n \; d_{n-1} \cdots d_1 \; d_0 & \\
\end{vmatrix}
\right. \\
n &\left\{
\begin{vmatrix}
d'_n d'_{n-1} \ldots d'_1 & & \\
& d'_n \; d'_{n-1} \ldots \; d'_1 & \\
& & d'_n \cdots d'_2 \; d'_1 \\
\end{vmatrix}
\right.
\end{aligned}
\right.
$$

is zero, where $d'_j = j d_j, j = 1, 2, \ldots, n$.

PROOF The representation (7-121) is valid even when $D(\,\cdot\,)$ has multiple roots. From it we can conclude that $D(\,\cdot\,)$ has multiple roots if, and only if, $D(\,\cdot\,)$ and $D'(\,\cdot\,)$ have a common root. A theorem[†] of Sylvester states that two polynomials, $A(\,\cdot\,)$ and $B(\,\cdot\,)$ say, have a common root if, and only if,

$$
0 =
\begin{vmatrix}
a_m & a_{m-1} & \cdots & a_0 & & \\
& a_m & a_{m-1} & \cdots & a_0 & \\
& & \cdots\cdots\cdots\cdots\cdots\cdots & & & \\
& & & a_m \; a_{m-1} & \cdots & a_0 \\
b_n & b_{n-1} & \cdots & b_0 & & \\
& b_n & b_{n-1} & \cdots & b_0 & \\
& & \cdots\cdots\cdots\cdots\cdots\cdots & & & \\
& & & b_n \; b_{n-1} & \cdots & b_0 \\
\end{vmatrix}
$$

where m and n are the degrees of $A(\,\cdot\,)$ and $B(\,\cdot\,)$, respectively. Applying this theorem to $D(\,\cdot\,)$ and $D'(\,\cdot\,)$ completes the proof. $\qquad\square$

---

† See, e.g., B. L. van der Waerden, *Modern Algebra*, rev. English ed., Ungar, New York (1953), p. 84.

Example 7-27 A partial-fraction expansion of $P(z) = (1 - \rho) (\alpha \sum_{i=1}^{k} z^i - 1)^{-1}$, where $\alpha \in (0, 1)$ is a known constant, is required in Example 8-7. If we write $P(z)$ in the form (7-116), then $N(z) \equiv 1 - \rho$ and $D(z) = \alpha \sum_{i=1}^{k} z^i - 1$.

To determine whether $D(\cdot)$ has multiple roots, write the determinant required in Proposition 7-11. It is

$$
\begin{array}{c|ccccccccc}
 & \multicolumn{4}{c}{\overbrace{\hspace{4cm}}^{k}} & \multicolumn{5}{c}{\overbrace{\hspace{4cm}}^{k-1}} \\
1 & \alpha & \alpha & \cdots & & \alpha & 1 & & & \\
2 & & \alpha & \cdots & & \alpha & \alpha & 1 & & \\
\vdots & & & & \cdots & & & & & \\
k-1 & & & & & \alpha & \alpha\ \alpha & \alpha & \cdots & \alpha\ 1 \\
k & k\alpha & (k-1)\alpha & \cdots & & \alpha & 0 & & & \\
k+1 & & k\alpha & \cdots & & 2\alpha & \alpha & & & \\
\vdots & & & & \cdots & & & & & \\
2k-1 & & & & & \alpha k & (k-1)\alpha & \cdots & 2\alpha\ 1 &
\end{array}
$$

Let d_{ij} be the (i, j)th element of the determinant. If the determinant were zero, there would exist row multipliers, λ_i say, such that

$$\sum_{j=1}^{2k-1} \lambda_i d_{ij} = 0 \qquad j = 1, 2, \ldots, 2k-1 \qquad (7\text{-}124)$$

By looking at columns $1, 2, \ldots, k$ in turn, it is clear that (7-124) holds for these values of j if, and only if,

$$\lambda_i = \begin{cases} 1 & i = k, k+1, \ldots, 2k-1 \\ -(k+i-1) & i = 1, 2, \ldots, k-1 \end{cases} \qquad (7\text{-}125)$$

It is equally clear that (7-125) makes (7-124) false for $j = k$; hence the determinant is not zero and all the roots of $D(\cdot)$ are distinct.

From (7-115),

$$D(z) = \left(\frac{1-z}{\delta_1}\right) \cdots \left(\frac{1-z}{\delta_k}\right)$$

where $\delta_1, \ldots, \delta_k$ are the k distinct roots of $D(\cdot)$; hence

$$D'(\delta_j) = \frac{-1}{\delta_j} \prod_{i \neq j} \left(\frac{1-\delta_j}{\delta_i}\right) \qquad j = 1, 2, \ldots, k \qquad (7\text{-}126)$$

Substituting (7-126) into (7-119) and using Proposition 7-10 yield

$$P(z) = \sum_{j=1}^{k} \frac{1-\rho}{1 - z/\delta_j} \prod_{i \neq j} \frac{\delta_i}{\delta_i - \delta_j} \qquad \qquad \square$$

Arguments Based on Analytic Functions

To introduce the class of problems treated in this section, consider the following example.

Example 7-28: M/D/c queue Assume customers arrive according to a Poisson process at rate λ and are served by c servers. All service times are the constant v; the order of service is arbitrary except no customer is kept waiting when there is an idle server.

Let $X(t)$ be the number of customers in the system at time t. The process $\{X(t); t \geq 0\}$ is not Markov, but there is a useful embedded Markov chain. Let $Y_n = X(nv)$, $n \in I$. Since the service times are constant, those customers in service at time nv must depart by time $(n + 1)v$, and no other customers can depart during $(nv, (n + 1)v]$. Let A_n denote the number of arrivals during $(nv, (n + 1)v]$; then

$$Y_{n+1} = A_n + (Y_n - c)^+ \qquad (7\text{-}127)$$

The memoryless property of exponential interarrival times suffices to prove A_n is independent of $Y_1, \ldots, Y_{n-1}$; hence (7-127) establishes that $\{Y_n; n \in I\}$ is a Markov chain with an infinite state space. It is readily checked (Exercise 7-109) that this chain is irreducible and consists of aperiodic states.

The random variables $A_1, A_2, \ldots$ defined above are i.i.d.; let $a_j = P\{A_1 = j\}, j \in I$. Use Theorem 7-13 to prove that the stationary vector π of $\{Y_n; n \in I\}$ satisfies (Exercise 7-109)

$$\pi_i = a_i \sum_{j=0}^{c} \pi_j + \sum_{j=c+1}^{\infty} \pi_j a_{i-j+c} \qquad i \in I \qquad (7\text{-}128)$$

Define $\hat{\pi}(z) = \sum_{i=0}^{\infty} z^i \pi_i$. Multiplying the ith equation in (7-128) by z^i and then summing over all $i \in I$, we obtain

$$\hat{\pi}(z) = \hat{A}(z) \left(\sum_{j=0}^{c} \pi_j + \sum_{j=c+1}^{\infty} z^{j-c} \pi_j \right)$$

$$= \frac{\sum_{j=0}^{c-1} \pi_j (z^j - z^c)}{1 - z^c / \hat{A}(z)}$$

where $\hat{A}(z) \triangleq \sum_{j=0}^{\infty} z^j a_j = e^{-\lambda v(1-z)}$.

Let $\pi_j^* \triangleq \pi_0 + \pi_1 + \cdots + \pi_j$; then

$$\hat{\pi}(z) = \frac{(1 - z) \sum_{j=0}^{c-1} \pi_j^* z^j}{1 - z^c e^{\lambda v(1-z)}} \qquad (7\text{-}129)$$

From (7-129) it can be established that π satisfies $\sum_{i=0}^{\infty} \pi_i = 1$ if, and only if, $\rho \triangleq \lambda v / c < 1$ (Exercise 7-110). It remains to determine the constants $\pi_0^*, \pi_1^*, \ldots, \pi_{c-1}^*$ appearing in (7-129). Observe that when $\pi_0, \pi_1, \ldots, \pi_{c-1}$ are obtained, π_{c+i} can be obtained from (7-128) by recursion: Setting $i = 0$ in (7-128) yields π_c, setting $i = 1$ yields π_{c+1}, and so forth. $\square$

A class of problems suggested by Example 7-28 is this: Suppose the PGF $\hat{\pi}(z)$ is the ratio of two functions, with the numerator containing unknown constants.

How may these constants be evaluated? Frequently, the constants can be evaluated by making the following observation.

Since any PGF is a polynomial which converges for all $z \in \mathbb{C} \triangleq \{z: |z| \leq 1 + \delta\}$ for some $\delta \geq 0$, if

$$\hat{\pi}(z) = \frac{f(z)}{g(z)}$$

then *any root of $g(z)$ which is in $\mathbb{C}$ must be a root of $f(z)$.*

In (7-129), there are c unknown constants in the numerator. If it could be established that the denominator contained c distinct roots in the region of convergence of $\hat{\pi}(\cdot)$, then the unknown constants would satisfy a system of c independent linear equations. Solving these equations would yield the constants. A typical procedure for proving that the denominator has the required number of roots involves this theorem.

Theorem 7-19: Rouché's theorem Let $f(\cdot)$ and $g(\cdot)$ be *analytic*† inside and on a contour $\mathbb{C}$ with $|g(z)| < |f(z)|$ on $\mathbb{C}$. Then $f(\cdot) + g(\cdot)$ have the same number of zeros *inside* $\mathbb{C}$.

PROOF See, for example, George F. Carrier, Max Krook, and Carl E. Pearson, *Functions of a Complex Variable*, McGraw-Hill, New York (1966), p. 61. □

Example 7-29: Example 7-28 continued As a consequence of Exercise 7-110, when $\rho < 1$, the PGF $\hat{\pi}(z)$ given by (7-129) converges whenever $|z| \leq 1$. For $\delta > 0$ and small enough, $e^{\lambda v \delta} < (1 + \delta)^c$. Hence, for some small positive δ, when $|z| = 1 + \delta$,

$$\left| e^{-\lambda v(1-z)} \right| \leq e^{-\lambda v + \lambda v |z|} = e^{\lambda v \delta} < (1 + \delta)^c \triangleq |z|^c$$

Rouché's theorem, with $f(z) = z^c$ and $g(z) = -e^{-\lambda v(1-z)}$, asserts that the number of roots of the denominator of (7-129) within $\mathbb{C}' = \{z: |z| < 1 + \delta\}$ is the number of solutions of $z^c = 0$ that lies in $\mathbb{C}'$, which is known to be c. Proposition 7-11 may be used to establish that the roots are distinct (see Exercise 7-111). □

EXERCISES

7-105 Derive 7-110.

7-106 Derive 7-113; explain why $q \neq 0$.

7-107 Let $A(z)$ be a polynomial of degree m and α_1 be a root with multiplicity r. Show

$$A(z) = (z - \alpha_1)^r Q(z)$$

where $Q(\cdot)$ is a polynomial of degree $m - r$ and $Q(\alpha_1) \neq 0$.

† A function is *analytic* throughout a region if it has a Taylor's series expansion there.

7-108: The M/E_k/1 queue Let the service-time distribution in an M/G/1 queue have a gamma density with shape parameter k and mean $1/\mu$ (which requires the scale parameter to be $k\mu$). In queueing theory, this distribution is called *Erlang-k* and denoted E_k. The corresponding density function $g(\cdot)$ is

$$g(t) = \frac{k\mu(k\mu t)^{k-1}e^{-\mu t}}{(k-1)!} \qquad t \geq 0, k \in I_+$$

(a) Use (7-64) to establish

$$\hat{\pi}(z) = \frac{(1-\rho)(1-z)}{z(1+\rho/k-\rho z/k)^k - 1} \qquad \rho \triangleq \lambda/\mu$$

for this model.

(b) Prove that the denominator of $\hat{\pi}(z)$ has no multiple roots.

(c) Expand $\hat{\pi}(z)$ in partial fractions. What can you say about the form of π_n?

7-109 Prove that the process $\{Y_n ; n \in I\}$ in Example 7-28 is irreducible and consists of aperiodic stages. Derive (7-128).

7-110 (Continuation) Use (7-129) to establish that π satisfies $\sum_{i=0}^{\infty} \pi_i = 1$ if, and only if, $\rho = \lambda v/c < 1$.

7-111 Prove that the c roots of the denominator of (7-129) are distinct and that the constants π_0^*, $\pi_1^*, \ldots, \pi_{c-1}^*$ are the unique solution of c independent linear equations.

7-112 In an M/D/c queue, let $R_t(j) = P\{$at most j of those customers present at t_0 are present at $t + t_0\}$, $0 < t < v$, $j \in I$, and $\hat{R}_t(z) = \sum_{j=0}^{\infty} z^j R_t(j)$. Prove

$$\hat{R}_t(z) = \frac{e^{\lambda t(1-z)}\hat{\pi}(z)}{1-z} \tag{7-130}$$

where $\hat{\pi}(z)$ is given by (7-129).

7-113 (Continuation) Substitute (7-129) into (7-130). Prove that $\hat{R}_t(z)$ can be expanded as a power series in $w \triangleq z^c e^{\lambda v(1-z)}$ for $|z| < z_1$, where z_1 is the smallest root, in absolute value, of the denominator of (7-129). Establish

$$R_t(kc+j) = \sum_{i=0}^{c-1} \pi_i^* \sum_{n=0}^{k} \frac{e^{\lambda(t+nv)}[-\lambda(t+nv)]^m}{m!} \tag{7-131}$$

where $m = (k-n)c + j - i$, $k \in I$, and $0 \leq j \leq c - 1$.

7-114 (Continuation) Let W_n be the waiting time of the nth arriving customer, and assume customers enter service FIFO. Set $W(t) = \lim_{n \to \infty} P\{W_n \leq t\}$; Example 6-4 shows it exists. Use (7-131) to establish

$$W(t) = R_x(kc+c-1)$$

$$= \sum_{j=0}^{c-1} \pi_{c-1-j}^* \sum_{n=0}^{k} \frac{e^{\lambda(t-nv)}[-\lambda(t-nv)]^{nc+j}}{(nc+j)!} \qquad t = kv + x, k \in I, 0 \leq x \leq v$$

7-115 Observe that 1 is a root of the denominator of (7-129); let $z_1, \ldots, z_{c-1}$ be the other roots. Set $v = 1$ and show that

$$\pi_0 = (-1)^{c+1}(c-\lambda)\prod_{i=1}^{c-1}\frac{z_i}{1-z_i}$$

$$\sum_{j=0}^{c-1}\pi_j = \frac{c-\lambda}{\prod_{i=1}^{c-1}(1-z_i)}$$

and

$$\sum_{j=1}^{\infty} j\pi_j = (c-\lambda)\sum_{i=1}^{c-1}\frac{1}{1-z_i} + \frac{c-(c-\lambda)^2}{2(c-\lambda)}$$

BIBLIOGRAPHIC GUIDE

Markov processes are named after the mathematician A. A. Markov who introduced the concept in 1907. Markov studied a process with a discrete parameter and a finite number of states. The first investigation of a Markov chain with a countably infinite number of states was made by A. N. Kolmogorov in 1936. The analysis of Markov chains via renewal theory was launched by W. Feller in 1950 in the first edition of Feller (1968). For the most part, we have followed Feller's treatment. The notion of the fundamental matrix and its implications is due to Kemeny and Snell (1960).

The analysis of $M/G/1$ queues by the method of an embedded Markov chain is due to D. G. Kendall (1954). The first solution of the $M/D/c$ queue, which used embedded chain methods, was given in Crommelin (1932); our treatment in Section 7-9 follows Prabhu (1965) for details. The (s, S) inventory models presented in this chapter were inspired by the models in Wagner (1962).

Most textbooks about stochastic processes include material on Markov chains. Kemeny and Snell (1960) uses only elementary mathematics to study finite chains; more advanced mathematics (including measure theory) is used for analyzing infinite chains in Kemeny, Snell, and Knapp (1966). A rigorous non-measure-theoretic treatment is given in Part I of Chung (1967). Isaacson and Madsen (1976) give an extensive treatment of the algebraic solution of Markov chains and relate the algebraic properties to the classification of an MC. Our favorite sources for matrix algebra are Hildebrand (1965) and Gantmacher (1959). Hildebrand (1968) is a good source for difference equation methods.

References

Chung, Kai Lai: *Markov Chains with Stationary Transition Probabilities*, Springer-Verlag, New York (1967).

Crommelin, C. D.: "Delay Probability Formulas When the Holding Times Are Constant," *P. O. Elect. Engrs. J.* **25:** 41–50 (1932).

Feller, William: *An Introduction to Probability Theory and Its Applications*, vol. I, 3rd ed., Wiley, New York (1968).

Gantmacher, F. R.: *The Theory of Matrices*, vols. I and II, Chelsea, New York (1959).

Hildebrand, Francis B.: *Methods of Applied Mathematics*, 2nd ed., Prentice-Hall, Englewood Cliffs, N.J. (1965).

————: *Finite-Difference Equations and Simulations*, Prentice-Hall, Englewood Cliffs, N.J. (1968).

Isaacson, D. L., and R. W. Madsen: *Markov Chains: Theory and Applications*, Wiley, New York (1976).

Kemeny, John G., and J. Laurie Snell: *Finite Markov Chains*, Van Nostrand, New York (1960).

————; ————; and Anthony W. Knapp: *Denumerable Markov Chains*, Van Nostrand, New York (1966).

Kendall, David G.: "Stochastic Processes Occurring in the Theory of Queues and Their Analysis by the Method of the Embedded Markov Chain" *Ann. Math. Stat.* **34:** 338–354 (1954).

Prabhu, N. J.: *Queues and Inventories*, Wiley, New York (1965).

Wagner, Harvey M.: *Statistical Management of Inventory Systems*, Wiley, New York (1962).

EIGHT

CONTINUOUS-TIME MARKOV CHAINS

Continuous-time Markov chains generalize birth-and-death processes by removing the restriction that all transitions are between adjacent states. In this chapter, the primary properties of continuous-time Markov chains are presented. As the name suggests, this process has the Markov property, and the concepts of Chapter 7 are used repeatedly.

The name also (correctly) suggests that the process evolves in continuous time. A continuous-time model of some phenomenon may be more appropriate than a discrete-time model when many changes of state may occur in a short time. By adding the material in this chapter to our knowledge of (discrete-time) Markov chains, we give ourselves the flexibility to choose a discrete-time or continuous-time model, as appropriate. Optimization issues for continuous-time Markov chain models are presented in Section 5-3 of Volume II.

8-1 DEFINITIONS AND EXAMPLES

A continuous-time Markov chain, abbreviated CTMC, is a Markov process with a countable state space. The states are labeled by consecutive nonnegative integers.

When certain regularity conditions (which are specified below) are met, a CTMC is conceptually straightforward. It is a process that goes from state to state according to a Markov chain. Each time a state is visited, the process stays there for a random time that is independent of the past behavior of the process and has an exponential distribution, which may have a state-dependent parameter.† The following definitions and theorems are designed to deduce the above properties from more fundamental concepts. We defer the discussion of some properties of CTMCs to the more general setting of Section 9-1.

Formal Definitions

Definition 8-1 The stochastic process $\{X(t); \ t \geq 0\}$ is a *continuous-time Markov chain* with state space‡ $S \subset I$ if each $X(t)$ assumes values only in S and

$$P\{X(t_{n+1}) = j \mid X(t_0) = i_0, \ldots, X(t_n) = i_n\} = P\{X(t_{n+1}) = j \mid X(t_n) = i_n\}$$

(8-1)

holds for $n \in I$, $0 \leq t_0 < t_1 < \cdots < t_{n+1}$, and all $i_0, \ldots, i_n, j \in S$. A *CTMC* is called *finite (infinite)* when S is *finite (infinite)*.

Since a stochastic process is specified by its fidi's (8-1) is equivalent to

$$P\{X(t + s) = j \mid X(u), \ u < s, \ X(s) = i\} = P\{X(t + s) = j \mid X(s) = i\} \quad (8\text{-}1')$$

for any $t \geq 0$ and $s > 0$. Equations (8-1) and (8-1') express the *Markov property*.

If the right-side of (8-1') does *not* depend on s, the CTMC is called *homogeneous*. Henceforth, *only homogeneous CTMCs are considered*. Define the functions $P_{ij}(\cdot)$ by

$$P_{ij}(t) \triangleq P\{X(t + s) = j \mid X(s) = i\} \qquad i, j \in S, \ t > 0 \tag{8-2}$$

The function $P_{ij}(\cdot)$ is called the *transition probability function from state i to state j*, customarily abbreviated to *transition function*. The matrix of transition functions $P(\cdot) = (P_{ij}(\cdot))$ is called the *transition matrix* of the CTMC.

By their nature, the transition functions have the properties

$$P_{ij}(t) \geq 0 \qquad t > 0, \quad i, j \in S \tag{8-3}$$

and

$$\sum_{j \in S} P_{ij}(t) \leq 1 \qquad t > 0, \qquad i \in S \tag{8-4}$$

The inequality in (8-4) allows for the possibility of infinitely many transitions in a finite time. From (8-1) and (8-2), one obtains (Exercise 8-1)

$$P_{ij}(t + s) = \sum_{i \in S} P_{ik}(t) P_{kj}(s) \qquad i, j \in S, \quad t, s \geq 0 \tag{8-5}$$

which is the *Chapman-Kolmogorov equation* for a CTMC.

---

† The semi-Markov process, which is described in Section 9-1, allows this r.v. to have a general distribution and to depend on the next state that is visited.

‡ We place S in I for convenience. The essential property is that S has a countable number of elements.

It is natural to make the following *assumptions* about the transition functions.

Definition 8-2 A transition matrix is called *standard* if for all $i, j \in \mathcal{S}$,

$$\lim_{t \downarrow 0} P_{ii}(t) = 1 \quad\text{and}\quad \lim_{t \downarrow 0} P_{ij}(t) = 0 \quad i \neq j \tag{8-6}$$

Henceforth all transition matrices are assumed to be standard. We use (8-6) to extend the definition in (8-2) to

$$P_{ii}(0) = 1 \quad\text{and}\quad P_{ij}(0) = 0 \quad i, j \in \mathcal{S} \tag{8-2'}$$

Let ω be a sample path of the CTMC $\{X(t); t \geq 0\}$, and suppose that at epoch $S(\omega)$ a transition from state i to state j occurs. It is possible to define $X[S(\omega)] = i$ or $X[S(\omega)] = j$; in either case, $\{X(t); t \geq 0\}$ conforms to Definitions 8-1 and 8-2. We always make the latter definition; formally, we *assume* that for each possible sample path ω,

$$X(t; \omega) = \liminf_{s \downarrow t} X(s; \omega) \quad t \geq 0 \tag{8-7}$$

The reason that a "lim inf" appears in (8-7) is that the limit may not exist (see Proposition 8-3). Proposition 8-2 indicates how to rule out such unpleasant behavior. As a result of (8-7), the graph of a typical sample path has the form shown in Figure 8-1.

Define a_i by

$$a_i \triangleq P\{X(0) = i\} \quad i \in \mathcal{S}, \quad \sum_{j \in \mathcal{S}} a_j = 1 \tag{8-8}$$

The row vector $\mathbf{a} = (a_i ; i \in \mathcal{S})$ is the *vector of initial* probabilities. Define

$$P_j(t) \triangleq P\{X(t) = j\} \quad j \in \mathcal{S} \tag{8-9}$$

and the row vector

$$\mathbf{P}(t) \triangleq (P_j(t); j \in \mathcal{S}) \quad t > 0$$

which is called the *vector of state probabilities at time t*. Clearly,

$$P_j(t) = \sum_{i \in \mathcal{S}} a_i P_{ij}(t) \quad t \geq 0 \tag{8-10a}$$

$X(t, \omega)$

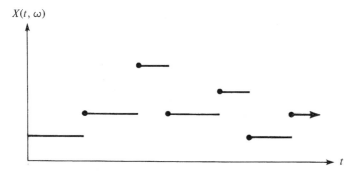

Figure 8-1 Part of a typical sample path.

In vector notation,

$$\mathbf{P}(t) = \mathbf{a}P(t) \qquad t \geq 0 \qquad\qquad (8\text{-}10b)$$

It is easy to show (Exercise 8-2) that the fidi's of a CTMC are determined by the vector of initial probabilities and the transition matrix, so these two objects specify† a CTMC. It is not trivial to prove [see, e.g., Chung (1967, sec. 2.4)] that given a matrix of functions satisfying (8-3) through (8-6), there exists a CTMC for which (8-2) and (8-8) hold (i.e., there is a CTMC for every transition matrix).

Examples

Example 8-1: Birth-and-death processes Every birth-and-death process is a CTMC. To verify this assertion, observe that parts (a) and (b) of the definition of a birth-and-death process, Definition 4-1, require a birth-and-death process to be a homogeneous CTMC. Part (c) of that definition specifies the form of the transition function, showing that a birth-and-death process is a special CTMC where all transitions are between adjacent states. Equations (4-2) through (4-4) imply (8-6), so the transition matrix is standard.

It is possible to choose the parameters of a birth-and-death process so that infinitely many transitions occur during a finite time (see Section 4-11); when this occurs, the sample paths of the process are not well defined. Thus, it is necessary to place some restrictions on the transition matrix of a CTMC so that such "explosions" do not occur.   ☐

Example 8-2: Start-up and shut-down processes Start-up and shut-down processes are defined in Section 4-6 and are shown there to satisfy axioms (a) and (b) for a birth-and-death process. These axioms are satisfied by any CTMC with $\mathcal{S} = I$. The remaining property of a start-up and shut-down process is the special form of the transition matrix required by (4-45) through (4-49), which implies that the process is a CTMC with a standard transition matrix.   ☐

Example 8-3 A model of a two-station message-forwarding system is shown in Figure 8-2. All messages are intended for location 3. A message that originates at location 2, called a *2-message*, uses a line from 2 to 3; a message that originates at location 1, a *1-message*, requires first a line from 1 to 2 and then a line from 2 to 3. A message that finds no outgoing line available at its arrival epoch is lost. However, a 1-message that arrives when a line from 1 to 2 is available is not lost even if, at that epoch, no lines are available from 2 to 3. At an epoch where a 1-message reaches location 2, if no lines from 2 to 3 are available, then the 1-message waits as long as required for a line from 2 to 3 to become available. During this waiting time, no other 1-customer can use the line from 1 to 2 belonging to the waiting message. Note that when a 1-customer is waiting, all arriving 2-customers are lost.

---

† Since several stochastic processes can have the same fidi's (see Section 3-2), more than one CTMC may have the given vector of initial probabilities and transition matrix.

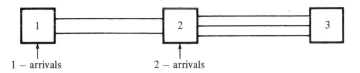

Figure 8-2 A two-station message-forwarding system.

This model may be used to describe a data collection system. Suppose that a company with headquarters in Chicago receives sales data from its regional offices in Dallas and Kansas City. The company has communication links from Dallas to Kansas City and from Kansas City to Chicago. We can interpret locations 1, 2, and 3 as Dallas, Kansas City, and Chicago, respectively.

Let c_i denote the number of lines from i to $i + 1$. Assume that i-messages arrive at location i according to a Poisson process with rate λ_i, that these processes are independent of one another and of the message processing times, and that the latter are i.i.d. and have an exponential distribution with mean $1/\mu$. Let a state of the system be the ordered triple (n_1, n_2, n_3), where n_i is the number of messages on the link from i to $i + 1$, $i = 1, 2$, and n_3 is the number of 1-messages at location 2 waiting for a line from 2 to 3. Clearly, each of these quantities is nonnegative, and the states can be numbered by the elements of I. Assume the latter has been done,† and let $X(t)$ be the state of the system at time t. The independence and exponential assumptions imply that (8-1) holds, so $\{X(t); t \geq 0\}$ is a CTMC. □

EXERCISES

8-1 Obtain (8-5) from (8-1) and (8-3).

8-2 Show that $\mathbf{a}$ and $P(t)$ determine the fidi's of a CTMC.

8-3 Recast Example 6-2 in continuous time by changing the geometric distributions to exponential distributions with the same mean. Show that the continuous time model is a CTMC.

8-4 In the machine repair model in Example 4-4, the TTYs are identical. Suppose that they are different. Assume that the times between failures for the ith TTY are i.i.d. exponential r.v.'s with mean $1/\lambda_i$ and that the repair times for the ith TTY are i.i.d. exponential r.v.'s with mean $1/\mu_i$. Formulate a CTMC to describe this model when there are two TTYs and one mechanic.

8-2 SAMPLE-PATH PROPERTIES

Strong Markov Property

Let the random variable S_n be the epoch of the nth transition of the CTMC $\{X(t); t \geq 0\}$. It is tempting to assert that the Markov property and time homogeneity

---

† The explicit form of the renumbering is not required. The reason the states have to be labeled by elements of I is to conform with Definition 8-1.

imply an analog of the MC property expressed by (7-5), namely that

$$P\{X(S_n + t) = j \mid X(S_n) = i\} = P\{X(t) = j \mid X(0) = i\} \qquad t > 0 \qquad (8\text{-}11)$$

In (8-1'), s is a *fixed* (i.e., not random) epoch and S_n is an r.v., so (8-11) does not necessarily follow from (8-1') and homogeneity. However, (8-11) is true for the following reason.

A *stopping time* (cf. Section 6-2) for a CTMC is an r.v., say A, such that the occurrence of the event $\{A \leq t\}$ is determined only by $X(u)$ for $u \leq t$.

The random variable S_n defined above is a stopping time because the event $\{S_n \leq t\}$ depends on only those transitions which occur no later than t. Similarly, the r.v. representing the epoch of the kth entry to state j is a stopping time for any $j \in S$ and positive integer k. The r.v. representing the epoch of the last entry to state j is not a stopping time because it depends on the evolution of the process after t.

Definition 8-3 A CTMC has the *strong Markov property* if for every stopping time A,

$$P\{X(A + t) = j \mid X(s), s < A; X(A) = i\} = P\{X(t) = j \mid X(0) = i\} \qquad t > 0$$

The importance of assuming (8-6) and (8-7) is shown by this theorem.

Theorem 8-1 If a CTMC has a standard transition matrix and (8-7) holds, then it has the strong Markov property.

PROOF See Chung (1967, p. 179). □

The validity of (8-11) is an immediate consequence of Theorem 8-1. The concept of the strong Markov property does not arise in discrete Markov chain theory because discrete chains make transitions at the fixed epochs 1, 2, ... which are (trivially) stopping times.

The following proposition shows that each entry to a fixed state is a re-generation epoch (cf. Definition 6-1).

Proposition 8-1 Let A be a stopping time for the CTMC $\{X(t); t \geq 0\}$, and fix some state $i \in S$. If $P\{X(A) = i\} = 1$ and A takes the value $a < \infty$, the fidi's of $\{X(t + a); t \geq 0\}$ are the same as the fidi's of $\{X(t); t \geq 0\}$ when $X(0) = 1$ (w.p.1).

PROOF See Chung (1967, p. 181). □

The epoch of the kth entry to some fixed state satisfies the hypothesis of Proposition 8-1. If we regard such an epoch as the "present," Proposition 8-1 asserts that given the present, the "future" is independent of the "past" and the process "starts from scratch" each time the fixed state is entered. Occasionally we say that the future is *conditionally independent* of the past, given the present.

Sample-Path Properties

Let S_n be the epoch of the nth transition of a particular CTMC, with $S_0 = 0$. Define the intertransition times $\tau_n = S_n - S_{n-1}, n \in I_+$.

> **Theorem 8-2** Let $\{X(t); t \geq 0\}$ be a CTMC. For each $i \in 8$, and $n \in I_+$,
>
> $$P\{\tau_n > t \mid X(S_{n-1}) = i\} = e^{-\alpha_i t} \qquad t \geq 0 \qquad (8\text{-}12)$$
>
> for some $\alpha_i \geq 0$ (possibly $\alpha_i = \infty$).

PROOF By homogeneity, we may take $n = 1$ in (8-12) without loss of generality. Fix i and write

$$
\begin{aligned}
H^c(t+s) &\triangleq P\{\tau_1 > t + s \mid X(0) = i\} \\
&= P\{S_1 > t + s \mid X(0) = i, S_1 > t\} P\{S_1 > t \mid X(0) = i\}
\end{aligned} \qquad (8\text{-}13)
$$

Use the strong Markov property and homogeneity, in turn, to obtain

$$
\begin{aligned}
P\{S_1 > t + s \mid X(0) = i, S_1 > t\} &= P\{S_1 > t + s \mid X(t) = i\} \\
&= P\{S_1 > s \mid X(0) = i\} \qquad (8\text{-}14) \\
&= H^c(s)
\end{aligned}
$$

Combining (8-13) and (8-14) yields

$$H^c(t+s) = H^c(s)H^c(t) \qquad t, s \geq 0 \qquad (8\text{-}15)$$

The only bounded solution† of (8-15) is the exponential function, and since probabilities are inherently bounded, (8-12) follows from (8-15). □

Times between Transitions

Let $W(t)$ be the time until the first transition after epoch t. A formal definition of $W(t)$ proceeds as follows. Set $N(0) = 0$ and define

$$N(t) \triangleq \sup\{n : S_n \leq t\} \qquad t > 0$$

where $N(t)$ is the number of transitions that occur by time t. Now define

$$W(t) \triangleq S_{N(t)+1} - t \qquad t \geq 0$$

which, by the definition of S_n, is equivalent to

$$W(t) = S_{N(t)} + \tau_{N(t)+1} - t \qquad t \geq 0 \qquad (8\text{-}16)$$

Using (8-16), Theorem 8-2, and the memoryless property of exponential r.v.'s, in

---

† See, e.g., E. Parzen, *Stochastic Processes*, Holden-Day, San Francisco (1962), p. 121.

turn, yields

$$P\{W(t) > s \mid X(t) = i\} = P\{\tau_{N(t)+1} > t - S_{N(t)} + s \mid X(t) = i\}$$

$$= P\{\tau_{N(t)+1} > t - S_{N(t)} + s \mid X(S_{N(t)}) = i, \tau_{N(t)+1} > t - S_{N(t)}\}$$

$$= e^{-\alpha_i s} \qquad s \geq 0 \tag{8-17}$$

Three different types of states can be identified from Theorem 8-2. If $\alpha_i = 0$, then $P\{\tau_n > t \mid X(S_n) = i\} = 1$ for all $t \geq 0$, so once state i is entered, it is never left. If $\alpha_i = \infty$, then $P\{\tau_n > 0 \mid X(S_n) = i\} = 0$, so state i is left immediately after being entered. If $0 < \alpha_i < \infty$, then $P\{0 < \tau_n < \infty \mid X(S_n) = i\} = 1$ and each visit to state i is for a positive and finite time (w.p.1).

Definition 8-4 State i is called *absorbing* if $\alpha_i = 0$, *instantaneous* if $\alpha = \infty$, and *stable* if $0 < a_i < \infty$.

Stable states are "natural" in most contexts, and absorbing states can be used to describe a terminal condition and are employed in first-passage-time analysis (see Exercise 8-7). Instantaneous states do not seem to appear in operations research models (see Exercise 8-6 for a pathological property of instantaneous states). The following proposition indicates how to rule out instantaneous states.

Proposition 8-2 (*a*) If the state space S is finite, then no state is instantaneous. (*b*) The sample paths are right-continuous† with probability 1 if, and only if, no state is instantaneous.

PROOF See Çinlar (1975, p. 244). □

A good reason for choosing (if possible) S so that no state is instantaneous is provided by part *b* of Proposition 8-2. Theorem 6-7 is a powerful tool for obtaining limiting probabilities. One of its hypotheses is that the sample paths are right-continuous and have left limits. The former is established by part *b* of Proposition 8-2, and the latter will hold if it can be shown that only finitely many transitions occur in any finite time (see Propositions 8-4 and 8-5 below).

Structure of a Continuous-Time Markov Chain

Define $Y_n = X(S_n)$, $n \in I$; hence Y_n is the nth state entered. The following theorem shows that $\{Y_n; n \in I\}$ is a Markov chain and that τ_n depends on Y_n but not on Y_k, $k < n$.

---

† That is, "lim inf " can be replaced by "lim" in (8-7).

Theorem 8-3 For $\{\tau_n\}$ and $\{Y_n\}$ defined as above,

$$P\{Y_{n+1} = j, \tau_{n+1} > t \mid Y_0, \ldots, Y_{n-1}, Y_n = i, \tau_1, \ldots, \tau_n\} = p_{ij} e^{-\alpha_i t} \qquad t \geq 0$$

$$(8\text{-}18)$$

for any $i, j \in S$. Also, (p_{ij}) has the properties

(a) $p_{ij} \geq 0$
(b) $p_{ii} = 0$
(c) $\sum_{j \in S} p_{ij} = 1$

PROOF Choose $i, j \in S$, $j \neq i$. The conditioning variables determine the CTMC $\{X(t); t \geq 0\}$ for $0 \leq t \leq S_n$, so the left-side of (8-18) is

$$P\{Y_{n+1} = j, \tau_{n+1} > t \mid X(t), t < S_n; X(S_n) = i\}$$

$$= P\{Y_{n+1} = j, \tau_{n+1} > t \mid X(S_n) = i\} \qquad \text{(strong Markov property)}$$

$$= P\{Y_1 = j, \tau_1 > t \mid X(0) = i\} \qquad \text{(homogeneity)}$$

$$= P\{\tau_1 > t \mid X(0) = i\} P\{Y_1 = j \mid Y_0 = i, \tau_1 > t\} \qquad (8\text{-}19)$$

The first probability on the right-side of (8-16) is $e^{-\alpha_i t}$ by Theorem 8-2. From the definition of $W(\cdot)$, $Y_1 = X[t + W(t)]$ for $t < S_1$, and the event $\{Y_0 = i, \tau_1 > t\}$ is equivalent to the event $\{X(u) = i, 0 \leq u \leq t\}$. Use the Markov property and homogeneity, in turn, to obtain

$$P\{Y_1 = j \mid Y_0 = i, \tau_1 > t\} = P\{X[t + W(t)] = j \mid X(u) = i, 0 \leq u \leq t\}$$

$$= P\{X[t + W(t)] = j \mid X(t) = i\}$$

$$= P\{X[W(0)] = j \mid X(0) = i\}$$

$$= P\{Y_1 = j \mid X(0) = i\}$$

which is independent of t, so it defines a function of i and j alone, p_{ij} say. Properties (a), (b), and (c) follow immediately. $\qquad \square$

Theorem 8-3 verifies the claim that a CTMC goes from state to state according to a Markov chain and remains in each state an exponentially distributed amount of time that depends on only the current state. The MC $\{Y_n; n \in I\}$ is called the *embedded chain* of the CTMC.

The construction of a sample path for a CTMC via Theorem 8-3 proceeds as follows. Assume there are no instantaneous states. Let $X(0; \omega)$ be chosen according to the vector of initial probabilities, say $X(0; \omega) = i_0$. Let $S_1(\omega)$ be obtained as a sample from the exponential distribution with mean $1/\alpha_{i_0}$, $Y_1(\omega)$ is obtained from the transition matrix $P = (p_{ij})$ of the embedded chain, and so forth, generating $Y_2(\omega)$, $Y_3(\omega)$, $\ldots$, $S_2(\omega)$, $S_3(\omega)$, $\ldots$. The sample path $X(\cdot; \omega)$ is defined by $X(t; \omega) = Y_n(\omega)$ for $t \in [S_n(\omega), S_{n+1}(\omega))$.

Pathological Behavior

If S is infinite, the construction above will fail if $S_\infty(\omega) \triangleq \lim_{n \to \infty} S_n(\omega) < \infty$, that is, if infinitely many transitions occur by a finite time. Here is the reason why.

Proposition 8-3 Let $Y_\infty(\omega) = \lim_{n \to \infty} Y_n(\omega)$. If S is infinite and no state is instantaneous, then (w.p.1)

$$S_\infty(\omega) < \infty \implies Y_\infty(\omega) = \infty$$

Consequently, the construction of $X(t; \omega)$ for $t > S_\infty(\omega)$ is not well defined if $S_\infty(\omega) < \infty$.

PROOF The proof is done by contradiction. Suppose $S_\infty(\omega) < \infty$ and there is a finite subset $\mathcal{F} \subset S$ such that $Y_n \in \mathcal{F}$ for all $n \in I$. Since $\mathcal{F}$ is finite, at least one state of $\mathcal{F}$, say j, must have been entered infinitely many times. Since instantaneous states are prohibited by assumption, the length of each visit to state j has mean $1/\alpha_j > 0$, and the strong law of large numbers implies that (w.p.1) the total time spent in state j is infinite. Hence $\mathcal{F}$ cannot be finite. ☐

The following equation is a consequence of Proposition 8-3:

$$P\{S_\infty < t\} > 0 \iff \sum_{j \in S} P\{X(t) = j\} < 1 \qquad t > 0 \tag{8-20}$$

The proof of (8-20) is requested in Exercise 8-7. Thus, if $P\{S_\infty < \infty\} = 0$, then (8-4) holds as an equality and the construction of sample paths via Theorem 8-3 is valid.

The method of proof for Proposition 8-3 can be used to establish the next proposition (see Exercise 8-8).

Proposition 8-4 If either (a) S is finite or (b) there is a positive number B such that $\alpha_i \leq B$ for all $i \in S$, then

$$P\{S_\infty < \infty\} = 0$$

Thus, pathologies occur only when the state space is infinite and the transition rates are unbounded. Unfortunately, some important processes have these properties, e.g., the $M/M/\infty$ queue. A necessary and sufficient condition for $P\{S_\infty < \infty\} = 0$ is given by Theorem II.19.3 in Chung (1967). This condition is difficult to apply; an easily applied sufficient (but not necessary) condition is described below.

Proposition 8-5 Let $\{X(t); t \geq 0\}$ be a CTMC, $\{Y_n; n \in I\}$ be its embedded MC, and α_i be defined by (8-12). If (a) $\alpha_i < \infty$ for all i and (b) $\{Y_n; n \in I\}$ consists entirely of persistent states, then

$$P\{S_\infty < \infty\} = 0 \tag{8-21}$$

PROOF Let j be any state and $S_n(j)$ be the time spent in state j after n transitions of the embedded chain. Since j is persistent, it will be entered infinitely many times as n tends to infinity. Since $\alpha_i < \infty$ and the lengths of the visits to j are i.i.d., $\lim_{n\to\infty} S_n(j)$ is the sum of infinitely many independent exponential r.v.'s, hence it is (w.p.1) infinite (use, e.g., Theorem 4-13 with $\lambda_i \equiv \alpha_i$). Since S_n is the time spent in all states after n transitions, $S_n \geq S_n(j)$ and

$$P\{S_\infty < \infty\} \leq P\{\lim_{n\to\infty} S_n(j) < \infty\} = 0 \qquad \square$$

Notice that (b) can be weakened; the essential condition is that a closed set of persistent states be reached after a finite number of transitions. Since condition (a) usually holds in applications, the latter condition is the crucial one to check.

EXERCISES

8-5 Let $\{X(t); t \geq 0\}$ be a CTMC, and let i and j be two members of the state space. Explain how to obtain the d.f. of the first-passage time from i to j, say $G_{ij}(\,\cdot\,)$, by constructing another CTMC, say $\{X'(t); t \geq 0\}$, where j is an absorbing state, and finding $P\{X'(t) = j \mid X(0) = i\}$.

8-6 Show that if an instantaneous state, i say, is entered at epoch t, then the time spent in state i during $[t, t + \epsilon]$ is positive, for any $\epsilon > 0$. *Hint*: Define $I_i(t)$ to be 1 if $X(t) = i$ and 0 otherwise; then prove

$$E\left(\int_t^{t+\epsilon} I_i(s)\, ds \mid X(t) = i\right) = \int_0^\epsilon P_{ii}(s)\, ds > 0$$

8-7 Derive (8-20).
8-8 Prove Proposition 8-4.

8-3 TRANSITION FUNCTION

Transition Rates

Usually a CTMC is specified in terms of its *transition rates*. This is justified by the next proposition.

Proposition 8-6 For any standard transition matrix, the limits

$$q_i = \lim_{t\to 0} \frac{1 - P_{ii}(t)}{t} \quad \text{and} \quad q_{ij} = \lim_{t\to 0} \frac{P_{ij}(t)}{t} \qquad i \neq j$$

exist, with $q_i \geq 0$ and $0 \leq q_{ij} < \infty$.

PROOF See Theorems II.2.4 and II.2.5 in Chung (1967). $\qquad \square$

These limits have the interpretation of transition rates at time 0. To see this, assume that the probability of more than one transition during $[0, t)$ is $o(t)$ for small t. Then $1 - P_{ii}(t) + o(t)$ is the probability that a transition occurs before t, and it equals $q_i t$. Similarly, $P_{ij}(t) + o(t)$ is the probability that a transition from state i to state j occurred by t, and it equals $q_{ij} t$.

The interpretation of the limits in Proposition 8-6 as transition rates makes it natural to suspect that

$$q_i = \sum_{j \neq i} q_{ij} \qquad i \in S \tag{8-22}$$

However, (8-22) does not necessarily hold because the general theory allows for transitions from state i to "state ∞." When (8-22) holds, the CTMC is said to be *conservative*, and *henceforth we consider only conservative processes.*

From Theorem 8-3, again by assuming that the probability of more than one transition in $(0, t)$ is $o(t)$,

$$P_{ii}(t) = e^{-\alpha_i t} + o(t) \Rightarrow q_i = \alpha_i \qquad i \in S \tag{8-23a}$$

and

$$P_{ij}(t) = p_{ij}(1 - e^{\alpha_i t}) + o(t) \Rightarrow q_{ij} = \alpha_i p_{ij} \qquad i \neq j \tag{8-23b}$$

These equations relate the transition structure described by Theorem 8-3 to the transition rates in a one-to-one way. Since the former defines the process when (8-18) holds, we expect that the transition rates determine the transition function when (8-21) holds. This is, in fact, true, as we show below.

Forward and Backward Kolmogorov Equations

A system of differential equations satisfied by the transition functions can be derived by dividing $[0, t + s)$ into $[0, t)$ and $[t, t + s)$ and then dividing $[0, t)$ into $[0, t - s)$ and $[t - s, t)$ for s small and positive (cf. Section 4-4 for the special case of birth-and-death processes). The Chapman-Kolmogorov equation (8-5) and Proposition 8-6 yield, in the former case,

$$P_{ij}(t + s) = \sum_{k \in S} P_{ik}(t)P_{kj}(s)$$

$$= P_{ij}(t)[1 - sq_j + o(s)] + \sum_{k \neq j} P_{ik}(t)[sq_{kj} + o(s)] \qquad i \in S \tag{8-24a}$$

and, in the latter,

$$P_{ij}(t) = \sum_{k \in S} P_{ik}(t - s)P_{kj}(s)$$

$$= P_{ij}(t - s)[1 - sq_i + o(s)] + \sum_{k \neq j} P_{ik}(t - s)[sq_{kj} + o(s)] \qquad i \in S \tag{8-24b}$$

When S is finite, the sum of the $o(s)$ terms is automatically $o(s)$. When S is infinite, the validity of (8-21) and (8-22) is a sufficient condition for the sum of the $o(s)$ terms to be $o(s)$ [see Chung (1967, sec. II.18)].

From (8-24a) and (8-24b), we obtain

$$\lim_{s \downarrow 0} \frac{P_{ij}(t+s) - P_{ij}(t)}{s} = \lim_{s \downarrow 0} \frac{P_{ij}(t) - P_{ij}(t-s)}{s}$$

Hence the common value is the derivative of $P_{ij}(\,\cdot\,)$, denoted by $\dot{P}_{ij}(\,\cdot\,)$, and

$$\dot{P}_{ij}(t) = -q_j P_{ij}(t) + \sum_{k \neq j} P_{ik}(t) q_{kj} \qquad i \in \mathcal{S} \qquad (8\text{-}25)$$

Equation (8-25) is called the *forward Kolmogorov equation*. Equation (8-25) has a simple flow-of-probability interpretation. When the initial state is i, the net rate of flow of probability into state j at time t is $\dot{P}_{ij}(t)$. The outward flow rate is the amount there, $P_{ij}(t)$, multiplied by the rate at which transitions leave state j, which is q_j; the inward flow rate is the sum over all other states k of the amount in state k, $P_{ik}(t)$, multiplied by the rate at which transitions from state k to state j occur, which is q_{kj}.

Define $q_{ii} = -q_i$ and the matrix $Q = (q_{ij})$. Then Q is the matrix of transition rates of the CTMC, where the negative sign associated with q_{ii} indicates that it is an outward transition rate. Sometimes Q is called the *Q-matrix* of the CTMC.

Let $P(t)$ be the matrix $(P_{ij}(t))$ and $\dot{P}(t)$ be the matrix $(\dot{P}_{ij}(t))$. Then (8-25) can be written as the matrix equation.

$$\dot{P}(t) = P(t)Q \qquad t > 0 \qquad (8\text{-}26a)$$

By dividing $(0, t+s]$ into $(0, s]$ and $(s, t+s]$ and dividing $(0, t-s]$ into $(0, s]$ and $(s, t-s]$, for s small and positive, letting $s \downarrow 0$ yields†

$$\dot{P}(t) = QP(t) \qquad t > 0 \qquad (8\text{-}26b)$$

which is the backward Kolmogorov equation (see Section 4-4 and Exercise 8-9).

The initial condition for (8-26a) and (8-26b) is (8-6), which in matrix notation is

$$P(0) = I \qquad (8\text{-}26c)$$

where I is the identity matrix whose dimension is the number of states (possibly infinite).

If (8-26a), (8-26b) and (8-26c) were relationships among scalars, $P(t) = e^{Qt}$ would be the unique solution. For matrices, *define*

$$e^{Qt} \triangleq \sum_{n=0}^{\infty} \frac{Q^n t^n}{n!} \qquad (8\text{-}27)$$

---

† Equation (8-22) is not needed for (8-26b) to be valid; see Chung (1967, sec. II.18).

It can be shown that (8-27) is the unique solution of both (8-26a) and (8-26b) subject to (8-26c). When S is finite, the proof of this fact is easy. One just has to show that the right-side of (8-27) satisfies the differential equations; this is left to you. When S is infinite, the proof is more delicate because convergence of the right-side must be established; the implicit conditions placed on Q by (8-21) and (8-22) imply that the right-side of (8-27) converges.†

The properties of a transition matrix given by (8-3) and (8-4) are not explicitly required in (8-26). It is easily shown that (8-27) represents a matrix with nonnegative elements and row sums of 1; see Exercise 8-10. When S is finite, it may appear from (8-27) that computing $P(t)$ is routine, but calculating e^{Qt} can be a difficult task.‡

The preceding results are summarized in the following theorem.

Theorem 8-4 Let $\{X(t); t \geq 0\}$ be a CTMC with the property (8-22), and let the associated Q-matrix be conservative. Then the transition functions satisfy the forward equations (8-26a), the backward equations (8-26b), and the boundary conditions (8-26c). Moreover, (8-27) is the unique solution to both systems of differential equations.

As a consequence of Theorem 8-4, the Q-matrix determines the transition function of the CTMC; for this reason, the Q-matrix is sometimes called the *generator* of the CTMC.

Digraph Representation

The digraph of a CTMC is formed by associating a node with each state. If $q_{ij} > 0$, a directed arc from node i to node j with weight q_{ij} is drawn. If $q_{ij} = 0$, or $i = j$, no arc from i to j is drawn. From (8-20) it is apparent that the digraph of a CTMC is the same as the digraph of the embedded MC with all arcs from a state to itself deleted.

EXERCISES

8-9 Derive (8-26a) and (8-26b).

8-10 Let Q be a standard, conservative Q-matrix. Show that the entries of e^{Qt} are nonnegative and the row sums are 1's. *Hint*: Show that $q_{ij} \geq 0$ for $i \neq j$ implies $e^{Qt} \geq 0$ for all sufficiently small t; then use the representation $e^{Qt} = e^{(Qt/n)^n}$.

8-11 Display the Q-matrix for Exercise 8-4. Write out the forward and the backward Kolmogorov equations.

---

† See Richard Bellman, *Introduction to Matrix Analysis*, McGraw-Hill, New York (1960), chap. 10.

‡ See Cleve Moler and Charles Van Loan, "Nineteen Dubious Ways to Compute the Exponential of a Matrix," *SIAM Rev.* **20**: 801–836 (1978).

8-4 MORE EXAMPLES

Example 8-4: Example 7-25 continued The inventory of trailers goes up when returns occur and goes down when demands occur. Assume that the returns and demands are generated by independent Poisson processes.

Costs are time homogeneous and are not discounted. A holding cost proportional to the amount of inventory is charged continuously. When there is a setup cost for returning items (to reflect the cost of packaging, preparing paperwork, etc.), there is an incentive to return several items together in order to spread the setup cost over several items. Suppose additional items can be obtained from the supplier very quickly; then in the model we assume that the supply lead time is zero. Suppose also that the ordering cost is proportional to the quantity ordered. There is no incentive to order large quantities, and the holding-cost rate provides an incentive to order small quantities, so one item should be ordered each time a customer arrives and the inventory is zero. This mode of operation is intuitively good, and it can be shown to be optimal.†

The policy described is characterized by two numbers, a and b say. When the inventory level reaches b, the amount $b - a$ is returned, i.e., the inventory is reduced to a. Let $1/\lambda$ be the mean time between returns, $1/\mu$ be the mean time between demands, $X(t)$ be the inventory level at epoch t, and let $X(0) = i_0$, $0 \le i_0 < b$, be given.

The assumption that the demand and return processes are Poisson and independent of each other implies that $\{X(t); t \ge 0\}$ is a CTMC; use the memoryless property of the interevent times to check that (8-1) holds. The state space is $\{0, 1, \ldots, b - 1\}$; there is no state b because a return in state $b - 1$ triggers an inventory reduction. (You might make b an instantaneous state; try it and see what happens to the analysis.)

The digraph of the CTMC is shown in Figure 8-3. Except for the arc from $b - 1$ to a, this is the digraph of a birth-and-death process. Each arc labeled with a λ (alternatively, μ) represents a transition caused by a return (alternatively, demand). The Q-matrix for this CTMC is

$$
Q = \begin{array}{c}
\begin{array}{c} \\ 0 \\ 1 \\ \vdots \\ a \\ \vdots \\ b-2 \\ b-1 \end{array}
\begin{array}{cccccccccc}
0 & 1 & 2 & \ldots & a-1 & a & a+1 & \ldots & b-3 & b-2 & b-1 \\
\left(\begin{array}{cccccccccc}
-\lambda & \lambda & 0 & \ldots & 0 & 0 & 0 & \ldots & 0 & 0 & 0 \\
\mu & -(\lambda+\mu) & \lambda & \ldots & 0 & 0 & 0 & \ldots & 0 & 0 & 0 \\
\cdots\cdots\cdots\cdots\cdots\cdots\cdots\cdots\cdots\cdots\cdots \\
0 & 0 & 0 & \ldots & \mu & -(\lambda+\mu) & \lambda & \ldots & 0 & 0 & 0 \\
\cdots\cdots\cdots\cdots\cdots\cdots\cdots\cdots\cdots\cdots\cdots \\
0 & 0 & 0 & \ldots & 0 & 0 & 0 & \cdots & \mu & -(\lambda+\mu) & \lambda \\
0 & 0 & 0 & \cdots & 0 & \lambda & 0 & \cdots & 0 & 0 & -\lambda
\end{array}\right)
\end{array}
\end{array}
$$

□

---

† Daniel P. Heyman, "Return Policies for an Inventory System with Positive and Negative Demands," *Nav. Res. Log. Quart.* **25**: 581–596 (1978).

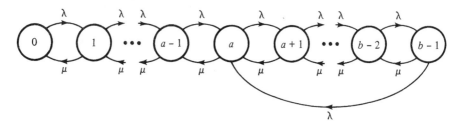

Figure 8-3 Digraph of the CTMC.

Example 8-5: Method of stages The queueing models for which we have obtained the limiting probabilities have a Poisson arrival process and, with the exception of the $M/G/1$ queue, also have exponentially distributed service times. The $M/G/1$ queue is analyzed by an embedded MC (Examples 7-18 and 7-19). This technique is not applicable to the $M/G/c$ queue with $c \geq 2$ (Exercise 7-4).

Although models with Poisson arrivals and exponential service times provide reasonably accurate descriptions of many congestion phenomena, they do not describe all situations. Therefore, richer classes of distributions are needed. Since exponential distributions often yield tractable models, it is natural to consider models with r.v.'s which are based on exponential r.v.'s, such as sums of exponential r.v.'s. One hopes that this larger class of distributions yields queueing models which are tractable. Indeed they are, and the *method of stages* is an important analytical approach.

Suppose an r.v. has a gamma distribution with mean $1/\mu$ and variance $1/(k\mu^2)$ with k a positive integer.† Such an r.v. has the same distribution as the sum of k independent exponential r.v.'s, each having mean $1/(k\mu)$. If this r.v. is, for example, a service time, it can be represented as k exponential *stages* (or *phases*) of service that are performed sequentially. The customer departs after the last stage is completed. Thus, a nonexponentially distributed r.v. is replaced by a series of exponential r.v.'s.

To be specific, consider a queueing system with Poisson arrivals and two servers in parallel, each having service times as above with $k = 2$. Let m denote the capacity of the queue.

Define the state of the queue to be (n, s_1, s_2), where n is the number of customers in the system (including customers in service) and s_i is the stage of the customer in server i. If server i is idle, $s_i = 0$. Let $X(t)$ be the state at time

---

† In queueing theory, this special form of the gamma distribution is called Erlang-k distribution, abbreviated E_k; see Section A-2. The method of stages is credited to the queueing theory pioneer A. K. Erlang.

t; with this state space, $\{X(t); t \geq 0\}$ is† a CTMC. The digraph is shown below for $m = 1$. □

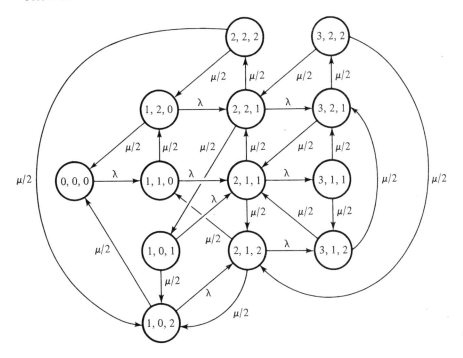

EXERCISES

8-12 Recall the definition of a Y-process given in Exercise 4-37. Show that Example 8-4 is a Y-process.

8-13 The assumption in Example 8-4 that deliveries are instantaneous may not always be tenable. Suppose deliveries have an exponentially distributed lead time. To protect against stockouts, orders are placed whenever the stock level is reduced from $c + 1$ to c; assume $0 < c < a$. Draw the digraph and display the Q-matrix for this model. Assume that the delivery times have an Erlang-2 distribution. Use the method of stages to construct a CTMC for this model, and draw its digraph.

8-14 Companies that provide service to a large number of customers located in a wide geographical area frequently offer, by telephone, information and ordering services from a centrally located facility. A typical example is a major airline that advertises a toll-free number to call for information and reservations. The calls to this toll-free number can be handled in a variety of ways; consider this way. There are i information clerks and r reservation clerks. When a call arrives, the caller identifies whether she or he wants information or is going to make a reservation. A call is routed to a clerk of the corresponding type, if one is available. If the only clerks available are of the other type, the call is routed to one of them. If all the clerks are busy, the call waits in a buffer that can hold b calls. A call that arrives when the buffer is full receives a busy signal.

† If you have studied the previous examples, you should readily identify this process as a CTMC without having to check the definition. A countable state space and independent exponentially distributed times between events are the key.

Assume that reservation clerks process reservations at a faster rate than information clerks; make the corresponding assumption for information calls. After receiving information, some customers may choose to make a reservation. In this case, they make it with the clerk who provided the information, who is occupied for an additional time, which is distributed as above.

State the assumptions required for the number of calls in the system to be modeled as a CTMC. Draw the digraph. What are the probabilities of interest in this model? How difficult would they be to compute if $i = 200$, $r = 300$, and $b = 100$?

8-15 A store operates with two checkout counters. Counter 1 serves anyone, but counter 2 serves only customers buying no more than 10 items. Since counter 1 is closer to the front door, a customer who is eligible to be served at counter 2 goes there if, and only if, counter 1 is busy.

Assume that customers who purchased more (no more) than 10 items arrive according to a Poisson process with rate α_1 (α_2) and the service times at counter i are i.i.d. exponentially distributed r.v.'s with mean $1/\mu_i$, $i = 1, 2$.

Let $N_i(t)$ be the number of customers present at counter i at time t.

(a) Show that $\{N_1(t); t \geq 0\}$ is a birth-and-death process with birth rates $\lambda_0 = \alpha_1 + \alpha_2$, $\lambda_n = \alpha_1$ for $n \in I_+$, and death rate μ_1 for every state in I_+.

Let $p_1(0) = \lim_{t \to \infty} P\{N_1(t) = 0\}$. Since $\alpha_2 p_1(0)$ is the long-run arrival rate at counter 2 (why?), it is tempting to assert that in the steady state, $N_2(t)$ fluctuates exactly as an M/M/1 queue with arrival rate $\lambda_2 = \alpha_2 p_1(0)$ and service rate μ_2.

(b) Explain why this assertion is false.

(c) Show that $\{(N_1(t), N_2(t)); t \geq 0\}$ is a CTMC, and draw the digraph.

8-5 ALGEBRAIC SOLUTION OF FINITE CONTINUOUS-TIME MARKOV CHAINS

Let the states of the CTMC $\{X(t); t \geq 0\}$ be $\{0, 1, \ldots, s\}$ with $s < \infty$, and let Q denote the Q-matrix. It can easily be shown that $\lambda_0 = 0$ is an eigenvalue of Q; see Exercise 8-16. Let λ_i, $1 \leq i \leq s$, denote the remaining eigenvalues, and assume they are distinct. Applying Theorem 7-2 to (8-27) yields

$$P(t) = e^{Qt} = \sum_{n=0}^{\infty} \frac{t^n}{n!} \sum_{k=0}^{s} \lambda_k^n Z_k$$

$$= Z_0 + \sum_{k=1}^{s} e^{\lambda_k t} Z_k \tag{8-28}$$

where

$$Z_k = \frac{\prod_{r \neq k} (Q - \lambda_r I)}{\prod_{r \neq k} (\lambda_k - \lambda_r)} \qquad k = 0, 1, \ldots, s$$

Although (8-27) may not be a good way† to compute $P(t)$, it does provide important qualitative information.

Proposition 8-7 Re $\lambda_k \leq 0$, $k = 1, 2, \ldots, s$.

PROOF Each matrix Z_k has at least one nonzero element, the (i, j)th say, and $0 \leq P_{ij}(t) \leq 1$. If‡ Re $\lambda_k > 0$, then $|e^{\lambda_k t} Z_k(i, j)| \to \infty$ as $t \to \infty$. Since the

---

† See Moler and Van Loan, op. cit.

‡ Recall that when z is the complex number $z = x + iy$, $|e^z| = e^x$.

eigenvalues are distinct, this forces $|P_{ij}(t)| \to \infty$, which is absurd; hence Re $\lambda_k \le 0$. □

It follows from Proposition 8-7 that $P(t) \to Z_0$ exponentially fast, with the rate controlled by the eigenvalue with smallest positive magnitude.

EXERCISE

8-16 Let **1** denote a column vector with $s + 1$ elements that are all 1's. Show that $Q\mathbf{1} = \mathbf{0}$, where **0** is the column vector consisting entirely of 0's.

8-6 LIMITING AND STATIONARY PROBABILITIES

Limiting and stationary probabilities are similar to their counterparts for Markov chains.

Definitions

Definition 8-5 Let $\{X(t); t \ge 0\}$ be a CTMC with state space $\mathcal{S}$ and Q-matrix Q. The vector $\mathbf{\Pi} = (\Pi_j ; j \in \mathcal{S})$ is a *stationary* (or *invariant*) *probability vector* of the CTMC if

$$\Pi Q = 0 \qquad (8\text{-}29a)$$

$$\sum_{j \in \mathcal{S}} \Pi_j = 1 \qquad (8\text{-}29b)$$

$$\Pi_j \ge 0 \qquad j \in \mathcal{S} \qquad (8\text{-}29c)$$

From (8-26a) and (8-26c),

$$P(t) = \int_{0^+}^{t} \dot{P}(x)\, dx + I = Q \int_{0^+}^{t} P(x)\, dx + I \qquad (8\text{-}30)$$

where the integral in (8-30) is interpreted componentwise (i.e., the integral of a matrix of functions is the matrix of integrals). If a stationary probability vector is chosen as the vector of initial probabilities, then combining (8-10b) $[\mathbf{P}(t) = \mathbf{a}P(t)]$, (8-29a), and (8-30) yields

$$\mathbf{P}(t) = \mathbf{\Pi} \qquad t \ge 0 \qquad (8\text{-}31)$$

so the state probability vector remains at $\mathbf{\Pi}$ for all time. Equation (8-31) is the continuous-time counterpart of (7-46).

Definition 8-6 Let $\{X(t); t \ge 0\}$ be a CTMC with state space $\mathcal{S}$ and vector of initial probabilities **a**. The vector $\mathbf{p} = (p_j; j \in \mathcal{S})$ is the *steady-state* (or *limiting*) *distribution* of the CTMC if

$$\lim_{t \to \infty} P\{X(t) = j\} = p_j \qquad j \in S \qquad (8\text{-}32)$$

for any choice of **a**.

It is readily established that Definition 8-6 is equivalent to

$$\lim_{t \to \infty} P_{ij}(t) = p_j \qquad i, j \in S \qquad (8\text{-}32')$$

From (8-31) and (8-32) it follows that a steady-state distribution must be a stationary vector and can exist only when there is a unique solution to (8-29). The results below establish that a *sufficient* condition for the limit in (8-32') to exist is that the embedded Markov chain be irreducible and consist of ergodic states. By using more complicated arguments,† it is possible to include null-persistent states. The hypothesis of irreducibility can always be weakened, as in Corollary 7-9.

From Theorem 8-3, for each fixed $j \in S$, each epoch that the process enters state j is a regeneration epoch. In particular, this means that times between successive entries to state j are i.i.d. random variables.

Obtaining Limiting Probabilities

Proposition 8-8 Let μ_{jj} denote the mean time between entries to state j for the CTMC $\{X(t); t \geq 0\}$. If

(a) $\mu_{jj} < \infty$
(b) $q_j < \infty$
(c) $P\{X(t) = j \text{ for some } t \mid X(0) = i\} = 1$ for all i

all hold, then

$$\lim_{t \to \infty} P_{ij}(t) = \frac{1}{q_j \mu_{jj}} \qquad j \in S \qquad (8\text{-}33)$$

PROOF From Theorem 8-3, $1/q_j$ is the mean time spent in state j between successive entries to state j, with each entry being a regeneration epoch. From (c), a first entry into state j occurs (w.p.1). Equation (8-33) is an immediate consequence of Theorem 6-7. ☐

To see that the hypotheses of Proposition 8-8 are not necessary for the limit in (8-33) to exist, take j as an absorbing state. Then $q_j = 0$ and since a second visit to j never occurs, $\mu_{jj} = \infty$. Obviously, if part (c) of Proposition 8-8 holds, $\lim_{t \to \infty} P_{ij}(t) = 1$.

Theorem 8-3 shows that the embedded MC controls the sequence of states visited by a CTMC. Proposition 8-8 shows that when the initial state is

---

† Combine Theorem 1 in Rupert G. Miller, Jr., "Stationary Equations in Continuous Time Markov Chains," *Trans. Am. Math. Soc.* **109**: 35–44 (1963), with Theorem II.16.2 in Chung (1967).

j and $0 < q_j \mu_{jj} < \infty$, $P_{jj}(t)$ converges to a positive limit. For MCs, the corresponding property is possessed by ergodic states (Proposition 7-3 and Definition 7-3). These considerations motivate the following definition.

Definition 8-7 A CTMC is called *irreducible* if its embedded MC is irreducible. State j of a CTMC is called *ergodic* if $0 < q_j \mu_{jj} < \infty$.

The difficulty in using (8-33) to calculate p_j is that one must be able to check part (c) of Proposition 8-8 and compute μ_{jj}. If the embedded chain has a limiting probability vector π, then π_j is the proportion of transitions that enter state j. Since $1/q_j$ is the mean length of each visit to state j, the proportion of time the CTMC spends in state j should be proportional to π_j/q_j; thus

$$p_j = \frac{\pi_j/q_j}{\sum_{i \in 8} \pi_i/q_i} \qquad j \in 8 \qquad (8\text{-}34)$$

The argument above is now made precise.

Theorem 8-5 Let $\{X(t); t \geq 0\}$ be a CTMC with properties

(a) $q_j < \infty$, $j \in 8$
(b) it† is irreducible
(c) the embedded Markov chain has a stationary probability vector [i.e., a solution of (7-45)], π say
(d) $\sum_{i \in 8} \pi_i/q_i < \infty$

Then the limiting distribution is given by (8-34).

PROOF From (b), (c), and Theorem 7-11, π is unique, so (8-34) is well defined, and all states in the embedded MC are ergodic. From Theorem 7-9 each $\pi_i > 0$, and from Exercise 7-78 the expected number of visits to state i between visits to state j is π_i/π_j. The mean length of each visit to state i is $1/q_i$. Hence

$$\mu_{jj} = \frac{1}{\pi_j} \sum_{i \in 8} \frac{\pi_i}{q_i} \qquad j \in 8 \qquad (8\text{-}35)$$

[which theorems establish (8-35)?] and (d) ensures that $\mu_{jj} < \infty$. Since the embedded chain is irreducible, state j will be entered after a finite number of transitions from any initial state; from (a) the time between transitions is finite (w.p.1), hence part (c) of Proposition 8-8 holds. Thus, (8-33) is valid; substituting (8-35) into (8-33) yields (8-34). ☐

When 8 is finite, parts (c) and (d) of Theorem 8-5 are automatically satisfied. Since part (a) is always assumed to hold and easy to check in any case, verifying part (b) is all that is required to assert that (8-34) provides the limiting probabilities.

† Note that (b) $\Rightarrow q_j > 0$ for all j.

Since a CTMC is described by its Q-matrix, it is preferable to express the limiting probabilities as a solution to (8-29). Theorem 8-6 is the analog of Theorem 7-11.

Theorem 8-6: Characterization of the limiting distribution Let $\{X(t); t \geq 0\}$ be an irreducible CTMC. If Π is a solution of (8-29) such that

$$\sum_{i \in S} \Pi_i q_i < \infty \tag{8-36}$$

then Π is the unique solution of (8-29) and

$$p_j \triangleq \lim_{t \to \infty} P\{X(t) = j \,|\, X(0) = i\} = \Pi_j \qquad i, j \in S \tag{8-37}$$

PROOF Write (8-29a) as

$$\Pi_j q_j = \sum_{i \neq j} \Pi_i q_{ij} \qquad i, j \in S \tag{8-38}$$

and set

$$x_j \triangleq \Pi_j q_j \qquad j \in S \tag{8-39}$$

Substitute (8-39) into (8-38) and use (8-23) to obtain

$$x_j = \sum_{i \neq j} \frac{x_i}{q_i} q_{ij} = \sum_{i \in S} x_j p_{ij} \qquad j \in S$$

That is, x_j is the limiting probability of being in state j for the embedded MC. From Theorem 8-5, p_j is given by (8-34); substituting (8-39) into (8-34) yields (8-37). Since limits are unique, Π must be the only solution of (8-29). $\square$

When S is finite or q_i is a bounded function of i, (8-36) is automatically satisfied, so one just has to solve (8-29) to obtain the limiting distribution. Exercise 4-21 illustrates that (8-29) may have a unique solution for which (8-36) fails. Exercise 8-17 asks you to prove that for an irreducible CTMC, $p_j > 0$ for some $j \in S$ only if (8-36) holds.

Observe that the left-side of (8-36) is the long-run rate at which transitions occur. With this interpretation, it is natural to expect that (8-36) is valid in most applied models. In the special case of a birth-and-death process, (8-36) reduces to the necessary and sufficient condition for the existence of limiting probabilities given in Theorem 4-4 (Exercise 8-20).

It is often inconvenient to treat (8-29b) explicitly in solving (8-29); fortunately, we do not have to.

Proposition 8-9 Equations (8-29) have a solution if, and only if, there exists a vector $\mathbf{y}$ such that

$$\mathbf{y}Q = 0 \tag{8-40a}$$

$$\sum_{j \in S} y_j < \infty \tag{8-40b}$$

$$y_j \geq 0 \qquad j \in S \tag{8-40c}$$

Furthermore, the solution of (8-29) is obtained from **y** by

$$\Pi_j = \frac{y_j}{\sum_{i \in S} y_i} \tag{8-41}$$

PROOF This is left as Exercise 8-18. $\qquad\square$

The next theorem may simplify solving (8-29) and (8-40). It is the continuous-time analog of Theorem 7-13 and generalizes Proposition 4-2.

Theorem 8-7: Alternate characterization of the stationary distribution Consider a CTMC with state space S and Q-matrix Q. A vector Π satisfying (8-29b) and (8-29c) solves (8-29a) if, and only if, Π solves

$$\sum_{i \in A} \Pi_i \sum_{j \in B} q_{ij} = \sum_{j \in B} \Pi_j \sum_{i \in A} q_{ji} \tag{8-42}$$

for any sets A and B such that $A \cup B = S$ and $A \cap B$ is empty.

PROOF The proof is analogous to the proof of Theorem 7-13 and is left as Exercise 8-19. $\qquad\square$

Theorem 8-7 has the interpretation that a CTMC is stationary whenever the rate of flow of probability into a set of states equals the rate of flow of probability out of that set. For this reason, (8-42) and (8-29a) are called the *steady-state balance equations*.

Examples

Example 8-6: Example 8-4 continued In this example, the CTMC is irreducible and $S < \infty$, so the limiting probability vector is the unique solution of (8-29). To obtain it, we use Proposition 8-8 and Theorem 8-7. Put (8-40a) in the form (8-42) with $A = \{0, 1, \ldots, k\}$, for $k = 0, 1, \ldots, b - 2$ in turn. For each k, this means that the net flow of probability across a vertical line immediately to the right of node k in the digraph of this CTMC (Figure 8-3) is zero. The resulting equations are

$$\lambda y_k = \mu y_{k+1} \qquad\qquad k = 0, 1, \ldots, a - 1 \tag{8-43a}$$

and

$$\lambda y_k = \mu y_{k+1} + \lambda y_{b-1} \qquad k = a, a + 1, \ldots, b - 2 \tag{8-43b}$$

Let $\rho = \lambda/\mu$ and temporarily assume $\rho \neq 1$. Solve (8-43a) in terms of y_0 to obtain

$$y_k = \rho^k y_0 \qquad\qquad k = 0, 1, \ldots, a \tag{8-44a}$$

Similarly, solve (8-43b) in terms of y_{b-1} by iterating it for $k = b - 2$,

$b - 3, \ldots, a$, in turn, to obtain

$$y_k = \frac{(1 - \rho^{b-k})y_{b-1}}{\rho^{b-1-k}(1 - \rho)} \qquad k = a, a + 1, \ldots, b - 1 \qquad (8\text{-}44b)$$

Equations (8-44a) and (8-44b) must agree at $k = a$, hence

$$\rho^a y_0 = \frac{(1 - \rho^{b-a})y_{b-1}}{\rho^{b-1-a}(1 - \rho)}$$

or

$$y_{b-1} = \frac{(1 - \rho)\rho^{b-1}y_0}{1 - \rho^{b-a}} \qquad (8\text{-}45)$$

Substituting (8-45) into (8-44b) yields

$$y_k = \frac{(\rho^k - \rho^b)y_0}{1 - \rho^{b-a}} \qquad k = a, a + 1, \ldots, b - 1 \qquad (8\text{-}44c)$$

and all the y_k are expressed in terms of y_0. From (8-44a) and (8-44c) it is evident that if $y_0 > 0$, then $y_k > 0$ for $k = 1, 2, \ldots, b - 1$, so $\mathbf{y}$ satisfies (8-40). To obtain Π_j from (8-41), it is necessary to compute $\sum y_k$. From (8-43a) and (8-43b),

$$\sum_{k=0}^{b-1} y_k = \sum_{k=0}^{a-1} y_k + \sum_{k=a}^{b-1} y_k$$

$$= y_0 \left[\frac{1 - \rho^a}{1 - \rho} + \frac{(\rho^a - \rho^b)/(1 - \rho) - (b - a)\rho^b}{1 - \rho^{b-a}} \right] \qquad (8\text{-}46)$$

$$= y_0 \frac{1 - \rho^{b-a} - (1 - \rho)(b - a)\rho^b}{(1 - \rho)(1 - \rho^{b-a})}$$

Substituting (8-44a), (8-44c), and (8-46) into (8-41) yields

$$\Pi_j = \frac{(1 - \rho)(1 - \rho^{b-a})}{1 - \rho^{b-a} - (1 - \rho)(b - a)\rho^b} \times \begin{cases} \rho^j & j = 0, 1, \ldots, a \\ \dfrac{\rho^j - \rho^b}{1 - \rho^{b-a}} & j = a, a + 1, \ldots, b - 1 \end{cases}$$

$\square$

Example 8-7: $M/E_k/1$ queue The steady-state distribution of the number of customers in an $M/E_k/1$ queue is derived as a special case of the $M/G/1$ queue in Exercise 7-108. The purpose of this example is to demonstrate the importance of the structure of Q in solving (8-29) and the use of Theorem 8-6.

Recall that in this model, customers arrive according to a Poisson process, with rate λ say, and the service times are i.i.d. random variables with density function

$$g(t) = \frac{k\mu(k\mu t)^{k-1}e^{-k\mu t}}{(k - 1)!}$$

with k a positive integer. By using the method of stages described in Example 8-5, the service time of a customer is represented as the sum of k i.i.d. stages, each one having an exponential distribution with mean $(k\mu)^{-1}$.

Following Example 8-5, we can represent the number of customers in an $M/E_k/1$ queue as a CTMC by considering states of the form (n, s), where n is the number of customers in the system and s is the stage of the customer in service. We use the convention that $s = 1$ when $n = 0$, so $s \in \{1, 2, \ldots, k\}$. A customer in stage s has completed $s - 1$ stages. When there is a single server, a scalar state will suffice. If there are n customers in the system and the customer in service has completed $s - 1$ stages of service, then each of the $n - 1$ customers in queue has k stages of service remaining and the customer in service has $k - s + 1$ stages remaining, for a total of $k(n - 1) + k - s + 1$ stages in the system. When

$$i \triangleq k(n - 1) + k - s + 1 = kn - s + 1$$

the pair (n, s) and i provide equivalent information (why can this not be done for two or more servers?). The smallest integer at least as large as i/k is the number of customers in the system.

Let $X(t)$ denote the number of stages in the system at time t; checking that $\{X(t); t \geq 0\}$ is a CTMC is left to you. The digraph of this CTMC is shown in Figure 8-4. At each arrival epoch, the state increases by k, and at each epoch where a stage of service is completed, the state decreases by 1. From the digraph it is easy to see that the CTMC is irreducible.

Let

$$p_i = \lim_{t \to \infty} P\{X(t) = i\} \qquad i \in I$$

and

$$P_j = \lim_{t \to \infty} P\{j \text{ customers present at } t\} \qquad j \in I$$

Then

$$P_j = \sum_{i=1+(j-1)k}^{jk} p_i$$

and, in particular, $P_0 = p_0$.

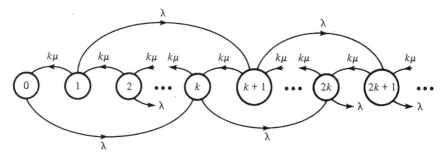

Figure 8-4 Digraph of the $M/E_k/1$ queue.

According to Theorem 8-6, if a vector satisfying (8-29) and (8-46) can be found, it is the limiting distribution. Use Theorem 8-7 with $A = \{0, 1, \ldots, i\}$ for each $i \in I_+$ to write (8-29) as†

$$\lambda \sum_{j=i-k}^{i-1} p_j = k\mu p_i \qquad i \in I_+ \tag{8-47}$$

where $p_j \triangleq 0$ for $j < 0$. One way to solve (8-47) is to treat it as a difference equation. Regard the equations when $i \geq k$ as the general form and the equations for $i < k$ and (8-29b) as boundary conditions. The general equation is

$$kp_i - \rho(p_{i-1} + \cdots + p_{i-k}) = 0 \tag{8-48}$$

where $\rho = \lambda/\mu$. The theory of difference equations [e.g., Hildebrand (1968, sec. 1.7)] asserts that‡

$$p_i = y^i \tag{8-49}$$

solves (8-48) for some suitable values of y. Substituting (8-49) into (8-48) shows that y must satisfy

$$ky^k - \rho(y^{k-1} + \ldots + y + 1) = 0 \tag{8-50}$$

A polynomial of degree k has k roots, $y_1, \ldots, y_k$ say, and each can be used in (8-49). Thus, the general solution of (8-49) is

$$p_i = \sum_{n=1}^{k} c_n y_n^i \qquad i \in I \tag{8-51}$$

where $c_1, \ldots, c_k$ are constants that are chosen to satisfy the boundary conditions. There are k boundary equations, which are all linear in the c_n's. It is possible to show that the boundary equations are independent; hence the constants can be obtained by solving a system of linear equations. By using generating-function methods, the constants can be obtained directly, as shown below. Observe that (8-51) demonstrates that p_i, and hence P_j, is a sum of k geometric terms.

Let $\hat{P}(z) \triangleq \sum_{i=0}^{\infty} z^i p_i$. We obtain $\hat{P}(z)$ by multiplying the ith equation in (8-47) by z^{i-1} and summing over $i \in I_+$. Observe [from Figure 8-4 or by writing (8-47) for $i = 1, 2, \ldots, k + 1$] that the left-side of (8-47) contains p_0 in each of the first k equations and in no other equation. Then p_1 appears in only the second through $(k + 1)$th equations, and so forth. Hence

$$\lambda(1 + z + \cdots + z^{k-1})(p_0 + zp_1 + \cdots) = \frac{k\mu[\hat{P}(z) - p_0]}{z}$$

---

† You should be able to write (8-47) directly from Figure 8-4 by a flow-of-probability argument.

‡ Or, guess a solution of the form (8-49) based on the particular difference equations solved in Chapter 7.

is obtained by taking a weighted sum on both sides of (8-47). Collecting terms and simplifying yield

$$\hat{P}(z) = \frac{p_0}{1 - (\rho/k)(z + z^2 + \cdots + z^k)} \tag{8-52}$$

From $\hat{P}(1) = 1$ we obtain $p_0 = 1 - \rho$, which conforms to the general result for $M/G/1$ queues obtained in Section 7-6. Now expand $\hat{P}(z)$ by partial fractions. First, compare the denominator of (8-52) with (8-50), and observe that the roots of the denominator of (8-52), $z_1, z_2, \ldots, z_k$ say, are the reciprocals of the roots of (8-48); that is, we can write $z_n = 1/y_n$. Then, using the result of Example 7-27, we can write (8-52) as

$$\hat{P}(z) = \sum_{n=1}^{k} \frac{c_n}{1 - zy_n}$$

where

$$c_n = (1 - \rho) \prod_{\substack{m=1 \\ m \neq n}}^{k} \frac{y_m}{y_m - y_n}$$

In principle, the $E_j/E_k/c$ queue can be analyzed as a CTMC by the method of stages, for any j, k, and c. However, for multiple-server queues, writing the balance equations is tedious and solving them is a major undertaking. Shapiro (1966) describes the $M/E_k/2$ queue, and Mayhugh and McCormick (1968) describe the $M/E_k/c$ queue.† The $E_j/E_k/c$ queue is treated by Yu (1977), and tables have been prepared by Hillier and Lo (1981).‡

EXERCISES

8-17 Let $\{X(t); t \geq 0\}$ be an irreducible CTMC. Prove that (8-36) implies $p_j \triangleq \lim_{t \to \infty} P\{X(t)=j\} > 0$ for some j in the state space.

8-18 Prove Proposition 8-9.

8-19 Prove Theorem 8-7.

8-20 Show that when the CTMC is a birth-and-death process, (8-36) becomes (4-40).

8-21 The queueing system $M/H_2/2/2$ has Poisson arrivals, two servers, and no waiting positions; customers who arrive when both servers are busy are lost. The service times are i.i.d., and their distribution is a mixture of two different exponential distributions; thus the density function of the service times is given by $g(t) = p\mu_1 e^{-\mu_1 t} + (1 - p)\mu_2 e^{-\mu_2 t}$.

Let $N(t)$ be the number of customers present at epoch t.

(a) Explain why $\{N(t); t \geq 0\}$ is not a CTMC.

---

† Saul Shapiro, "The M-Server Queue with Poisson Input and Gamma Distributed Service of Order Two," *Oper. Res.* **14**: 685–694 (1966); J. O. Mayhugh and R. E. McCormick, "Steady State Solution of the Queue $M/E_k/r$," *Manage. Sci.* **14**: 692–712 (1968).

‡ Oliver S. Yu, "The Steady-State Solution of a Heterogeneous-Server Queue with Erlang Service Times," *Algorithmic Methods in Probability*, Marcel F. Neuts (ed.), North-Holland, New York (1977); F. S. Hillier and F. D. Lo, *Queueing Tables and Graphs*, North-Holland, New York (1981).

(b) Do something similar to the method of stages to construct a CTMC from which $P\{N(t) = n\}$ can be extracted.

(c) Prove

$$p_n \triangleq \lim_{t \to \infty} P\{N(t) = n\} = \frac{a^n/n!}{1 + a + a^2/2} \tag{8-53}$$

for $n = 0, 1, 2$ and $a = \lambda/\mu$, where λ is the customer arrival rate and $1/\mu \triangleq p/\mu_1 + (1 - p)/\mu_2$ is the mean service time. Observe that (8-53) is identical to the corresponding solution of Exercise 4-24 where the service times are exponentially distributed.

8-22 The queueing system $M/E_2/2/2$ is the same as the system described in Exercise 8-21 except the service times have an Erlang-2 distribution. Use the method of stages to establish that (8-53) holds for the $M/E_2/2/2$ queue. *Hint*: First establish that for a two-server system, the method of stages can be employed, using the status (n, k), where n is the number of customers in the system and k is the number in the second stage of service.

8-23 (Continuation) In an $M/E_2/2/2$ system which has reached the steady state, let $B(\cdot)$ be the d.f. of the remaining service time of the customer in service, given that exactly one customer is in service. Let $b(\cdot)$ be the associated density function. Establish

$$b(t) = \mu e^{-2\mu}(1 + 2\mu t) \qquad t \geq 0 \tag{8-54}$$

Show that $b(t)$ given by (8-54) is the density of the equilibrium-excess r.v. in a renewal process where the times between renewals have an E_2 distribution with mean $1/\mu$.

8-24 (Continuation) In an $M/E_2/2/2$ system, let $R_n(t_1; t_2)$ be the steady-state probability that there are n customers in service, and the remaining service time of the customer in server i (if any) is no larger than t_i, $i = 1, 2, n = 0, 1, 2$. Let $r_n(\cdot; \cdot)$ be the associated joint density function. Prove

$$r_n(t_1; t_2) = p_n b(t_1)b(t_2) \qquad n = 0, 1, 2 \tag{8-55}$$

where p_n is given by (8-53) and $b(\cdot)$ is given by (8-54).

8-25 Repeat Exercises 8-23 and 8-24 for the $M/H_2/2/2$ system [note that the form of (8-54) must be changed].

8-26 In an $M/M/2$ queue, assume that a customer who arrives when both servers are free always goes to server 1. What is the steady-state probability that server 1 is busy?

8-7 ANOTHER CONSTRUCTION OF THE CONTINUOUS-TIME MARKOV CHAIN

The analytic theory of CTMCs can be greatly simplified by assuming that there is a number Λ such that $q_i \leq \Lambda$ for all $i \in \mathcal{S}$. Such CTMCs are called *uniformizable*. By Proposition 8-4, this assumption rules out explosions, but it does even more. A uniformizable CTMC can be described as a process where transition epochs are generated by a homogeneous Poisson process and the transitions from state to state are governed by a (discrete-time) Markov chain. (Why is this different from the conclusion of Theorem 8-3?) This construction of a CTMC is useful in some Markov decision problems; see Section 5-3 of Volume II.

Let $\{X(t); t \geq 0\}$ be a uniformizable CTMC with Q-matrix Q and state space $\mathcal{S}$. Choose

$$\Lambda \geq \max_{i \in \mathcal{S}} q_i$$

and let $\{N(t); t \geq 0\}$ be a Poisson process with rate Λ. Let $\{Z(t); t \geq 0\}$ be a process with state space $\mathcal{S}$ which is constructed as follows: $Z(t)$ makes transitions at each epoch where $\{N(t); t \geq 0\}$ makes a jump, and the transitions form a Markov chain with transition matrix $P = (p_{ij})$, where

$$
p_{ij} = \begin{cases} \dfrac{q_{ij}}{\Lambda} & i \neq j \\[2mm] 1 - \dfrac{q_i}{\Lambda} & i = j \end{cases}
\tag{8-56}
$$

Notice that if $q_i < \Lambda$, there are transitions from state i to itself in the embedded chain; the state of $\{Z(t); t \geq 0\}$ does not change when these transitions occur.

Let $P_{ij}(t) = P\{Z(t) = j \mid Z(0) = i\}$; a flow-of-probability argument yields

$$
\dot{P}_{ij}(t) = -\Lambda P_{ij}(t)(1 - p_{ij}) + \Lambda \sum_{k \neq i} P_{ik}(t) p_{kj}
\tag{8-57}
$$

Substituting (8-56) into (8-57) yields

$$
\dot{P}_{ij}(t) = -q_j P_{ij}(t) + \sum_{k \neq i} P_{ik}(t) q_{kj}
$$

which is identical to (8-26); hence $Z(t)$ has the same distribution as $\{X(t), t \geq 0\}$. Since $\{Z(t); t \geq 0\}$ is a Markov process by its construction (check this for yourself), the two processes have the same fidi's.

8-8* REVERSIBLE MARKOV CHAINS

Imagine that a movie is made of each sample path of a Markov chain (discrete or continuous time). The intuitive concept of a reversible process is that it is impossible to tell whether the movie is being shown in the ordinary direction or in reverse.

Reversible Markov chains are an important tool in examining networks of queues. In this section, reversibility is used to prove that, in the steady state, the departures from an $M/M/c$ queue form a Poisson process.

Definition of Reversibility

In order to capture fully the notion that "time goes backward," the definitions we used must be extended to include negative values for the time parameters. For an MC, let $\mathbf{i}$ denote the set of *all* integers, and replace I by $\mathbf{i}$ in Definition 8-1. For a CTMC, removing the condition that $t_0 \geq 0$ in Definition 8-1 defines the CTMC $\{X(t); -\infty < t < \infty\}$. The following definition of reversibility applies to discrete- and continuous-time Markov chains.†

---

† Recall that $A \sim B$ means A and B have the same distribution.

Definition 8-8: Reversibility Fix $\mathfrak{J}$ as either $\mathfrak{i}$ or $(-\infty, \infty)$. In the former case, write $X_t = X(t)$, $t \in \mathfrak{i}$. The Markov chain $\{X(t); t \in \mathfrak{J}\}$ is *reversible* if for any $\tau \in \mathfrak{J}$ and $n \in I_+$,

$$[X(t_1), \ldots, X(t_n)] \sim [X(\tau - t_1), \ldots, X(\tau - t_n)] \tag{8-58}$$

where $t_1 \leq t_2 \leq \cdots \leq t_n$ are elements of $\mathfrak{J}$.

Definition 3-3 states that $\{X(t); t \in \mathfrak{J}\}$ is *stationary* if the distribution of $[X(s + t_1), \ldots, X(s + t_n)]$ does not depend on s.

Proposition 8-10 If the MC $\{X(t); t \in \mathfrak{J}\}$ is reversible, then it is stationary.

PROOF Choosing $\tau = 0$ in (8-58) yields

$$[X(t_1), \ldots, X(t_n)] \sim [X(-t_1), \ldots, X(-t_n)] \tag{8-59}$$

Choosing $\tau = s$ and setting $t_i' = s + t_i$ yields

$$[X(s + t_1), \ldots, X(s + t_n)] \sim [X(-t_1), \ldots, X(-t_n)]$$

Combining this with (8-59) establishes the desired conclusion. □

Characterization of Reversibility

The following theorem provides a test for reversibility of a CTMC in terms of the Q-matrix. An analogous theorem holds for an MC (Exercise 8-27). The intuitive basis for the theorem is this. By Proposition 8-10, only stationary processes can be reversible. When the CTMC is stationary, $P\{X(t) = i\} = \Pi_i$, where Π is a stationary probability vector of the chain. The rate of flow of probability from i to j is $\Pi_i q_{ij}$. In order for the process to look the same when time is reversed, the rate of flow of probability from j to i must be at the same rate, hence $\Pi_i q_{ij} = \Pi_j q_{ji}$.

Theorem 8-8 Consider a stationary CTMC with state space $\mathcal{S}$ and Q-matrix Q that is irreducible and consists of ergodic states. This CTMC is reversible if, and only if, for all $i, j \in \mathcal{S}$,

$$\Pi_i q_{ij} = \Pi_j q_{ji} \tag{8-60}$$

where Π is the stationary probability vector.

PROOF Let $\{X(t); -\infty < t < \infty\}$ be a stationary CTMC that is irreducible and consists of ergodic states. Stationarity implies that $P\{X(t) = j\}$ is the same for all values of t, so denote it by P_j. Setting $t_1 = t$, $t_2 = t + s$, and $\tau = 2t + s$ in (8-58) yields

$$P\{X(t) = i, X(t + s) = j\} = P\{X(t + s) = i, X(t) = j\}$$

So it follows that

$$P\{X(t + s) = j \mid X(t) = i\} \frac{P_i}{s} = P\{X(t + s) = i \mid X(t) = j\} \frac{P_j}{s}$$

Letting $s \downarrow 0$ and using the definition of q_{ij} yield

$$P_i q_{ij} = P_j q_{ji} \tag{8-61}$$

Summing both sides of (8-61) over all $j \in \mathcal{S}$ yields $\mathbf{PQ} = \mathbf{0}$; hence $\mathbf{P}$ is a stationary probability vector.

Now assume (8-60) holds. From the Markov property and stationarity, it is sufficient to prove

$$P\{X(t) = j, X(0) = i\} = P\{X(t) = i, X(0) = j\} \qquad i, j \in \mathcal{S}, \quad t > 0 \tag{8-62}$$

We establish (8-62) by proving that for any sequence of states which are visited during $(0, t)$, the probability that these transitions occur and that $X(t) = j$ and $X(0) = i$ is identical to the probability that the same states are visited in reverse order and $X(t) = i$ and $X(0) = j$.

Iterate (8-60) to obtain

$$\Pi_i q_{i i_1} \Pi_{i_1} q_{i_1 i_2} \cdots \Pi_{i_n} q_{i_n j} = \Pi_j q_{j i_n} \cdots \Pi_{i_2} q_{i_2 i_1} \Pi_{i_1} q_{i_1 i}$$

The hypotheses placed on the CTMC imply $\Pi > 0$; canceling like terms on both sides yields

$$\Pi_i q_{i i_1} \cdots q_{i_n j} = \Pi_j q_{j i_n} \cdots q_{i_1 i} \tag{8-63}$$

From Theorem 8-3 and (8-18), the probability that $X(0) = i$, states $i_1, \ldots,$ i_n are visited in turn prior to t, and $X(t) = j$ is

$$\Pi_i q_{i i_1} q_{i_1 i_2} \cdots q_{i_n j} f(t)$$

where $f(t)$ is the convolution of exponential functions with parameters $\alpha_1,$ $\alpha_{i_1}, \ldots, \alpha_{i_n}, \alpha_j$. Note that the order of the subscripts does not affect $f(t)$. Similarly, the probability that $X(0) = j$, states $i_n, \ldots, i_1$ are visited in turn prior to t, and $X(t) = i$ is

$$\Pi_j q_{j i_n} q_{i_n i_{n-1}} \cdots q_{i_1 i} f(t)$$

From (8-63), the two probabilities above are equal, hence (8-62) is true. $\square$

There is a sufficient condition for reversibility in terms of the (undirected) *graph* of a CTMC. The graph of a CTMC is formed by having one node for each state, and an arc connects nodes i and j if $q_{ij} + q_{ji} > 0$. A graph is *connected* if there is a sequence of arcs connecting every pair of nodes. A connected graph is a *tree* if removing *any* arc breaks the graph into two pieces.

Proposition 8-11 If a CTMC has a stationary distribution and its graph is a tree, then the CTMC is reversible.

PROOF If there is no arc between i and j, then $q_{ij} = q_{ji} = 0$ and (8-60) is trivially true. If there is an arc between i and j, the tree property implies that removing this arc separates the nodes into two disjoint sets, A and B say, with $i \in A$ and $j \in B$. The stationary distribution satisfies (8-42), which simplifies to (8-60) by the tree property. $\square$

Example 8-8: Birth-and-death processes Let $\{X(t); t \geq 0\}$ be a birth-and-death process possessing a stationary distribution, and extend it in the obvious way to $\{X(t); -\infty < t < \infty\}$. Since (8-60) above reduces to (4-34), such a birth-and-death process is reversible. Figure 4-2 shows that the graph of a birth-and-death process is a tree whenever it is connected, so Proposition 8-11 is clearly satisfied. □

Departure Process of the M/M/c Queue

Let the birth-and-death process in Example 8-8 represent an M/M/c queue. Interpret $t = 0$ as a time when the queue is in the steady state. That is, we imagine that the first customer arrived at "$t = -\infty$" and the queue has evolved infinitely long when time 0 is reached. In the notation of Example 4-10, the birth rates and death rates are $\lambda_n \equiv \lambda$ and $\mu_n = n\mu$ for $0 \leq n \leq c$ and $\mu_n = c\mu$ for $n > c$. Let $R(t)$ denote the number of departures during $(0, t]$.

The following theorem is a direct consequence of Example 8-8.

Theorem 8-9: Burke's output theorem† Let $\{X(t); -\infty < t < \infty\}$ be an M/M/c queue with arrival rate λ and traffic intensity $\rho \triangleq \lambda/(c\mu) < 1$. Then the departure process $\{R(t); t \geq 0\}$ is Poisson with rate λ.

PROOF The condition $\rho < 1$ ensures that a stationary distribution exists (Example 4-10), and Example 8-8 shows that $\{X(t); -\infty < t < \infty\}$ is reversible. Reversibility implies that the process is probabilistically the same when time runs backward as when time runs forward. *When time runs backward, departure epochs become arrival epochs and vice versa.* Thus, the departure epochs in reversed time are probabilistically identical to the departure epochs in forward time and probabilistically identical to the arrival process in forward time, which is Poisson at rate λ. □

Observe that the proof of Theorem 8-9 does not use the special structure of the death rates. The crucial elements of the proof are the Poisson arrival process and the reversibility of the process which records the number of customers present; see Exercise 8-28. The statement of Theorem 8-9 and its method of proof are used in Chapter 13.

EXERCISES

8-27 Let $\{X_n; n \in \mathfrak{i}\}$ be a MC. Prove that it is reversible if, and only if, $\pi_i p_{ij} = \pi_j p_{ji}$ for all i and j in the state space, where (p_{ij}) is is the transition matrix and $\boldsymbol{\pi}$ is a stationary distribution.

† The first proof of this theorem is due to P. J. Burke, "The Output of a Queueing System," *Oper. Res.* **4**: 699–704 (1956). Burke's proof is based on renewal theory. The method of proof that uses the reversibility property of a birth-and-death process is due to Edgar Reich, "Waiting Times When Queues Are in Tandem," *Ann. Math. Stat.* **28**: 768–773 (1957).

8-28 Show that the MCs (extended in the obvious way to $n < 0$) in Exercise 7-18 are reversible, and find the stationary distributions. *Hint*: Use (8-60) to simplify the computations required to obtain Π.

8-29 Repeat Exercise 8-28 for the MCs in Exercise 7-19. *Hint*: Conjecture that the MCs are reversible, use Theorem 8-8 to verify the conjecture, and simplify by obtaining the stationary probabilities.

8-30 Let $\{X(t); -\infty < t < \infty\}$ be the number of customers present at time t in an M/M/c queue, as in Theorem 8-9. Prove that for $t > 0$, $X(t)$ is independent of the departure epochs during $(0, t)$.

BIBLIOGRAPHIC GUIDE

A standard reference work for CTMCs is part II of Chung (1967). Our presentation was greatly influenced by that book. Chapter 8 of Çinlar (1975) was also an important influence. The notion of a uniformizable CTMC appears in Chapter 10 of Feller (1971) and is exploited in Keilson (1979). Reversible Markov chains were first considered by A. N. Kolmogorov in 1935. The treatment given here is based on Kendall (1975), Whittle (1975), and Kelley (1979). Reversible processes are also considered in Keilson (1979).

References

Chung, Kai Lai: *Markov Chains with Stationary Transition Probabilities*, Springer-Verlag, New York (1967).

Çinlar, Erhan: *Introduction to Stochastic Processes*, Prentice-Hall, Englewood Cliffs, N.J. (1975).

Feller, William: *An Introduction to Probability Theory and Its Applications*, vol. 2, Wiley, New York (1971).

Hildebrand, Francis B.: *Methods of Applied Mathematics*, 2d ed., Prentice-Hall, Englewood Cliffs, N.J. (1965).

Keilson, Julian: *Markov Chain Models—Rarity and Exponentiality*, Springer-Verlag, New York (1979).

Kelley, F. P.: *Markov Processes and Reversibility*, Wiley, New York (1979).

Kendall, David G.: "Some Problems in Mathematical Genealogy," *Perspectives in Probability and Statistics: Papers in Honour of M. S. Bartlett*, Academic Press, London (1975), pp. 325–345.

Whittle, P.: "Reversibility and Acyclicity," *Perspectives in Probability and Statistics: Papers in Honour of M. S. Bartlett*, Academic Press, London (1975), pp. 217–224.

NINE

MARKOV PROCESSES

In this chapter we examine four stochastic processes with a Markov property. The semi-Markov process generalizes the discrete-time and continuous-time Markov chains by allowing the times between transitions to be arbitrarily distributed nonnegative random variables which may depend on the current state and the next state. The Markov renewal process is based on the same structure as the semi-Markov process; the former is concerned with the state at time t, and the latter is concerned with the number of changes of state during $(0, t]$. The Markov renewal process generalizes the renewal processes studied in Chapter 5 by not requiring that successive interrenewal times be identically distributed.

The random-walk process generalizes the simple-random-walk process described in Section 7-8, by permitting the step sizes to have an arbitrary distribution. The diffusion process is obtained from the simple-random-walk process by letting the step size and the time between steps approach zero so that a continuous-time process whose state space is the real line is obtained.

The greater generality achieved by these processes encompasses important operations research models which are not amenable to analysis with the methods of earlier chapters.

9-1 SEMI-MARKOV PROCESSES

The semi-Markov process, abbreviated SMP, is a generalization of a discrete-time Markov chain where the times between transitions are allowed to be r.v.'s which depend on the current, and possibly the next, state. It contains the continuous-time Markov chain as a special case.

We start with a state space S. As usual, we adopt the convention that $S \subset I$, and the states are numbered consecutively, starting with zero. The basic data that govern an SMP are specified in this definition.

Definition 9-1 Let $Q(\cdot) = (Q_{ij}(\cdot))$, where for each $i, j \in S$, $Q_{ij}(\cdot)$ has these properties:
(a) $Q_{ij}(t) = 0$ for $t < 0$
(b) $\sum_{j \in S} Q_{ij}(\infty) = 1$
(c) $Q_{ij}(t)$ is monotone in t
We call $Q(\cdot)$ a *matrix of transition distributions*.

We use $Q(\cdot)$ to construct a stochastic process $\{Z(t); t \geq 0\}$ which makes transitions among the states of S.† In order to define $\{Z(t); t \geq 0\}$ properly, we introduce a process which describes the transitions. In what follows, interpret Y_n as the state of the Z-process at the epoch of its nth transition and X_n as the elapsed time between the nth and $(n + 1)$th transitions.

In the next definition, the row vector $\mathbf{a} = (a_k; k \in S)$ is the *vector of initial probabilities*.

Definition 9-2 The (X, Y)-*process* is any two-dimensional stochastic process $\{(X_n, Y_n); n \in I\}$ such that for all $i, j \in S$

$$P\{Y_0 = k\} = a_k \qquad k \in S \quad (a_k \geq 0, \sum_{k \in S} a_k = 1) \qquad (9\text{-}1)$$

and

$$P\{Y_{n+1} = j, X_n \leq x \mid Y_0, X_0, Y_1, X_1, \ldots, X_{n-1}, Y_n = i\}$$
$$= P\{Y_{n+1} = j, X_n \leq x \mid Y_n = i\} = Q_{ij}(x) \qquad x \geq 0 \qquad (9\text{-}2)$$

Observe that (9-2) asserts that given Y_n, X_n is independent of $X_0, X_1, \ldots, X_{n-1}$. Summing (9-2) over all $j \in S$ yields

$$P\{X_n \leq x \mid Y_n = i\} = \sum_{j \in S} Q_{ij}(x) \triangleq H_i(x) \qquad x \geq 0, i \in S \qquad (9\text{-}3)$$

Definition 9-1 implies that $H_i(\cdot)$ is a proper d.f. for each i. We call $H_i(x)$ the *holding-time* distribution function in state i, and when $Y_n = i$, X_n is the *holding-time* in state i.

The structure of the Y-process is given by this proposition.

Proposition 9-1 The process $\{Y_n; n \in I\}$ is a Markov chain with state space S and transition matrix $P = (p_{ij})$, where

$$p_{ij} = Q_{ij}(\infty) \qquad i, j \in S \qquad (9\text{-}4)$$

---

† This is not meant to suggest the Q-matrix of a CMTC. It is unfortunate that Q is the standard notation for both matrices.

PROOF Use (9-2) to obtain

$$P\{Y_{n+1} = j \mid Y_0, \ldots, Y_n = i\} = \lim_{x \to \infty} P\{Y_{n+1} = j, X_n \le x \mid Y_n = i\} = Q_{ij}(\infty)$$

It remains to check that P is a stochastic matrix; parts (a) and (c) of Definition 9-1 imply that each $p_{ij} \ge 0$, and part (b) states that the row sums equal 1. $\square$

From (9-2), (9-4), and the definition of conditional probability, if $p_{ij} > 0$, then

$$P\{X_n \le x \mid Y_n = i, \quad Y_{n+1} = j, \quad X_k, Y_k (0 \le k < n)\}$$

$$= P\{X_n \le x \mid Y_n = i, \quad Y_{n+1} = j\} = \frac{Q_{ij}(x)}{p_{ij}} \triangleq F_{ij}(x) \qquad i, j \in \mathbb{S} \quad (9\text{-}5)$$

If $p_{ij} = 0$, then $F_{ij}(\cdot)$ can be defined arbitrarily. Thus, the distribution of X_n depends on the current and next states. Parts (a) and (c) of Definition 9-1 and (9-5) establish that each $F_{ij}(\cdot)$ is a proper d.f. of a nonnegative r.v.

Several processes can be defined in terms of the (X, Y)-process. Define the process $\{S_n; n \in I\}$ by

$$S_0 = 0 \quad \text{and} \quad S_n = X_0 + \cdots + X_{n-1} \qquad n \in I_+ \qquad (9\text{-}6)$$

Here S_n is interpreted as the epoch of nth transition of the (X, Y)-process. Define $\{N(t); t \ge 0\}$ by

$$N(t) = \sup\{n : S_n \le t\} \qquad (9\text{-}7)$$

It represents the number of transitions that occur by time t. Finally, we define a semi-Markov process.

Definition 9-3: Semi-Markov process The process $\{Z(t); t \ge 0\}$, defined by

$$Z(t) = Y_{N(t)}$$

is a *semi-Markov process*. The process $\{Y_n; n \in I\}$ is called the *embedded Markov chain* of $\{Z(t); t \ge 0\}$.

Thus, $Z(t)$ is the state of the Y-process at the epoch $S_{N(t)}$, that is, at the last transition epoch.

Let us recapitulate the constructive definition of an SMP given above. Start with a state space $\mathbb{S}$, an initial probability vector $\mathbf{a}$, and a matrix of transition functions $Q(\cdot)$, where each $Q_{ij}(\cdot)$ can be written $Q_{ij}(\cdot) = p_{ij} F_{ij}(\cdot)$. The matrix $P = (p_{ij})$ is stochastic, and each $F_{ij}(\cdot)$ is the d.f. of a nonnegative r.v. A process evolves as follows: Transitions occur from state to state according to a Markov chain; when the process is in state i, the next state visited is j with probability p_{ij}. When the successive states are i and j, the d.f. of the time to make the transition is $F_{ij}(\cdot)$. The process which records the state at each time is the semi-Markov process.

Examples

Example 9-1† It is important to be able to predict the length of stay of patients in each section of a hospital in order to provide sufficient, but not wasteful, amounts of resources such as beds, special equipment, nurses, etc. Since care units frequently are formed by collecting patients with similar diagnoses and required facilities, the movement of patients with the same disease type among the care units is statistically stable.

Consider all patients with a particular disease, specifically, heart disease (acute myocardial infarction or related diseases, to be more specific). Let the state of a patient be the care unit in which he or she resides. All coronary patients enter the coronary care unit, which we take to be state 0. After that, patients are transferred among the following states: (1) postcoronary care, (2) intensive care, (3) medical unit, (4) surgical unit, and (5) ambulatory unit. Finally, each patient enters one of the following states and stays there: (6) extended-care unit, (7) home, and (8) death.

An analysis of the histories of 555 patients indicated that movement of patients from state to state is adequately modeled by a Markov chain and that the length of stay in each state is adequately described by an r.v. which depends on the current state and the next state to which the patient will be transferred, but not on previous states. Thus, the sojourn of a patient can be described by an SMP. □

Example 9-2: (s, S) inventory model with purchase lead times‡ Consider an inventory system, operating with an (s, S) policy, where the inventory level is observed continuously and demands occur according to a Poisson process. When a demand reduces the inventory to s, an order for $S - s$ items is placed; the order is received after a random amount of time which is independent of all past and future events. Only one order may be outstanding at a time. Assume that demands which occur when no items are on hand are backlogged. When an order is received, if the inventory is no larger than s, another order is placed immediately.

Let $I(t)$ be the inventory at time t (a negative value indicates the number of backlogged demands), λ be the rate of the Poisson demand process, and $G(\cdot)$ be the common d.f. of the purchase lead times. A graph of a typical sample path of $I(t)$ is shown in Figure 9-1.

The process $\{I(t); t \geq 0\}$ is *not* an SMP because when there is a purchase order outstanding, the evolution of the process is not completely determined by the current inventory level; the time that the order has been outstanding

† This example is taken from Edward P. C. Kao, "Modeling the Movement of Coronary Patients within a Hospital by Semi-Markov Processes," *Oper. Res.* **22**: 683–699 (1974).

‡ Examples 9-2 and 9-3, and their subsequent analysis, are due to A. J. Fabens, "The Solution of Queueing and Inventory Models by Semi-Markov Processes," *J. Royal. Stat. Soc.* **B23**: 113–127 (1961).

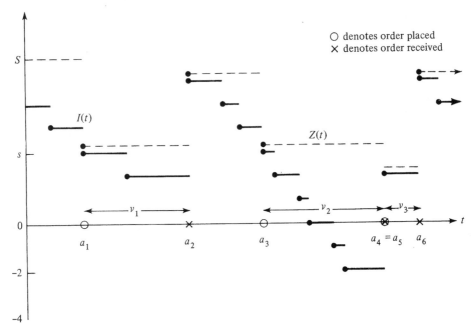

Figure 9-1 A section of a sample path of $I(t)$; $s = 3$, $S = 7$.

has an influence [unless $G(t) = 1 - e^{-\lambda t}$]. However, there is an *embedded* SMP, $\{Z(t); t \ge 0\}$ say. It is formed by recording the value of $I(\cdot)$ at only those epochs where an order is either placed or received. These points are connected so that the resulting sample paths are right-continuous step functions. The sample path of $\{Z(t); t \ge 0\}$ is also shown in Figure 9-1.

A formal definition of $\{Z(t); t \ge 0\}$ is made as follows. Let a_{2i-1} be the epoch at which the ith order is placed and a_{2i} be the epoch at which the ith order is received, $i \in I_+$. Naturally, $a_{2i-1} \le a_{2i}$, and $a_{2i+1} = a_{2i}$ is possible; for example, $a_4 = a_5$ in Figure 9-1. Set $a_0 = 0$, and for every $t \ge 0$, define

$$t^* \triangleq \max\{a_j : a_j \le t, j \in I\}$$

Then

$$Z(t) \triangleq I(t^*), \qquad t \ge 0 \qquad\qquad (9\text{-}8)$$

The memoryless property of the exponential times between demands and the assumption that the lead times are i.i.d. cause $\{Z(t); t \ge 0\}$ to be an SMP; the detailed verification that the definition is satisfied is left as Exercise 9-3. This embedded SMP is used to obtain limiting (as $t \to \infty$) probabilities for $I(t)$ in Example 9-4 and Exercise 9-13. $\qquad\square$

Example 9-3: A bulk-service M/G/1 queue Consider an M/G/1 queue in which customers are served in batches of fixed size K. If less than K customers are present at an epoch when the server is available, they are not taken into service until enough more customers arrive to make a batch of size K. If at least K customers are present at a service completion epoch, a batch of size K is formed and taken into service. The ordinary $M/G/1$ queue is the special case $K = 1$.

Let $X(t)$ denote the number of customers present at time t, λ be the arrival rate, V_i be the service time of the ith batch, and $G(\cdot)$ be the d.f. of the service time of a batch. A typical sample path of $X(t)$ is shown in Figure 9-2.

This queueing model is interesting for the following reason. Let $I(t)$ be the inventory level defined in Example 9-2, and choose $K = S - s$. Then $X(\cdot)$ and $I(\cdot)$ are related by

$$X(t) = S - I(t) \qquad t \geq 0 \tag{9-9}$$

Comparing Figures 9-1 and 9-2 shows that the epochs where an order is placed correspond to epochs where a batch enters service, that purchase lead times correspond to service times, and that epochs where an order is received correspond to service completion epochs.

The analysis of the $M/G/1$ queue in Chapter 7 indicates that $\{X(t); t \geq 0\}$ is not an SMP; but a process embedded at service completion epochs and

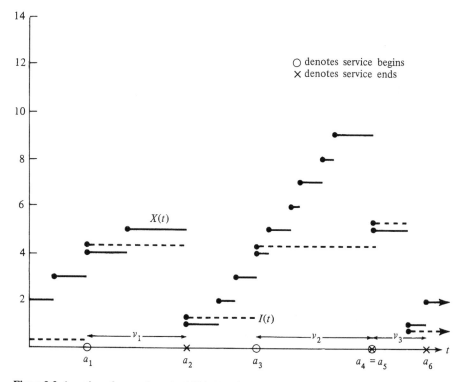

Figure 9-2 A section of a sample path of $X(t)$, $K = 4$.

arrival epochs where a batch enters service is an SMP. Let $\{Z(t); t \geq 0\}$ be that process; it is defined formally as follows. Let a_{2i-1} be the epoch where the ith batch enters service and a_{2i} be the epoch where the ith batch completes service, and set $a_0 = 0$. Define, for every $t \geq 0$,

$$t^* \triangleq \max\{a_j : a_j \leq t, j \in I\}$$

Then

$$Z(t) \triangleq X(t^*) \qquad t \geq 0 \qquad (9\text{-}10)$$

The correspondence between $\{X(t); t \geq 0\}$ and $\{I(t); t \geq 0\}$ and Exercise 9-3 establish that $\{Z(t); t \geq 0\}$ is an SMP. $\qquad\qquad \square$

Transition Function

Let $\{Z(t); t \geq 0\}$ be an SMP with state space S, and define the functions $P_{ij}(\cdot)$ by

$$P_{ij}(t) \triangleq P\{Z(t) = j \mid Z(0) = i\} \qquad i, j \in S, \quad t \geq 0 \qquad (9\text{-}11)$$

The function $P_{ij}(\cdot)$ is called the *transition probability function from state i to state j*, customarily abbreviated to *transition function*. The matrix of transition functions $(P_{ij}(\cdot))$ is called the *transition matrix* of the SMP.

By analogy with the characteristics of a standard transition matrix of a CTMC, *assume*

$$P_{ii}(0) = 1 \quad \text{and} \quad P_{ij}(0) = 0 \qquad i \neq j \qquad (9\text{-}12)$$

The following equation is a generalization of the backward equations of a Markov chain. Recall from (9-3) that

$$H_i(x) \triangleq P\{X_n \leq x \mid Y_n = i\}$$

Proposition 9-2 The transition functions of an SMP, with state space S and matrix of transition distributions $Q(\cdot)$, satisfy

$$P_{ij}(t) = \sum_{k \in S} \int_0^t P_{kj}(t - x)\, dQ_{ik}(x) \qquad\qquad i \neq j, \quad t \geq 0 \qquad (9\text{-}13a)$$

$$P_{ii}(t) = 1 - H_i(t) + \sum_{k \in S} \int_0^t P_{ki}(t - x)\, dQ_{ik}(x) \qquad t \geq 0 \qquad (9\text{-}13b)$$

PROOF Condition on the length of the first transition time X_0 and the first state visited Y_1. For $i \neq j$, if $X_0 = x$ and $Y_1 = k$, then $Z(t) = j$ if, and only if, the process is in state j exactly $t - x$ time units after entering state k. This probability equals the probability that an SMP with initial state k and matrix of transition distributions $Q(\cdot)$ is in state j at time $t - x$, which is $P_{kj}(t - x)$, and is independent of Y_0 and X_0 by (9-2). Removing the condition on X_0 and Y_1 yields (9-13a). When $i = j$, we must add to the above probability the probability that no transitions occur by time t; from (9-3), this is $1 - H_i(t)$. □

Pathological Behavior

As in the case of CTMCs, some parameter values cause $P\{N(t) = \infty\} > 0$. Under these circumstances, it is possible that (9-13) has many solutions. For a complete discussion of these issues, refer to Pyke (1961a) and Feller (1964). A sufficient condition that rules out these pathologies is given below.

Proposition 9-3 Assume the embedded MC $\{Y_n; n \in I\}$ has a nonempty set of ergodic states, $\mathcal{E}$ say. If for each $i \in \mathcal{S}$,

$$\lim_{n \to \infty} P\{Y_n \in \mathcal{E} \mid Y_0 = i\} = 1 \qquad \text{and} \qquad H_i(0) < 1$$

then

$$P\{N(t) < \infty\} = 1 \qquad t \geq 0 \tag{9-14}$$

PROOF Fix $i \in \mathcal{S}$ and let $j \in \mathcal{E}$ be the first state in $\mathcal{E}$ that is reached (possibly $j = i$). Let n_k be the stage of the kth visit to state j, that is, the kth smallest n such that $Y_n = j$ (possibly $n_1 = 0$). Let S_{n_k} be the epoch of the kth entry to state j, where S_n is defined by (9-6) to be $X_0 + \cdots + X_{n-1}$. Define

$$T_1 = S_{n_1}, \qquad T_k = S_{n_{k+1}} - S_{n_k} \qquad k \in I_+$$

where T_k is the time between the kth and $(k + 1)$th entries of state j. Since the SMP probabilistically starts from scratch each time state j is entered, $T_2, T_3, \ldots$ are i.i.d., hence $\{T_i\}_1^\infty$ generates a (possibly delayed) renewal process. Exercise 5-16 establishes $P\{\sum_{k=1}^\infty T_k = \infty\} = 1$. Observe that

$$\sum_{k=1}^\infty T_k = \lim_{k \to \infty} S_{n_k} = \sum_{n=0}^\infty X_n$$

which proves $P\{\sum_{n=0}^\infty X_n = \infty\} = 1$. Combining this with the definition of $N(t)$ given by (9-7), $N(t) = \sup\{n : S_n \leq t\}$, completes the proof. □

Theorem 7-11 may be used to test whether a closed set of persistent states is eventually entered when $\mathcal{S}$ is infinite. When $\mathcal{S}$ is finite, the proof of Corollary 7-4 shows that this condition is always satisfied. In fact, when $\mathcal{S}$ is finite, (9-14) always holds (Exercise 9-7). *Henceforth it is assumed that (9-14) holds.*

First-Passage Times

First-passage times for SMPs are analogous to first-passage times for Markov chains. A convenient way to define them starts by defining $N_j(t)$ to be the number of times state j is entered by time t; formally,

$$N_j(t) \triangleq \# \{n : Y_n = j, \ 0 < n \le N(t)\} \tag{9-15}$$

where $\# \{ \cdot \}$ stands for the number of points in the set $\{ \cdot \}$. Let T_{ij} denote the *first-passage time from state i to state j*. Then

$$T_{ij} \triangleq \inf\{t : N_j(t) > 0 \,|\, Z(0) = i\} \qquad i, j \in \mathcal{S} \tag{9-16}$$

Note that when $p_{ii} > 0$, a visit to state i contains a geometrically distributed number of first passages from i to itself. From (9-16), for $i, j \in \mathcal{S}$,

$$G_{ij}(t) \triangleq P\{T_{ij} \le t\} = P\{N_j(t) > 0 \,|\, Z(0) = i\} \qquad t \ge 0 \tag{9-17}$$

Conditioning on the first transition yields the relation between the $G_{ij}(\cdot)$'s and $Q_{ij}(\cdot)$'s:

$$G_{ij}(t) = Q_{ij}(t) + \sum_{k \ne j} \int_0^t G_{kj}(t - x) \, dQ_{ik}(x) \qquad i, j \in \mathcal{S} \tag{9-18}$$

This is a generalization of (7-23). The transition functions and the first-passage-time distribution functions are related by

$$P_{ij}(t) = \int_0^t G_{ij}(t - u) \, dP_{jj}(u) \qquad i \ne j \tag{9-19a}$$

and

$$P_{ii}(t) = \int_0^t G_{ii}(t - u) \, dP_{ii}(u) + 1 - H_i(t) \tag{9-19b}$$

The proof of (9-19) is left to you. Various consequences of these equations are explored in Exercises 9-8 and 9-9.

Let $\mu_{ij} = E(T_{ij})$ and $\eta_i = E(X_0 \,|\, Y_0 = i)$; the latter is the expected duration of a transition starting in state i. In terms of (9-3), $\eta_i = \int_0^\infty [1 - H_i(u)] \, du$.

Proposition 9-4 The expected first-passage times of an SMP satisfy

$$\mu_{ij} = \eta_i + \sum_{k \ne j} \mu_{kj} p_{ik} \qquad i \in \mathcal{S} \tag{9-20}$$

PROOF Let $\eta_{ij} = E(X_0 \,|\, Y_0 = i, \ Y_1 = j)$; then $\eta_i = \sum_{j \in \mathcal{S}} \eta_{ij} p_{ij}$. Condition on the first transition of the embedded MC to obtain

$$\mu_{ij} = \sum_{k \ne j} (\eta_{ik} + \mu_{kj}) p_{ik} + \eta_{ij} p_{ij}$$

$$= \eta_i + \sum_{k \ne j} \mu_{kj} p_{ik} \qquad\qquad \square$$

Classification of States

States of an SMP have classifications similar to the states of a Markov chain (Section 7-5) and for the same purposes. In fact, the classification of a state of an SMP is determined primarily by its classification in the embedded Markov chain.

Definition 9-4 State i is called *persistent* if $P\{T_{ii} < \infty\} = 1$; otherwise, it is called *transient*. State i is called *null* if $E(T_{ii}) = \infty$ and *nonnull* if $E(T_{ii}) < \infty$. A persistent, nonnull state is called *ergodic*.

Definition 9-5 States i and j are said to *communicate*, denoted $i \leftrightarrow j$, if either $P\{T_{ij} < \infty\}P\{T_{ji} < \infty\} > 0$ or $i = j$. A nonempty set of states C is called a *communicating class* if (a) $i \leftrightarrow j$ for all i, $j \in C$ and (b) $i \in C$ and $P\{T_{ij} < \infty\} > 0$ implies $j \in C$. If a communicating class consists of a single state, that state is called *absorbing*. If the only communicating class is the set of all states, the SMP is called *irreducible*.

This is the main classification result.

Theorem 9-1 Each of the following statements is valid for an SMP if, and only if, it is valid for the embedded MC:

(i) State i is persistent.
(ii) State i is transient.
(iii) State i communicates with state j.

PROOF From (9-15) and (9-16) we obtain

$$P\{T_{ij} < \infty\} = P\{Y_n = j \text{ for some } n \in I_+ \mid Y_0 = i\} \tag{9-21}$$

The properties claimed in the theorem depend on only $P\{T_{ij} < \infty\}$. Equation (9-21) demonstrates that these probabilities are identical to the corresponding first-passage probabilities for the embedded MC. □

Here is an immediate consequence of Theorem 9-1.

Corollary 9-1 Each of the following statements is valid for an SMP if, and only if, it is valid for the embedded MC:

(i) A set of states is a communicating class.
(ii) State i is absorbing.
(iii) The process is irreducible.

Notice that we have not asserted that state i is ergodic in an SMP if, and only if, it is ergodic in the embedded MC. The reason is that both implications are false in general. Counterexamples are given in Exercises 9-8 and 9-9. The next theorem gives a simple sufficient condition for an ergodic state of the embedded MC to be an ergodic state of the SMP.

Theorem 9-2 Let i be a state of a given SMP, and let C denote the communicating class to which it belongs. State i is ergodic if it is an ergodic state of the embedded MC and $\sum_{j \in C} \eta_j < \infty$.

PROOF Since i is an ergodic state of the embedded MC, it is persistent there; by Theorem 9-1, it is a persistent state of the SMP. Since persistence is a class property, the set C is well defined. It only remains to show $E(T_{ii}) < \infty$.

Since i is an ergodic state of the embedded MC, a first passage from i to i passes through states in C only. Let v_{ji} be the expected number of visits to state j, between consecutive visits to state i, for $j \in C$. By Exercise 7-78, $v_{ji} = \pi_j/\pi_i$, where π is the stationary probability vector of the MC formed by the states of C; furthermore, $\pi_i > 0$. Thus

$$E(T_{ii}) = \sum_{j \in C} v_{ji} \eta_j \leq \sum_{j \in C} \frac{\eta_j}{\pi_i} < \infty \qquad (9\text{-}22)$$

$\square$

Observe that the equality in (9-22) provides a way of computing $E(T_{ii})$ from the stationary probabilities of the embedded MC.

As in the special case of discrete Markov chains, periodic states may occur. Recall that state j is *periodic with period* ξ if (w.p.1) T_{jj} assumes values in only $\{k\xi : k \in I\}$. Periodic states are not fundamentally different from aperiodic states, but they require special care in stating limiting results. The analog of Theorem 7-5 is Proposition 9-5 below.

Proposition 9-5 Let i and j be different states of an SMP. If $i \leftrightarrow j$, then either both states are aperiodic or both states are periodic with the same period.

PROOF See Çinlar (1975), proposition 10.2.23. $\square$

Limiting Probabilities

Recall that for the special case of a CTMC, the limiting probabilities are given in (8-34), which can be written as

$$p_j = \frac{\pi_j \eta_j}{\sum_{k \in S} \pi_k \eta_k}$$

The heuristic argument to obtain (8-34) does not use the properties of the exponential holding-time d.f., so it is natural to suspect that it holds for the SMP. It does! Here is the first step in the proof.

Proposition 9-6 For an SMP with state space S, if $\mu_{jj} < \infty$, then

$$\lim_{t \to \infty} P_{ij}(t) = \frac{G_{ij}(\infty)\eta_j}{\mu_{jj}} \qquad i, j \in S \qquad (9\text{-}23)$$

If T_{jj} is an arithmetic r.v., then (9-23) is valid only when t is a multiple of the span of T_{jj}.

PROOF For any SMP, entries into any fixed state $j \in S$ are regeneration epochs. Hence, for a given SMP $\{Z(t); t \geq 0\}$ with $Z(0) = i$, the process $\{Z_j(t); t \geq 0\}$, defined by

$$Z_j(t) = \begin{cases} 1 & \text{if } Z(t) = j \\ 0 & \text{if } Z(t) \neq j \end{cases}$$

is a (possibly delayed) regenerative process. Since η_j is the mean time that state j is occupied during a first passage from state j to itself, (9-23) is a consequence of part (b) of Corollary 6-7. $\square$

With the aid of Proposition 9-6, it is easy to prove this important result.

Theorem 9-3 Let $\{Z(t); t \geq 0\}$ be an irreducible SMP consisting of ergodic states, and let π be the stationary probability vector of the embedded MC. Then for all $i, j \in S$,

$$p_j \triangleq \lim_{t \to \infty} P\{Z(t) = j \mid Z(0) = i\} = \frac{\pi_j \eta_j}{\sum_{k \in S} \pi_k \eta_k} \qquad (9\text{-}24)$$

If T_{jj} is an arithmetic r.v., then (9-24) is valid only when t is a multiple of the span of T_{jj}.

PROOF Since the SMP is irreducible, the only communicating class is the set of all states S. Thus, the equality in (9-22) can be written

$$\pi_j \mu_{jj} = \sum_{k \in S} \pi_k \eta_k \qquad (9\text{-}25)$$

Since $i \leftrightarrow j$ and the states are ergodic, $G_{ij}(\infty) = 1$. Substituting (9-25) into (9-23) yields (9-24). $\square$

Theorem 9-3 does not completely characterize the limiting behavior of an SMP because the remaining length of the transition in progress at time t typically depends on $Z(t)$. Furthermore, the next state to be visited typically depends on the remaining length of the transition in progress.

Let $B_{jk}(t, x; i)$ be the probability that at time t the current state is j, the remaining time of the transition in progress is no longer than x, and the next transition is to state k, given that the initial state is i:

$$B_{jk}(x, t; i) \triangleq P\{Z(t) = j, Y_{N(t)+1} = k, S_{N(t)+1} \leq t + x \mid Z(0) = i\} \qquad (9\text{-}26)$$

Theorem 9-4 Let $\{Z(t); t \geq 0\}$ be an SMP which satisfies the hypotheses of Theorem 9-3, and let $B_{jk}(x, t; i)$ be defined by (9-26). Then

$$B_{jk}(x) \triangleq \lim_{t \to \infty} B_{jk}(x, t; i) = p_j p_{jk} \frac{1}{\eta_j} \int_0^x F_{jk}^c(u) \, du \qquad (9\text{-}27)$$

for all $i \in S$, $x > 0$. If T_{jj} is an arithmetic r.v., then (9-27) is valid only when t is a multiple of the span of T_{jj}.

PROOF Consider the case $i = j$; the epochs when state j is entered are the renewal epochs of a renewal process where the times between renewals are i.i.d. replicas of T_{jj}. Suppress the parameter i for notational convenience. Use the renewal argument (see page 118) to write

$$B_{jk}(x, t \mid T_{jj})$$

$$= \begin{cases} B_{jk}(x, t - T_{jj}) & \text{if } T_{jj} \leq t \\ P\{Z(t) = j, \ Y_{N(t)+1} = k, \ t \leq S_{N(t)} \leq t + x \mid Z(0) = j, \ T_{jj}\} & \text{if } T_{jj} > t \end{cases}$$

Let $w(t)$ denote the probability above, and let $W(t) = \int_t^\infty w(s) \, dG_{jj}(s)$. Unconditioning on T_{jj} yields

$$B_{jk}(x, t) = w(t) + \int_0^t B_{jk}(x, t - s) \, dG_{jj}(s) \tag{9-28}$$

For $Z(0) = j$ and $T_{jj} = s > t$, $Z(t) = j$ if, and only if, there are no transitions during $[0, t]$; moreover, $P\{X_0 \leq t\} = 0$. Hence,

$$w(t) = \int_t^\infty P\{t < X_0 \leq t + x, \ Y_1 = k \mid Z(0) = j, \ T_{jj} = s\} \, dG_{jj}(s)$$

$$= \int_0^\infty P\{t < X_0 \leq t + x, \ Y_1 = k \mid Z(0) = j, \ T_{jj} = s\} \, dG_{jj}(s)$$

$$= P\{t < X_0 \leq t + x, \ Y_1 = k \mid Z(0) = j\} \tag{9-29}$$

The solution of the renewal-type equation (9-28) is

$$B_{jk}(x, t) = w(t) + \int_0^t w(t - s) \, dM_{jj}(s) \tag{9-30}$$

where $M_{jj}(\cdot)$ is the renewal function corresponding to the distribution function $G_{jj}(\cdot)$. From (9-29), $w(\infty) = 0$. Applying the key renewal theorem (Theorem 5-8) to (9-30) [how do we know $w(\cdot)$ is directly Riemann-integrable?] and using (9-29) yield

$$B_{jk}(x) = \frac{1}{\mu_{jj}} \int_0^\infty P\{s < X_0 \leq s + x, \ Y_1 = k \mid Z(0) = j\} \, ds$$

$$= \frac{1}{\mu_{jj}} \int_0^\infty p_{jk}[F_{jk}^c(s) - F_{jk}^c(s + x)] \, ds$$

$$= \frac{p_{jk}}{\mu_{jj}} \int_0^x F_{jk}^c(s) \, ds \tag{9-31}$$

Substituting $\mu_{jj} = \eta_j/p_{jj}$ [from (9-23)] into (9-31) yields (9-26). $\qquad \square$

Summing (9-26) over $k \in S$ yields this corollary.

Corollary 9-4 With the hypothesis of Theorem 9-3, for $i, j \in S$

$$B_j(x) \triangleq \lim_{t \to \infty} P\{Z(t) = j, \, S_{N(t)+1} - t \le x \,|\, Z(0) = i\}$$

$$= \frac{p_j}{\eta_j} \int_0^x H_j^c(u) \, du \tag{9-32}$$

Recognizing $\int_x^\infty [H_j^c(u) \, du / \eta_j]$ as the equilibrium-excess distribution (see Theorem 5-10) of a renewal process where the interevent times have distribution $H_j(\cdot)$ provides an intuitive basis for (9-32). Since the equilibrium excess and deficit r.v.'s have the same distribution, it is natural to conjecture the next proposition.

Proposition 9-7 With the hypothesis of Theorem 9-3, for $i, j \in S$

$$B_j^*(x) \triangleq \lim_{t \to \infty} P\{Z(t) = j, \, t - S_{N(t)} \le x \,|\, Z(0) = i\}$$

$$= \frac{p_j}{\eta_j} \int_0^x H_j^c(u) \, du \tag{9-33}$$

PROOF This is left as Exercise 9-10. □

Example

Example 9-4: Example 9-3 continued The transition matrix P of the embedded Markov chain is given by (the unwritten entries are all zero)

$$
P =
\begin{array}{c}
\\
0 \\
1 \\
2 \\
\vdots \\
K-1 \\
K \\
K+1 \\
K+2 \\
\vdots
\end{array}
\left(
\begin{array}{ccccccccc}
0 & 1 & 2 & \cdots & K-1 & K & K+1 & \cdots \\
 & & & & & 1 & & \\
 & & & & & 1 & & \\
 & & & & & 1 & & \\
\multicolumn{8}{c}{\cdots\cdots} \\
 & & & & & 1 & & \\
c_0 & c_1 & c_2 & \cdots & c_{K-1} & c_K & c_{K+1} & \cdots \\
0 & c_0 & c_1 & \cdots & c_{K-2} & c_{K-1} & c_K & \cdots \\
0 & 0 & c_0 & \cdots & c_{K-3} & c_{K-2} & c_{K-1} & \cdots \\
\multicolumn{8}{c}{\cdots\cdots\cdots}
\end{array}
\right)
$$

where $c_k \triangleq \int_0^\infty e^{-\lambda t}[(\lambda t)^k / k!] dG(t)$ is the probability that k customers arrive during the service time of a batch. The embedded MC is irreducible and aperiodic; it is left to you to establish that the states are ergodic when $\rho \triangleq \lambda v_G < 1$, where v_G is the mean service time of a batch.

The unique stationary (and limiting) probabilities of the embedded chain satisfy $\pi = \pi P$, which in detail is

$$\pi_j = c_0 \pi_{K+j} + c_1 \pi_{K+j-1} + \cdots + c_j \pi_K \qquad j \ne K \tag{9-34a}$$

and

$$\pi_K = \sum_{i=0}^{K-1} \pi_i + c_0 \pi_{2K} + c_1 \pi_{2K-1} + \cdots + c_K \pi_K \tag{9-34b}$$

Observe that once $\pi_0, \ldots, \pi_{K-1}$ are obtained, π_K can be determined from (9-34a) with $j = 0$, then π_{K+1} can be determined from the equation with $j = 1$, and so forth. The probabilities $\pi_0, \ldots, \pi_{K-1}$ can be obtained by generating-function methods. When $K \geq 2$, the probabilities $\pi_0, \ldots, \pi_{K-1}$ can be obtained via the complex-variable approach used in the analysis of the $M/D/c$ queue in Section 7-9. It will obscure the point of this example if we pause to present the details. Since this model with $K = 1$ is the ordinary $M/G/1$ queue, it deserves more attention.

For $K = 1$, form the generating function $\hat{\pi}(z) \triangleq \sum_{j=0}^{\infty} z^j \pi_j$ by multiplying the jth equation in (9-34) by $z^j (|z| \leq 1)$ and summing over $j \in I$. The result is

$$\hat{\pi}(z) = z\pi_0 + \frac{1}{z} \sum_{j=0}^{\infty} z^{j+1} \pi_{j+1}(c_0 + c_1 z + c_2 z^2 + \cdots)$$

$$= z\pi_0 + \frac{\hat{c}(z)[\hat{\pi}(z) - \pi_0]}{z} \tag{9-35}$$

$$= \frac{[z^2 - \hat{c}(z)]\pi_0}{z - \hat{c}(z)}$$

where

$$\hat{c}(z) \triangleq \sum_{k=0}^{\infty} z^k c_k = \tilde{G}(\lambda - \lambda z)$$

and $\tilde{G}(\cdot)$ is the LST of $G(\cdot)$. From $\lim_{z \uparrow 1} \hat{\pi}(z) = 1$,

$$\pi_0 = \frac{1 - \rho}{2 - \rho} \tag{9-36}$$

is obtained.

Now we will relate $\{\pi_j\}$ and $\{p_j\}$; we do not restrict our attention to $K = 1$. For $j < K$, a transition out of state j always enters state K and represents a time interval when $K - j$ customers arrive; hence $\eta_j = (K - j)/\lambda$, $0 \leq j < K$. For $j \geq K$, each transition out of state j represents a time interval when a batch is served; hence $\eta_j = \nu_G$, $j \geq K$. Once the $\{\pi_j\}$ are determined, the limiting probabilities $\{p_j\}$ can be obtained from (9-24).

Since the process $\{X(t); t \geq 0\}$ regenerates whenever $Z(t) = K$, the limiting probabilities

$$P_j \triangleq \lim_{t \to \infty} P\{X(t) = j \mid X(0) = i\} \qquad j \in I$$

exist and are independent of i. Because the SMP is embedded, it is not necessarily true that $P_j = p_j$ for all j. The relation between $\{P_j\}$ and $\{p_j\}$ is

$$P_j = \sum_{k=0}^{j} \frac{p_k/\eta_k}{\lambda} = \sum_{k=0}^{j} \frac{p_k}{K-k} \qquad j = 0, 1, \ldots, K-1 \qquad (9\text{-}37a)$$

and

$$P_j = \sum_{k=K}^{j} \frac{p_k}{\nu_G} \int_0^\infty \frac{(\lambda x)^{j-k} e^{-\lambda x} G^c(x)}{(j-k)!} \, dx \qquad j = K, K+1, \ldots \qquad (9\text{-}37b)$$

which we now derive.

Observe that if $X(t) = j < K$, the last transition epoch of the SMP corresponds to a service completion epoch. Suppose the last service completion occurs at time $t - x$ with $X(t - x) = k$; then $Z(t) = k$ and $S_{N(t)} = t - x$. Let $\beta_{kj}(x)$ be the probability that $j - k$ customers arrive during a time interval of length x, given that less than $K - j$ arrive. Thus

$$\beta_{kj}(x) = \frac{(\lambda x)^{k-j}/(k-j)!}{\sum_{i=0}^{K-j-1} (\lambda x)^i/(i!)}$$

Therefore, for $j < k$ and $s < t$,

$$P\{X(t) = j \mid Z(t) = k, t - S_{N(t)} = x\} = \beta_{kj}(x)$$

Uncondition, let $t \to \infty$, and use Proposition 9-7 to obtain†

$$P_j = \sum_{k=0}^{j} \int_0^\infty \beta_{kj}(x) \, dB_k^*(x)$$

$$= \sum_{k=0}^{j} \frac{p_k}{\eta_k} \int_0^\infty \beta_{kj}(x) \, dH_k^c(x) \qquad j = 0, 1, \ldots, K-1$$

where $B_k^*(\,\cdot\,)$ is given by (9-33). Explicit evaluation of the integral (which is left as Exercise 9-11) yields (9-37a).

When $X(t) = j \geq K$, the most recent transition epoch of the embedded SMP corresponds to an epoch where a service begins (which may also be an epoch where a service ends). Let $\alpha_{kj}(x)$ be the probability that $k - j$ customers arrive in an interval of length x. Arguing as before and using Corollary 9-4 yield

$$P_j = \sum_{k=K}^{j} \int_0^\infty \alpha_{kj}(x) \, dB_k(x) \qquad j = K, K+1, \ldots$$

where $B_k(\,\cdot\,)$ is given by (9-32). Explicit evaluation of the integral, using (9-32), yields (9-37b).

An interesting special case of this example is $K = 1$; then $\{X(t); t \geq 0\}$ is the number of customers present in an M/G/1 queue. In this case, (9-37) can be used to prove that $\{P_{jj}\}$ is identical to the limiting probabilities of the embedded Markov chain described in Example 7-18. The details are left as Exercise 9-12. □

---

† The Helly-Bray theorem (Proposition A-11) may be used to justify the integral in the first line.

EXERCISES

9-1 Show that a CTMC is equivalent to an SMP with $Q_{ij}(x) = (\lambda_{ij}/\lambda_i)(1 - e^{-\lambda_i x})$, $i \neq j$, and $Q_{ii}(x) \equiv 0$, where $\lambda_{ij} \geq 0$, $i \neq j$, and $\lambda_i = \sum_{j \neq i} \lambda_{ij}$.

9-2 Show that the state of an alternating renewal process [this is denoted $I(\cdot)$ in Section 5-6] is an SMP where $S = \{0, 1\}$, $Q_{01}(\cdot) = F(\cdot)$, $Q_{10} = G(\cdot)$, and $Q_{ij}(x) \equiv 0$ otherwise.

9-3 Prove that the process $\{Z(t); t \geq 0\}$ defined in Example 9-2 is an SMP.

9-4 Show that for a CTMC, (9-13a) and (9-13b) are equivalent to the backward Kolmogorov equation.

9-5 Let $\tilde{P}_{ij}(s)$ be the LST of $P_{ij}(\cdot)$, $\tilde{Q}_{ij}(s)$ be the LST of $Q_{ij}(\cdot)$, and $\tilde{P}(s)$ and $\tilde{Q}(s)$ be the corresponding matrices. Let $\tilde{H}(s)$ be the matrix whose (i, i)th element is the LST of $H_i(\cdot)$ and whose off-diagonal elements are all zero. When S is finite, show that

$$\tilde{P}(s) = [I - \tilde{Q}(s)]^{-1}[I - \tilde{H}(s)] \qquad \text{Re } s > 0$$

From this equation, deduce that (9-12) has a unique solution.

9-6 Explain how to compute $P(t)$ recursively from (9-13) when all transition times are a multiple of some constant, Δ say. How would you use this scheme to approximate $P(t)$ in the general case?

9-7 Prove that (9-14) is valid whenever the state space is finite. *Hint:* Let $F(\cdot)$ be a d.f., with $F(0) = 0$, such that $F(\cdot)$ is an upper bound for every $F_{ij}(\cdot)$. Prove $P\{N(t) \geq n\} \leq F_n(t) \to 0$ as $n \to \infty$ for every $t > 0$, where $F_n(\cdot)$ is the n-fold convolution of $F(\cdot)$ with itself.

9-8 Consider the birth-and-death process with state space I and parameters $\lambda_0 = 1$ and $\lambda_n = \mu_n = 2^n$, $n \in I_+$. Show that this SMP has ergodic states but the states of the embedded MC are persistent-null. You may use the fact that positive limiting probabilities exist if, and only if, the states are ergodic.

9-9 Consider the birth-and-death process with state space I, $\lambda_0 = 1$, $\lambda_n = 2^{n-1}$, and $\mu_n = 2^n$. Show that this SMP has states which are not ergodic but that the states of the embedded MC are ergodic.

9-10 Prove Proposition 9-7.

9-11 Complete the derivation of (9-37). Prove that the integral in (9-37b) equals $(1 - \sum_{i=0}^{j-k} c_i)/\lambda$, where c_i is the probability that i customers arrive during the service time of a batch.

9-12 Let $K = 1$ in Example 9-4. Let $\{\xi_i\}$ denote the limiting distribution of the embedded Markov chain for the M/G/1 queue derived in Example 7-18, and let $\{P_j\}$ be as in Example 9-4. Prove $P_j = \xi_j$, $j \in I$. *Hint:* For $\{\pi_j\}$ as in Example 9-4, first prove $\pi_j = d\xi_j$ for $j \neq 1$, and then prove $\pi_j = d(\xi_0 + \xi_1)$ for some constant d.

9-13 Interpret Example 9-4 in terms of the inventory model in Example 9-2.

9-14 Use the embedded SMP in Example 9-4 to obtain the limiting probabilities for the M/M/1 queue.

9-15 Recall the machine breakdown model in Examples 7-2 and 7-6. Suppose that repairs always take two days, that a perfectly running machine stays in that mode for an exponentially distributed time with mean $1/u$, and that a machine with a minor defect works for a time that is uniformly distributed over $[0, 2/d]$. Let g be the probability that a perfectly running machine fails completely and ϕ be the probability that a repair produces a perfectly running machine. What are the limiting probabilities for this model?

9-2* MARKOV RENEWAL PROCESSES

The Markov renewal process, abbreviated MRP, is closely related to the semi-Markov process. The SMP records the state of the (X, Y)-process defined in Section 9-1; the MRP records the number of visits to each state in S of the

(X, Y)-process. The MRP is a generalization of a renewal process in which different distributions of the times between renewals are chosen according to a Markov chain. The notation introduced in Section 9-1 is used throughout this section.

Let $N_k(t)$ be the number of visits that the (X, Y)-process makes to state k during $(0, t]$. The MRP is $\{(N_k(t); k \in S); t \geq 0\}$; its state at time t is a vector which records the number of visits made to each $k \in S$. Here is the formal definition.

Definition 9-6: Markov renewal process Let the (X, Y)-processes be given by Definition 9-2. Define the functions

$$U_{nk}(t) = \begin{cases} 1 & \text{if } Y_n = k, X_0 + \cdots + X_{n-1} \leq t, n \in I_+ \\ 0 & \text{otherwise} \end{cases} \tag{9-38}$$

for $k \in I$ and $t \geq 0$, and

$$N_k(t) = \sum_{n=0}^{\infty} U_{nk}(t) \qquad k \in S, \quad t \geq 0 \tag{9-39}$$

The process $\{(N_k(t); k \in S); t \geq 0\}$ is called a *Markov renewal process*.

Observe that when S consists of a single state, an MRP is a renewal process. For each k, $\{N_k(t); t \geq 0\}$ is the (possibly delayed) renewal process with generic interrenewal time T_{kk}.

Markov Renewal Function

The renewal function plays a central role in renewal theory. The analogous quantity for an MRP is defined now.

Definition 9-7: Markov renewal function The matrix of functions $M(\cdot) = (M_{ij}(\cdot))$, where

$$M_{ij}(t) = E[N_j(t) \mid Y_0 = i] \qquad i, j \in S, \quad t \geq 0 \tag{9-40}$$

is the *Markov renewal function*.

From (9-38), (9-39), and (9-40),

$$M_{ij}(t) = E\left(\sum_{n=0}^{\infty} (U_{nj}(t) \mid Y_0 = i) \right) = \sum_{n=0}^{\infty} E[U_{nj}(t) \mid Y_0 = i]$$

$$\tag{9-41}$$

$$= \sum_{n=1}^{\infty} P\{Y_n = j, S_n \leq t \mid Y_0 = i\}$$

From (9-2), $P\{Y_1 = j, S_1 \le t \mid Y_0 = i\} = Q_{ij}(t)$; by conditioning on the first transition,

$$P\{Y_2 = j, S_2 \le t \mid Y_0 = i\} = \sum_{k \in \mathcal{S}} \int_0^t P\{Y_2 = j, X_1 \le t - u \mid Y_1 = k\} \, dQ_{ik}(u)$$

$$= \sum_{k \in \mathcal{S}} \int_0^t Q_{kj}(t - u) \, dQ_{ik}(u)$$

$$= \sum_{k \in \mathcal{S}} \int_0^t Q_{ik}(t - u) \, dQ_{kj}(u)$$

This motivates the following definition of matrix convolution.

Definition 9-8 Let $A(\cdot) = (a_{ij}(\cdot))$ and $B = (b_{ij}(\cdot))$ be square matrices of the same size, with $A(t) = B(t) = 0$ for $t < 0$, and let $\mathcal{S}$ be the set of row indices. The matrix $C(\cdot) = (c_{ij}(\cdot))$, where

$$c_{ij}(t) = \sum_{k \in \mathcal{S}} \int_0^t a_{ik}(t - u) \, db_{kj}(u) \qquad i, j \in \mathcal{S} \tag{9-42}$$

is the *convolution of $A(\cdot)$ and $B(\cdot)$* if all the integrals exist. We write $C(t) = A * B(t)$. When $A(\cdot) = B(\cdot)$, the *n-fold convolution of $A(\cdot)$ with itself* is denoted by $A_n(\cdot)$ and defined by

$$A_1(\cdot) = A(\cdot) \qquad A_n(t) = A_{n-1} * A(t) \qquad n = 2, 3, \ldots \tag{9-43}$$

From (9-43) and (9-2), the nth term in the right-side of (9-41) is the (i, j)th element of $Q_n(t)$; hence

$$M(t) = \sum_{n=1}^{\infty} Q_n(t) \qquad t \ge 0 \tag{9-44}$$

Equation (9-44) is the MRP version of (5-22). The MRP version of the renewal equation (5-26) can be obtained via the renewal argument (Exercise 9-18), or from (9-44), as follows. Using the results of Exercise 9-17 and (9-43), we have

$$M(t) = Q(t) + \sum_{n=2}^{\infty} Q_n(t)$$

$$= Q(t) + \sum_{n=1}^{\infty} [Q_n * Q(t)]$$

$$= Q(t) + \left(\sum_{n=1}^{\infty} Q_n \right) * Q(t)$$

$$\tag{9-45}$$

$$= Q(t) + M * Q(t) = Q(t) + Q * M(t) \qquad t \ge 0$$

Equation (9-45) is called the *Markov renewal equation.*

Let $\tilde{Q}_{ij}(s)$ be the LST of $Q_{ij}(\,\cdot\,)$, $\tilde{M}_{ij}(s)$ be the LST of $M_{ij}(\,\cdot\,)$, $\tilde{Q}(s) = (\tilde{Q}_{ij}(s))$, and $\tilde{M}(s) = (\tilde{M}_{ij}(s))$. The MRP version of (5-9) is given by this theorem.

Theorem 9-5 For an MRP with matrix of transition distributions $Q(\,\cdot\,)$ and Markov renewal function $M(\,\cdot\,)$, the LSTs $\tilde{Q}(\,\cdot\,)$ and $\tilde{M}(\,\cdot\,)$ satisfy

$$[I - \tilde{Q}(s)]\tilde{M}(s) = \tilde{Q}(s) \qquad \tilde{M}(s) \geq 0 \qquad \text{Re } s > 0 \qquad (9\text{-}46)$$

When the state space is finite, the unique solution of (9-46) is

$$\tilde{M}(s) = [I - \tilde{Q}(s)]^{-1}\tilde{Q}(s) = \sum_{n=1}^{\infty} [\tilde{Q}(s)]^n = [I - \tilde{Q}(s)]^{-1} - I \qquad (9\text{-}47)$$

When the state space is infinite, $\tilde{M}(s)$ is the smallest solution of (9-46), in the sense that if $\xi(s)$ satisfies $[I - \tilde{Q}(s)]\xi(s) = \tilde{Q}(s)$, $\xi(s) \geq 0$, then $\tilde{M}_{ij}(s) \leq \xi_{ij}(s)$ for all i and j, Re $s > 0$.

PROOF Since the LST of $Q * M(t)$ is $\tilde{Q}(s)\tilde{M}(s)$, (9-46) follows from (9-45) by taking LSTs on both sides and rearranging terms.

For† $s > 0$, $0 \leq \tilde{Q}_{ij}(s) = p_{ij} \int_0^\infty e^{-st} \, dF_{ij}(t) < p_{ij}$ and $\sum_{j \in S} \tilde{Q}_{ij}(s) < 1$. Thus $[\tilde{Q}(s)]^n \to 0$ as $n \to \infty$, and by Theorem 6-3, when the state space is finite, $[I - \tilde{Q}(s)]^{-1} \triangleq \sum_{n=0}^{\infty} [\tilde{Q}(s)]^n$ exists. Multiplying (9-46) on the left by $[I - \tilde{Q}(s)]^{-1}$ yields (9-47).

Suppose the state space is infinite and $\tilde{M}(s)$ and $\tilde{\xi}(s)$ are distinct solutions of (9-46). Iterating (9-46) n times yields (suppress the dependence on s for notational clarity)

$$\xi = \tilde{Q} + \tilde{Q}\xi = \tilde{Q} + \tilde{Q}(\tilde{Q} + \tilde{Q}\xi)$$

$$= \tilde{Q} + \tilde{Q}^2 + \tilde{Q}^2(\tilde{Q} + \tilde{Q}\xi)$$

$$= \tilde{Q} + \tilde{Q}^2 + \cdots + \tilde{Q}^n + \tilde{Q}^{n+1}\xi$$

Letting $n \to \infty$ yields (since $\tilde{Q}$ and ξ are nonnegative)

$$\xi = \tilde{M} + \lim_{n \to \infty} \tilde{Q}^{n+1}\xi \geq \tilde{M} \qquad\qquad \square$$

Corollary 9-5 When the state space is finite,

$$\tilde{Q}(s) = \tilde{M}(s)[I + \tilde{M}(s)]^{-1} \qquad \text{Re } s > 0 \qquad (9\text{-}48)$$

PROOF From (9-47), $I + \tilde{M}(s) = [I - Q(s)]^{-1}$, which establishes that $[I + \tilde{M}(s)]^{-1}$ exists. Rearranging (9-46) yields $\tilde{Q}(s)[I + \tilde{M}(s)] = \tilde{M}(s)$, and (9-48) follows immediately. $\square$

Corollary 9-5 implies that $Q(\,\cdot\,)$ can be deduced from $M(\,\cdot\,)$. Since $Q(\,\cdot\,)$ and $a(\,\cdot\,)$ determine the transitions of an SMP, *to each finite-state MRP there corresponds exactly one SMP with the same (X, Y)-process.*

† Re $s > 0$ when s is complex.

Generalized Markov Renewal Equations

Let $Q(\,\cdot\,)$ be the matrix of transition distributions of an MRP with state space S; $D(\,\cdot\,) = (D_{ij}(\,\cdot\,))$, a given matrix of functions with $i, j \in S$; and $A(\,\cdot\,) = (A_{ij}(\,\cdot\,))$ be an unknown matrix of functions. A *generalized Markov renewal equation* is

$$A(t) = D(t) + Q * A(t) \qquad t \geq 0 \tag{9-49}$$

In the scalar (i.e., one-state) case of renewal theory, the unique bounded solution of (9-49) is $A(t) = D(t) + M * D(t)$, by Theorem 5-6 and Exercise 5-23. In the matrix case, more structure is required to solve (9-49).

Let $\mathcal{B}$ be the set of functions $b_i(\,\cdot\,)$ from $S \times \mathbb{R}_+$ to $\mathbb{R}$ such that, for each $i \in S$, $b_i(t)$ is bounded in finite intervals and, for each $t \in \mathbb{R}_+$, $b_i(t)$ is bounded in i.† Let $\mathcal{B}_+$ denote the subset of functions in $\mathcal{B}$ that assume only nonnegative values.

> **Proposition 9-8** Let $Q(\,\cdot\,)$ be the matrix of transition distributions of an MRP and $M(\,\cdot\,)$ be the renewal function. Then
>
> *(a)* $Q(\,\cdot\,) \in \mathcal{B}_+$ and $M(\,\cdot\,) \in \mathcal{B}_+$
>
> and
>
> *(b)* $A(\,\cdot\,) \in \mathcal{B}_+ \Rightarrow Q * A(\,\cdot\,) \in \mathcal{B}_+$

PROOF This is left as Exercise 9-19. □

> **Theorem 9-6** Let $D(\,\cdot\,) \in \mathcal{B}_+$. Every solution of (9-49) with $A(\,\cdot\,) \in \mathcal{B}_+$ is of the form
>
> $$A(t) = D(t) + M * D(t) + C(t) \tag{9-50}$$
>
> where $C(\,\cdot\,)$ satisfies
>
> $$C(t) = Q * C(t) \qquad t \geq 0 \tag{9-51}$$
>
> In particular,
>
> $$A(t) = D(t) + M * D(t) \tag{9-52}$$
>
> is a solution.

PROOF Assume $A(\,\cdot\,) \in \mathcal{B}_+$. Iterate (9-49) n times to obtain

$$
\begin{aligned}
A(t) &= D(t) + Q * [D(t) + Q * A(t)] \\
&= D(t) + Q * D(t) + Q_2 * [D(t) + Q * A(t)] \\
&= D(t) + [Q * D(t) + Q_2 * D(t) + \cdots + Q_n * D(t)] + Q_{n+1} * A(t)
\end{aligned}
$$

Since $D(\,\cdot\,) \geq 0$, the term in brackets increases with n and approaches $M*D(t)$ as $n \to \infty$. The properties of $Q(\,\cdot\,)$ given in Definition 9-1 imply $Q_{n+1} * A(t)$

† We use the notation $\mathbb{R}_+ = [0, \infty)$ and $\mathbb{R} = (-\infty, \infty)$.

decreases with n; hence it has a limit, $C(t)$ say, as $n \to \infty$. Thus

$$C(t) \triangleq \lim_{n \to \infty} Q_{n+1} * A(t) = Q * \lim_{n \to \infty} Q_n * A(t) = Q * C(t)$$

which establishes (9-51). Since $C(\cdot) \equiv 0$ satisfies (9-51), (9-52) is a particular solution. □

From Theorem 9-6 it is evident that (9-52) is the unique solution (9-49) if, and only if, $C(\cdot) \equiv 0$ is the only solution of (9-51). Fortunately, that is the typical case.

Theorem 9-7 The only solution of (9-51) is $C(\cdot) \equiv 0$ if, and only if, $P\{N(t) < \infty\} = 1$ holds.

PROOF See Çinlar (1975), Proposition 10.3.14. □

Key Renewal Theorem for MRPs

Call an MRP *irreducible* if the SMP with the same matrix of transition distributions is irreducible. There is no difficulty in extending the key renewal theorem (Theorem 5-9) to irreducible MRPs with a finite state space.

Theorem 9-8 Let $M(\cdot)$ be the Markov renewal function of an irreducible MRP with matrix of transition functions $Q(\cdot)$ and finite state space $\mathcal{S} = \{0, 1, \ldots, S\}$. For each $j \in \mathcal{S}$, let $h_j(\cdot)$ be a directly Reimann-integrable function. Then

$$\lim_{t \to \infty} \sum_{j=0}^{S} \int_0^t h_j(t - u) \, dM_{ij}(u) = \frac{1}{\mu_{jj}} \sum_{j=0}^{S} \int_0^\infty h_j(u) \, du \qquad (9\text{-}53)$$

for each $i \in \mathcal{S}$, where μ_{jj} is given by (9-22). If the states are periodic, (9-53) holds only when t is a multiple of the span.

PROOF From the proof of Proposition 9-6 and the comment following Definition 9-6, for each i, $M_{ij}(\cdot)$ is a (possibly delayed) renewal function, where μ_{jj} is the mean time between renewals. Since $S < \infty$, $G_{ij}(\infty) = 1$, for all i, $j \in \mathcal{S}$, and thus the key renewal theorem implies

$$\int_0^t h_j(t - u) \, dM_{ij}(u) \to \frac{1}{\mu_{jj}} \int_0^\infty h_j(u) \, du \qquad j \in \mathcal{S} \text{ as } t \to \infty$$

Since $S < \infty$, the limit on the left-side of (9-53) can be brought inside the sum. □

Note that the last step of the proof of Theorem 9-8 is not valid when $\mathcal{S}$ is infinite. In that case, a more restrictive property than direct Reimann integrability must be assumed; see Çinlar (1975), sec. 10.4, for a complete discussion.

An Example

Example 9-5: Example 9-1 continued In Example 9-1, states 0 to 5 are transient, and states 6, 7, and 8 are absorbing. For $j \leq 5$, $M_{0j}(\infty)$ is the mean number of visits that a patient makes to care unit j before entering an absorbing state. This quantity is easily obtained from (9-44) or (9-47); see Exercise 9-20.

It is more interesting to model the simultaneous movement of many patients. Let $\Phi(t)$ be a deterministic function which denotes the cumulative number of patients who have arrived by time t. By its nature, $\Phi(\cdot)$ is a nondecreasing function, and it is natural to assume that $\Phi(\cdot)$ has an LST. Assume that each patient moves from state to state according to the same SMP and that the SMPs are independent from one patient to another.

At time t, the mean number of visits to state j made by those customers (if any) who arrived at time $\tau < t$ is $M_{0j}(t - \tau)$. The mean number of visits to state j at time t made by all customers, $V_j(t)$ say, is given by

$$V_j(t) = \int_0^t M_{0j}(t - \tau) \, d\Phi(\tau) \tag{9-54a}$$

The LST version of (9-54a) is

$$\tilde{V}_j(s) = \tilde{M}_{0j}(s)\Phi(s) \tag{9-54b}$$

Let $\mathbf{V}(t) = (V_0(t), \ldots, V_8(t))$ and $\boldsymbol{\Phi}(t) = [\Phi(t), 0, \ldots, 0]$. From (9-54b) and (9-47),

$$\tilde{\mathbf{V}}(s) = \tilde{\boldsymbol{\Phi}}(s)\tilde{\mathbf{M}}(s) = \tilde{\boldsymbol{\Phi}}(s)\,[I - \tilde{Q}(s)]^{-1}\tilde{Q}(s) \tag{9-55}$$

Observe that (9-55) remains true if patients can first enter the system in any state, where $\Phi_j(t)$ is the number of patients who have arrived (from the outside) at state j by time t, and $\boldsymbol{\Phi}(t) = [\Phi_0(t), \ldots, \Phi_8(t)]$. □

EXERCISES

9-16 Let $G_{jj}^{(n)}(\cdot)$ denote $G_{jj}(\cdot)$ convoluted with itself n times, $n \in I_+$. Show that $M_{jj}(t) = \sum_{n=1}^{\infty} G_{jj}^{(n)}(t)$ and $M_{ij}(t) = G_{ij} * M_{jj}(t)$, $i \neq j$. If the corresponding SMP is irreducible and all states are persistent, prove

$$\lim_{t \to \infty} \frac{M_{ij}(t)}{t} = \frac{1}{\mu_{jj}} \qquad i, j \in S$$

9-17 For $A(\cdot)$ and $B(\cdot)$ as in Definition 9-8, prove† (a) $A * B(t) = B * A(t)$, (b) $\sum_{n=1}^{\infty} A_n(t) * A(t) = (\sum_{n=1}^{\infty} A_n) * A(t)$, and (c) $\mathcal{L}\{A * B(t)\} = \mathcal{L}\{A(t)\}\mathcal{L}\{B(t)\}$, where the term on the right is a product of two matrices.

9-18 Derive (9-45) by the renewal argument.

† This is a situation in which the type of integral used is important. For Riemann-Stieltjes integrals, $B * A(t)$ exists; with Lebesque integrals, it may not exist. See, e.g., Feller (1971, p. 143).

9-19 Prove Proposition 9-8.

9-20 In Example 9-1, let P be the transition matrix of the embedded Markov chain of the SMP.

(a) Show $M(\infty) = \sum_{n=1}^{\infty} P^n$, where P^n is the product of P with itself n times. Observe that P can be written

$$P = \begin{pmatrix} P_1 & P_2 \\ 0 & I \end{pmatrix}$$

where P_1 is a 7×7 matrix, P_2 is a 7×3 matrix, and the other matrices are sized conformally.

(b) Prove $M_{ij}(\infty) = (I - P_1)^{-1}$, $0 \le i, j \le 6$.

(c) Prove $\lim_{t \to \infty} P_{ij}(t) = (I - P_1)^{-1} P_2$, $0 \le i \le 6$, $7 \le j \le 9$. Notice that only the embedded MC is important in this exercise; the solution can be obtained by MC methods if the relevant features of Example 9-1 are described by a Markov chain.

9-21 Construct an MRP such that $N(t) = \sum_{j=0}^{\infty} N_j(t)$ is the number of departures during $(0, t]$ from an M/G/1 queue. Display the matrix $Q(\cdot)$ for the MRP.

9-22 MRPs with Rewards Suppose a reward structure is imposed on the transitions of an MRP [more precisely, on the (X, Y)-process] as follows. For a transition that starts at state i and ends x time units later at state j, let $r_{ij}(x; u)$ be the cumulative reward received u time units after the transition begins ($0 \le u \le x$). Let $r_{ij}(x) \triangleq r_{ij}(x; x)$; it denotes the reward for the completed transition. Let $R_i(t)$ be the sum of the rewards received during $(0, t]$. For sample path ω, if $N(t; \omega) = n$ and $Y_n(\omega) = i_n$, $n \in I$, then

$$R_i(t; \omega) = r_{i, i_1}[X_0(\omega)] + r_{i_1, i_2}[X_1(\omega)] + \cdots + r_{i_{n-1}, i_n}[X_{n-1}(\omega)]$$
$$+ r_{i_n, i_{n+1}}[X_n(\omega), t - S_n(\omega)]$$

Prove that for each $i \in \mathcal{S}$,

$$E[R_i(t)] = a_i(t) + \sum_{j \in \mathcal{S}} \int_0^t E[R_j(t - s)] \, dQ_{ij}(s)$$

$$= a_i(t) + \sum_{j \in \mathcal{S}} \int_0^t a_j(t - s) \, dM_{ij}(s)$$

where

$$a_i(t) = \sum_{j \in \mathcal{S}} \left[\int_t^{\infty} r_{ij}(x; t) \, dF_{ij}(x) + \int_0^t r_{ij}(x) \, dF_{ij}(x) \right]$$

9-23 (Continuation) Suppose that rewards are discounted continuously at continuous interest rate $\beta > 0$. Then

$$\tilde{r}_i(\beta) = \sum_{j \in \mathcal{S}} \int_0^{\infty} \int_0^x e^{-\beta t} \, dr_{ij}(x; t) \, dQ_{ij}(x)$$

is the expected present worth of a completed reward about to be earned when state i is entered. Show

$$\tilde{R}_i(\beta) = \int_0^{\infty} e^{-\beta t} \, dE[R_i(t)] = \sum_{j \in \mathcal{S}} \tilde{M}_{ij}(\beta) \tilde{r}_j(\beta) \qquad i \in \mathcal{S}$$

9-24 (Continuation) When the SMP is irreducible and consists of ergodic states, prove that for $r_i \triangleq \tilde{r}_i(0)$ (assumed absolutely convergent),

$$\lim_{t \to \infty} \frac{R_i(t)}{t} = \lim_{t \to \infty} \frac{E[R_i(t)]}{t} = \frac{\sum_{j \in \mathcal{S}} \pi_j r_j}{\sum_{j \in \mathcal{S}} \pi_j \eta_j}$$

for each $i \in \mathcal{S}$, the first equality holding w.p.1.

9-3* RANDOM WALKS ON THE LINE

In the simple random walk described in Section 7-8, the step sizes are r.v.'s taking on the values ± 1. In this section, the steps have an arbitrary distribution. Our main use of this increased generality is an analysis of the delays in the $GI/G/1$ queue and some related storage models.

Definitions

Definition 9-9: Random-walk process Let Y_1, Y_2, ... be i.i.d. random variables with the property

$$P\{Y_1 > 0\} > 0 \qquad P\{Y_1 < 0\} > 0 \qquad \text{and} \qquad P\{-\infty < Y_1 < \infty\} = 1$$
(9-56)

Fix a number x_0, and define S_n by

$$S_0 = x_0 \qquad S_n = S_{n-1} + Y_n \qquad n \in I_+$$
(9-57)

The stochastic process $\{S_n; n \in I\}$ is called a *random walk starting at* x_0. When $x_0 = 0$, the random walk is said to *start at the origin*; in this case,

$$S_n = Y_1 + Y_2 + \cdots + Y_n \qquad n \in I_+$$
(9-58)

An immediate consequence of (9-57) is that $\{S_n; n \in I\}$ is a time homogeneous Markov process. The state space is discrete or continuous according as Y_1 is discrete or continuous. The essential qualitative feature of a random walk is upward *and* downward fluctuations; condition (9-56) avoids trivialities. For most of the theory, it is sufficient to deal with only random walks that start at the origin.

The study of random walks focuses on the extreme values taken on by the random walk. Toward this end, we make these definitions.

Definition 9-10 Let $\{S_n; n \in I\}$ be a random walk. The random variable T_1 defined by

$$\{T_1 = n\} \Leftrightarrow \{S_j \leq S_0, j = 1, 2, \ldots, n - 1, S_n > S_0\}$$
(9-59)

is called the *first ascending ladder epoch*. If the event on the right-side of (9-59) does not occur, set $T_1 = \infty$. If $T_1 < \infty$, the random variable S_{T_1} is called the *first ascending ladder height*. The pair (T_1, S_{T_1}) is the *first ascending ladder point*.

Thus T_1 is the first epoch where the random walk reaches a value which is (strictly) larger than all previous values, and S_{T_1} is the record-setting value. An example of a random walk in which $P\{T_1 = \infty\} > 0$ is the simple random walk with $P\{Y_1 = -1\} > P\{Y_1 = 1\}$; see Exercise 9-25.

Use (9-57) to write (9-59) as

$$\{T_1 = n\} \Leftrightarrow \{Y_1 \leq 0, \cdots, Y_1 + \cdots + Y_{n-1} \leq 0, Y_1 + \ldots + Y_n > 0\} \quad (9\text{-}59a)$$

Hence, $\{T_1 = n\}$ is independent of S_0. If $T_1 < \infty$, the process $\{S_n; n \geq T_1\}$ is stochastically the same as $\{S_n; n \in I\}$ except for the first value (which is S_{T_1} rather than S_0). The process $\{S_n; n \geq T_1\}$ has a first ascending ladder height, T_1' say. From (9-57) and (9-59),

$$\{T_1' = n\} \Leftrightarrow \{Y_{T_1+1} \leq 0, \ldots,$$

$$Y_{T_1+1} + \cdots + Y_{T_1+n-1} \leq 0, Y_{T_1+1} + \cdots + Y_{T_1+n} > 0\} \quad (9\text{-}59b)$$

Equations (9-59a) and (9-59b) and the assertion that the Y_k's are i.i.d. yield

$$P\{T_1 = n\} = P\{T_1' = n\} \qquad n \in I \quad (9\text{-}60)$$

Since $T_2 \triangleq T_1 + T_1'$ is the second epoch where a record high is reached, it is the *second* ascending ladder epoch of $\{S_n; n \in I\}$. In the same manner, we define the third, fourth, etc. ascending ladder epochs; let T_k denote the kth one. Define $T_0 = 0$, and if $T_n < \infty$, define

$$X_n = S_{T_n} - S_{T_{n-1}} \qquad n \in I_+ \quad (9\text{-}61)$$

Thus

$$S_{T_n} = x_0 + X_1 + \cdots + X_n \qquad n \in I \quad (9\text{-}62)$$

The pair (T_n, S_{T_n}) is the nth *ascending ladder point*; the first element is the epoch at which it occurs, and the second element is the value of the random walk at that epoch.

By reversing the inequalities in (9-59), we obtain the definition of the *first descending ladder epoch*, denoted by T_1^-; specifically,

$$\{T_1^- = n\} \Leftrightarrow \{S_j \geq S_0, j = 1, 2, \ldots, n-1, S_n < S_0\} \quad (9\text{-}63)$$

The random variable $X_1^- \triangleq S_{T_1^-}$ is the *first descending ladder height*; T_k^- and $S_{T_k^-}$ are defined recursively, as above. Since the ladder epochs do not depend on $x_0 \triangleq S_0$, without loss of generality we can choose $x_0 = 0$. Multiplying each Y_k by -1 interchanges ascending ladder points and descending ladder points, so it is sufficient to develop results for only ascending ladder points.

When the Y_k's have a continuous d.f., it does not matter whether the inequalities in (9-59) and (9-63) are strong or weak because $P\{S_j = S_0\} = 0, j \in I_+$. When the d.f. of the Y_k's has mass points, it is technically important, but conceptually trite, to distinguish between strong and weak inequalities. Define the random variable T_1^w by

$$\{T_1^w = n\} \Leftrightarrow \{S_j < S_0, j = 1, 2, \ldots, n-1, S_n \geq S_0\} \quad (9\text{-}64)$$

it is called the *first ascending weak ladder epoch*. Exercises 9-26 and 9-27 ask you to derive some simple relationships between T_1 and T_1^w which allow us to obtain results for the latter from corresponding results for the former.

Examples

Example 9-6: $GI/G/1$ queue Let T_n be the arrival epoch of the nth customer, with $0 = T_0 \le T_1 \le T_2 \le \cdots$. Define the interarrival times U_n by $U_n = T_n - T_{n-1}$, $n \in I_+$, and let V_n denote the service time of customer n. Assume that the customers are served in order of arrival.

Let D_n denote the delay of customer n; that is, the time between the customer's arrival and entry into service. Assume that no customers are present when customer 0 arrives, so $D_0 \triangleq 0$.

Equation (2-15) holds for each sample path; hence it is a valid relation among random variables. Thus

$$D_n = (D_{n-1} + V_{n-1} - U_n)^+ \qquad n \in I_+$$

Defining $Y_n = V_{n-1} - U_n$, $n \in I_+$, yields the representation

$$D_n = (D_{n-1} + Y_n)^+ \qquad n \in I_+ \qquad D_0 = 0 \qquad (9\text{-}65)$$

If the "positive part" operator were not present in (9-65), (9-65) would be a version of (9-57); thus, it appears that $\{D_n; n \in I\}$ is related to a random walk. To make the connection, let S_n be defined by (9-57), with $S_0 = 0$. From Figure 9-3, it is readily seen that the index of the kth customer (excluding customer 0) who finds the server free on arrival is the kth descending ladder epoch. This fact can also be deduced from (9-65). Let K_n (an r.v.) be the number of descending ladder epochs which are less than or equal to n. From Figure 9-3 it is apparent that

$$D_n = S_n - S_{K_n} \qquad n \in I_+ \qquad (9\text{-}66)$$

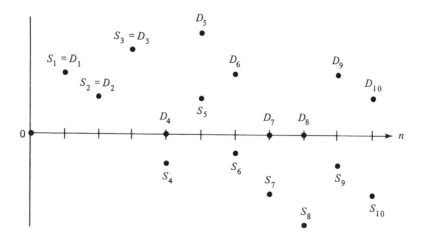

Figure 9-3 A sample path of a $GI/G/1$ queue.

and (9-66) can be deduced analytically from (9-65) by induction. (This is left for you to do.) From the definition of K_n,

$$S_{K_n} = \min\{S_0, S_1, \ldots, S_n\} \tag{9-67}$$

Substituting (9-67) into (9-66) yields

$$D_n = S_n - \min\{S_k : 0 \leq k \leq n\} \qquad n \in I_+ \tag{9-68}$$

In some settings, (9-68) is the most useful representation of D_n; later in this section, the following representation is used. From (9-68),

$$\begin{aligned} D_n &= \max\{S_n - S_k : 0 \leq k \leq n\} \\ &= \max\{Y_1 + \cdots + Y_n, \ldots, Y_{n-1} + Y_n, Y_n, 0\} \end{aligned} \tag{9-69}$$

Since† the Y_k's are i.i.d., $Y_{n-(k-1)} + \cdots + Y_n \sim Y_1 + \cdots + Y_k \triangleq S_k$, hence (9-69) implies

$$D_n \sim \max\{S_k : 0 \leq k \leq n\} \qquad n \in I_+ \tag{9-70}$$

Notice that (9-70) is the only equation whose validity depends on the assumption that $Y_1, Y_2, \ldots$ are i.i.d. □

Example 9-7: Cash balance model Let S_n denote the cash balance of a firm at the end of day n; the initial cash balance is $x_0 \triangleq S_0$. Let U_n and V_n denote the receipts and disbursements on day n, and set $Y_n \triangleq U_n - V_n$. If the firm maintains a line of credit, then S_n may be negative; a negative value is interpreted as a loan (see Section 2-4 for further commentary on this model). Clearly, S_n is given by (9-57). For large firms with many customers and suppliers, in a stable economic environment it may be reasonable to assume that $Y_1, Y_2, \ldots$ are i.i.d., in which case $\{S_n ; n \in I\}$ is a random walk.

Suppose the firm does not use a line of credit, but covers negative balances by transferring funds from another account (an interest-bearing account kept for that purpose).‡ Let D_n be the cash balance on day n with this policy. Then D_n satisfies (9-65) for $n \in I_+$, with $D_0 = x_0$. Consequently, D_n satisfies (9-70):

$$D_n \sim \max\{0, Y_1, Y_1 + Y_2, \ldots, Y_1 + Y_2 + \cdots + Y_n\} \qquad □$$

Renewal Analysis of Ladder Epochs

Let $F(x; n) = P\{T_1 = n, S_{T_1} \leq x\}, n \in I_+, x \geq 0$. Then

$$F(x) \triangleq \sum_{n \in I_+} F(x; n) = P\{S_{T_1} \leq x\} \qquad x \geq 0 \tag{9-71}$$

---

† Recall that the notation $X \sim Y$ means that X and Y have the same distribution.

‡ See Section 8-4 in Vol. II for optimization in a model of cash balance decisions.

and

$$f_n \triangleq \lim_{x \to \infty} F(x; n) = P\{T_1 = n\} \qquad n \in I_+ \qquad (9\text{-}72)$$

Furthermore, since S_{T_1} is finite if, and only if, $T_1 < \infty$,

$$F(\infty) = P\{S_{T_1} < \infty\} = P\{T_1 < \infty\} \qquad (9\text{-}73)$$

Hence, $F(\cdot)$ is the (possibly defective) d.f. of each X_n defined by (9-61). The number of ascending ladder heights in $(S_0, S_0 + t]$ is now seen to be the number of "renewals" in $(0, t]$ of the renewal process generated by $\{X_n\}_1^\infty$. Let $M(t)$ be the mean number of those ladder heights; then

$$M(t) = \sum_{n=1}^{\infty} F_n(t) \qquad t > 0 \qquad (9\text{-}74)$$

where $F_n(\cdot)$ is the n-fold convolution of $F(\cdot)$ with itself.

If $F(\infty) < 1$, each ascending ladder epoch has probability $1 - F(\infty) > 0$ of being the last one; hence, the probability that exactly k ascending ladder epochs occur is $[1 - F(\infty)][F(\infty)]^k$, $k \in I$. Since each ascending ladder height is positive and finite, $M(\infty)$ is the mean number of ascending ladder epochs that occur; thus,

$$M(\infty) = \frac{F(\infty)}{1 - F(\infty)} \qquad (9\text{-}75)$$

and

$$M(\infty) < \infty \Leftrightarrow F(\infty) < 1 \qquad (9\text{-}76)$$

Let $M^w(t)$ be the mean number of ascending weak ladder heights in $[S_0, S_0 + t]$. Exercise 9-28 asks you to prove that

$$M^w(t) = \frac{M(t)}{1 - \delta} \qquad t \geq 0 \qquad (9\text{-}77)$$

where $\delta < 1$ is the probability that the first ascending ladder epoch of either type is a weak one (δ is formally defined in Exercise 9-26).

The renewal function $M(t)$ has the following unexpected interpretation.

Proposition 9-9: Duality lemma Let $\{S_n; n \in I\}$ be a random walk starting at the origin, and define N_t by

$$N_t = \#\{n : 0 < S_n \leq t, S_k > 0 \quad \text{for} \quad k = 1, 2, \ldots, n\}$$

Then $E(N_t) = M(t)$, where $M(t)$ is given by (9-74).

PROOF Use the argument employed in deriving (9-70) from (9-69) to obtain

$$P\{S_n > S_j \quad \text{for} \quad j = 0, 1, \ldots, n-1 \quad \text{and} \quad S_n \leq t\}$$
$$= P\{Y_n > 0, \ Y_n + Y_{n-1} > 0, \ \ldots, \ Y_n + \cdots + Y_1 > 0 \quad \text{and} \quad S_n \leq t\}$$
$$= P\{Y_1 > 0, \ Y_1 + Y_2 > 0, \ \ldots, \ Y_1 + \cdots + Y_n > 0 \quad \text{and} \quad S_n \leq t\}$$
$$= P\{S_j > 0 \quad \text{for} \quad j = 1, 2, \ldots, n \quad \text{and} \quad S_n \leq t\} \qquad (9\text{-}78)$$

The first term is the probability that n is an ascending ladder epoch and $S_n \leq t$; the last term is the probability that $S_j > 0$ for $1 \leq j \leq n$ and $S_n \leq t$. Hence,

$$P\{n \text{ is a ladder epoch}, S_n \leq t\} = P\{S_j > 0 \quad \text{for} \quad j = 1, 2, \ldots, n, S_n \leq t\}$$

$$(9\text{-}79)$$

for $n \in I_+$ and $t > 0$. Summing the right-side of (9-79) over $n \in I_+$ yields $E(N_t)$. If n is a ladder epoch, it is the first, or second, or third, or ...; hence summing the left-side of (9-79) over $n \in I_+$ yields $\sum_{n=1}^{\infty} F_n(t)$. $\quad\square$

Classification of Random Walks

Theorem 9-9 below proves that there are two types of random walks, those that oscillate and those that drift to infinity. Then Theorem 9-10 classifies random walks based on the sign of $E(Y_1)$. Throughout this subsection we assume, without loss of generality, that $S_0 = 0$.

Theorem 9-9: Classification of random walks For a random walk $\{S_n; n \in I\}$ there are three possibilities:
(a) $P\{T_1 < \infty\} = P\{T_1^- < \infty\} = 1$, in which case

$$E(T_1) = \infty \qquad \text{and} \qquad E(T_1^-) = \infty \qquad (9\text{-}80)$$

and

$$P\{\lim_{n \to \infty} |S_n| < \infty\} = 1 \qquad (9\text{-}81)$$

(b) $P\{T_1 < \infty\} < 1$ and $P\{T_1^- < \infty\} = 1$, in which case

$$E(T_1^-) = \frac{F(\infty)}{(1 - \delta)[1 - F(\infty)]} < \infty \qquad (9\text{-}82)$$

and

$$P\{\lim_{n \to \infty} S_n = -\infty\} = 1 \qquad (9\text{-}83)$$

Moreover, (w.p.1) the random walk achieves a finite maximum, L say, with $L \geq 0$ and

$$P\{L \leq t\} = [1 - F(\infty)][1 + M(t)] \qquad t \geq 0 \qquad (9\text{-}84)$$

(c) $P\{T_1 < \infty\} = 1$ and $P\{T_1^- < \infty\} < 1$, in which case the obvious analogs of (9-82), (9-83), and (9-84) hold.

PROOF It is easily checked that (9-78) remains valid if the "greater than" signs are changed to "greater than or equal to." Hence

$$P\{S_n \geq S_j \quad \text{for} \quad j = 0, 1, \ldots, n\} = P\{S_j \geq 0 \quad \text{for} \quad j = 0, 1, \ldots, n\}$$

$$= P\{T_1^- > n\} \qquad (9\text{-}85)$$

The leftmost term in (9-85) is the probability that n is an ascending weak ladder epoch. Summing the outer terms of (9-85) over $n \in I$ yields

$$M^w(\infty) = \sum_{n=0}^{\infty} P\{T_1^- > n\} = E(T_1^-) \qquad (9\text{-}86)$$

Use (9-76) and (9-86) to obtain

$$P\{T_1^- < \infty\} < 1 \Rightarrow E(T_1^-) = \infty \Leftrightarrow P\{T_1 < \infty\} = 1 \qquad (9\text{-}87a)$$

Multiply each Y_k by -1 so the roles of the ascending and descending ladder epochs are interchanged, repeat the above argument, and conclude

$$P\{T_1 < \infty\} < 1 \Rightarrow E(T_1) = \infty \Leftrightarrow P\{T_1^- < \infty\} = 1 \qquad (9\text{-}87b)$$

From (9-87) we can conclude that

$$P\{T_1^- < \infty\} < 1 \qquad \text{and} \qquad P\{T_1 < \infty\} < 1$$

is impossible; hence, the three listed cases are exhaustive. Furthermore, (9-80) is a direct consequence of (9-87). A consequence of (a) is that there is no last ascending or descending ladder epoch, which implies (9-81).

Suppose (b) holds. Then (9-82) is obtained by combining (9-75), (9-77), and (9-86). The strong law of large numbers asserts

$$P\{\lim_{n\to\infty} \frac{S_n}{n} = E(X_1^-)\} = 1$$

Since $E(X_1^-) < 0$, (9-83) follows. From $P\{T_1 < \infty\} < 1$, there is a last ascending ladder epoch (w.p.1); hence

$$L \triangleq \sup\{S_0, S_1, S_2, \ldots\} < \infty \qquad (\text{w.p.1})$$

Each ascending ladder epoch has probability $1 - F(\infty)$ of being the last one. If there were n ascending ladder epochs, the d.f. of L is $F_n(t)$ (for $n \in I_+$) and $\{L \le t\}$ certainly occurs if $n = 0$. Hence

$$P\{L \le t\} = [1 - F(\infty)]\left[1 + \sum_{n=1}^{\infty} F_n(t)\right] \qquad t > 0$$

which with (9-74) yields (9-84). $\square$

The behavior of a random walk described by (9-81) is called *oscillating*, and the behavior described by (9-83) is called *drifting* (to $-\infty$). The strong law of large numbers and (9-58) show that $E(Y_1) = 0$ is necessary for (9-81) to hold; the next theorem asserts that $E(Y_1) = 0$ is also sufficient for (9-81) to hold.

Theorem 9-10 Let $\{S_n; n \in I\}$ be a random walk.
(a) If $E(Y_1) = 0$, then the random walk is oscillating.
(b) If $E(Y_1) < 0$, then the random walk drifts to $-\infty$.
(c) If $E(Y_1) > 0$, then the random walk drifts to $+\infty$.

PROOF The strong law of large numbers asserts that (w.p.1)

$$\lim_{n \to \infty} \frac{S_n}{n} = E(Y_1) \tag{9-88}$$

This establishes (b) and (c) immediately.

Suppose $E(Y_1) = 0$. Let m be the jth ascending ladder epoch; then

$$\frac{S_m}{m} = \frac{(X_1 + \cdots + X_j)/j}{(T_1 + \cdots + T_j)/j} \tag{9-89}$$

From Theorem 9-9, as $n \to \infty$, either $j \to \infty$ or the number of descending ladder epochs approaches ∞. In the latter case, multiply each Y_k by -1, so $j \to \infty$ always. This implies $P\{T_1 < \infty\} = 1$.

From (9-88), the left-side of (9-89) goes to zero as $m \to \infty$ (w.p.1). Since each $X_i > 0$, the numerator of the right-side of (9-89) is positive for all j; hence the denominator must become infinite as $j \to \infty$. This implies (via the strong law of large numbers) that $E(T_1) = \infty$.

The statements of Theorem 9-9 imply that T_1 is proper with an infinite mean if, and only if, the random walk is oscillating; this establishes (a). $\square$

Analysis of the *GI/G/*1 Queue

Theorems 9-9 and 9-10 are now used to continue the analysis of the $GI/G/1$ queue that started in Example 9-6. Let

$$D_n(t) \triangleq P\{D_n \le t\} \qquad n \in I \qquad \text{and} \qquad H(t) \triangleq P\{Y_1 \le t\}$$

An immediate consequence of (9-65) is

$$D_n(t) = \int_{-\infty}^{t} D_{n-1}(t - x) \, dH(x) \qquad t \ge 0 \tag{9-90}$$

Let $1/\lambda = E(U_1)$, $1/\mu = E(V_1)$, and $\rho = \lambda/\mu$. Then, from the definition $Y_n = V_{n-1} - U_n$,

$$E(Y_1) = \frac{\rho - 1}{\lambda} \tag{9-91}$$

Consider the limiting behavior as $n \to \infty$. According to Theorem 9-9, there are three possible cases.

If $\rho < 1$, by Theorem 9-10 the S_n-process is drifting to $-\infty$. From (9-70) and (9-84), $D(t) \triangleq \lim_{n \to \infty} D_n(t)$ is a proper d.f. and $D(t) = P\{L \le t\}$. Conditioning on Y_1 yields

$$\{L \le t\} \Leftrightarrow \{Y_1 = y \le t \quad \text{and} \quad \max(0, Y_2, Y_2 + Y_3, \ldots) \le t - y\}$$

Since $\max\{0, Y_2, Y_2 + Y_3, \ldots\} \sim L$, unconditioning on Y_1 yields

$$D(t) = \int_{-\infty}^{t} D(t - y) \, dH(y) \qquad t \ge 0 \qquad (\rho < 1) \tag{9-92}$$

If $\rho = 1$, by Theorem 9-10 the S_n-process is oscillating, and hence

$$D(t) = P\{L \le t\} = 0 \qquad t \ge 0 \qquad (\rho \ge 1) \tag{9-93}$$

Since there are infinitely many descending ladder epochs of the S_n-process, and since each one corresponds to a customer who finds the server free, the server becomes free infinitely often. From (9-80), the mean time between these epochs is infinite.

If $\rho > 1$, Theorem 9-10 asserts that the S_n-process is drifting to $+\infty$, and hence (9-93) holds. Since there is a last descending ladder epoch, eventually the server is working continually (w.p.1).

Equations (9-90), (9-92), and (9-93) were obtained first in Lindley (1952); (9-90) and (9-92) are often called the *Lindley equations*. For $\rho < 1$, the existence of the proper distribution function $D(\,\cdot\,)$ is ensured by the regenerative properties† of $\{D_n; n \in I\}$, as shown in Example 6-5. The additional contribution of the above analysis is the integral equation (9-92). Unfortunately, explicit solutions of (9-92) are not obtained often. The solution can be found for the $M/G/1$ and $GI/M/1$ models (see Exercises 9-34 and 9-35) from the formal solution of (9-92) by the analysis outlined in Exercises 9-30 through 9-33.

The nature of the difficulty in solving (9-92) can be appreciated by using (9-70) and (9-84) to put (9-92) in the form

$$D(t) = [1 - F(\infty)][1 + M(t)] \qquad t \ge 0 \tag{9-94}$$

Thus, solving (9-92) is equivalent to first determining $F(\,\cdot\,)$ and then calculating $M(\,\cdot\,)$. The latter is a standard problem in renewal theory and is no more difficult than inverting the LST given by (5-9); the former is the major source of difficulty.

EXERCISES

9-25 Let $\{S_n; n \in I\}$ be a simple random walk with $p = P\{Y_1 = 1\}$ and $q = 1 - p = P\{Y_1 - 1\}$. Prove that

$$P\{T_1 < \infty\} = \begin{cases} 1 & \text{if } p \ge q \\ \dfrac{p}{q} & \text{if } p < q \end{cases}$$

9-26 Define δ by

$$\delta \triangleq \sum_{n=1}^{\infty} P\{S_1 < S_0, \ldots, S_{n-1} < S_0, S_n = S_0\}$$

and let W_1 be the first ascending weak ladder height. Prove that $0 \le \delta < 1$,

$$\delta = P\{W_1 = 0\} = P\{T_1^w < T_1\}$$

---

† Theorems 9-9 and 9-10 provide a convenient proof that the mean length of a regenerative cycle is finite when $\rho < 1$; see Example 6-5 and Exercise 6-28.

and

$$P\{W_1 \le t\} = \delta + (1 - \delta)F(t) \qquad t \ge 0$$

9-27 Show that the probability that k ascending weak ladder epochs occur between an adjacent pair of (strict) ascending ladder epochs is $(1 - \delta)\delta^k$.

9-28 Derive (9-77).

9-29 For the $GI/G/1$ queue, what interpretation can you provide for $D_0 = x_0 > 0$? If $D_0 > 0$, prove that (9-92) still holds. *Hint*: Look at Example (5-19). Show that if D_0 has distribution function $D(\cdot)$, then $D(t) = P\{D_n \le t\}$ for all $n \in I$.

9-30 Let $\{S_n; n \in I\}$ be a random walk starting at the origin. Let $M_n(t) = P\{S_1 > 0, \dots, S_n > 0, S_n \le t\}$ for $t > 0$ and $M_n(t) = 0$ for $t \le 0$, $n \in I_+$. Show that the renewal function $M(\cdot)$ given by (9-74) can be expressed as $M(t) = \sum_{n=1}^{\infty} M_n(t)$.

Let $Q_n(t) = P\{S_1 > 0, \dots, S_{n-1} > 0, S_n \le t\}$ for $t \le 0$ and $Q_n(t) = 0$ for $t > 0$, $n \in I_+$. What is the interpretation of $Q_n(t)$? Show that

$$Q_{n+1}(t) = \int_0^{\infty} M_n(y) \, dH(t - y) \qquad\qquad t \le 0, \quad n \in I_+ \qquad (9\text{-}95a)$$

$$M_{n+1}(t) = \int_0^{\infty} M_n(y) \, d[H(t - y) - H(-y)] \qquad t > 0, \quad n \in I_+ \qquad (9\text{-}95b)$$

and

$$M_n(t) + Q_n(t) = \int_0^{\infty} M_{n-1}(t - y) \, dG(y) \qquad -\infty < t < \infty, \quad n \in I_+ \qquad (9\text{-}96)$$

9-31 (Continuation) Let $Q(t) = \sum_{n=1}^{\infty} Q_n(t)$. Prove

$$Q(t) = \int_0^{\infty} M(y) \, dH(t - y) + Q_1(t) \qquad\qquad t \le 0 \qquad (9\text{-}97a)$$

and

$$M(t) = \int_0^{\infty} M(y) \, dH(t - y) - Q(0) + M_1(t) + Q_1(0) \qquad t > 0 \qquad (9\text{-}97b)$$

The asymmetry in (9-97a) and (9-97b) permits $Q(\cdot)$ to be obtained when $H(t) = Ce^{\alpha t}$ for $t \le 0$. When $H(\cdot)$ has a density, one need not distinguish between strict and weak ladder epochs; by multiplying each Y_k by -1, the roles of $Q(\cdot)$ and $F(\cdot)$ are reversed. The procedure outlined below yields $F(\cdot)$ whenever one tail of $H(\cdot)$ is exponential.

9-32 Suppose $H(t)$ has a derivative for all t, and for $t < 0$, $H(t) = Ce^{\alpha t}$, for some constant C. Show

$$q(t) \triangleq \frac{d}{dt} Q(t) = C'e^{\alpha t} \qquad t \le 0 \qquad (9\text{-}98)$$

and exhibit C'.

Let $Y_n^* = -Y_n$, $S_n^* = -S_n$, $n \in I_+$, $S_0^* = S_0 = 0$; and let $a *$ denote a quantity for a starred process. Argue that $Q(-t) = F^*(-t)$, $t \ge 0$. Show $H^*(t) = 1 - \psi e^{\alpha t}$, $t \ge 0$, where $\psi = 1 - H^*(0)$.
Assume $E(Y_1^*) \ge 0$. Deduce, with the aid of (9-98), that

$$F^*(t) = 1 - e^{-\alpha t} \qquad \text{and} \qquad M^*(t) = \alpha t \qquad t \ge 0 \qquad (9\text{-}99)$$

Use (9-97a) to derive

$$Q^*(t) = H^*(t) + \alpha \int_{-\infty}^{t} H(y) \, dy \qquad t < 0 \qquad (9\text{-}100)$$

and

$$Q^*(0) = 1 - \alpha E(Y_1^*) \qquad (9\text{-}101)$$

9-33 (Continuation)   Assume $E(Y_1^*) < 0$. Show

$$\frac{d}{dt} F^*(t) = (\alpha - x)e^{-\alpha t} \qquad t \geq 0 \tag{9-102}$$

for some number x, $0 < x < \alpha$. Prove

$$\frac{d}{dt} M^*(t) = (\alpha - x)e^{-xt} \qquad t > 0 \tag{9-103}$$

Explain why $Q^*(0) = 1$ and that this implies that x is the unique positive solution of

$$\int_{-\infty}^{\infty} e^{xy} \, dH(y) = 1 \tag{9-104}$$

Show

$$F(\infty) = 1 - \frac{x}{\alpha} \tag{9-105}$$

9-34   The $GI/M/1$ Queue   Let $A(t) = P\{U_0 \leq t\}$, $1/\lambda = E(U_0)$,   and $P\{V_0 \leq t\} = 1 - e^{-\mu t}$,   with $\rho \triangleq \lambda/\mu < 1$. Show†

$$H(t) = \begin{cases} 1 - \tilde{A}(\mu)e^{-\mu t} & t \geq 0 \\[2mm] A^c(t) - e^{-\mu t} \displaystyle\int_{-t}^{\infty} e^{-\mu z} \, dA(z) & t \leq 0 \end{cases}$$

Use (9-94) and the results in Exercise 9-33 to obtain

$$D(t) = 1 - \frac{\mu - x}{x} e^{-xt} \qquad t \geq 0$$

9-35   The $M/G/1$ Queue   Let $P\{U_0 \leq t\} = 1 - e^{-\lambda t}$, $G(t) = P\{V_0 \leq t\}$, and $v = E(V_0)$, with $\rho \triangleq \lambda v < 1$. Show

$$H^*(t) \triangleq P\{Y_1^* \leq t\} = \begin{cases} 1 - \tilde{G}(\lambda)e^{-\lambda t} & t \geq 0 \\[2mm] \lambda e^{-\lambda t} \displaystyle\int_{-t}^{\cdot} e^{-\lambda z} G^c(z) \, dz & t \leq 0 \end{cases}$$

Observe that $H^*(\cdot)$ satisfies the hypothesis in Exercise 9-32. Use the results of Exercise 9-32 to obtain

$$Q^*(t) = \lambda \int_{-t}^{\infty} G^c(z) \, dz \qquad t < 0$$

$$Q^*(0) = \rho = 1 - F(\infty)$$

and

$$F(t) = \lambda \int_{t}^{\infty} G^c(z) \, dz \qquad t \geq 0$$

Use these results, (9-74), and (9-94) to obtain this version of the Pollaczek-Khintchine formula (see Exercise 7-60):

$$D(t) = (1 - \rho) + \sum_{n=1}^{\infty} \rho^n B_n(t) \qquad t \geq 0$$

where $B(t) = \int_t^{\infty} G^c(z) \, dz/v$ and $B_n(t)$ is the n-fold convolution of $B(\cdot)$ with itself.

† Recall the notation $\tilde{A}(s) = \int_0^{\infty} e^{-st} \, dA(t)$ and $A^c(t) = 1 - A(t)$, where $A(\cdot)$ is the d.f. of a non-negative random variable.

9-36 For the model in Example 9-7, explain how to compute the probability that no loans are required during the first N days, where N is given.

9-4* DIFFUSION PROCESSES

A *diffusion process* (on the real line) is a Markov process with a continuous time parameter, a continuous state space contained in $(-\infty, \infty)$, and continuous sample paths. Many particle-flow phenomena occurring in physics and engineering are naturally modeled by diffusion processes. In operations research, diffusion processes are primarily used as approximations to discrete processes. These approximations sometimes can be solved explicitly when the motivating discrete model is intractable. For example, diffusion process approximations for the $GI/G/1$ and $GI/G/c$ queues are given in Section 13-2.

Diffusion processes are more complicated conceptually and technically than the other Markov processes treated in this book. The discussion in this section is less rigorous and complete than in other sections of this chapter.

The following terminology is used. Let $\{X(t); t \geq 0\}$ be a diffusion process. Define

$$F(t, x; t_0, x_0) = P\{X(t_0 + t) \leq x \mid X(t_0) = x_0\} \qquad t, t_0 \geq 0$$

which is the *distribution function* of the process. Assume that

$$f(t, x; t_0, x_0) \triangleq \frac{\partial}{\partial x} F(t, x; t_0, x_0)$$

exists; it is called the *transition density* of the process. When $F(\,\cdot\,, \,\cdot\,; t_0, \,\cdot\,)$ does not depend on t_0, we suppress t_0 from the notation.

By analogy with the standard transition matrix of a CTMC, we *assume*

$$\lim_{t \downarrow 0} F(t, x; t_0, x_0) = \begin{cases} 0 & \text{if } x < x_0 \\ 1 & \text{if } x \geq x_0 \end{cases}$$

Brownian Motion as a Limit of Simple Random Walks

Let $\{S_n; n \in I\}$ be a simple random walk, as described in Section 7-8. Let $p = P\{S_{n+1} = S_n + 1\}$ and $q = 1 - p = P\{S_{n+1} = S_n - 1\}$. For any integers i and j, define

$$P_{ij}(n) = P\{S_n = j \mid S_0 = i\} \qquad n \in I \tag{9-106}$$

By conditioning on the event at epoch n,

$$P_{ij}(n) = pP_{i, j-1}(n - 1) + qP_{i, j+1}(n - 1) \qquad n \in I_+ \tag{9-107}$$

In the simple random walk, each change of state has magnitude 1, and the state changes at epochs 1, 2, ... Suppose each change of state has magnitude $\Delta x > 0$ and the state changes at epochs $\Delta t, 2 \Delta t, \ldots$. This changes the scale of the process but does not affect the stochastic structure. As $\Delta x \downarrow 0$ and $\Delta t \downarrow 0$, we expect that the resulting process has continuous sample paths and is a diffusion; that is indeed true.

For $\Delta x > 0$ and $\Delta t > 0$ and fixed, let t be some multiple of Δt and let x and x_0 be multiples of Δx. Let

$$f(t, x; x_0) = P\{S_t = x \mid S_0 = x_0\} \tag{9-108}$$

From (9-107),

$$f(t, x; x_0) = p f(t - \Delta t, x - \Delta x; x_0) + q f(t - \Delta t, x + \Delta x; x_0) \tag{9-109}$$

Let us investigate (9-109) as $\Delta x \downarrow 0$ and $\Delta t \downarrow 0$.

We cannot send Δx and Δt to zero arbitrarily; the rates are controlled by the following considerations. Suppose $S_0 = 0$. For each $\Delta t > 0$, if t is a multiple of Δt, in time t there are $n(t) = t/\Delta t$ transitions; hence

$$E(S_{n(t)}) = \frac{(p - q)\Delta x \; t}{\Delta t} \tag{9-110a}$$

and

$$\text{Var}(S_{n(t)}) = \frac{(\Delta x)^2[1 - (p - q)^2]t}{\Delta t} = \frac{4pqt(\Delta x)^2}{\Delta t} \tag{9-110b}$$

In order for $E(S_{n(t)})$ to be finite, p and q must vary with Δt so that $(p - q)\Delta x/\Delta t$ approaches some limit, c say. For $\text{Var}(S_{n(t)})$ to be positive and finite, $(\Delta x)^2/\Delta t$ approaches some limit, D^2 say. To effect this, set

$$\Delta x = D\sqrt{\Delta t} \tag{9-111}$$

and

$$p = \frac{1 + \Delta x(c/D^2)}{2} \qquad q = \frac{1 - \Delta x(c/D^2)}{2} \tag{9-112}$$

Substituting (9-111) and (9-112) into (9-110a) and (9-110b) yields

$$E(S_{n(t)}) = ct \tag{9-113a}$$

and as $\Delta x \downarrow 0$,

$$\text{Var}(S_{n(t)}) \to D^2 t \tag{9-113b}$$

where c and D^2 can be chosen arbitrarily.

Treat x, x_0, and t as continuous variables, and assume that $f(\cdot, \cdot; \cdot)$ is sufficiently smooth to have a Taylor series expansion up to $o(\Delta t)$ terms. Write (9-109) as

$$f = p\left[f - \Delta t \dot{f} - \Delta x f' + \frac{(\Delta x)^2 f''}{2}\right]$$
$$+ q\left[f - \Delta t \dot{f} + \Delta x f' + \frac{(\Delta x)^2 f''}{2}\right] + o(\Delta t) \tag{9-114}$$

where the suppressed arguments are $(t, x; x_0)$, a dot stands for $\partial/\partial t$, a prime stands for $\partial/\partial x$, and a double prime stands for $\partial^2/\partial x^2$. Divide both sides of (9-114) by Δt, rearrange, and use (9-111) and (9-112) to obtain

$$\dot{f} = cf' + \frac{D^2}{2} f'' + \frac{o(\Delta t)}{\Delta t}$$

Letting $\Delta t \downarrow 0$ [hence $\Delta x \downarrow 0$ by (9-111)] yields

$$\frac{\partial}{\partial t} f(t, x; x_0) = -c \frac{\partial}{\partial x} f(t, x; x_0) + \frac{D^2}{2} \frac{\partial^2}{\partial x^2} f(t, x; x_0) \qquad t \geq 0 \quad (9\text{-}115)$$

The partial differential equation (9-115) occurs in the physical theory of heat and is known to physicists as the Fokker-Planck equation. In the stochastic process literature, it is called the *forward Kolmogorov equation of Brownian motion*.

The solution of (9-115) can be obtained by probabilistic reasoning. For Δx, $\Delta t > 0$, there are $t/\Delta t$ i.i.d. displacements by time t. The mean and variance of the sum of these displacements are given by (9-113). By the central-limit theorem, as $\Delta t \downarrow 0$, the distribution of the total displacement approaches the normal d.f., hence

$$f(t, x; x_0) = \left(D\sqrt{2\pi t}\right)^{-1} e^{-(x - x_0 - ct)^2/(2D^2 t)} \qquad t > 0 \qquad (9\text{-}116)$$

should solve (9-115). The truth of this assertion can be established by direct substitution.

Discussion

We have just shown that the equation satisfied by the transition function of a random walk which jumps by the amount Δx every Δt units of time approaches (9-115) when Δt and Δx go to zero and are related by (9-111). The central-limit theorem asserts that the solution of (9-109) approaches (9-116), and this can be checked by direct calculation. That is all that is proved; what does it suggest?

Suppose $\Delta t = 1/k$, $k \in I_+$, and let $N(t)$ be the largest integer less than or equal to $t/\Delta t$. Define $S^{(k)}(t)$ by $S^{(k)}(t) = S_{N(t)}^{(k)}$, where $\{S_n^{(k)}; n \in I\}$ is the random walk having parameters p and q given by (9-112) with $\Delta t = 1/k$. Figure 9-4 illustrates $S^{(4)}(t)$.

It seems reasonable that as $k \to \infty$,

1. $\{S^k(t); t \geq 0\}$ approaches some process, $\{B(t); t \geq 0\}$ say, with state space $(-\infty, \infty)$.
2. $\{B(t); t \geq 0\}$ is Markov.
3. $\{B(t); t \geq 0\}$ has continuous sample paths.
4. The transition density of $\{B(t); t \geq 0\}$ is given by (9-116).

All these statements are true; their proofs are beyond the scope of this book† and may be found in Billingsley (1968) and Breiman (1968). Thus, $\{B(t); t \geq 0\}$ is a diffusion process. It is called *Brownian motion* (abbreviated BM) or the *Wiener process*.

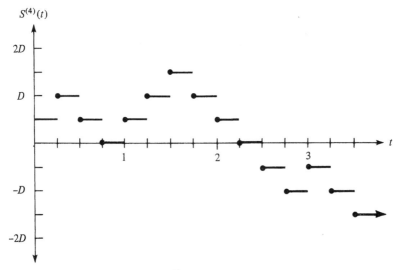

Figure 9-4 A partial sample path of $S^{(4)}(t)$ with $x_0 = 1$.

From statements 1, 2, and 4,

$$F(t, x; x_0) \triangleq P\{B(t + s) \leq x \mid B(s) = x_0\}$$

$$= \int_{-\infty}^{x} f(t, y; x_0)dy = \Phi\left(\frac{x - x_0 - ct}{D\sqrt{t}}\right) \qquad (9\text{-}117)$$

where $\Phi(\cdot)$ is the standard normal d.f. It is left to you to show $F(t, x; x_0)$ also satisfies (9-115). Thus, BM is specified by two parameters, c and D^2, where

$$c = \frac{E[B(t + s) - B(s)]}{t} \qquad t > 0, \quad s \geq 0 \qquad (9\text{-}118a)$$

and is called the *drift* and

$$D^2 = \frac{\text{Var}[B(t + s) - B(s)]}{t} \qquad t > 0, \quad s \geq 0 \qquad (9\text{-}118b)$$

and is called the *diffusion coefficient*.

Phenomena that are Markov, and where displacements occur rapidly and have small magnitude, are candidates for a BM process model. The BM is determined from observations or hypotheses via (9-118). The ease of doing the latter is what makes a BM convenient to use; the Markov property and continuous (but, in fact, nondifferentiable) sample paths may require extensive justification.

---

† Notice that the first statement assumes a notion of convergence of stochastic processes. To develop this notion, topology is required to describe when two processes have sample paths that are close, and measure theory is needed to discuss the probabilities assigned to sets of sample paths. A discussion at a lower level of abstraction does not seem possible.

Boundary Conditions for Brownian Motion

The ideas of reflecting and absorbing barriers for random walks (in Section 7-8) carry over to BM processes. The challenge is to represent these barriers as boundary conditions on (9-115).

Consider a BM with drift c and diffusion coefficient D^2. When boundaries are imposed, (9-118) holds when s is not close to the boundary and t is small, so these parameters maintain their interpretation. Suppose the initial position of the BM is $x_0 > 0$ and the reflecting barrier is at zero. Convince yourself that placing the barrier at zero involves no loss of generality.

From (7-102b), the boundary equation of the random walk with jumps of size Δx every Δt is

$$f(t, 0; x_0) = q[f(t - \Delta t, \Delta x; x_0) + f(t - \Delta t, 0; x_0)] \tag{9-119}$$

Expanding f in a Taylor's series around the point $(t, 0; x_0)$ transforms (9-119) to

$$\dot{f} = \left(1 - \frac{1}{2q}\right) \frac{1}{\Delta t} f + \frac{1}{2} \frac{\Delta x}{\Delta t} f' + \frac{1}{4} \frac{(\Delta x)^2}{\Delta t} f'' + \frac{o(\Delta t)}{\Delta t} \tag{9-120}$$

Using (9-111) and (9-112), we observe that as $\Delta t \downarrow 0$, the right-side of (9-120) becomes infinite unless

$$f = -\frac{f'}{2} \Delta x \frac{2q}{2q - 1} = \frac{f'}{2} \frac{1 - c \, \Delta x/D^2}{c/D^2} \rightarrow \frac{D^2 f'}{2c} \qquad \text{as} \qquad \Delta t \downarrow 0$$

Thus, the boundary condition for a reflecting barrier at zero is

$$cf(t, 0; x_0) = \frac{D^2}{2} \frac{\partial}{\partial x} f(t, x; x_0)\big|_{x=0} \qquad x_0, t > 0 \tag{9-121}$$

For the distribution function $F(\,\cdot\,,\,\cdot\,;\,\cdot\,)$ defined in (9-117), the boundary condition is derived from (9-121); it is

$$F(t, 0; x_0) = 0 \qquad x_0, t > 0 \tag{9-122}$$

It is not difficult† to solve (9-115) subject to (9-122) for $F(t, x; x_0)$; the solution is

$$F(t, x; x_0) = \Phi\left(\frac{x - x_0 - ct}{D\sqrt{t}}\right) - e^{2cx/D^2} \Phi\left(\frac{-x - x_0 - ct}{D\sqrt{t}}\right) \tag{9-123}$$

Exercise 9-39 asks you to establish that the boundary condition for an absorbing barrier at $b > x_0$ is

$$f(t, b; x_0) = 0 \qquad t > 0 \tag{9-124}$$

---

† A nice derivation is given in G. F. Newell, *Applications of Queueing Theory*, Springer-Verlag, New York (1974), chap. 6.

The solution of (9-115) subject to (9-124) is†

$$f(t, x; x_0) = \phi\left(\frac{x - ct}{D\sqrt{t}}\right) - e^{2(b - x_0)c/D^2} \phi\left(\frac{x - 2(b - x_0) - ct}{D\sqrt{t}}\right) \qquad x_0, x \le b$$

where $\phi(\cdot)$ is the standard normal density function.

Kolmogorov Equations for General Diffusion Processes

In the BM process, (9-118) asserts that the mean and variance of $B(t + s) - B(s)$ depend on only t. Now we introduce diffusion processes where this need not hold.

Let $\{X(t); t \ge 0\}$ be a diffusion process. Assume that for any $\delta > 0$,

$$P\{|X(t_0 + \Delta t) - X(t_0)| > \delta : X(t_0) = x_0\} = o(\Delta t) \tag{9-125}$$

Equation (9-125) states that the probability of a change of state larger than δ during a time Δt is small compared to Δt. Even with (9-125), it is possible that $E[X(t_0 + \Delta t) - X(t_0) | X(t_0) = x_0] = \infty$. To avoid such unpleasant situations, define these truncated moments:

$$\alpha(t, x_0) = E[X(t_0 + \Delta t) - X(t_0): X(t_0) = x_0 | X(t_0 + \Delta t) - X(t_0)| < \delta] + o(\Delta t) \tag{9-126}$$

and

$$\beta(t, x_0) = \text{Var}[X(t_0 + \Delta t) - X(t_0): X(t_0) = x_0 | X(t_0 + \Delta t) - X(t_0)| < \delta] + o(\Delta t) \tag{9-127}$$

with $\beta(t, x_0) \ge 0$. These are called the *infinitesimal mean* and *variance* of the process, respectively.

The generalizations of the backward Kolmogorov equation in Exercise 9-37 and the forward Kolmogorov equation (9-115) are given in this theorem.

Theorem 9-11 Let $\{X(t); t \ge 0\}$ be a diffusion process with transition density $f(\cdot, \cdot; \cdot, \cdot)$. If (9-125), (9-126), and (9-127) are valid and the indicated partial derivatives exist, then $f(\cdot, \cdot; \cdot, \cdot)$ satisfies the backward equation

$$-\frac{\partial f(t, x; t_0, x_0)}{\partial t_0} = \alpha(t_0, x_0) \frac{\partial f(t, x; t_0, x_0)}{\partial x_0} + \frac{\beta(t_0, x_0)}{2} \frac{\partial^2 f(t, x; t_0, x_0)}{\partial x_0^2} \tag{9-128}$$

and the forward equation

$$\frac{\partial f(t, x; t_0, x_0)}{\partial t} = -\frac{\partial[\alpha(t, x)f(t, x; t_0, x_0)]}{\partial x} + \frac{1}{2} \frac{\partial^2[\beta(t, x)f(t, x; t_0, x_0)]}{\partial x^2} \tag{9-129}$$

† See, e.g., Cox and Miller (1965, p. 221).

Furthermore, the d.f. of the process also satisfies (9-128).

PROOF See, e.g., Bharucha-Reid (1960, chap. 3). □

The BM process is the special case where $\alpha(t, x) = c$ and $\beta(t, x) = D^2$. Another important special case is when $\alpha(t, x) = -\alpha x$ for some given $\alpha > 0$ and $\beta(t, x) = D^2$; this is called the *Ornstein-Uhlenbeck process*. In this process, the infinitesimal mean is proportional to the state and of opposite sign. A variant of the Ornstein-Uhlenbeck process is used as an approximation to the $GI/G/c$ queue in Section 13-2.

First-Passage Times

Let $\{X(t); t \geq 0\}$ be a diffusion process. Since it is Markov, the first-passage time between states x and y can be defined generically by

$$T_{xy} = \inf\{t > 0: X(t) = y \mid X(0) = x\}$$

Let

$$G_{xy}(t) \triangleq P\{T_{xy} \leq t\} \quad \text{and} \quad g_{xy}(t) = \frac{d}{dt} G_{xy}(t)$$

Here are two ways to get the first-passage-time d.f. from the d.f. of the process. If an absorbing barrier is placed at $y > x$. then

$$P\{T_{xy} > t\} = P\{X(t) < y \mid X(0) = x\} \qquad t \geq 0, \, y > x$$

Hence,

$$G_{yx}(t) = 1 - \int_{-\infty}^{y} f(t, u; x) \, du \qquad t \geq 0, \, y > x \tag{9-130}$$

Example 9-8: Brownian motion without barriers When $\{X(t); t \geq 0\}$ is a **BM**, substituting (9-124) into (9-130) yields

$$G_{xy}(t) = 1 - \Phi\left(\frac{y - ct}{D\sqrt{t}}\right) + e^{2(y-x)c/D^2} \Phi\left(\frac{2x - y - ct}{D\sqrt{t}}\right) \qquad t \geq 0, \, y > x$$

$$\tag{9-131}$$

From (9-131), we obtain, for $y > x$,

$$P\{T_{xy} < \infty\} = \begin{cases} 1 & \text{if } c \geq 0 \\ e^{2(y-x)c/D^2} & \text{if } c < 0 \end{cases} \tag{9-132}$$

□

The second way to obtain the first-passage-time d.f. exploits the continuity of the sample paths. Assume (without loss of generality) that $x_0 < x$, and choose y

such that $x_0 < y < x$. Since y must be reached before x,

$$f(t, x; x_0) = \int_0^t g_{x_0 y}(u) f(t - u, x; y) \, du \qquad x_0 < y < x \qquad (9\text{-}133)$$

It is convenient to obtain the Laplace transform of the first-passage-time density (which coincides with the LST of the associated d.f.) from (9-133). Let

$$\tilde{f}(s, x; x_0) = \int_0^\infty e^{-st} f(t, x; x_0) \, dt \qquad \text{and} \qquad \tilde{g}_{xy}(s) = \int_0^\infty e^{-st} g_{xy}(t) \, dt$$

Taking Laplace transforms on both sides of (9-133) and rearranging terms yield

$$\tilde{g}_{x_0 y}(s) = \frac{\tilde{f}(s, x; x_0)}{\tilde{f}(s, x; y)} \qquad x_0 < y < x \qquad (9\text{-}134)$$

In (9-134), x is a parameter of no significance, and the right-side is the ratio of a function of x_0 and a function of y.

Example 9-9: Brownian motion with a reflecting barrier Let $\{X(t); t \geq 0\}$ be a BM with $X(0) = x_0 \geq 0$ and a reflecting barrier at zero. Thus, the transition density satisfies (9-115) with boundary condition (9-121). The associated d.f. satisfies

$$F(0, x; x_0) = \begin{cases} 0 & \text{if } x < x_0 \\ 1 & \text{if } x \geq x_0 \end{cases} \qquad (9\text{-}135)$$

When $x_0 > y \geq 0$, the reflecting barrier at zero has no effect on $T_{x_0 y}$, and $y < 0$ certainly cannot be reached from $x_0 > 0$. Hence, the first-passage-time d.f. can be obtained from (9-131) when $x_0 > y$.

When $0 \leq x_0 < y$, (9-134) provides a convenient way to obtain the Laplace transform of the first-passage-time d.f. The first step in employing (9-134) is to obtain $\tilde{f}(s, x; x_0)$. Taking Laplace transforms on both sides of (9-115) yields

$$s\tilde{f}(s, x; x_0) = -c \frac{d}{dx} \tilde{f}(s, x; x_0) + \frac{D^2}{2} \frac{d^2}{dx^2} \tilde{f}(s, x; x_0) \qquad (9\text{-}136)$$

The solution of (9-136) subject to (9-121) and (9-135), for $x > x_0$, is (see Exercise 9-45)

$$\tilde{f}(s, x; x_0) = \frac{2e^{-\psi x_0}}{\psi D^2} (\xi_1 e^{\xi_1 x} - \xi_2 e^{\xi_2 x}) - \frac{2}{D^2} (e^{\xi_1(x - x_0)} - e^{\xi_2(x - x_0)}) \qquad (9\text{-}137)$$

where $\xi_1, \xi_2 = c(1 \pm \sqrt{1 + 2sD^2/c^2})/D^2$ and $\psi = \xi_1$ if $c \geq 0$ and $\psi = \xi_2$ if $c < 0$.

Set $r = \sqrt{1 + 2sD^2/c^2}$ and $\alpha = -cr/D^2$; substituting (9-137) into (9-134) yields (for any c)

$$\tilde{g}_{x_0 y}(s) = e^{c(y - x_0)/D^2} \frac{(r - 1)e^{\alpha x_0} + (r + 1)e^{-\alpha x_0}}{(r - 1)e^{\alpha y} + (r + 1)e^{-\alpha y}} \qquad 0 \leq x_0 < y \qquad (9\text{-}138)$$

Notice that the parameter x does not appear in (9-138), and (9-138) is the ratio of a function of x_0 and the same function of y. It is possible to simplify (9-138) by using the hyperbolic trigonometric functions

$$\sinh z \triangleq \frac{e^z - e^{-z}}{2} \qquad \text{and} \qquad \cosh z \triangleq \frac{e^z + e^{-z}}{2}$$

where z may be complex. Then

$$\tilde{g}_{x_0 y} = e^{c(y - x_0)/D^2} \; \frac{r \cosh \alpha x_0 - \sinh \alpha x_0}{r \cosh \alpha y - \sinh \alpha y} \qquad 0 \le x_0 < y \qquad (9\text{-}139)$$

A primary use of (9-139) is to obtain the moments of $T_{x_0 y}$ by differentiation with respect to s. The first moment is given by

$$E(T_{x_0 y}) = \frac{D^2}{2c^2} e^{-2cy/D^2} \left[1 - e^{-2c(y - x_0)/D^2} - \frac{2c(y - x_0)}{D^2} e^{2cy/D^2} \right] \qquad (9\text{-}140)$$

when $x_0 < y$. Obtaining higher moments is tedious.

Equations (9-139) and (9-140) are used in Exercise 13-9. □

EXERCISES

9-37 Derive the backward Kolmogorov equation of Brownian motion,

$$\frac{\partial}{\partial t} f(t, x; x_0) = c \frac{\partial}{\partial x_0} f(t, x; x_0) + \frac{D^2}{2} \frac{\partial^2}{\partial x_0^2} f(t, x; x_0) \qquad (9\text{-}141)$$

from random-walk equations. Verify that (9-116) solves (9-141).

9-38 Show that by setting $\tau = (c^2/D^2)t$ and $\xi = -(c/D^2)x$, (9-115) can be put into the dimensionless form

$$\frac{\partial}{\partial \tau} f(\tau, \xi; \xi_0) = -\frac{\partial}{\partial \xi} f(\tau, \xi; \xi_0) + \frac{1}{2} \frac{\partial^2}{\partial \xi^2} f(\tau, \xi; \xi_0) \qquad (9\text{-}142)$$

Show that the "$\partial/\partial \xi$ term" can be eliminated by setting $g(\tau, \xi; \xi_0) = e^{\tau/2 - \xi} f(\tau, \xi; \xi_0)$.

9-39 Derive (9-124).

9-40 For a BM process with a reflecting barrier at zero with $x_0 \ge 0$, show that

$$\lim_{t \to \infty} F(t, x; x_0) = 1 - e^{2cx/D^2} \qquad x \ge 0, c < 0$$

and that the limit is zero otherwise.

9-41 Let $\{B(t); t \ge 0\}$ be a BM process and $\{B^*(t); t \ge 0\}$ be the same process except that a reflecting barrier at zero is imposed; assume $B(0) = B^*(0) \ge 0$. Show that $B^*(t)$ can be represented as $B^*(t) = B(t) - \sup_{s \le t}\{B(s)\}$.

9-42 Let $\{X(t); t \ge 0\}$ be a birth-and-death process with constant birth rate λ and constant death rate μ. Let $P_{ij}(t) = P\{X(t) = j \mid X(t) = i\}$; it is given by (4-126), which is cumbersome to work with. The similarities between birth-and-death processes and random walks, on one hand, and between random walks and BM processes, on the other, suggest that a suitably chosen BM might provide a good approximation to a given birth-and-death process. Replace the discrete variables i and j by the continuous variables x_0 and x, respectively, and $P_{ij}(t)$ by $f(t, x; x_0)$. Show that $f(\cdot, \cdot; \cdot)$ satisfies the equations of a BM with a reflecting barrier at zero, having $c = \lambda - \mu$ and $D^2 = \lambda + \mu$. How do you account for these values for c and D^2?

9-43 (Continuation) How would you use the BM in Exercise 9-42 to approximate (a) $P\{X(t) = j \mid X(0) = i\}$ and (b) $E[X(t) \mid X(0) = i]$? What properties does your answer have in the steady state?

9-44 Compare (9-132) to the answer to Exercise 7-96 when $c = p - q$ and $D^2 = 4pq$.

9-45 In Example 9-9, let $h(x) = \tilde{f}(s, x; x_0)$. Establish

$$sh(x) - \tilde{f}(0, x; x_0) = -ch'(x) + \frac{D^2}{2} h''(x)$$

Let $\tilde{h}(\xi) = \int_0^\infty e^{-\xi x} h(x) dx$; show

$$\tilde{h}(\xi) = \frac{\xi h(0) - (2/D^2) e^{-\xi x_0}}{(\xi - \xi_1)(\xi - \xi_2)}$$

where ξ_1 and ξ_2 are defined in the text. Invert this Laplace transform to obtain

$$(\xi_1 - \xi_2) h(x) = h(0)(\xi_1 e^{\xi_1 x} - \xi_2 e^{\xi_2 x}) - \begin{cases} 0 & \text{if } x < x_0 \\ \dfrac{2(e^{\xi_1(x - x_0)} - e^{\xi_2(x - x_0)})}{D^2} & \text{if } x \geq x_0 \end{cases}$$

Explain why $\int_0^\infty h(x)\, dx = 1/s$; use it to obtain $h(0) = [2/(\psi D^2)] e^{-\psi x_0}$.

9-46 Let $\{X(t); t \geq 0\}$ be a BM with a reflecting barrier at $b > 0$. For $0 < x \leq b$, use (9-138) to show that the LST of the d.f. of the first-passage time from x to 0 is given by

$$\tilde{g}_x(s; b) = e^{-cx/D^2} \frac{r \cosh \alpha(b - x) - \sinh \alpha(b - x)}{r \cosh \alpha b - \sinh \alpha b}$$

Hint: No calculations are required. Use this result to show that the LST of the d.f. given by (9-141) is

$$\tilde{g}_{xy}(s) = \begin{cases} e^{-\xi_2(y - x)} & \text{if } c \geq 0 \\ e^{-\xi_1(y - x)} & \text{if } c < 0 \end{cases}$$

Use this formula to derive $E(T_{xy}) = (y - x)/c$, when $y > x$ and $c > 0$.

BIBLIOGRAPHIC GUIDE

P. Lévy and W. L. Smith independently presented papers introducing semi-Markov processes at the International Congress of Mathematicians held at Amsterdam in 1954. Both authors coined this name for the new class of processes. Also in 1954, L. Takács introduced a process which is essentially an SMP. The classification of SMPs and the term *Markov renewal process* first appeared in Pyke (1961a). Our presentation is based on Pyke (1961a, 1961b) and Çinlar (1975, Chap. 10). Çinlar (1969) offers a comprehensive survey, and the generalized Markov renewal equations first appeared there.

The treatment of random walks in Section 9-3 follows Feller (1971, Chap. 12). A complete discussion of random walks is given in Spitzer (1964). First-passage times are explored in Kemperman (1961).

The derivation of the forward Kolmogorov equation for Brownian motion by taking limits of random walks was done first by A. Einstein in 1905; see Einstein (1956). This approach is exploited in Cox and Miller (1965, Chap. 5), which contains many specific results for one-dimensional BM processes. Other treatments of diffusion processes at this level of mathematical detail include Bharucha-Reid (1960, Chap. 3) and Karlin and Taylor (1975, Chap. 7). The rigorous justification of the passage to the limit in Einstein's approach is based on weak convergence of probability measures, particularly Donsker's theorem. This is more advanced material and may be found in Billingsley (1968) and Breiman (1968). A measure-theoretic treatment of BM where the measure theory is rather unobtrusive appears in Wong (1971). Diffusion processes are used in inventory and cash-balance models in Harrison and Taylor (1978) and Constantinides and Richard (1978). Many financial models employ a diffusion process to describe stock price movements. An early paper by Osborne (1959) proposed this type of model.

References

Bharucha-Reid, A. T.: *Elements of the Theory of Markov Processes and Their Applications*, McGraw-Hill, New York (1960).

Billingsley, Patrick: *Convergence of Probability Measures*, Wiley, New York (1968).

Breiman, Leo: *Probability*, Addison-Wesley, Reading, Mass. (1968).

Çinlar, E.: "Markov Renewal Theory," *Adv. Appl. Probab.* **1**: 123–187 (1969).

———: *Introduction to Stochastic Processes*, Prentice-Hall, Englewood Cliffs, N.J. (1975).

Constantinides, George M., and Scott F. Richard: "Existence of Optimal Simple Policies for Discounted-Cost Inventory and Cash Management in Continuous Time," *Oper. Res.* **26**: 620–636 (1978).

Cox, D. R., and H. D. Miller: *The Theory of Stochastic Processes*, Methuen, London (1965).

Einstein, A.: *Investigations on the Theory of Brownian Movement*, Dover, New York (1956).

Feller, William: *An Introduction to Probability Theory and Its Applications*, vol. 2, Wiley, New York (1971).

———: "On Semi-Markov Processes," *Proc. Nat. Acad. Sci.* **51**: 653–659 (1964).

Harrison, J. Michael, and Allison J. Taylor: "Optimal Control of a Brownian Motion Storage System," *Stochastic Process. Appl.* **6**: 179–194 (1978).

Karlin, Samuel, and Howard M. Taylor: *A First Course in Stochastic Processes*, Academic, New York (1975).

Kemperman, J. H. B.: *The Passage Problem for a Stationary Markov Chain*, University of Chicago Press, Chicago (1961).

Lévy, Paul: "Systèmes Semi-Markoviens à au plus une infinité dénombrable d'états possibles," *Proc. Int. Congr. Math.* **2**: 294, **3**: 416–426 (1954).

Lindley, D. V.: "The Theory of Queues with a Single Server," *Proc. Cambridge Philos. Soc.* **48**: 277–289 (1952).

Osborne, M. F. M.: "Brownian Motion in the Stock Market," *Oper. Res.* **7**: 145–173 (1959). See also pp. 806–811 for historical comments.

Pyke, Ronald: "Markov Renewal Processes: Definitions and Preliminary Properties," *Ann. Math. Stat.* **33**: 1231–1242 (1961a).

———: "Markov Renewal Processes with Finitely Many States," *Ann. Math. Stat.* **33**: 1243–1259 (1961b).

Smith, W. L.: "Regenerative Stochastic Processes," *Proc. Roy. Soc. London* **A232**: 6–31 (1955). [The abstract of this paper appears in *Proc. Int. Congr. Math.* **2**: 304–305 (1954)].

Spitzer, Frank: *Principles of Random Walk*, Van Nostrand, Princeton, N.J. (1964).

Takács, Lajos: "Some Investigations concerning Recurrent Stochastic Processes of a Certain Type," *Mag. Tud. Akad. Kut. Intez. Kozl.* **3**: 115–128 (1954). [An English summary appears in *Math Rev.* **17**: 866 (1956).]

Wong, Eugene: *Stochastic Processes in Information and Dynamical Systems*, McGraw-Hill, New York (1971).

TEN

STATIONARY PROCESSES AND ERGODIC THEORY

A stationary stochastic process has the property that its fidi's are unaffected by a shift in the time parameter. That is, the origin of the time axis plays no essential role. This mode of behavior often is associated with the "steady state" or "statistical equilibrium"; these notions are made precise by a limit theorem. Another fundamental result of the theory is the ergodic theorem, which states that the sample mean of a stationary process converges, as the sample size approaches infinity, to the mean of an arbitrary r.v. of the process.

Sections 10-1 and 10-2 concern general stationary stochastic processes. Many operations research models are regenerative processes, and stationary regenerative processes are introduced in Section 10-3. Further applications of stationary regenerative processes are found in Chapters 11, 12, and 13.

10-1* DEFINITION AND EXAMPLES

Definition 3-3, of a stationary process, is repeated here.

Definition 10-1: Stationary stochastic process The stochastic process $\{X(t); t \in \mathcal{T}\}$ is *stationary* if for any $n \in I_+$, when $t_1 < t_2 < \cdots < t_n$ and $s + t_1, \ldots, s + t_n$ are elements of $\mathcal{T}$, then

$$[X(t_1), \ldots, X(t_n)] \sim [X(s + t_1), \ldots, X(s + t_n)] \tag{10-1}$$

is valid.

Equation (10-1) asserts that the fidi's are unaffected by a shift in the time parameter. In particular, when $n = 1$, (10-1) asserts $X(t) \sim X(s + t)$, hence each $X(t)$ has the same distribution.

Example 10-1: I.i.d random variables Let X_0, X_1, ... be i.i.d. random variables with distribution function $F(\cdot)$. Then

$$P\{X_{i_1} \leq x_1, \ldots, X_{i_n} \leq x_n\} = F(x_1) \cdots F(x_n)$$

for any $i_1, \ldots, i_n \in I$, which verifies (10-1). $\square$

Stationary Homogeneous Markov Chains

In order to prove that a given process is stationary, (10-1) must be shown to hold. The special structure of a homogeneous Markov chain leads to a simple criterion. In the proposition below, interpret $\mathcal{T}$ as either I or $[0, \infty)$, so both discrete- and continuous-time chains are included.

Proposition 10-1 The homogeneous Markov chain $\{X(t); t \in \mathcal{T}\}$ is stationary if, and only if, for every state i, $P\{X(t) = i\}$ is the same for all $t \in \mathcal{T}$.

PROOF Let $P_{ij}(\cdot)$ be the transition function of the process. For $0 \leq t_1 < \ldots < t_n$,

$$P\{X(t_1) = i_1, \ldots, X(t_n) = i_n\}$$
$$= P\{X(t_1) = i_1\} P_{i_1 i_2}(t_2 - t_1) \cdots P_{i_{n-1} i_n}(t_n - t_{n-1})$$

and

$$P\{X(t_1 + s) = i_1, \ldots, X(t_n + s) = i_n\}$$
$$= P\{X(t_1 + s) = i_1\} P_{i_1 i_2}(t_2 - t_1) \cdots P_{i_{n-1} i_n}(t_n - t_{n-1})$$

These two probabilities are equal for all choices of $i_1, \ldots, i_n$ if, and only if, $P\{X(t_1) = i_1\} = P\{X(t_1 + s) = i_1\}$. $\square$

Hence, a Markov chain can be shown to be stationary by establishing that the one-dimensional distributions are time-invariant. Proposition 10-1 can be extended to cover Markov processes with an uncountable state space (Exercise 10-1).

Examples

Example 10-2: Markov chains If the initial distribution of an MC is a stationary probability vector π, (7-44) shows that all one-dimensional distributions are identical; hence Proposition 10-1 applies and such an MC is a stationary process. Similarly, if a CTMC has an initial probability vector that is a stationary probability vector, then (8-31) and Proposition 10-1 establish that the CTMC is a stationary process. $\square$

Example 10-3 Let A_t^* be the deficit at t of an equilibrium renewal process (see Sections 5-5 and 5-6 for definitions). The interpretation of an equilibrium renewal process as a renewal process that is "started at random" motivates the following modification of the construction given in Section 5-6. With this modification, $\{A_t^*; t \geq 0\}$ is stationary.

Let $\{N^*(t); t \geq 0\}$ be an equilibrium renewal process with interrenewal times $X_1^*, X_2, X_3, \ldots$ and renewal epochs $S_1^*, S_2^*, \ldots$. Let A_0^* be an r.v. with density $F^c(\cdot)/v$, where $v < \infty$ is the mean of the distribution function $F(\cdot)$, which is the common d.f. of $X_2, X_3, \ldots$. Set $S_0^* = -A_0^*$ and interpret it as a renewal epoch prior to time 0, as shown in Figure 10-1. The length of the renewal in progress at time 0, say L_0^*, is larger than A_0^*; construct it as follows. Let Z be an r.v. with distribution function $F(\cdot)$, and let L_0^* have the d.f. of Z given $Z > A_0^*$. Then the time to the first renewal is given by $X_1^* = L_0^* + S_0^*$. Thus,

$$G^c(y) \triangleq P\{X_1^* > y\} = \int_0^\infty P\{L_0^* > y + u \,|\, A_0^* = u\} \, dP\{A_0^* \leq u\}$$

$$= \int_y^\infty \frac{F^c(y + u)}{F^c(u)} \frac{F^c(u)}{v} \, du$$

$$= \frac{1}{v} \int_y^\infty F^c(u) \, du$$

which agrees with the choice of $G(\cdot)$ made for the equilibrium renewal process in Definition 5-4.

Define A_t^* by

$$A_t^* = t - S_{N^*(t)}^* \qquad t \geq 0 \tag{10-2}$$

Observe that for $t \geq S_1^*$, (10-2) agrees with the definition of A_t on page 129; for $0 \leq t < S_1^*$, $A_t = t - S_0^*$ and (5-41a) does not hold.

It is clear that $\{A_t^*; t \geq 0\}$ is a Markov process with state space $[0, \infty)$. By Exercise 10-1, showing that

$$A_t^c(z) \triangleq P\{A_t^* > z\}$$

is the same for all $t \geq 0$ establishes that $\{A_t^*; t \geq 0\}$ is stationary.

For $t \leq z$,

$$A_t^c(z \,|\, A_0^* = u) = \begin{cases} 0 & \text{if } u \leq z - t \\ P\{L_0^* > u + t\} & \text{if } u > z - t \end{cases}$$

hence

$$A_t^c(z) = \int_{z-t}^\infty \frac{F^c(u + t)}{F^c(u)} \frac{F^c(u)}{v} \, du$$

$$= \frac{1}{v} \int_z^\infty F^c(u) \, du \qquad t \leq z \tag{10-3a}$$

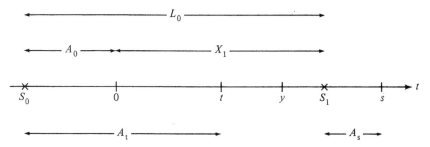

Figure 10-1 Renewal process near time 0.

For $t > z$, $A_t^* > z$ if either $X_1^* > t$ or there is a renewal at some epoch $u < t - z$ and the elapsed time until the next renewal is at least $t - u$. Thus

$$A_t^c(z) = G^c(t) + \int_0^{t-z} F^c(t - u) \, dM_G(u)$$

$$= \frac{\int_t^\infty F^c(u) \, du + \int_z^t F^c(u) \, du}{v}$$

$$= \frac{1}{v} \int_z^\infty F^c(u) \, du \qquad t > z \qquad (10\text{-}3b)$$

where (5-54) is used to obtain the second equality. Hence, $\{A_t^* ; \; t \geq 0\}$ is stationary. This result is important in the analysis in Section 10-3. ☐

Covariance Function

Suppose $\{X(t); \; t \geq 0\}$ is stationary, and let $\mu = E[X(0)]$ and $\sigma^2 = \text{Var}[X(0)]$, both of which are finite. Then

$$\text{Cov}[X(t), X(s)] = E([X(t) - \mu][X(s) - \mu])$$

$$= E([X(|t - s|) - \mu][X(0) - \mu])$$

$$\triangleq c(|t - s|) \qquad (10\text{-}4)$$

where the second equality follows from the stationarity property. In particular,

$$c(0) = \text{Var}[X(t)] = \sigma^2 \qquad (10\text{-}5)$$

As a consequence of (10-4), the covariance of any two r.v.'s of a stationary process depends on only the time difference between them. The reverse implication need not hold; for example, if either $E[X(0)]$ or $\text{Var}[X(0)]$ fails to exist, $\text{Cov}[X(t), X(s)]$ is not defined. Although we do not use them in this book, it may be useful to be aware of weakly stationary processes defined below.

Definition 10-2: Weakly stationary process The stochastic process $\{X(t); \; t \in \mathcal{T}\}$ is *weakly* (or *wide-sense*) *stationary* if $E[X(t)]$ and $\text{Var}[X(t)]$ are finite and $\text{Cov}[X(t), X(s)]$ depends on only $|t - s|$.

To distinguish them from weakly stationary processes, processes satisfying Definition 10-1 often are called *strictly stationary*. This emphasis is not required in this book.

EXERCISES

10-1 Let $\{X(t); t \in \mathcal{T}\}$ be a Markov process whose state space is an interval of the real line. Prove that this process is stationary if the d.f. of $X(t)$ is the same for all $t \in \mathcal{T}$.

10-2 Let B_t be the excess r.v. of an equilibrium renewal process at time t. Prove that $\{B_t; t \geq 0\}$ is stationary.

10-3 A process $\{X(t); t \in \mathcal{T}\}$ is said to have *stationary increments* if for $s, t \in \mathcal{T}$ with $0 \leq s < t$, the distribution of $X(t) - X(s)$ depends on only $t - s$. Prove that an equilibrium renewal process has stationary increments.

10-2* ERGODIC THEOREMS FOR STATIONARY PROCESSES

Let $\{X_n; n \in I\}$ be a stationary process and $v = E(X_0)$; assume $|v| < \infty$. For X_0, $X_1, \ldots$ i.i.d., the strong law of large numbers asserts

$$\lim_{N \to \infty} \frac{X_0 + X_1 + \cdots + X_{N-1}}{N} = v \quad \text{(w.p.1)} \qquad (10\text{-}6)$$

We show that (10-6) is valid for a large class of stationary processes. The discrete-parameter result (10-6) is then used to prove

$$\lim_{T \to \infty} \frac{1}{T} \int_0^T X(t)\, dt = v \quad \text{(w.p.1)} \qquad (10\text{-}7)$$

when $\{X(t); t \geq 0\}$ is stationary.

Discrete-Parameter Processes

The proof of (10-6) has two steps. Let

$$X_N^{\#} = \frac{X_0 + X_1 + \cdots + X_{N-1}}{N} \qquad N \in I_+ \qquad (10\text{-}8)$$

The first step establishes the existence of a random variable $X^{\#}$ such that

$$\lim_{N \to \infty} X_N^{\#} = X^{\#} \quad \text{(w.p.1)} \qquad (10\text{-}9)$$

Clearly, $E(X^{\#}) = v$. The second step identifies a class of stationary processes such that $X^{\#}$ is actually a constant; in this class (10-9) implies (10-6).

The proof of (10-9) consists of establishing these two lemmas.

Lemma 10-1 Let $\{X_n; n \in I\}$ be a stationary process, a and b be any pair of rational numbers with $a < b$,

$$X^* = \lim_{N \to \infty} \sup X_N^{\#} \qquad (10\text{-}10)$$

and

$$X_* = \lim_{N \to \infty} \inf X_N^{\#} \qquad (10\text{-}11)$$

Define the event A by

$$A = \{\omega : X^*(\omega) > b > a > X_*(\omega)\} \qquad (10\text{-}12)$$

where ω is a sample path of $\{X_n ; n \in I\}$. Then

$$P\{A\} = 0 \Rightarrow (10\text{-}9) \text{ is valid} \qquad (10\text{-}13)$$

for some random variable $X^{\#}$.

Lemma 10-2 Let A be defined by (10-12) and $I(A)$ be the indicator function of A. If $|E(X_0)| < \infty$, then

$$E[(X_0 - b)I(A)] \geq 0 \qquad (10\text{-}14a)$$

and

$$E[(a - X_0)I(A)] \geq 0 \qquad (10\text{-}14b)$$

are valid.

Adding (10-14a) and (10-14b) yields

$$E[(a - b)I(A)] \geq 0 \Rightarrow E[I(A)] = 0 \Leftrightarrow P\{A\} = 0$$

Thus Lemmas 10-1 and 10-2 establish this important theorem.

Theorem 10-1: Ergodic theorem For any stationary process $\{X_n ; n \in I\}$ with $|E(X_0)| < \infty$, (10-9) is valid.

Proof of Lemma 10-1

For each sample path ω, $\lim_{N \to \infty} X_N^{\#}(\omega)$ exists unless $X^*(\omega) > X_*(\omega)$, that is, unless there are rational numbers $a < b$ such that $X^*(\omega) > b > a > X_*(\omega)$. Alternatively, (10-9) fails only when $\omega \in A$ for some pair of rational numbers.

Let $\{(a_i, b_i)\}$ denote the (countable) set of all pairs of rational numbers with $a_i < b_i$, and let

$$A_i = \{\omega : X^* > b_i > a_i > X_*\}$$

Then $A = \bigcup_{i \in I} A_i$, and so

$$P\{A\} \leq \sum_{i \in I} P\{A_i\}$$

Thus, if $P\{A_i\} = 0$ for all $i \in I$, then $P\{A\} = 0$ and (10-9) holds w.p.1. $\qquad \square$

Proof of Lemma 10-2

Let

$$Y_i = X_i - b \qquad\qquad i \in I$$

$$S_{i,k} = Y_i + \cdots + Y_{i+k-1} \qquad k \in I_+, i \in I$$

and

$$M_{i,n} = \max\{0, S_{i,1}, \ldots, S_{i,n}\} \qquad n \in I_+, i \in I$$

With this notation, $X_N^* = S_{0,N}/N + b$, hence $X_N^* > b \Leftrightarrow S_{0,N} > 0$; this yields

$$I(A) > 0 \Rightarrow M_{0,n} > 0 \qquad \text{all } n \text{ large enough} \tag{10-15}$$

Let $I(M_{0,n} > 0)$ denote the indicator function of the event $\{M_{0,n} > 0\}$. Since $M_{0,n}$ is nondecreasing in n and nonnegative, $\lim_{n \to \infty} I(M_{0,n} > 0)$ exists and [from (10-15)] equals 1 if, and only if, $I(A) > 0$. Thus,

$$I(A) = I(A) \lim_{n \to \infty} (M_{0,n} > 0) \tag{10-16}$$

The hypothesis $|E(X_0)| < \infty$ implies $|E(Y_0)| < \infty$. From (10-16),

$$E[Y_0 I(A)] = E[Y_0 I(A) \lim_{n \to \infty} I(M_{0,n} > 0)]$$
$$= \lim_{n \to \infty} E[Y_0 I(A) I(M_{0,n} > 0)] \tag{10-17}$$

The second equality is obtained by applying the monotone convergence theorem to the positive and negative parts of Y_0.

Observe that

$$M_{1,n} = \max\{0, S_{1,1}, \ldots, S_{1,n}\} \geq S_{1,k} \qquad k = 1, \ldots, n-1$$

hence

$$Y_0 + M_{1,n} \geq Y_0 + S_{1,k} = S_{0,k+1} \qquad k = 1, \ldots, n-1 \tag{10-18}$$

Alternatively,

$$Y_0 \geq S_{0,k} - M_{1,n} \qquad k = 2, \ldots, n \tag{10-18a}$$

Since $M_{1,n} \geq 0$ and $Y_0 = S_{0,1}$,

$$Y_0 \geq Y_0 - M_{1,n} = S_{0,1} - M_{1,n} \tag{10-18b}$$

Combining (10-18a) and (10-18b) yields

$$Y_0 \geq S_{0,k} - M_{1,n} \qquad k = 1, \ldots, n \tag{10-19}$$

Taking the maximum over $k = 1, \ldots, n$ on both sides of (10-19) yields

$$Y_0 \geq \max_{1 \leq k \leq n} S_{0,k} - M_{1,n}$$
$$= M_{0,n} - M_{1,n} \qquad (\text{if } M_{0,n} > 0) \tag{10-20}$$

Therefore,

$$E[Y_0 I(M_{0,n} > 0)I(A)] \geq E[M_{0,n} - M_{1,n})I(M_{0,n} > 0)I(A)]$$

$$= E[M_{0,n}I(A)] - E[M_{1,n}I(M_{0,n} > 0)I(A)]$$

$$\geq E[M_{0,n}I(A)] - E[M_{1,n}I(A)] = 0 \qquad (10\text{-}21)$$

The last equality holds because the stationarity of $\{X_n ; n \in I\}$ implies $M_{0,n} \sim M_{1,n}$.

Since $Y_0 \triangleq X_0 - b$, combining (10-21) with (10-17) yields

$$E[(X_0 - b)I(A)] \geq 0 \qquad (10\text{-}14a)$$

To establish (10-14b), set $Z_i = a - X_i$ and repeat the arguments above. $\square$

Convergence to a Constant

Example 10-1 and (10-1) illustrate a stationary process for which the limiting random variable $X^{\#}$ takes on the value v with probability 1. This is not always the case, as shown by the following example.

Example 10-4 Let X_0 be an r.v. with finite mean, and define $X_n = X_0$ for $n = 1, 2, \ldots$. The process $\{X_n ; n \in I\}$ is easily seen to be stationary, and $X_N^{\#} = X_0$ for each $N \in I$. Thus $X^{\#} = X_0$ and is a constant only when X_0 is a constant. $\square$

To ensure that $P\{X^{\#} = v\} = 1$, processes such as the one in Example 10-4 have to be ruled out. This is done by introducing the concept of *metric transitivity*, which is defined below.

Let $\mathbf{x} = (x_0, x_1, \ldots)$ be a sample path of the process $\{X_n ; n \in I\}$. We call Ξ a *shift operator* if

$$\Xi\mathbf{x} = (x_1, x_2, \ldots)$$

A set of sample paths A is called *shift-invariant* (or *measure-preserving*) if

$$\mathbf{x} \in A \Rightarrow \Xi\mathbf{x} \in A$$

Definition 10-3 A stationary process $\{X_n ; n \in I\}$ is *metrically transitive* (or *ergodic*) if for every shift-invariant set A, $P\{X_0, X_1, \ldots) \in A\}$ is either 0 or 1.

Example 10-5: Example 10-4 continued In Example 10-4, suppose $P\{X_0 = 0\} = P\{X_0 = 1\} = \frac{1}{2}$. It is easily checked that $A_0 = \{0, 0, \ldots)\}$ and $A_1 = \{(1, 1, \ldots)\}$ are shift-invariant sets. Since

$$P\{(X_0, X_1, \ldots) \in A_0\} = P\{X_1 = 0\} = \frac{1}{2}$$

$\{X_n ; n \in I\}$ is not metrically transitive. $\square$

Example 10-6 Take X_0, A_0, and A_1 as in Example 10-5, but suppose X_0, X_1, ... are i.i.d. Then

$$P\{(X_0, X_1, \ldots) \in A_0\} = \lim_{n \to \infty} (\tfrac{1}{2})^n = 0 = P\{(X_0, X_1, \ldots) \in A_1\}$$

A moment's reflection should convince you that A_0 and A_1 are the only shift-invariant sets that conceivably could have positive probability, so $\{X_n ; n \in I\}$ is metrically transitive. □

Examples 10-5 and 10-6 are typical in the sense that $X^\#$ is a constant when the process is metrically transitive.

Theorem 10-2 Let $\{X_n ; n \in I\}$ be a stationary process that is metrically transitive, and suppose $v = E(X_0)$ is finite. Then

$$P\{X^\# = v\} = 1 \tag{10-22}$$

PROOF Let α be a real number, and look at the set of sample paths

$$A = \left\{ \mathbf{x} : \lim_{n \to \infty} \frac{x_0 + \cdots + x_{n-1}}{n} \le \alpha \right\}$$

Since $\lim_{n \to \infty} (x_0 + \cdots + x_{n-1})/n = \lim_{n \to \infty} (x_1 + \ldots + x_n)/n$ for every sequence of numbers $\{x_n\}$, A is shift-invariant. For this choice of A,

$$P\{(X_0, X_1, \ldots) \in A\} = P\left\{ \lim_{n \to \infty} \frac{X_0 + \cdots + X_{n-1}}{n} \le \alpha \right\}$$

$$= P\{X^\# \le \alpha\} = 0 \quad \text{or} \quad 1 \tag{10-23}$$

because A is shift-invariant and the process is metrically transitive. Since (10-23) is the d.f. of a constant and $E(X^\#) = v$, (10-22) is proved. □

It is often difficult to prove metric transitivity from the fundamental properties of many stationary processes which arise in operations research. For example, combining Exercises 9-29 and 10-1 proves that the waiting-time process of a $GI/G/1$ queue with $\rho < 1$ is stationary when the d.f. of D_0 is given by (9-92). We know of no proof that $\{D_n ; n \in I\}$ is metrically transitive. For this reason, often one finds that the property of metric transitivity is a hypothesis.

Role of the Ergodic Theorem in Simulation

The classical method of analyzing simulation experiments relies on the ergodic theorem to obtain point estimates. Rather than describe a general situation in abstract notation, we discuss a particular case, the $GI/G/1$ queue.

Suppose you wish to find the steady-state probability that a customer is delayed for more than 10 units of time. Suppose, further, that solving (9-92) numerically is unattractive, and you decide to simulate the delays on a computer.

After the simulation is run, the data are analyzed according to the following rationale. After the initial transient period, the delay process is stationary. Choose N large and define

$$X_n = \begin{cases} 1 & \text{if } D_{N+n} > 10 \\ 0 & \text{otherwise} \end{cases}$$

for $n \in I$. If $N + K$ delays are observed in the simulation, with K large, the ergodic theorem implies

$$\hat{p} = \frac{X_0 + \cdots + X_K}{K + 1}. \tag{10-24}$$

is a good estimate of the desired probability in the sense that as $K \to \infty$, $\hat{p}$ approaches the desired quantity.

The practical problem with this method of analysis is choosing N and K large enough. The theoretical problem is verifying that the simulated process does indeed approach a stationary process; by Theorem 10-2, one should further verify metric transitivity. The practical difficulties can be overcome by discarding (10-24) as a point estimate when the process is regenerative (see Section 6-4). The theoretical problem is discussed in Section 10-3.

Continuous-Parameter Processes

Let $\{X(t); t \geq 0\}$ be stationary. The analog of (10-9) is that there exists a random variable, $X^{\#}$ say, such that

$$\lim_{T \to \infty} \frac{1}{T} \int_0^T X(t)\, dt = X^{\#} \qquad \text{(w.p.1)} \tag{10-25}$$

The first issue is the meaning of $\int_0^T X(t)\, dt$. Let Ω be the set of all sample paths of a stochastic process (not necessarily stationary) and $(\Omega, \mathcal{F}, \mathscr{P})$ be the probability space. For each $\omega \in \Omega$, the sample path $x(t; \omega)$ is a function of t, and $\int_0^T x(t; \omega)\, dt$ has its usual meaning.† The numerical values of these integrals are functions of ω; that is, they are r.v.'s The random variable $\int_0^T X(t)\, dt$ assumes the value $\int_0^T x(t; \omega)\, dt$ on sample path ω. Conditions for the latter to be a real number are given below.

Proposition 10-2 For any stochastic process $\{X(t); t \geq 0\}$, if

(a) $\int_a^b E(|X(t)|)\, dt < \infty$ for real numbers $a < b$

then $\int_a^b x(t; \omega)\, dt$ is finite (w.p.1). If the process is stationary, then (a) may be replaced by

(b) $v = E(|X(0)|) < \infty$.

---

† To be rigorous, Lebesgue integration should be used. The idea of what is going on will not be lost if Riemann integrals are used. The two integrals agree when the Riemann integral exists.

PROOF Write†

$$\infty > \int_a^b E(|X(t)|)\, dt = \int_a^b \left[\int_\Omega |x(t;\, \omega)|\, d\mathscr{P}(\omega) \right] dt$$

By Fubini's theorem, the double integral equals $\int_\Omega \int_a^b |x(t;\, \omega)|\, dt\, d\mathscr{P}(\omega)$, which implies

$$\infty > \int_a^b |x(t;\, \omega)|\, dt \geq \left| \int_a^b x(t;\, \omega)\, dt \right| \qquad \text{(w.p.1)}$$

which establishes the first assertion.

When the process is stationary and (b) holds,

$$\int_a^b E(|X(t)|)\, dt = v(b - a) < \infty$$

and the second assertion follows from the first. $\qquad\qquad\qquad\square$

This is the analog of Theorem 10-1.

Theorem 10-3 Let $\{X(t);\, t \geq 0\}$ be stationary with $v = E(|X(0)|) < \infty$. Then (10-25) is valid for some random variable $X^{\#}$.

PROOF Proposition 10-2 ensures that $\int_0^T X(t)\, dt/T$ is finite for all $T > 0$. For each $n \in I$, define

$$X_n \triangleq \int_n^{n+1} X(t)\, dt \qquad \text{and} \qquad Y_n \triangleq \int_n^{n+1} |X(t)|\, dt \qquad (10\text{-}26)$$

The processes $\{X_n;\, n \in I\}$ and $\{Y_n;\, n \in I\}$ are stationary and satisfy the hypothesis of Theorem 10-1 (this requires some checking). Hence

$$\lim_{N \to \infty} \frac{1}{N} \int_0^N X(t)\, dt = \lim_{N \to \infty} \frac{1}{N} \sum_{n=0}^N X_n < \infty \qquad (10\text{-}27a)$$

and

$$\lim_{N \to \infty} \frac{1}{N} \int_0^N |X(t)|\, dt = \lim_{N \to \infty} \frac{1}{N} \sum_{n=0}^N Y_n < \infty \qquad (10\text{-}27b)$$

and existence and finiteness being w.p.1. From (10-27b),

$$\frac{Y_N}{N} = \frac{N+1}{N} \frac{1}{N+1} \sum_{n=0}^N Y_n - \frac{1}{N} \sum_{n=0}^{N-1} Y_n \to 0 \qquad \text{(w.p.1)} \qquad (10\text{-}28)$$

as $N \to \infty$. Let $\lfloor T \rfloor$ denote the integer part of T. Then

$$\frac{1}{T} \int_0^T X(t)\, dt = \frac{\lfloor T \rfloor}{T} \frac{1}{\lfloor T \rfloor} \int_0^{\lfloor T \rfloor} X(t)\, dt + \varepsilon_T \qquad (10\text{-}29)$$

† The notation $\int_\Omega f(\omega)\, d\mathscr{P}(\omega)$ is used when the range of integration is a set.

where

$$|\varepsilon_T| = \left| \frac{1}{\lfloor T \rfloor} \int_{\lfloor T \rfloor}^{T} X(t) \, dt \right| \leq \int_{\lfloor T \rfloor}^{\lfloor T \rfloor + 1} |X(t) \, dt \to 0 \qquad (10\text{-}30)$$

as $T \to \infty$ (w.p.1) by (10-28). Since $\lim_{T \to \infty} (\lfloor T \rfloor / T) = 1$, the desired result follows from (10-27a), (10-29), and (10-30). $\qquad \square$

Corollary 10-3 If $\{X_n; \, n \in I\}$ defined by (10-26) is metrically transitive, then $P\{X^\# = v\} = 1$.

PROOF It follows directly from Theorem 10-2. $\qquad \square$

10-3* STATIONARY REGENERATIVE PROCESSES

It may be difficult to prove that a given stochastic process is stationary by appealing to Definition 10-1, because checking that all the fidi's conform to (10-1) can be burdensome. It is often desirable to use an indirect method based on some structural properties of the process at hand. In this section we show that (1) by picking suitable initial conditions, every regenerative process (for which the mean time between regenerations is finite) is a stationary process and (2) an important subclass of regenerative processes are stationary after they have run for a sufficient time.

The theoretical contribution of these results is that an interesting class of stationary processes is identified. This has practical consequences of the following sort. In many models, it is clear from the outset that only steady-state results can be obtained in easily computable form. Instead of writing the time-dependent equations, then proving the existence of limits, and finally solving the limiting equations, it is easier to consider the steady state directly and use its properties in the analysis. The point is illustrated by this example.

Example 10-7: Machine repair model with general repair times In this example, Example 4-4 is extended to arbitrarily distributed repair times. The price that is paid for this generalization is less specific results. Let m be the number of machines and λ be the instantaneous failure rate of each working machine. There are n mechanics, each of whom requires an average of $1/\mu$ units of time to repair a failed machine. Let $X(t)$ be the number of inoperative machines at time t.

Assume that at time 0 the system is in "the steady state," i.e., that $\{X(t); t \geq 0\}$ *is stationary.* Let $N_F(t)$ and $N_R(t)$ be the number of failures and the number of repairs during $(0, t]$, respectively. Manifestly,

$$X(t) = X(0) + N_F(t) - N_R(t) \qquad t > 0 \qquad (10\text{-}31)$$

Since $X(t)$ is stationary, $E[X(t)] = E[X(0)]$. Hence (10-31) implies

$$E[N_F(t)] = E[N_R(t)] \qquad t > 0 \qquad (10\text{-}32)$$

The failure rate at time t is $[m - X(t)]\lambda$, hence

$$E[N_F(t)] = \lambda \int_0^t (m - E[X(u)]) \, du$$

$$= \lambda t(m - E[X(0)]) \qquad t > 0 \qquad (10\text{-}33)$$

Let v_S be the average number of machines undergoing repair (not including those waiting for repair). By stationarity, $v_S \triangleq E[\min\{X(t), n\}] = E[\min\{X(0), n\}]$, so it is independent of t. Argue intuitively that (a rigorous proof is given in Section 11-3)

$$E[N_R(t)] = v_S \mu t \qquad t > 0 \qquad (10\text{-}34)$$

Let $v_Q = E[X(0)] - v_S$, which is the mean number of machines waiting for repair, and $a = \lambda/\mu$. Substituting (10-33) and (10-34) into (10-32) and rearranging give

$$v_Q = m - \frac{v_S(a + 1)}{a} \qquad (10\text{-}35)$$

which is a quantitative relation between the average number of busy servers and the average number of waiting machines.

Notice that (10-32), (10-33), and (10-34) are consequences of the assumption that $\{X(t); t \geq 0\}$ is stationary. A rigorous analysis of this model would include a justification of this assumption.

In (10-35), suppose m, λ, and μ are fixed by physical and/or institutional constraints, but n can be chosen freely. The servers are being used efficiently if $r \triangleq v_S/n$ is close to 1 (from below), and they are effective in keeping the machines running if $q \triangleq v_Q/m$ is close to zero. For fixed values of r and q, (10-35) yields

$$n = \frac{1 - q}{r} \frac{a}{a + 1} m \qquad (10\text{-}36)$$

as the number of servers required. [What happens to v_Q/m and v_S/n when (10-36) yields a fractional value of n that is rounded up or down?] $\qquad \square$

Stationary Version of a Regenerative Process

Example 10-2 shows that Markov chains (which are regenerative processes) are stationary if we choose as the initial distribution what would be the limiting distribution from an arbitrary initial state. That result is now shown to hold for any regenerative process satisfying the hypotheses of Theorem 6-7, which guarantee that the limiting distribution exists.

Given a regenerative process $\{X(t); t \geq 0\}$, let $S_1, S_2, \ldots$ be the regeneration epochs, $X_1 = S_1$, $X_n = S_n - S_{n-1}$, $n = 2, 3, \ldots,$

$$F(x) \triangleq P\{X_1 \leq x\} \qquad \text{and} \qquad v \triangleq E(X_1) \qquad (10\text{-}37)$$

Let $N(t)$ be the number of regenerations by time t and A_t be the deficit at t of the renewal process $\{N(t); t \geq 0\}$. Formally,

$$N(t) = \sup\{n: S_n \leq t, \quad n \in I\} \qquad t \geq 0 \tag{10-38}$$

and

$$A_t = t - S_{N(t)} \qquad t \geq 0 \tag{10-39}$$

Let $N^*(t)$ and A_t^* be the corresponding r.v.'s in an equilibrium renewal process. From Definition 6-1a, $X(t)$ has the representation

$$X(t) = Z_{N(t)}(A_t) \qquad t \geq 0 \tag{10-40}$$

where $Z_n(\,\cdot\,)$ is the nth segment of $\{X(t); t \geq 0\}$.

Using Example 10-3 for motivation and notation, we define $X^*(t)$ by

$$X^*(t) = Z_{N_*(t)}(A_t^*) \qquad t \geq 0 \tag{10-41}$$

In particular, $X^*(0) = Z_0(A_0^*)$; this is meant to reflect starting a regenerative process "at random." To support this interpretation, observe that (10-41) and the fact that $Z_1(\,\cdot\,), Z_2(\,\cdot\,), \ldots$ are stochastically identical imply†

$$P\{X^*(t) \leq x \,|\, A_t^* = y\} = P\{Z_0(y) \leq x \,|\, S_1 > y\}$$
$$= P\{X(y) \leq x \,|\, S_1 > y\} \tag{10-42}$$

From (10-42) we obtain

$$
\begin{aligned}
P\{X^*(0) \leq x\} &= \int_0^\infty P\{Z_0(y) \leq x \,|\, A_0^* = y\}\, dP\{A_0^* \leq y\} \\
&= \frac{1}{v} \int_0^\infty P\{X(y) \leq x \,|\, S_1 > y\} F^c(y)\, dy \\
&= \frac{1}{v} \int_0^\infty P\{X(y) \leq x,\, S_1 > y\}\, dy \\
&= \lim_{t \to \infty} P\{X(t) \leq x\}
\end{aligned}
\tag{10-43}
$$

where the last equality follows from Corollary 6-7a.

Definition 10-4 Let $X(t)$ be defined by (10-40) and assume $v = E(S_1) < \infty$. Let $X^*(t)$ be defined by (10-41). The process $\{X^*(t); t \geq 0\}$ is the *stationary version* of $\{X(t); t \geq 0\}$. If S_1 is arithmetic, t is restricted to be a multiple of the span.

It should be clear to you that $\{X^*(t); t \geq 0\}$ is a delayed regenerative process. The justification for the adjective stationary in Definition 10-4 is

---

† Since conditional probability functions may not be unique, (10-42) holds w.p.1 with respect to the probability measure of S_1.

Theorem 10-4 If $\{X(t); t \geq 0\}$ is a regenerative process with finite mean cycle length, then its stationary version is a stationary process.

PROOF Definition 10-1 requires that

$$P\{X^*(t_1) = x_1, \ldots, X^*(t_n) = x_n\} = P\{X^*(t_1 + s) = x_1, \ldots, X^*(t_n + s) = x_n\}$$

hold for any $n \in I_+$, $s > 0$, $0 \leq t_1 \leq \cdots \leq t_n$, and any numbers $x_1, \ldots, x_n$.

The process segments $Z_0(\cdot)$, $Z_1(\cdot)$, ... are independent and stochastically identical, and $\{A_t^*; t \geq 0\}$ is stationary. (See Example 10-3.) Therefore,

$$P\{X^*(t_j) \leq S_j; j = 1, \ldots, n\}$$

$$= \int_0^\infty P\{X^*(t_j) \leq x_j; j = 1, \ldots, n \mid A_{t_1} = y\} \, dP\{A_{t_1} \leq y\}$$

$$= \int_0^\infty P\{Z_0(y) \leq x_1\} P\{X^*(t_j - t_1 + y) \leq x_j; j = 2, \ldots, n \mid Z_0(y) \leq x_1\} F^c(y) \frac{dy}{v}$$

$$= \int_0^\infty P\{X^*(t_j) \leq x_j; j = 1, \ldots, n \mid A_{t_1 + s} = y\} \, dP\{A_{t_1 + s} \leq y\}$$

$$= P\{X^*(t_j + s) \leq x_j; j = 1, \ldots, n\} \qquad \square$$

Theorem 10-4 partially characterizes the steady state of a regenerative process. By assuming that the proper initial conditions prevail, i.e., that we are working with the stationary version, it is legitimate to apply arguments based on stationarity. Since the process $\{X(t); t \geq 0\}$ in Example 10-7 is regenerative with finite mean times between regenerations (see Exercise 10-4), Theorem 10-4 establishes (10-32), (10-33), and (10-34) for the stationary version of the given process.

Asymptotic Stationarity of Regenerative Processes

Theorem 10-4 does not assert that stationarity accompanies letting $t \to \infty$. However, (10-42) suggests that it does occur; that is the content of Theorem 10-5 below.

Theorem 10-5 Let $\{X(t); t \geq 0\}$ be a regenerative process such that $v < \infty$ and either
(a) the sample paths are in the set† $D[0, \infty)$ or
(b) $F(\cdot)$ has a density on some interval is valid.
Let $\{X^*(t); t \geq 0\}$ be its stationary version; then

$$\lim_{t \to \infty} P\{X(t + t_j) \leq x_j; j = 1, \ldots, n\} = P\{X^*(t_j) \leq x_j; j = 1, \ldots, n\} \quad (10\text{-}44)$$

---

† Recall that $D[0, \infty)$ is the set of real-valued functions which are continuous from the right and have limits from the left.

for any $0 \leq t_1 < t_2 < \cdots < t_n$, $x_1, \ldots, x_n$, and $n \in I_+$. When X_1 is arithmetic, (10-44) holds when t is a multiple of the span.

PROOF Let

$$g_t(y) \triangleq P\{X(t + t_j) \leq x_j; j = 1, \ldots, n \mid A_t = y\}$$

and

$$g^*(y) \triangleq P\{X^*(t_j) \leq x_j; j = 1, \ldots, n \mid A_0^* = y\}$$

Extend the argument leading to (10-42) (how?) to obtain

$$g_t(y) = g^*(y) \qquad y \geq 0 \tag{10-45}$$

It has been shown [see Miller (1972)] that either (a) or (a') implies that $g^*(y)$ is directly Riemann-integrable. Define the renewal function $M(t) = E[N(t)]$. From Theorem 5-10, the key renewal theorem, and (10-44),

$$
\begin{aligned}
P\{X(t + t_j) \leq x_j ; j = 1, \ldots, n\} &= \int_0^t g_t(y)\, dP\{A_t \leq y\} \\
&= \int_0^t g^*(y) F^c(y)\, dM(t - y) \\
&\to \int_0^\infty g^*(y) F^c(y)\, \frac{dy}{\nu} \qquad (t \to \infty) \\
&= P\{X^*(t_j) \leq x_j; j = 1, \ldots, n\} \quad \square
\end{aligned}
$$

Combining Theorems 10-4 and 10-5 yields this conclusion: *A regenerative process satisfying the hypotheses of Theorem 10-5 is asymptotically stationary.* As $t \to \infty$, such a process achieves the steady state, which means that its state probabilities can be computed from the stationary version. The process representing the number of failed machines given in Example 10-7 satisfies hypothesis (a) because its sample paths are step functions, so Theorems 10-4 and 10-5 justify (10-32), (10-33), and (10-34).

Example 10-8: *GI/G/c* queue Example 6-5 gives a simple condition for the process representing the number of customers in the queue to satisfy the hypothesis of Theorem 10-5. The queue-length process therefore achieves the steady state. Since regeneration epochs are those epochs where an arriving customer finds all servers free, the regeneration cycles are busy cycles in queueing terminology. At any instant in the steady state, the time since the start of the current busy cycle is distributed as the equilibrium-deficit r.v. of the busy cycles. The conditional distribution of the state at that instant is given by (10-42). In the special case of the *M/G/1* queue, a formula for the right-side of (10-42) has been obtained.† $\square$

† Donald P. Gaver, Jr., "Embedded Markov Chain Analysis of a Waiting Line Process in Continuous Time," *Ann. Math. Stat.* **30**: 698–720 (1959); see especially Sec. 3.

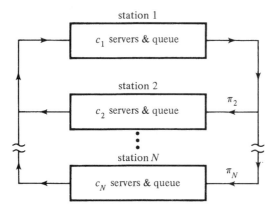

Figure 10-2 A schematic description of
a queueing network.

Example 10-9: Semi-Markov processes Let $\{Z(t); \ t \geq 0\}$ be an irreducible
SMP with ergodic states. Recall that entries into any fixed state are regeneration epochs. The assumption of ergodicity implies that the mean time
between regenerations is finite and hypothesis (a) holds. Thus, Theorems 10-4
and 10-5 imply that such a SMP is asymptotically stationary. □

Example 10-10: Closed network of queues† A *closed network* of queues is a
system with a fixed number of jobs that are served at different stations. We
discuss networks of queues in Chapter 12. Figure 10-2 portrays the special
structure where a customer goes to station i, with probability π_i, after finish-
ing service at station 1 and then returns to station 1. Each station may have
several servers.

A closed network of queues is a popular model for studying a multiprogramming computer system. In a multiprogramming computer system,
several jobs may be handled simultaneously by utilizing the central processing unit (CPU) to process one job while other jobs are served by peripheral devices (e.g., tapes, disks, etc.). Each program (job) is executed in stages;
a portion of the required information is brought from peripheral devices into
core memory to enable the CPU to process it. The results are then sent out
through some peripheral devices, and the process is repeated until the job is
completed. The utilization of the system's elements is improved by allowing
several programs to reside simultaneously in the system while being processed. The system is typically nondeterministic, so some interference among
the jobs is unavoidable and queues may be formed at the CPU and the
peripheral devices.

Let M be the number of jobs (customers) in a closed network of queues;

† This example is taken from B. Avi-Itzhak and D. P. Heyman, "Approximate Queueing
Models for Multiprogramming Computer Systems," *Oper. Res.* **21**: 1212–1230 (1973).

interpret M as the maximum number of programs that can simultaneously reside in the computer. If the computer is almost always fully loaded, then most of the time there will be M jobs in the computer, and the closed network of queues is a useful model for studying its operation. In Figure 10-2, let station 1 represent the CPU and stations 2, 3, ..., N represent various groups of peripheral devices.

Assume that the ith station consists of c_i identical servers operating in parallel with a common queue and that service times at a server of the ith station are i.i.d. exponential r.v.'s with mean $1/\mu_i$. There is no service-time dependence among stations or among servers of the same station, and each station has enough waiting positions so that stations do not interfere with one another. The queue discipline at each station is arbitrary, except that it does not change the service time of any job or cause a customer to wait when there is an idle server.

The assumption of exponentially distributed service times allows us to describe the position of all the jobs by a finite state CTMC (how?). This process is a special case of an SMP; by Example 10-9 it is asymptotically stationary, so the steady state is meaningful.

Let a_i be the average number of busy servers at station i, in the steady state. In the steady state, the average flow (jobs per time unit) into a given station must equal the average flow out of the station. Therefore

$$a_1 \mu_1 \pi_i = a_i \mu_i \qquad i = 2, 3, \ldots, N$$

This relationship can be simplified by defining $\gamma_i \triangleq \mu_1 \pi_i / \mu_i$, $i = 2, 3, \ldots, N$, and $\gamma_1 \triangleq 1$; hence

$$a_i = \gamma_i a_1 \qquad i = 1, 2, \ldots, N \tag{10-46}$$

From (10-46) we observe that in the steady state, the average utilization of the servers at station i, that is, a_i/c_i, is obtained when a_1 is found; this is done in Chapter 12. Additional features of this model are explored in Chapters 11 and 13. $\qquad \square$

EXERCISE

10-4 Let W_n be the waiting time of the nth arriving customer in a $GI/G/c$ queue. Assume the "regularity conditions" in Example 6-5 hold, and prove that $\{W_n; n \in I\}$ is asymptotically stationary.

BIBLIOGRAPHIC GUIDE

The brief treatment of stationary processes in Sections 10-1 and 10-2 is based on Doob (1953). The proof of Theorem 10-1 is taken from Karlin and Taylor (1975). A comprehensive treatment of stationary processes is given in Cramér and Leadbetter (1967). The material in Section 10-3 is based on Miller (1972).

References

Cramér, Harold, and M. R. Leadbetter: *Stationary and Related Stochastic Processes*, Wiley, New York (1967).

Doob, J. L.: *Stochastic Processes*, Wiley, New York (1953).

Karlin, Samuel, and Howard M. Taylor: *A First Course in Stochastic Processes*, 2d ed., Academic, New York (1975).

Miller, Douglas R.: "Existence of Limits in Regenerative Processes," *Ann. Math. Stat.* **43**: 1275–1282 (1972).

PART

B

OPERATING CHARACTERISTICS OF STOCHASTIC SYSTEMS

In this part of the book, we use the material in Part A to study the operating characteristics of many models. A large part of the literature on operating characteristics is concerned with queueing models. Many of the results about queueing models can be recast in terms of models of inventories, dams, cash balances, etc.; if that were done, much of the intuitive content and historical perspective would be lost. Thus, we choose to present this material in its original setting, queueing models, and to comment occasionally on other interpretations.

Many important aspects of queueing theory are discussed in Part A, and we now give a short guide to them. Examples of queueing phenomena are introduced in Chapter 2, and our notation for queueing models is given in Section 3-3. Queues with Poisson arrivals and exponential service times are extensively studied in Chapter 4. The queue with infinitely many servers and Poisson arrivals is discussed in Section 5-7. In Section 6-4, the theory of regenerative processes is used to obtain simple sufficient conditions for the existence of steady-state distributions for various operating characteristics of the $GI/G/c$ queue. The busy period of the $M/G/1$ queue is obtained in Section 6-5. The embedded Markov chain method for examining the $M/G/1$ queue, and its variants, is presented in Sections 7-1 and 7-6. The embedded Markov chain analysis of the $M/D/c$ queue is given in Section 7-9. The method of stages for treating queues with Erlang service or interarrival times is described in Section 8-4. It is shown in Section 8-8 that in the steady state, the departures from an M/M/c queue form a Poisson process. A formulation of the M/G/1 queue as a semi-Markov process is given in Section 9-1, and Lindley's integral equation for the delay distribution of the GI/G/1 queue is obtained in Section 9-3. The concept of the steady state of a queueing system is explored in Section 10-3.

Most of the analyses listed focus on obtaining explicit formulas for interesting probabilities in a specific model. In this part of the book, we emphasize the qualitative properties of queueing models and show how they may be exploited to obtain quantitative properties in particular settings. We also show how the methods of analysis developed in Part A can be combined to solve complicated models that probe fundamental properties of congestion systems.

SYSTEM PROPERTIES

This chapter describes important properties that are shared by a large number of specific models. Because of their wide applicability, we call them *system properties*. Examples of system properties include "Poisson arrivals see time averages" and the queueing formula $L = \lambda W$. We show that the system properties can be used in structured models to obtain specific conclusions. In this way, we pinpoint when, and why, special assumptions lead to sharper conclusions.

11-1 WORK-IN-SYSTEM (VIRTUAL-DELAY) PROCESS

Models of queues, inventories, dams, and other storage systems have in common the following. They are models of systems where some storable quantity which we call *work* arrives, is stored for a while, and then is released. Therefore, the stochastic process describing the work in the system plays a central role in several kinds of models.

An informal description of the work-in-system process uses this interpretation of congestion phenomena. A single server works at a fixed rate. Customers arrive at a queue in front of the server and bring with them random quantities of work; this accounts for random service times. A customer stays in queue until all work that arrived earlier has been completed, is in service while the work is done, and then departs. The *work in the system* at time t is the remaining work to be done for the customer in service plus the sum of the work requirements of the customers in the queue. If work requirements are expressed as the time that the server requires to process the work and customers are served in order of arrival, then the work in the system at time t is the delay (i.e., time spent in queue) that would be suffered were a customer to arrive at time t. For this reason, the work in the system is also called the *virtual delay*; the modifier *virtual* emphasizes that perhaps no customer will arrive at t. (Suppose customers arrive at integer times and t is not an integer; then the work in system at t is perfectly well defined, but it does not represent the delay of any customer.)

To fix these ideas, consider an example. Suppose the single server is an olive oil press that presses ψ olives per hour. Customers bring baskets of olives to be pressed, and the random variable Υ_i denotes the number of olives brought by customer i. The work in the system at time t is the number of uncrushed olives. Observe that the service time of customer i is $V_i \triangleq \Upsilon_i/\psi$; by choosing the right set of units we can always take $\psi = 1$, and the service time of a customer is numerically equal to the work requirement. Indeed, we always do this.

Formal Definition

Assume there is a single server who works at unit rate whenever work is available and is idle otherwise. Let the random variable $S(t)$ denote the amount of work that is brought to the system during $(0, t]$ and $W(0)$ represent the work in the system at time 0. The stochastic process $\{S(t); t \geq 0\}$ is called the *offered work process* and is given. The only assumptions we place on it are that the sample paths are step functions which are continuous from the right and have limits from the left [i.e., are in $D[0, \infty)$]. The former condition is a convention for when to count an arrival. A typical graph of a sample path of the offered work process is shown in Figure 11-1a.

The usual way to construct $S(t)$ is by

$$S(t) = V_1 + V_2 + \cdots + V_{A(t)} \qquad t > 0 \qquad (11\text{-}1)$$

where an empty sum is taken as zero; in (11-1), $A(t)$ represents the number of arrivals during $(0, t]$, and V_i is the service time of customer i. Notice that no assumptions about $\{A(t); t \geq 0\}$ and $\{V_i; i \in I_+\}$ are made [except for the indirect effect of the conditions on the sample paths of $S(\cdot)$]. The service times may depend on the interarrival times and/or the delays, the arrival process may be nonstationary, etc.

Let the random variable $Y(t)$ be the total time during $(0, t]$ that the server is idle, and let $W(t)$ be the work in the system at time t. Since the server works at unit rate, the amount of work completed during $(0, t]$ is $t - Y(t)$. The amount of work present at t is the amount of work present at time 0 plus the amount of work offered, less the amount of work completed. Hence for each sample path,

$$W(t) = W(0) + S(t) + [t - Y(t)] \qquad t \geq 0 \qquad (11\text{-}2)$$

Assume that the server is idle at t if, and only if, $W(t) = 0$. Defining the indicator function

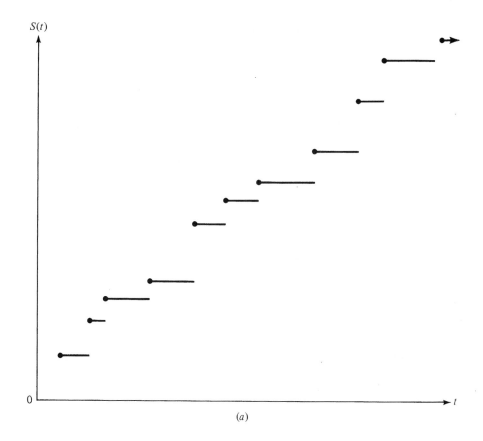

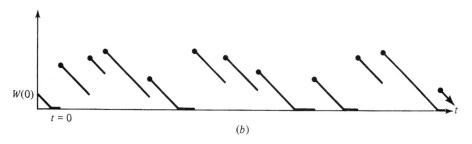

Figure 11-1 Sample paths of (a) $S(\cdot)$ and (b) $W(\cdot)$.

$$\theta[W(t)] = \begin{cases} 1 & \text{if } W(t) = 0 \\ 0 & \text{if } W(t) > 0 \end{cases}$$

allows us to represent $Y(t)$ by

$$Y(t) = \int_0^t \theta[W(u)] \, du \tag{11-3}$$

Substituting (11-3) into (11-2) yields the sample-path relation

$$W(t) = W(0) + S(t) + \int_0^t \theta[W(u)]\, du - t \qquad t \geq 0 \tag{11-4}$$

The graph of $W(\cdot)$ which corresponds to the graph of $S(\cdot)$ given in Figure 11-1a is given in Figure 11-1b.

From a modeling viewpoint, (11-4) is a conclusion; from a mathematical viewpoint, it defines $W(\cdot)$ in terms of $S(\cdot)$.

Definition 11-1 Let the process $\{S(t); t \geq 0\}$, having sample paths in $D[0, \infty)$ that are nondecreasing step functions, represent the work offered to a single server who operates at unit rate. A process $\{W(t); t \geq 0\}$ satisfying (11-4) is the *work-in-system* (or *virtual-delay*) *process*.

To justify Definition 11-1, one must show that some stochastic process satisfies (11-4). Questions of existence and uniqueness are resolved in Beneš (1963, Sec. 3.2).

Examples

Example 11-1: Single-server queues Let T_n be the epoch when the nth customer arrives and D_n be the delay of that customer; assume customers are served in order of arrival. Then $D_n = W(T_n)$ and the departure epoch of customer n is $T_n + D_n + V_n = T_n + W(T_n^+)$. Thus, the delay process can be analyzed by examining the virtual-delay process at arrival epochs. $\square$

Example 11-2: Production and inventory control model† In order to meet a random demand process, the following production scheme is sometimes used (cf. Sections 3-1 and 3-2 in Volume II). Items are produced and placed in inventory at rate ψ whenever the inventory is no larger than the *base-stock level M*; when the inventory reaches M, production is stopped until the next demand reduces the inventory below M, and then production resumes. Demands that cannot be satisfied by current inventory are backlogged, creating negative stock levels.

The demand process is assumed to be given by (11-1). Interpret $A(t)$ as the number of customers who arrive during $(0, t]$ and V_i as $V_i = \Upsilon_i/\psi$, where Υ_i is the quantity requested by customer i. That is, V_i is the time required to produce the units demanded by customer i, and we can scale time so that $\psi = 1$ (replace t by t/ψ).

Let $X(t)$ denote the inventory level at time t. A typical sample path of $X(\cdot)$ is shown in Figure 11-2a. Let $W(t) \triangleq M - X(t)$; it denotes the *deficit* below the base-stock level. In Figure 11-2b, we display the sample path of

---

† This example is due to D. P. Gaver, Jr., "Operating Characteristics of a Simple Production Inventory—Control Model," *Oper. Res.* **9**: 635–649 (1961).

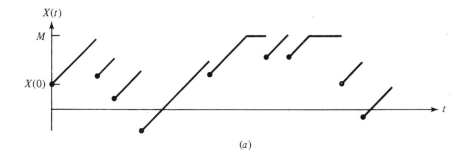

(a)

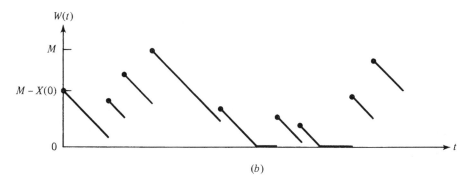

(b)

Figure 11-2 (a) A typical sample path of $X(\cdot)$; (b) corresponding sample path of $W(\cdot)$.

$W(\cdot)$ that corresponds to the sample path of $X(\cdot)$ given in Figure 11-2a. Notice that a backlog situation $[X(\cdot) < 0]$ corresponds to $W(\cdot) > M$.

The important feature of Figure 11-2b is that it looks just like Figure 11-1b. It is easy to verify that $W(\cdot)$ is given by (11-4). From $W(\cdot)$, we can obtain, for example, the distribution of the inventory level (and its mean), the proportion of time that the production facility is idle, the probability that a demand is not fully satisfied, and the distribution of time required to fully satisfy a backlogged demand (see Exercise 11-1). $\qquad \square$

Example 11-3: Reservoir model Let $S(t)$, given by (11-1), represent the amount of rainwater that is deposited in a reservoir by time t. Interpret $A(t)$ as the number of rainstorms by time t and V_i as the volume of water dropped into the reservoir by storm i. Assume that water in the reservoir is depleted (this may include evaporation) at a constant rate, and choose units so that this rate is unity. Let $W(t)$ represent the amount of water in the reservoir at time t. If we assume that rainwater is the only relevant source of water and that the duration of rainstorms is small enough to be regarded as a point, then $W(\cdot)$ is clearly seen to be given by (11-4). In this context, $Y(t)$ is interpreted as the "dry time" during $[0, t]$. $\Box$

Example 11-4: $G/G/1$ queue with limited waiting times† In some situations, a customer will not (or cannot) wait indefinitely for service. For example, switching machines used in the long-distance telephone network will "time out" a call if the call is delayed longer than a threshold amount. This scheme prevents an overload in one part of the network tying up machines in the rest of the network.

To model this phenomenon, we associate with customer n the random variable K_n, which represents the time customer n is willing (or permitted) to wait in queue. If D_n (the delay of customer n) is no larger than K_n, customer n will enter service. Let $Z_n = Z(D_n)$ be the maximum time customer n is willing (or permitted) to spend in the server given that the delay is D_n. If V_n (the service time of customer n) is larger than Z_n, customer n receives Z_n units of service and then exits before completing service. Assume that $K_1, K_2, \ldots$ are i.i.d.; let

$$R(x) = P\{K_1 \leq x\}$$

and

$$F_n(x, y) = P\{Z_n(y) \leq x \mid D_n = y\}$$

These d.f.'s may be defective. The interpretations of the defects are

$$1 - R(\infty) = P\{\text{customer } n \text{ will wait indefinitely}\}$$

and

$$1 - F_n(\infty, y) = P\{\text{customer } n \text{ will not leave until service}$$
$$\text{is completed} \mid D_n = y\}$$

Indeed, for the canonical queueing model where customers do not leave the system until service is completed, we have $R(x) \equiv 0$ and $F_n(x, y) \equiv 0$.

Two interesting special cases of this model are (a) that a customer who starts service will always stay until service is completed and (b) that the total time (waiting plus service) that a customer n is willing to spend in the system is a random variable, η_n say, with $\eta_1, \eta_2, \ldots$ being i.i.d. and having distribution function $K(\cdot)$. In case (a), $R(\cdot)$ is arbitrary and $F(x, y) \equiv 0$. In case (b), $R(x) = K(x)$ and

$$F(x, y) = \frac{P\{y < \eta_1 \leq y + x\}}{P(\eta_1 > y)} = \frac{K(y + x) - K(y)}{1 - K(y)}$$

---

† The formulation and analysis of this model are taken from Sec. 4.10 of Gnedenko and Kovalenko (1968).

The telephone switching situation described above belongs to case (a) with $R(x) = 0$ for x larger than the length of a time-out interval and $R(x) = 1$ otherwise.

How does this model fit into the framework of the work-in-system process? In (11-1), V_n is the service-time requirement of customer n. In this model, customer n may spend less time than V_n in service. Let V_n' denote the *actual time* customer n spends in service, and interpret V_n as the time required for customer n to complete service. The workload process is

$$S'(t) = V_1' + V_2' + \cdots + V_{A(t)}' \qquad t > 0 \qquad (11\text{-}1')$$

and $W(\,\cdot\,)$ is defined by (11-4) with $S'(\,\cdot\,)$ replacing $S(\,\cdot\,)$. Observe that $W(t)$ depends only on events occurring before t.

To find the distribution of V_n', we identify three mutually exclusive and collectively exhaustive cases. Suppose $W(T_n) \triangleq D_n = y$.

1. If $K_n < y$, then customer n joins the queue but does not enter service and $V_n' = 0$.
2. If $K_n \geq y$ and $Z_n = Z(y) < V_n$, then customer n joins the queue, enters service but leaves before service is complete and $V_n' = Z(y) < V_n$.
3. If $K_n \geq y$ and $Z_n \geq V_n$, then customer n completes service and $V_n' = V_n$.

Combining these cases and letting $G(x) = P\{V_1 \leq x\}$ yields

$$P\{V_n' \leq x \mid D_n = y\} = R(y^-) + R^c(y^-)F(x, y) + R^c(y^-)F^c(x, y)G(x)$$
$$= 1 - R^c(y^-)F^c(x, y)G^c(x) \qquad x \geq 0 \qquad (11\text{-}5)$$

where $F^c(x, y) = 1 - F(x, y)$.

Exercises 11-3, 11-4, and 11-5 develop further properties of this model. The steady-state delay distribution when the arrivals form a Poisson process is obtained later in Example 11-6. $\qquad\qquad\square$

Queues with Multiple Servers

Number the customers in order of arrival, and let T_n and V_n denote the arrival epoch and service time of the nth customer, respectively. Let $A(t)$ be the number of arrivals by t, and define the indicator function

$$\delta_n(t) \triangleq \begin{cases} 1 & \text{if customer } n \text{ is in service at } t \\ 0 & \text{otherwise} \end{cases}$$

Necessarily, $\delta_n(t) = 0$ for $t < T_n$.

The work performed for customer n by time t is $\int_0^t \delta_n(u)\, du$, and the total work performed by time t is

$$\sum_{n=1}^{A(t)} \int_0^t \delta_n(u)\, du = \int_0^t \sum_{n=1}^{A(t)} \delta_n(u)\, du$$

The work brought to the servers by time t is $S(t)$, which is given by (11-1).

Therefore, the work in the system at time t is given by

$$
\begin{aligned}
W(t) &= W(0) + S(t) - \sum_{n=1}^{A(t)} \int_0^t \delta_n(u) \, du \\
&= W(0) + \sum_{n=1}^{A(t)} \left[V_n - \int_0^t \delta_n(u) \, du \right]
\end{aligned}
\tag{11-6}
$$

The nth item in the sum on the right-side of (11-6) represents the amount of unfinished work associated with customer n at time t.

The effect of the number of servers c is implicit in (11-6) because c partially determines when customer n is in service. When $c > 1$, a customer may enter service when $W(\cdot) > 0$, so $W(\cdot)$ does not have the interpretation of virtual delay. Exercise 11-7 asks you to verify that (11-6) agrees with (11-4) when $c = 1$.

EXERCISES

11-1 In Example 11-2, show (a) $P\{X(t) \leq x\} = P\{W(t) \geq M - x\}$, (b) $P\{\text{production facility is idle at } t\} = P\{W(t) = 0\}$, (c) $P\{\text{demand } n \text{ cannot be satisfied immediately}\} = P\{W(T_n) + V_n > M\}$, and (d) if backlogged demands are served FIFO, then $P\{\text{demand } n \text{ is backlogged longer than } t\} = P\{W(T_n) + V_n > t + M\}$.

11-2 Let $W(t)$ be the water content at time t of a reservoir with finite capacity K and constant unit release rate. What is the analog of (11-4) for this model?

11-3 A version of the model in Example 11-4 can be used to describe a node in a data communication network (see Exercise 7-18) where the input buffer can hold a fixed number of packets and a message is placed in the buffer only if there is room for all its packets. Assume the number of packets in the buffer is known at all times (is this reasonable?). What are $G(\cdot)$ and $F(\cdot, \cdot)$ for this example?

11-4 In Example 11-4, $W(T_n)$ is not the time spent in queue by customer n (even when customers are served FIFO) because customer n may leave the queue without obtaining service. Let D'_n denote the time spent in queue by customer n and $H_n(x) = P\{W(T_n) \leq x\}$. Show
  (a) $P\{D'_n \leq x\} = H_n(x) + H_n^c(x)R(x)$
  (b) $P\{V'_n = V_n \mid V_n = x\} = \int_0^\infty F^c(x, y) \, dH_n(y)$
  (c) $P\{\text{customer } n \text{ receives no service}\} = \int_0^\infty F^c(x, y) \, dH_n(y)$
  (d) $P\{\text{customer } n \text{ starts, but does not finish, service}\} = \int_0^\infty R^c(y) \int_0^\infty F(z, y) \, dG(z) \, dH_n(y)$

11-5 In Example 11-4, assume that the arrivals form a renewal process with interarrival-time distribution function $A(\cdot)$ having mean $1/\lambda < \infty$ and that there exists a finite constant M such that $P\{K_1 \leq M\} = 1$ and $A^c(M) > 0$. Prove that $W(t)$ is regenerative and the mean number of arrivals in each regeneration cycle is finite. *Hint*: Prove that the mean time between regenerations is no larger than $[\lambda A^c(M)]^{-1}$. Show that when the arrivals form a Poisson process, $\{W(t); t \geq 0\}$ is a Markov process.

11-6 Consider the M/G/1 queue with arrival rate λ and service-time distribution function $G(\cdot)$, and let $W(t, x) = P\{W(t) \leq x\}$. Argue that

$$
W(t + \Delta t, x) = (1 - \lambda \, \Delta t)W(t, x + \Delta t) + \lambda \, \Delta t \int_0^x G(x - y) \, d_y \, W(t, y) + o(\Delta t)
$$

and then derive Takács' integrodifferential equation (assume that the derivatives exist)

$$
\frac{\partial W(t, x)}{\partial t} = \frac{\partial W(t, x)}{\partial x} - \lambda \left[W(t, x) - \int_0^x G(x - y) \, d_y \, W(t, y) \right]
$$

11-7 Show that (11-6) agrees with (11-4) when there is a single server.

11-8 For the M/M/c queue, what is $\lim_{t\to\infty} E[W(t)]$?

11-9 In a $GI/G/1$ queue, suppose at time zero, n customers are present, one of which has just started service. Prove that the expected time until the system is empty is finite when the traffic intensity is less than 1.

11-2 POISSON ARRIVALS "SEE" TIME AVERAGES

One reason that steady-state probabilities are important is that they can be interpreted as long-run proportions. A birth-and-death process representing a machine breakdown model is described in Section 4-8, and it is shown that (typically) the proportion of time j machines are broken does not equal the proportion of failed machines that find j other broken machines in the repair facility at their failure epoch. However, when the failures (i.e., customer arrivals) occur according to a Poisson process, these two averages are the same. In this section, that particular property of Poisson arrivals is shown to hold under very general conditions; we express it as "Poisson arrivals 'see' time averages."

The obvious use of this result is that it suffices to find either the time average or the limiting distribution at arrival epochs. A more subtle application is to use the equality of these two stationary probabilities to write the equations that determine them; this is illustrated in Example 11-6, which extends the analysis of Example 11-4.

Assumptions

The usual setting for the results to be established is a queueing model, so we use the language of queues for interpretations. Analogous statements hold for inventories, reliability models, etc. Let $X(t)$ denote the state of some queueing process at time t. Examples of $X(t)$ include

$$X_1(t) = \text{number of customers present at } t$$
$$X_2(t) = \text{number of busy servers at } t$$
$$X_3(t) = \text{virtual delay at } t$$
$$X_4(t) = \text{vector representing number of customers}$$
$$\text{at each node of a network of queues}$$

Let $A(t)$ denote the number of arrival epochs during $(0, t]$ and T_n denote the nth arrival epoch.

Assumption 11-1 The sample paths of $\{X(t); t \geq 0\}$ are (w.p.1) in $D[0, \infty)$.

Assumption 11-2: Nonanticipation For every $t > 0$, $X(t^-)$ depends on $\{A(t); t \geq 0\}$ through (at most) $\{A(u); 0 \leq u < t\}$.

Assumption 11-3 $\{A(t); t \geq 0\}$ is a Poisson process with finite rate $\lambda > 0$.

Assumption 11-1 is primarily a convention for defining the state of the process at a transition epoch. It ensures that the definition

$$X_n \triangleq \lim_{h \downarrow 0} X(T_n - h) \triangleq X(T_n^-) \tag{11-7}$$

makes sense. In some circumstances, it contributes to the structure of the X-process, e.g., when $\{X(t); t \geq 0\}$ is a CTMC (see Proposition 8-2).

Assumption 11-2 states that the past arrivals completely determine the current state. This assumption is valid in most queueing models. An example of a situation where this assumption fails to hold is given in Wolff (1982). To see that $X_1(t)$ defined above typically satisfies this assumption, let D_n be the delay in queue and V_n be the service time of customer n. Then $T_n + D_n + V_n$ is the departure epoch of customer n. Consequently, $X_1(t) = \#\{n: T_n \leq t < T_n + D_n + V_n\}$. If the event $\mathcal{E} = \{D_n + V_n > t - T_n\}$ depends on the arrivals only through $\{A(u); u < t\}$, then Assumption 11-2 is satisfied. When customers enter service in order of arrival, $\mathcal{E}$ depends on only $T_1, \ldots, T_n$ and $V_1, \ldots, V_n$, and Assumption 11-2 holds whenever the service times are independent of future arrival times. When customers enter service LIFO (last-in, first-out) or in random order, or according to the priority schemes described in Section 11-6, $\mathcal{E}$ depends on only the arrival epochs and service times which occur no later than t, so the same condition verifies Assumption 11-2. The corresponding analyses for $X_2(t)$, $X_3(t)$, and $X_4(t)$ are requested in Exercise 11-10.

Since $A(t)$ is the number of arrival epochs in $(0, t]$, the theorems below apply to the first customer in a batch if the arrivals form a compound Poisson process.

The formal statement of Assumption 11-2, is†

Assumption 11-2: Formal Version For every $t > 0$ and set B for which $P\{X(t) \in B\}$ is defined,

$$P\{X(t^-) \in B \,|\, A(u), \quad u \geq 0\} = P\{X(t^-) \in B \,|\, A(u), \quad 0 \leq u \leq t\}$$

This form is used in proving theorems.

Theorems

We give two versions of "Poisson arrivals 'see' time averages": first for finite t and then for limiting averages. For t finite, we wish to prove, for any set B such that $P\{X(t^-) \in B\}$ is defined,

$$P\{X(t^-) \in B\} = P\{X(t^-) \in B \,|\, t \text{ is an arrival epoch}\}$$

The main difficulty in proving this statement is that $P\{t \text{ is an arrival epoch}\} = 0$ and conditioning on an event having probability 0 is not allowed.‡ We avoid this

---

† Recall that $P\{ \cdot \,|\, A(u), u \in \mathcal{T}\}$ is an abbreviation for $P\{ \cdot \,|\, A(u_i)a_i, i = 1, \ldots, n\}$, where $n \in I_+$, each $u_i \in \mathcal{T}$, and each a_i is in the state space of the A-process.

‡ However, it is permissible to condition on the value of an r.v., for example, $Y = y$, when $P\{Y = y\} = 0$. See pages 505 and 506.

difficulty by observing that t is an arrival epoch if, and only if, $A(t) - A(t - h) \geq 1$ for any $h > 0$. Then

$$\lim_{h \downarrow 0} P\{X(t^-) \in B \,|\, A(t) - A(t - h) \geq 1\}$$

corresponds to our intuitive notion of $P\{X(t^-) \in B \,|\, t$ is an arrival epoch$\}$.

Theorem 11-1 Under Assumptions 11-1 through 11-3, for every $t \geq 0$ and any $h > 0$,

$$P\{X(t^-) \in B\} = P\{X(t^-) \in B \,|\, A(t) - A(t - h) \geq 1\} \qquad (11\text{-}8)$$

for every set B such that the left-side of (11-8) is defined.

PROOF For any $h > 0$, $P\{A(t) - A(t - h) \geq 1\} = 1 - e^{-\lambda h} > 0$. For notational ease, let

$$a = \{X(t^-) \in B\}$$

$$b = \{A(u); 0 \leq u \leq t\}$$

and

$$c = \{A(t) - A(t - h) \geq 1\}$$

The definition of conditional probability is

$$P\{a \,|\, c\} = \frac{P\{a, c\}}{P\{c\}} = P\{c \,|\, a\} \frac{P\{a\}}{P\{c\}} \qquad (11\text{-}9)$$

If we can show that a and c are independent, then (11-9) will imply (11-8). Assumption 11-2 implies $P\{a \,|\, b, c\} = P\{a \,|\, b\}$. The memoryless property of the exponential interarrival time implies $P\{c \,|\, b\} = P\{c\}$; that is, c and b are independent. Since independence is a symmetric relation, $P\{b \,|\, c\} = P\{b\}$. Summing over all possible values of b yields

$$P\{a \,|\, c\} = \sum_b P\{a \,|\, b, c\} P\{b \,|\, c\}$$

$$= \sum_b P\{a \,|\, b\} P\{b\} = P\{a\}$$

and so a and c are independent. $\square$

The limiting probability that i customers are present at an arrival epoch is obtained for the M/G/1 queue in Examples 7-18 and 7-19 and Exercise 7-61. Denote it by π_i; in the notation of this section, $\pi_i = \lim_{n \to \infty} P\{X_n = i\}$, $i \in I$. We want to establish

$$\pi_i = \lim_{t \to \infty} P\{X(t) = i\} \qquad (11\text{-}10)$$

This is not the same as letting $t \to \infty$ in (11-8); see Exercise 11-11. If the limiting probabilities in (11-10) are also long-run proportions, then

$$\lim_{N \to \infty} \frac{1}{N} \sum_{n=1}^{N} 1\,\{X_n = i\} = \lim_{T \to \infty} \frac{1}{T} \int_0^t 1\,\{X(t) = i\}\,dt \qquad \text{(w.p.1)} \qquad \text{(11-11)}$$

holds, where $1\{\,\cdot\,\}$ is the indicator function of the event in braces.

The proof of (11-10) and (11-11) given below uses the assumption that the regeneration points of $\{X(t); t \geq 0\}$ are arrival epochs. (The $M/G/c$ queue has this property; see Example 5-15.) In this setting, it is convenient to state the theorem in terms of the stationary version (see Section 10-3) of the $X(t)$- and X_n-processes and obtain (11-10) and (11-11) as corollaries.

Theorem 11-2† Let $\{X(t); t \geq 0\}$ be a nonnegative regenerative process with regeneration epochs $0 = S_0 \leq S_1 \leq S_2 \cdots$. Assume

(a) each S_i is an arrival epoch
(b) $X(S_i)$ has an a priori given value that is the same for all i

Let $0 = N_0 \leq N_1 \leq N_2 \cdots$ be the regeneration epochs of $\{X_n;\ n \in I\}$ and assume

(c) $E(N_1) < \infty$

Let $\{X^*(t); t \geq 0\}$ and $\{X_n^*;\ n \in I\}$ be the stationary versions of $\{X(t); t \geq 0\}$ and $\{X_n;\ n \in I\}$, respectively. If Assumptions 11-1 to 11-3 hold, then for all $x \geq 0$,

$$P\{X_0^* \leq x\} = P\{X^*(0) \leq x\} \qquad \text{(11-12)}$$

Before we prove this theorem, a few comments on the assumptions are in order. Many queueing systems regenerate when a customer arrives and the system is empty; this will satisfy (a). Condition (b) ensures that $\{X_n;\ n \in I\}$ is regenerative and is typically satisfied. Condition (c) is satisfied by $M/G/c$ queues with traffic intensity less than 1 (Example 5-15) and systems where the X_n-process is an ergodic Markov chain. Equation (11-17) below will establish $E(S_1) < \infty$, so $\{X^*(t);\ t \geq 0\}$ is defined. Note that setting $S_0 = 0$ implies there is a customer number 0 who arrives at time 0 so that customers $0, \dots, N_1 - 1$ arrive during the first cycle.

PROOF By its definition, $N_i = A(S_i)$; in particular, $N_1 = A(S_1)$. From (10-43),

$$P\{X^*(0) \leq x\} = \frac{1}{E(S_1)} \int_0^\infty P\{X(y^-) \leq x \mid S_1 > y\} P\{S_1 > y\}\,dy$$

$$= \frac{1}{E(S_1)} \int_0^\infty \sum_{n=0}^\infty P\{X(y^-) \leq x \mid A(y^-) = n, S_1 > y\} P\{A(y) = n,$$

$$S_1 > y\}\,dy \qquad \text{(11-13)}$$

† This theorem and its proof are due to Shaler Stidham, Jr., "Regenerative Processes in the Theory of Queues, with Applications to the Alternating Priority Queue," *Adv. Appl. Probab.* **4**: 542–577 (1972).

The obvious discrete-time version of (10-43) yields†

$$P(X_0^* \leq x) = \frac{1}{E(N_1)} \sum_{n=0}^{\infty} P\{X_n \leq x \mid N_1 > n\} P\{N_1 > n\}$$

$$= \frac{1}{E(N_1)} \sum_{n=0}^{\infty} \int_0^{\infty} P\{X_n \leq x \mid N_1 > n, T_{n+1} = y\} \, dP\{T_{n+1} \leq y, N_1 > n\}$$

$$(11\text{-}14)$$

Proving that the right-sides of (11-13) and (11-14) are equal will establish (11-12). The first step is to prove

$$P\{X_0^* \leq x\} = \frac{1}{E(N_1)} \int_0^{\infty} \sum_{n=0}^{\infty} P\{X(y^-) \leq x \mid A(y^-) = n,$$

$$S_1 > y\} \, dP\{T_{n+1} \leq y, N_1 > n\} \qquad (11\text{-}15)$$

Observe that $\{N_1 > n\} \Leftrightarrow \{S_1 > T_n\}$. Thus,

$$P\{X_n \leq x \mid T_n = y, N_1 > n\}$$

$$= P\{X(y^-) \leq x \mid T_{n+1} = y, N_1 > n\}$$

$$= P\{X(y^-) \leq x \mid T_{n+1} = y, S_1 > T_n\}$$

$$= P\{X(y^-) \leq x \mid T_{n+1} = y, S_1 > y\} \qquad (11\text{-}16)$$

$$= P\{X(y^-) \leq x \mid y \text{ is an arrival epoch}, A(y^-) = n, S_1 > y\}$$

$$= P\{X(y^-) \leq x \mid A(y^-) = n, S_1 > y\}$$

where the last equality follows from Theorem 11-1. Substituting (11-16) into (11-14) and interchanging the order of integration and summation (which is justified by Fubini's theorem, since probabilities are nonnegative) yield (11-15).

From Assumptions 11-2 and 11-3, the event $\{N_1 = n\}$ does not depend on $T_{n+1}, T_{n+2}, \ldots$; hence Wald's equation (Theorem 6-1) is applicable and yields

$$E(S_1) = \frac{E(N_1)}{\lambda} < \infty \qquad (11\text{-}17)$$

The proof is completed by proving

$$P\{T_{n+1} > y, N_1 > n\} = \lambda \int_y^{\infty} P\{A(u) = n, S_1 > u\} \, du \qquad (11\text{-}18)$$

because substituting (11-17) and (11-18) into (11-13) yields (11-15).

---

† The random variable A_t used in Section 10-3 is different from the random variable $A(t)$ used here.

Since S_1 is an arrival epoch, if $S_1 > T_n$, then $S_1 \geq T_{n+1}$; consequently

$$\{A(u) = n, S_1 > u\} \Leftrightarrow \{S_1 > u, T_n \leq u < T_{n+1}\}$$

Let $1\{\cdot\}$ be the indicator function of the event in brackets and $y \vee T_n$ denote $\max\{y, T_n\}$. Then

$$\int_y^\infty 1\{A(u) = n, S_1 > u\}\, du = \int_{y \vee T_n}^{T_{n+1}} 1\{S_1 > u\}\, du$$

$$= (T_{n+1} - y \vee T_n)1\{T_n > y\}1\{N_1 > n\}$$

Taking expectations on both sides and then bringing the expectation inside the integral yield

$$\int_y^\infty P\{A(u) = n, S_1 > u\}\, du = E(T_{n+1} - T_n \vee y \mid T_{n+1} > y, N_1 > n)$$

$$\times P\{T_{n+1} > y, N_1 > n\} \qquad (11\text{-}19)$$

The event $\{N_1 > n\}$ is independent of $X(t)$ for $t \leq T_n$, and the interarrival times are exponentially distributed (hence memoryless), so the expectation on the right-side of (11-19) is $1/\lambda$, which establishes (11-18). $\qquad\square$

Corollary 11-2a If N_1 is aperiodic, then $\pi(y) \triangleq \lim_{n \to \infty} P\{X_n \leq y\}$ exists and

$$\pi(y) = p(y) \triangleq \lim_{t \to \infty} P\{X(t) \leq y\} \qquad (11\text{-}20)$$

PROOF Since S_1 is a sum of exponentially distributed r.v.'s, it is not arithmetic; hence the regenerative property of $\{X(t); t \geq 0\}$ implies (Corollary 6-7a) that $p(y)$ exists. A similar argument establishes that $\pi(y)$ exists, and (11-20) follows from (11-12) and (10-44), which states that the right-sides of (11-12) and (11-20) are equal. $\qquad\square$

Corollary 11-2b If N_1 is aperiodic, then

$$\lim_{N \to \infty} \frac{1}{N} \sum_{n=1}^N 1\{X_n \leq y\} = \lim_{T \to \infty} \frac{1}{T} \int_0^T P\{X(t) \leq y\}\, dt \qquad (\text{w.p.1}) \qquad (11\text{-}21)$$

PROOF This is an immediate consequence of (11-20) and Theorem 6-9, which states that limiting probabilities coincide with asymptotic proportions in a regenerative process. $\qquad\square$

The proof of Theorem 11-2 given above uses the regenerative structure of the system in an essential way. The following generalization of Corollary 11-2b shows that the regenerative structure is needed only to prove that limits exist.

Theorem 11-3 Under Assumptions† 11-1 through 11-3, if the right-side of (11-21) is a constant, then (11-21) is valid.

† Wolff uses a slightly different form of Assumption 11-1. He requires the sample paths to be continuous from the left (not necessarily from the right). This difference is not essential for our purposes.

PROOF See Wolff (1982). The proof uses some martingale convergence theorems that are beyond the scope of this book. □

Example 11-5: M/G/1 queue In Example 11-1, we observed that the delay of customer n, denoted D_n, is the virtual delay at T_n, denoted $W(T_n^-)$. Take $W(\cdot)$ as $X(\cdot)$ in Theorem 11-2; it is easily verified that the hypotheses of the theorem are satisfied and that N_1 is aperiodic. Applying (11-20) yields

$$\lim_{t \to \infty} P\{W(t) \le y\} = \lim_{n \to \infty} P\{D_n \le y\} \qquad y \ge 0 \qquad (11\text{-}22)$$

The right-side of (11-22) is given by the Pollaczek-Khintchine formula in Section 7-6. In particular, $\lim_{t \to \infty} P\{W(t) = 0\} = 1 - \rho$. □

Example 11-6: Example 11-4 continued In the $GI/G/1$ queue with limited waiting times described in Example 11-4, assume the arrivals form a Poisson process with positive rate λ. In Exercise 11-5, you showed that if K_1, the maximum delay customer 1 will endure, is bounded, then the workload process regenerates when a customer arrives at a free server and the mean number of arrivals in a regeneration cycle is finite. Thus, hypotheses (a), (b), and (c) of Theorem 11-2 are satisfied. It is easy to check that Assumptions 11-1 through 11-3 are satisfied, and the solution of Exercise 11-5 shows $P\{N_1 = 1\}$, so N_1 is aperiodic by Exercise 6-11. Thus Corollary 11-2a applies; it implies

$$\lim_{n \to \infty} P\{W(T_n) \le x\} = \lim_{t \to \infty} P\{W(t) \le x\}$$
$$= P\{W^*(0) \le x\} \triangleq H(x) \qquad (11\text{-}23)$$

Furthermore, in Exercise 11-5 you showed that $\{W(t); t \ge 0\}$ is a Markov process.

Exercise 11-5 shows that the first limit in (11-23) is the delay distribution in the steady state. In the analysis below, the second limit is obtained. The fact that Poisson arrivals see time averages helps make $H(\cdot)$ an interesting d.f.

We now obtain $H(\cdot)$. Specifically, we prove that $H(\cdot)$ has mass at zero (which is the probability that a customer finds the server free) and a density for positive arguments. That is, when $W(t) \le m$ for all t,

$$H(x) = H(0) + \int_0^x h(u) \, du \qquad 0 \le x \le m \qquad (11\text{-}24)$$

where $H(0)$ and $h(\cdot)$ satisfy

$$h(x) = \lambda \int_0^x [1 - K(y)] [1 - F(x - y, y)] [1 - G(x - y)] h(y) \, dy$$
$$+ \lambda H(0)[1 - F(x,0)] [1 - G(x)] \qquad 0 \le x \le m \quad (11\text{-}25)$$

The functions $K(\cdot)$, $F(\cdot, \cdot)$, and $G(\cdot)$ are defined in Example 11-4.

Consider the stationary process $\{W^*(t); t \ge 0\}$; for any $\Delta > 0$, $H(\cdot)$ is the d.f. of $W^*(t)$ and $W^*(t + \Delta)$. Fix $\Delta > 0$. If there are no arrivals during $(t, t + \Delta]$, then

$$\{W^*(t) \le x + \Delta\} \Leftrightarrow \{W^*(t + \Delta) \le x\}$$

This occurs with probability $1 - \lambda\Delta + o(\Delta)$. If one arrival occurs during $(t, t + \Delta]$, say at z, then using (11-5) yields, for $0 \le y \le x + \Delta$,

$$P\{W^*(t + \Delta) \le x \mid W(t) = y, \text{ an arrival at } z\} = 1 - J(x - y + z, y - z)$$

where

$$J(x, y) \triangleq K^c(y^-)F^c(x, y)G^c(x)$$

If $y > x + \Delta$, the conditional probability above is zero. The probability of one arrival during $(t, t + \Delta]$ is $\lambda\Delta + o(\Delta)$, and given that it occurs, the arrival epoch is uniformly distributed over the interval (by Corollary 5-13). The probability that two or more arrivals occur during $(t, t + \Delta)$ is $o(\Delta)$.

Removing the conditioning on the number of arrivals and invoking the stationary and Markov properties of the W^*-process yield

$$H(x) = (1 - \lambda\Delta)H(x + \Delta)$$

$$+ \lambda\Delta \int_0^{x+\Delta} \int_t^{t+\Delta} [1 - J(x - y + z, y - z)] \, dz \, dH(y) + o(\Delta)$$

From the mean value theorem for integrals (Proposition A-8),

$$\int_t^{t+\Delta} [1 - J(x - y + z, y - z)] \, dz = 1 - J(x - y + \xi, y - \xi)$$

for some $\xi \in [0, \Delta]$. Hence

$$H(x + \Delta) = H(x) + \lambda\Delta \int_0^{x+\Delta} J(x - y + \xi, y - \xi) \, dH(y) + o(\Delta) \qquad (11\text{-}26)$$

Since $J(\,\cdot\,,\,\cdot\,)$ is a product of probabilities, it is no larger than 1, and the integral in (11-26) is no larger than 1. Hence (11-26) implies $H(\,\cdot\,)$ has a right derivative, $h(\,\cdot\,)$ say. Dividing both sides of (11-26) by Δ and letting $\Delta \downarrow 0$ yield

$$h(x) = \lambda H(0)J(x, 0) + \lambda \int_0^x J(x - y, y)h(y) \, dy$$

which is (11-25). By repeating the analysis above on the time interval $(t - \Delta, t)$, $h(\,\cdot\,)$ can be shown to be the left derivative for $x > 0$. Thus, (11-24) and (11-25) are established.

For the special case where the maximum tolerable delay is a constant and service times are exponentially distributed, an explicit solution for the steady-state delay distribution is requested in Exercise 11-12. □

EXERCISES

11-10 Show that the X_2-, X_3-, and X_4-processes defined on p. 391 satisfy Assumption 11-2.

11-11 Why does π_i differ from $\lim_{t \to \infty} P\{X(t) = i \mid t$ an arrival epoch$\}$?

11-12 In Example 11-5, suppose (a) service times have an exponential d.f. with mean $1/\mu$, (b) customers will wait in queue for, at most, a length of time m, and (c) customers do not defect from the server. What are $K(\,\cdot\,)$ and $F(\,\cdot\,,\,\cdot\,)$ for this model? Show that

$$H(0) = \frac{1 - \rho}{1 - \rho e^{-(\mu - \lambda)m}}$$

and

$$h(x) = \rho(\mu - \lambda) \frac{e^{-(\mu - \lambda)x}}{1 - \rho e^{-(\mu - \lambda)m}} \qquad 0 \le x \le m$$

where $\rho = \lambda/\mu$. Hint: Solve (11-25), ignoring the condition $x \le m$, and argue that this function agrees with the desired function for $x \le m$.

11-13 A proof of Theorem 11-3 given by Wolff uses the following device. Introduce observers who arrive according to a Poisson process with rate γ; the arrival epochs of the observers are independent of the system. We use this device to prove that the expected number of customers who arrive during $(0, t]$ and find the system in some state less than or equal to y is the arrival rate multiplied by the expected amount of time during $(0, t]$ that the state is less than or equal to y. Let $\Lambda(t)$ be the number of arrivals plus the number of observers that appear by time t and τ_i be epoch of the ith event in $\{\Lambda(t); t \ge 0\}$. Define the 0–1 variables δ_i, C_i, O_i, and $a_i(t)$ by

$$\delta_i = 1 \Leftrightarrow X(\tau_i) \le y \qquad \text{for some fixed } y \ge 0$$

$$C_i = 1 \Leftrightarrow \tau_i \text{ is a customer arrival epoch}$$

$$O_i = 1 \Leftrightarrow \tau_i \text{ is an observer arrival epoch}$$

and

$$a_i(t) = 1 \Leftrightarrow \tau_i \le t$$

Prove that for $t > 0$,

$$\gamma E\left(\int_0^t 1\{X(u) \le y\} \, du \right) = \frac{\gamma}{\lambda + \gamma} \sum_{i=1}^{\infty} E[\delta_i a_i(t)]$$

and

$$E\left(\sum_{n=1}^{A(t)} 1\{X(T_n) \le y\} \right) = E\left(\sum_{i=1}^{A(t)} \delta_i C_i \right) = \lambda E\left(\int_0^t 1\{X(u) \le y\} \, du \right)$$

11-3 THE QUEUEING FORMULA $L = \lambda W$: APPLICATIONS AND GENERALIZATIONS

Let $X(t)$ denote the number of customers in a queueing system at time t, $A(t)$ denote the number of arrivals by time t, and W_n denote the time spent in the system by the nth arriving customer. Assume that $\{X(t); t \ge 0\}$ and $\{W_n ; n \in I_+\}$ are stationary stochastic processes (cf. Section 10-3), and let

$$L = E[X(0)] \qquad W = E(W_1)$$

and λ be the arrival rate. One of the most celebrated results in queueing theory is that under very weak conditions,

$$L = \lambda W \qquad (11\text{-}27)$$

By direct calculation of L and W, (11-27) was obtained for the $M/M/c$ system in Exercise 4-52 and for the $M/G/1$ system in Example 7-19.

The many implications of (11-27) are due to the flexibility one has in choosing "the system" to which (11-27) is applied. Before we prove (11-27), some of its applications are illustrated.

Example 11-7 The direct calculations of W mentioned above assume customers enter service in the order of their arrival. Consider any other queue discipline which does not change the expected value of the arrival and departure epochs, i.e., which yields the same value of $E[X(t)]$, for all t, as does FIFO. Examples of such disciplines are LIFO and random order of service. Since these disciplines do not affect L or λ, (11-27) implies they result in the same value of W as does FIFO. This is the stochastic version of Theorem 2-3. $\qquad \square$

Example 11-8 Let $1/\mu$ be the mean service time in a $GI/G/c$ queue. Let L_s be the average number of customers who are either in queue or in service, and let W_s be the mean waiting time (i.e., time spent in queue or in service). If we regard the queue and the servers as "the system," (11-27) yields

$$L_s = \lambda W_s \qquad (11\text{-}28)$$

Let L_q be the average number of customers in the queue and W_q be the average time spent in the queue. If we regard the queue as "the system," (11-27) yields

$$L_q = \lambda W_q \qquad (11\text{-}29)$$

Since $W_s = W_q + 1/\mu$, and $L_s = L_q + L_b$, where L_b is the average number of busy servers, subtracting (11-29) from (11-28) yields

$$L_b = \frac{\lambda}{\mu} \triangleq a \qquad (11\text{-}30)$$

For the $M/M/c$ queue, (11-30) is obtained by direct calculation in (4-44). When $c = 1$, $L_b = P\{X(0) > 0\}$, which is the probability the server is busy, and (11-30) is the stochastic version of Theorem 2-6. $\qquad \square$

Overview of the Proof of Equation (11-27)

The proof of (11-27) is done in two steps. First, a sample-path version is established; this is a deterministic result that extends Theorem 2-4. Second, ergodic theory is used to establish that the averages on the right-side of (11-27) exist and to provide the deterministic version with an interpretation as an equation among expected values.

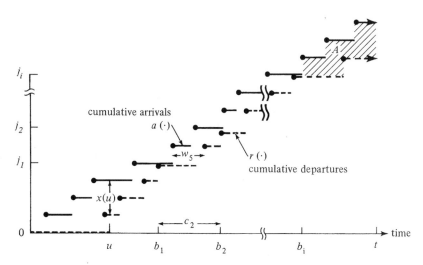

Figure 11-3 A sample path of a queue.

Aside from issues involving the existence of limits, here is the essential content of Theorem 11-4. The limits†

$$l \triangleq \lim_{t \to \infty} \frac{1}{t} \int_0^t x(u) \, du \qquad \lambda = \lim_{t \to \infty} \frac{a(t)}{t} \qquad w = \lim_{N \to \infty} \frac{1}{N} \sum_{n=1}^{N} w_n$$

are related by

$$l = \lambda w \qquad (11\text{-}31)$$

The intuitive background and the fundamental issue in the proof of (11-31) are shown in the following argument.

In Figure 11-3, the vertical distance between the cumulative arrival and cumulative departure curves is the number present, when customers are served FIFO, and the horizontal distances between the curves are the waiting times (for a more complete discussion, see Section 2-3).

Let c_k denote the number of customers served in the kth busy cycle, $j_i = c_1 + \cdots + c_i$, and b_i be the epoch where the ith busy cycle ends. In Figure 11-3, b_i is the ith epoch where $r(\cdot)$ reaches $a(\cdot)$ from below, and $j_i = a(b_i)$.

Let $\lambda_i = j_i/b_i$; it is the arrival rate during $(0, b_i]$. Theorem 2-4 states

$$\frac{1}{b_i} \int_0^{b_i} x(u) \, du = \frac{\lambda_i}{j_i} \sum_{n=1}^{j_i} w_n \qquad (11\text{-}32)$$

and is established in Section 2-3 by expressing the unshaded area between $a(\cdot)$ and $r(\cdot)$ in Figure 11-3 in two ways. Letting $i \to \infty$ in (11-32) establishes (11-31) as $t \to \infty$ through the values $b_1, b_2, \dots$.

† Lowercase letters denote realizations of the r.v.'s represented by the corresponding capital letters.

Suppose, as in Figure 11-3, $b_i < t < b_{i+1}$. In Figure 11-3, the crosshatched area A is given by

$$A = \int_{b_i}^{t} [a(u) - r(u)]\, du \qquad b_i < t < b_{i+1}$$

Let $\lambda(t) = a(t)/t$; then

$$\frac{1}{t} \int_0^t x(u)\, du = \frac{t}{b_i} \frac{1}{t} \int_0^{b_i} x(u)\, du + \frac{A}{t}$$

$$= \lambda(t) \frac{j_i}{a(t)} \frac{1}{j_i} \sum_{n=1}^{j_i} w_n + \frac{A}{t} \tag{11-33}$$

From (11-32) and (11-33), (11-31) holds as $t \to \infty$ if the relevant limits exist and

$$\frac{t}{b_i} \to 1 \qquad \frac{j_i}{a(t)} \to 1 \qquad \text{and} \qquad \frac{A}{t} \to 0$$

It would not be difficult to establish the first two conditions; the crux of the proof of (11-31) is to establish the third condition.

Proof of Equation (11-27): Little's Theorem

The first proof of Theorem 11-4 was given in Little (1961) and proceeded along the lines shown above. We use the proof given in Stidham (1974), which employs the following notation.

Number the customers in the order of their arrival, and for $n \in I_+$, let t_n be the arrival epoch and r_n be the departure epoch. Define the waiting time by

$$w_n = r_n - t_n \qquad n \in I_+ \tag{11-34}$$

and define

$$a(t) = \sup\{n : t_n \le t\} \qquad t \ge 0 \tag{11-35}$$

$$r(t) = \#\{n : r_n \le t\} \qquad t \ge 0 \tag{11-36}$$

and

$$x(t) = a(t) - r(t) \qquad t \ge 0 \tag{11-37}$$

It is important to note that these quantities are not random variables; they are realizations of random variables.

Theorem 11-4 If the limits

$$\lambda = \lim_{t \to \infty} \frac{a(t)}{t} \qquad \text{and} \qquad w = \lim_{N \to \infty} \frac{1}{N} \sum_{n=1}^{N} w_n$$

exist and are finite, then

$$l = \lim_{t \to \infty} \frac{1}{t} \int_0^t x(u)\, du$$

exists and

$$l = \lambda w \tag{11-31}$$

PROOF For each $n \in I_+$, let

$$1_n(t) = \begin{cases} 1 & t_n \leq t < r_n \\ 0 & \text{otherwise} \end{cases} \tag{11-38a}$$

Then

$$w_n = \int_{t_n}^{r_n} 1_n(u) \, du \quad \text{and} \quad x(t) \doteq \sum_{n=1}^{\infty} 1_n(t) \tag{11-38b}$$

If $\lambda = 0$, the queueing system is trite, but special arguments need to be employed; this case is left to you. Henceforth, assume $\lambda > 0$. Define the sets

$$\alpha(t) = \{n : t_n \leq t\} \quad \text{and} \quad \delta(t) = \{n : t_n + w_n \leq t\}$$

They are the customers who arrive by t and who depart by t, respectively. Since $w_n \geq 0$, $\delta(t) \subset \alpha(t)$ for each $t \geq 0$. Let

$$l_a(t) = \sum_{n=1}^{a(t)} w_n \qquad l(t) = \int_0^t x(u) \, du \qquad l_d(t) = \sum_{n=1}^{r(t)} w_n$$

If each customer was paid for waiting at a unit rate, $l_a(t)$ is the amount paid to those who arrived by t, $l(t)$ is the amount paid by t, and $l_d(t)$ is the amount paid to those who left by t. From this heuristic argument,

$$l_a(t) \geq l(t) \geq l_d(t) \qquad t \geq 0 \tag{11-39}$$

A formal proof of (11-39) proceeds as follows. From Proposition 2-1, $\lambda > 0$ implies $\alpha(t)$ is a finite set. From (11-38b),

$$\begin{aligned} l(t) &= \int_0^t \sum_{n=1}^{\infty} 1_n(u) \, du = \int_0^t \sum_{n=1}^{a(t)} 1_n(u) \, du \\ &= \sum_{n=1}^{a(t)} \int_0^t 1_n(u) \, du \end{aligned} \tag{11-40}$$

Also from (11-38b), $w_n \geq \int_0^t 1_n(u) \, du$, with equality holding only for $n \in \delta(t)$. Since $\delta(t) \subset \alpha(t)$, (11-40) implies (11-39).

We use (11-39) to obtain (11-31) by manipulating limits. From Proposition 2-1, $\lambda > 0$ implies

$$\lim_{t \to \infty} a(t) = \infty \quad \text{and} \quad \lim_{n \to \infty} t_n = \infty$$

and consequently,

$$w = \lim_{t \to \infty} \frac{1}{a(t)} \sum_{n=1}^{a(t)} w_n$$

Since λ and w are both assumed to be finite,

$$\lambda w = \lim_{t \to \infty} \frac{a(t)}{t} \frac{1}{a(t)} \sum_{n=1}^{a(t)} w_n$$

$$= \lim_{t \to \infty} \frac{1}{t} \sum_{n=1}^{a(t)} w_n = \lim_{t \to \infty} \frac{l_a(t)}{t} \tag{11-41}$$

Suppose we prove

$$\lim_{t \to \infty} \frac{l_a(t)}{t} = \lim_{t \to \infty} \frac{l_d(t)}{t} \tag{11-42}$$

Then (11-39), (11-41), and (11-42) imply (11-31). The first step in proving (11-42) is to establish

$$\lim_{n \to \infty} \frac{w_n}{t_n} = 0 \tag{11-43}$$

Write

$$\frac{w_n}{n} = \frac{\sum_{i=1}^{n} w_i - \sum_{i=1}^{n-1} w_i}{n}$$

$$= \frac{1}{n} \sum_{i=1}^{n} w_i - \frac{n-1}{n} \frac{1}{n-1} \sum_{i=1}^{n-1} w_i$$

From the assumption that $w < \infty$ we obtain $\lim_{n \to \infty} (w_n/n) = w - w = 0$. Proposition 2-1 asserts $t_n/n \to \lambda$, hence (11-43) holds true.

A consequence of (11-43) is that for any $\varepsilon > 0$, there exists N such that $n > N$ implies $w_n < \varepsilon t_n$. Choose t so large that $a(t) > N$. Then

$$l_a(t) \geq l_d(t) = \sum_{\substack{n \in \delta(t) \\ n \leq N}} w_n + \sum_{\substack{n \in \delta(t) \\ n > N}} w_n$$

$$\geq \sum_{\substack{n \in \delta(t) \\ n \leq N}} w_n + \sum_{\substack{n : t_n \leq t/(1 + \varepsilon) \\ n > N}} w_n$$

$$= \sum_{\substack{n \in \delta(t) \\ n \leq N}} w_n + \sum_{n \in \alpha[t/(1 + \varepsilon)]} w_n - \sum_{\substack{n \in \alpha[t/(1 + \varepsilon)] \\ n \leq N}} w_n$$

The first and third terms on the right-side are finite sums, even as $t \to \infty$, so they are $o(t)$ as $t \to \infty$. The middle term is $l_a[t/(1 + \varepsilon)]$. Hence,

$$\lim_{t \to \infty} \frac{l_a(t)}{t} \geq \lim_{t \to \infty} \frac{l_d(t)}{t}$$

$$\geq \lim_{t \to \infty} \frac{l_a[t/(1 + \varepsilon)]}{t}$$

$$= \lim_{t \to \infty} \frac{t/(1 + \varepsilon)}{t/(1 + \varepsilon)} \frac{l_a[t/(1 + \varepsilon)]}{t}$$

$$= (1 + \varepsilon)^{-1} \lim_{t \to \infty} \frac{l_a(t)}{t} \tag{11-44}$$

Since ε is arbitrarily small, (11-44) implies (11-42), which completes the proof.

$\square$

Equation (11-31), $l = \lambda w$, is a deterministic relation that holds for each realization of a queueing process, while (11-27), $L = \lambda W$, is a relation among expected values. To derive (11-27) from (11-31), interpret $x(t)$ as $X(t; \omega)$ and w_n as $W_n(\omega)$, where ω is a sample path of the queueing process.† Then (11-31) can be written

$$\lim_{t \to \infty} \frac{1}{t} \int_0^t X(u; \omega)\, du = \lim_{t \to \infty} \frac{A(t; \omega)}{t} \lim_{n \to \infty} \frac{1}{n} \sum_{i=1}^n W_i(\omega) \qquad (11\text{-}31a)$$

where $A(t)$ is the number of arrivals during $(0, t]$. By using an obvious notation, (11-31a) is

$$l(\omega) = \lambda(\omega)w(\omega) \qquad (11\text{-}31b)$$

The following line of reasoning in Little (1961) obtains (11-27) from (11-31b) and illustrates the power that can be gained by employing the general framework of stochastic processes. Let Ω be the set of all sample paths and $\{\Omega, \mathcal{F}, \mathcal{P}\}$ be a probability space on which $\{X(t); t \geq 0\}$ and $\{W_n; n \in I_+\}$ are defined.

Corollary 11-4 Let‡

$$L = \int_\Omega l(\omega)\, d\mathcal{P}(\omega) \qquad \text{and} \qquad W = \int_\Omega w(\omega)\, d\mathcal{P}(\omega)$$

If there is a constant λ such that (w.p.1)

$$\lambda(\omega) = \lambda \qquad (11\text{-}45)$$

then

$$L = \lambda W \qquad (11\text{-}46)$$

PROOF Integrating both sides of (11-31b) yields

$$L = \int_\Omega \lambda(\omega)w(\omega)\, d\mathcal{P}(\omega) = \lambda \int_\Omega w(\omega)\, d\mathcal{P}(\omega) = \lambda W \qquad \square$$

From Theorem 6-10, when $\{W_n; n \in I_+\}$ and $\{X(t); t \geq 0\}$ are regenerative, the number of customers served in a regenerative cycle is an aperiodic r.v. with finite mean, and the times between regeneration epochs are not arithmetic, then

$$w(\omega) = \lim_{n \to \infty} E(W_n) \triangleq E(W) \qquad \text{and} \qquad l(\omega) = \lim_{t \to \infty} E[X(t)] \triangleq E(X)$$

hold w.p.1. If $\{A(t); t \geq 0\}$ is regenerative (in particular, if it is a renewal process

---

† The precise construction of the sample paths and the probability space to which they belong is neglected because those who are concerned with such issues can easily fill in the details and those who are not should find the concepts intuitive.

‡ L and W are the means of the random variables $l(\cdot)$ and $w(\cdot)$.

with rate λ), then (11-45) holds and (11-45) asserts

$$E(X) = \lambda E(W) \tag{11-46a}$$

Let an asterisk denote the stationary version of a process. Then from (10-43), the limiting probabilities of a process are the initial probabilities of its stationary version, and so

$$E(W) = E(W_1^*) \quad \text{and} \quad E(X) = E[X^*(0)]$$

and the interpretation in (11-27) is obtained. When $\{W_n; n \in I_+\}$ and† $\{U_n; n \in I_+\}$ are stationary processes, the ergodic theorems in Section 10-2 imply that $w(\omega)$ and $\lambda(\omega)$ exist. Theorem 11-4 implies that $l(\omega)$ exists. Therefore

$$\int_\Omega w(\omega) \, d\mathscr{P}(\omega) = E(W_1) \quad \text{and} \quad \int_\Omega l(\omega) \, d\mathscr{P}(\omega) = E(X)$$

Let $E(U_1) = 1/\lambda$; Proposition 2-1 and Theorem 10-2 imply that (11-45) holds when $\{U_n; n \in I_+\}$ is metrically transitive (Definition 10-3). Thus, (11-46) implies (11-27) under these conditions.

The above results are summarized in this theorem.

Theorem 11-5 (a) Assume that $\{W_n; n \in I_+\}$, $\{X(t); t \geq 0\}$, and $\{A(t); t \geq 0\}$ are regenerative processes; the W-process is aperiodic; the A-process is non-arithmetic; and the mean lengths of the regeneration cycles are finite. If λ and $E(W)$ are finite, then

$$L = \lambda W \tag{11-46}$$

and

$$E(X) = \lambda E(W) \tag{11-46a}$$

are valid.

(b) Assume that $\{W_n; n \in I_+\}$ and $\{U_n; n \in I_+\}$ are stationary processes, with the latter metrically transitive and $E(U_1) = 1/\lambda < \infty$. If $E(W) < \infty$, then (11-46a) is valid.

If you have not done so already, notice that the stochastic hypotheses in Theorem 11-5 are used only to show that $w(\omega)$ and $\lambda(\omega)$ exist and that (11-44) holds. Any other hypotheses that accomplish this could be used to obtain the same conclusions.

Examples

Example 11-9 In the $GI/G/c/K$ queue, at most K customers can be in the system at a time. Let B be the steady-state probability that a customer is denied access to the facility, $1/\mu$ be the mean service time, and L_b be the

---

† Recall that U_n is the time between the $(n-1)$th and nth arrivals.

average number of busy servers. Exercise 11-15 asks you to prove formally that the rate at which customers gain access to the servers is $\lambda(1 - B)$, where λ is the rate at which customers attempt to enter the facility. Applying Theorem 11-5 to the "system" formed by the servers, we obtain $L_b = \lambda(1 - B)/\mu$, or

$$\lambda = \lambda B + L_b \mu \tag{11-47}$$

In telephony, (11-47) is called the equation of *conservation of load*. The left-side is the arrival rate, and the first term on the right-side is the arrival rate of those customers who are denied access to the facility. By conservation of flow, the second term must be the arrival rate of the accepted customers. The form of the second term makes it natural to interpret it as the departure rate from the facility; by conservation again, this is correct because in the steady state, the arrival rate of accepted customers equals the departure rate from the facility.

Rewrite (11-47) as $B = 1 - L_b/a$, where $a = \lambda/\mu$. Since $L_b \leq c$, we obtain

$$B \geq 1 - 1/\rho \tag{11-48}$$

where $\rho = a/c$. When the facility is so heavily loaded that most of the servers are busy most of the time (that is, L_b is close to c), the right-side of (11-48) is a good approximation of B. For the $M/M/c/K$ queue, the approximation performs well† when $\rho > 1.5$. ☐

Example 11-10 The formula $L = \lambda W$ has interpretations in contexts other than queueing models. Figure 11-3 can be interpreted in an inventory context. Then $a(\cdot)$ is the cumulative number of items put into inventory, $r(\cdot)$ is the cumulative number of items taken out of inventory, and $x(t)$ is the inventory at t. Theorems 11-4 and 11-5 assert that the average inventory level equals the rate at which items enter inventory multiplied by the average time an item spends in inventory.

Another nonqueueing context is cash balances. Checks deposited in a bank cannot be drawn against (or earn interest) until the check is cleared by the writer's bank. Theorems 11-4 and 11-5 assert that the average daily balance of uncleared checks L is the check deposit rate λ (in units of dollars per day), multiplied by the average number of days to clear a check W. Utilities and large retail stores attempt to make L small by reducing W. They establish accounts in many banks and deposit checks in the same bank on which they are written. A large firm incurs costs when it opens a new account; $L = \lambda W$ can provide guidance in deciding whether opening a new account is worthwhile. ☐

Other uses of $L = \lambda W$ are given in Sections 11-5 and 11-6.

---

† See Matthew J. Sobel, "Simple Inequalities for Multiserver Queues," *Manage. Sci.* **26**: 951–954 (1980).

An Existence Theorem for $E(W)$

To apply Theorem 11-5, $E(W) < \infty$ has to be established. The following theorem is easy to apply and difficult to prove.

Theorem 11-6 In a $GI/G/c$ queue, if for some fixed $k \in I_+$,

(a) $E(U_1) < \infty$ and $E(V_1^{k+1}) < \infty$

and

(b) $\rho \triangleq \dfrac{E(V_1)}{cE(U_1)} < 1$

then

$$E(W^k) < \infty$$

PROOF See J. Kiefer and J. Wolfowitz, "On the Characteristics of the General Queueing Process, with Applications to Random Walk," *Ann. Math. Stat.* **27**: 147–161 (1956). ☐

A Generalization of $L = \lambda W$

An essential part of the proof of Theorem 11-4 is the representation of w_n and $x(t)$ given in (11-38b) via the indicator function in (11-38a). Maxwell (1970) observed that interesting generalizations of $L = \lambda W$ could be obtained by replacing $1_n(t)$ by an arbitrary integrable function, $f_n(t)$ say; a rigorous treatment and several examples are given in Brumelle (1971, 1972). The version given here is due to Heyman and Stidham (1980).

It may reduce confusion if the usually suppressed ω's are used. Let Ω denote the set of appropriate sample paths of the queue and $(\Omega, \mathcal{F}, \mathcal{P})$ be the corresponding probability space. Number the customers in order of arrival, and associate with customer n a real-valued integrable function f_n defined on $\Omega \times [0, \infty)$. For each $\omega \in \Omega$, define

$$g_n(\omega) = \int_0^\infty f_n(\omega; u)\, du \qquad n = 1, 2, \ldots$$

and

$$h(\omega; t) = \sum_{n=1}^\infty f_n(\omega; t) \qquad t \geq 0$$

When they exist, define

$$G(\omega) = \lim_{N \to \infty} \frac{1}{N} \sum_{n=1}^N g_n(\omega)$$

and

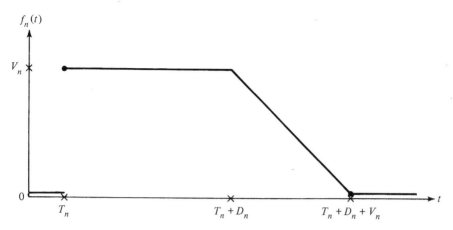

Figure 11-4 Graph of $f_n(t)$ = remaining work of customer n at time t.

$$H(\omega) = \lim_{t \to \infty} \frac{1}{t} \int_0^t h(\omega; u) \, du$$

Thus, $g_n(\omega)$ and $G(\omega)$ are generalizations of w_n and w, and $h(t; \omega)$ and $H(\omega)$ are generalizations of $x(t)$ and l. The analog of (11-31b) that is proved below is

$$H(\omega) = \lambda(\omega)G(\omega) \tag{11-49}$$

Let

$$H \triangleq \int_\Omega H(\omega) \, d\mathscr{P}(\omega) \qquad \text{and} \qquad G \triangleq \int_\Omega G(\omega) \, d\mathscr{P}(\omega) \tag{11-50}$$

Then the analog of Corollary 11-4 is: if $P\{\lambda(\omega) = \lambda\} = 1$, then

$$H = \lambda G \tag{11-51}$$

Example 11-11: Pollaczek-Khintchine equation† Consider an M/G/1 queue where $\rho < 1$ and customers are served FIFO. Let $f_n(t)$ denote the remaining work of customer n at time t. A typical graph of $f_n(t)$ is shown in Figure 11-4.

With this choice of $f_n(\cdot)$, $h(t)$ is the work in the system at t. Since the system regenerates, H is the expected virtual delay in the steady state. From "Poisson arrivals 'see' time averages," H is also the expected delay (i.e., wait in queue) in the steady state, denoted $E(D^*)$. Since g_n is the area between $f_n(\cdot)$ and the axis of abscissas, $g_n = D_n V_n + V_n^2/2$. Since customers are served FIFO, D_n and V_n are independent, and so $G = E(D^*)/\mu + E(V_1^2)/2$. Thus, (11-51) yields

$$E(D^*) = \rho E(D^*) + \lambda E(V_1^2)/2$$

from which

† This example is taken from Brumelle (1971).

$$E(D^*) = \frac{\lambda E(V_1^2)}{2(1-\rho)} \qquad \rho < 1 \tag{11-52}$$

$\square$

Proof of Equation (11-49)

The only reason (11-49) does not follow directly from the proof of Theorem 11-4 is that the analog of (11-43) is not necessarily true. In order to effect the analogous step, assumptions (a) and (b) are placed in Theorem 11-7.

Theorem 11-7 Fix $\omega \in \Omega$, and assume that for each $n \in I_+$,
(a) there exists s_n with $0 \le s_n < \infty$ such that

$$t \notin [t_n, t_n + s_n] \Rightarrow f_n(t) = 0$$

with

(b) $\displaystyle\lim_{n \to \infty} \frac{s_n}{n} = 0$

(c) $f_n(t) \ge 0 \qquad$ for $\qquad t \ge 0$

(d) $g_n < \infty$

If G and λ exist, with $0 < \lambda < \infty$ and $G < \infty$, then

$$H(\omega) = \lambda G(\omega) \tag{11-49}$$

is valid.

PROOF The proof is analogous to the proof of Theorem 11-4. The outline of the proof is given; filling in the details is left to you. Define the sets

$$\alpha(t) = \{n : t_n \le t\} \qquad \text{and} \qquad \delta(t) = \{n : t_n + s_n \le t\}$$

and the functions

$$G_a(t) = \sum_{n \,\varepsilon\, \alpha(t)} g_n \qquad F_n(t) = \int_0^t f_n(u)\, du \qquad G_d(t) = \sum_{n \,\varepsilon\, \delta(t)} g_n$$

By mimicking the proof of (11-39),

$$G_a(t) \ge \int_0^t h(u)\, du \ge G_d(t) \tag{11-53}$$

is readily established. Since λ and G are both assumed to be finite, repeating the derivation of (11-41) yields

$$\lambda G = \lim_{t \to \infty} \frac{G_a(t)}{t} \tag{11-54}$$

Combining (11-53) and (11-54) shows that the proof can be completed by establishing

$$\lim_{t \to \infty} \frac{G_a(t)}{t} = \lim_{t \to \infty} \frac{G_d(t)}{t}$$

This can be done by mimicking the proof of (11-42). □

Corollary 11-7 Let H and G be defined by (11-50). If $P\{\lambda(\omega) = \lambda\} = 1$, then $H = \lambda G$ is valid.

PROOF Mimic the proof of Corollary 11-4. □

Theorem 11-8 Assume, in addition to assumptions (a) through (d) of Theorem 11-6, that either

(e) $\{U_n; n \in I_+\}$ is regenerative with finite mean cycle time and nonarithmetic

or

(f) $\{U_n; n \in I_+\}$ is stationary and metrically transitive

and that either

(g) $\{g_n; n \in I_+\}$ is regenerative with finite mean cycle time and aperiodic

or

(h) $\{g_n; n \in I_+\}$ is stationary

Then $H = \lambda G$ is valid.

PROOF Mimic the proof of Theorem 11-5. □

Assumption (c) is stronger than required. If $\int_0^\infty |f_n(u)| \, du < \infty$ and the positive and negative parts of G are finite, then $H(\omega) = \lambda(\omega)G(\omega)$ is valid (see Exercise 11-20).

Example 11-12: Example 11-11 continued Consider a $GI/G/c$ queue with arrival rate λ and mean service time $1/\mu$. Assume $a = \lambda/\mu < c, P\{U_1 > V_1\} > 0$, and $E(D^*) < \infty$. Choose $f_n(\cdot)$ as in Example 11-11. Using a regenerative argument that should be familiar, we get

$$H = \lim_{t \to \infty} E[W(t)] \triangleq E[W^*(\infty)]$$

We have, as in Example 11-11, $g_n = D_n V_n + V_n^2/2$. When D_n is independent of V_n (for example, when customers are served FIFO or LIFO),

$$G = E(D^*)E(V_1) + \frac{E(V_1^2)}{2}$$

and $H = \lambda G$ yields

$$E[W^*(\infty)] = aE(D^*) + \frac{\lambda E(V_1^2)}{2} \tag{11-55}$$

When D_n and V_n are dependent (e.g., when the queue discipline serves next that customer in the queue with the smallest service time),

$$G = \lim_{n \to \infty} E(D_n V_n) + \frac{E(V_1^2)}{2}$$

Since $E(D^*)$ exists even when D_n and V_n are dependent (because that does not affect the regenerative properties), we may write the limit above as $\int_0^\infty E(D^* | V = v) \, dG(v)$. Thus, $H = \lambda G$ yields

$$E[W^*(\infty)] = \lambda \int_0^\infty E(D^* | V = v) \, dG(v) + \frac{E(V_1^2)}{2} \qquad (11\text{-}56)$$

The following inequalities† follow from (11-55):

$$E[W^*(\infty)] \geq (\leq)E(D^*) \Leftrightarrow E(D^*) \leq (\geq)\frac{aE(V_1^2)}{2(1-a)}$$

If $c = 1$, then $a = \rho$, so $E[W^*(\infty)] = E(D^*)$ if, and *only if*, $E(D^*)$ is given by the Pollaczek-Khintchine equation (11-52). Since $a < c$, (11-55) also implies

$$E[W^*(\infty)] - cE(D^*) \leq \frac{\lambda E(V_1^2)}{2} = \frac{aE(V_1^2)}{2E(V_1)} < \frac{cE(V_1^2)}{2E(V_1)} \qquad \square$$

Conservation laws in priority queues are shown to be consequences of $H = \lambda G$ in Section 11-5.

Additional Comments

The $H = \lambda G$ equations, (11-49) and (11-51), are relations among averages; the left-side is a customer average, and the right-side is the product of two time averages. Exercise 11-21 shows how to use $H = \lambda G$ to relate the steady-state probability that, on arrival, a customer finds j other customers in the facility (a customer average) to the asymptotic probability that there are j customers in the facility at time t (a time average).

The heuristic argument leading to Little's theorem shows that it can be interpreted as a conservation law which expresses the area between $a(\cdot)$ and $r(\cdot)$ in two ways. The theorems in this section are natural tools to employ in seeking conservation laws or relations between customers averages and time averages.

EXERCISES

11-14 In an M/G/1 queue, let $1/\mu$ be the mean service time, λ be the arrival rate, $E(D^*)$ be the expected delay in the steady state, and L be the average number of customers in the facility. Some

---

† These inequalities are due to Kneale T. Marshall and Ronald W. Wolff, "Customer Average and Time Average Queue Lengths and Waiting Times," *J. Appl. Prob.*, **8**: 535–542 (1971).

people believe $E(D) = L/\mu$; why is that assertion false in general? (Assume both sides exist and are finite—that is not the issue.) Why is it true for exponentially distributed service times? (Do not use the explicit formulas.) Use the assertion and $L = \lambda W$ to obtain L and $E(D^*)$.

11-15 In Example 11-9, let $J_n = 1$ if customer n gains access to the facility, and let $J_n = 0$ otherwise. Assume $B = \lim_{n \to \infty} P\{J_n = 0\}$ exists. The number of arrivals who reach the facility is given by $N(t) = \Sigma_1^{A(t)} J_n$. Prove

$$\lim_{t \to \infty} \frac{A(t)}{t} = \lambda \Rightarrow \lim_{t \to \infty} \frac{N(t)}{t} = \lambda(1 - B)$$

11-16 The M/G/1/K queue is the special case of Example 11-9 in which the arrivals form a Poisson process and there is one server. In Exercise 7-62, the limiting probability that j customers are present at a departure epoch, π_j say, is obtained. Let $X(t)$ denote the number of customers in the system at time t and $p_j = \lim_{t \to \infty} P\{X(t) = j\}$ (how do we know p_j exists?), $j = 0, 1, \ldots, K$. Let λ be the arrival rate, $1/\mu$ be the mean service time, and $\rho = \lambda/\mu$. Show

$$p_j = \frac{\pi_j}{\rho + \pi_0} \qquad j = 0, 1, \ldots, K$$

11-17 Example 7-25 describes an inventory system where demands and returns occur according to independent Poisson processes with rates λ and μ, respectively. When the inventory level reaches b, returns are not added to the inventory. Let p_0 and p_b be the limiting probabilities that the inventory is empty and full, respectively. Prove, without explicit calculations, that

$$\lambda = \lambda\pi_0 + (1 - \pi_b)\mu$$

Give an intuitive explanation of this equation.

11-18 In Example 10-10, let C be a generic r.v. representing the time elapsing between the epoch a particular job joins the queue at station 1 and the next such epoch; it is called the *cycle time* of a job. Let C_i be the cycle time conditional on the job's passing through station i, $i = 2, 3, \ldots, N$. Show

$$E(C) = \frac{M}{a_1\mu_1}$$

and

$$E(C_i) = \frac{L_1 + L_i/\pi_i}{a_1\mu_1} \qquad i = 2, 3, \ldots, N$$

where L_i is the average number of jobs (in queue and in service) at station i; π_i and a_i are defined in Example 10-10.

11-19 Verify that the conditions of Theorem 11-6 hold in Examples 11-11 and 11-12.

11-20 Discard assumptions (c) and (d) in Theorem 11-6. Let $f_n^+(t) = \max\{0, f_n(t)\}$ and $f_n^-(t) = \max\{0, -f_n(t)\}$. Affix a plus to g_n, $h(\cdot)$, G, and H when they are defined in terms of f_n^+; affix a minus similarly. Assume (a), (b), and $\int_0^\infty |f_n(t)| \, dt < \infty$ hold. Show that if λ, G^+, and G^- exist for each ω, with $0 < \lambda < \infty$, $G^+ < \infty$, and $G^- < \infty$, then (11-49) is valid.

11-21 In the GI/M/c/K queue, a Markov chain embedded at arrival epoch provides the steady-state probabilities at arrival epochs, π_j say (Exercise 7-10). Suppose we want the limiting time-average probabilities, p_j say. Provide an intuitive argument for the validity of

$$\lambda\pi_{j-1} = \min\{j, c\}\mu p_j \qquad j = 1, 2, \ldots, K \tag{11-57}$$

where $1/\lambda$ is the mean time between arrivals and $1/\mu$ is the mean service time.

A rigorous proof of (11-57) can be obtained via Theorem 11-7. Let $X(t)$ be the number of customers in the system at time t, and let $I_j(t)$ be 1 if $X(t) = j$ and 0 otherwise. Fix j and choose

$$s_n = \inf\{t : t \ge 0, X(t_n + t) = j - 1\}I_j(t)$$

and

$$f_n(t) = \begin{cases} I_j(t) & \text{if } t_n \le t < t_n + s_n \\ 0 & \text{otherwise} \end{cases}$$

(a) Show that assumptions (a) through (e) and (g) of Theorem 11-7 are satisfied.

(b) Show that

$$g_n = \int_{t_n}^{s_n + t_n} I_j(t) \, dt \triangleq \gamma_n(j)$$

Interpret $\gamma_n(j)$ in terms of the queueing process.

(c) Let $A_j(k)$ be the number of arrivals among the first k arrivals who find $j - 1$ customers in the system. Show

$$G = \lim_{k \to \infty} \frac{A_j(k)}{k} \lim_{k \to \infty} \frac{1}{A_j(k)} \sum_{n=1}^{A_j(k)} \gamma_n(j)$$

$$= \frac{\pi_{j-1}}{\mu \min\{j, c\}}$$

(d) Apply Theorem 11-8 and obtain (11-56).

11-22 In the $GI/G/c$ queue, let $R(t)$ be the remaining service time of the customers in service at time t [if there are none, set $R(t) = 0$]. Prove

$$E(R) \triangleq \lim_{t \to \infty} E[R(t)] = \frac{E(V_1^2)}{2E(V_1)}$$

where V_1 is the service time of customer 1 and $\rho < 1$ is the traffic intensity. Show further that when $c = 1$,

$$E(R \mid R > 0) = \frac{E(V_1^2)}{2E(V_1)}$$

Give a heuristic argument for the $M/G/1$ queue.

11-23 The purpose of this exercise is to show that assumption (b) in Theorem 11-6 is necessary. Choose $t_n = n$ and

$$f_n = \begin{cases} 1 & \text{if } t \in [k, k + 1/n] \text{ for some } k \in \{n, n + 1, \dots, 2n - 1\} \\ 0 & \text{otherwise} \end{cases}$$

Then $\lambda = g_n = G = 1$ and $s_n/n \to 2$. Let $H_k = \int_k^{k+1} h(t) \, dt$. Show that if H exists, $H = \lim_{k \to \infty} H_k$ and for $k \in I_+$,

$$H_{2k-1} = \frac{1}{k} + \frac{1}{k+1} + \cdots + \frac{1}{2k-1}$$

and

$$H_{2k} = \frac{1}{k+1} + \frac{1}{k+2} + \cdots + \frac{1}{2k}$$

Accept the fact that each of these series converges to $\ln 2 = 0.6931$, and observe that (11-49) is violated.

11-24 Let $V(x) = P\{W(\infty) \le x\}$ in a $GI/G/1$ queue with $\rho < 1$. Accept the fact that when the interarrival times and the service times have continuous densities, $v(x) \triangleq dV(x)/dx$ exists for $x > 0$. For each $x > 0$, let $D(t; x)$ be the number of times $W(t)$ crosses x from above during $(0, t]$. Prove

$$v(x) = \lim_{t \to \infty} \frac{D(t; x)}{t} \qquad x > 0 \qquad \text{(w.p.1)} \tag{11-58}$$

Hint: Choose $\Delta x > 0$, define

$$f_n(t) = \begin{cases} 1 & \text{if } W(t) \in [x, x + \Delta x] \quad \text{and} \quad T_n \le t < T_{n+1} \\ 0 & \text{otherwise} \end{cases}$$

and use $H = \lambda G$.

11-25 (Continuation) In an M/G/1 queue with $\rho < 1$, let $U(t; x)$ be the number of times $W(t)$ crosses x from below during $(0, t]$. Show that

$$\lim_{t \to \infty} \frac{U(t; x)}{t} = \lambda \int_0^x G^c(x - y) \, dV(x) \qquad x > 0 \qquad \text{(w.p.1)} \tag{11-59}$$

where λ is the arrival rate and $G^c(x) = P\{V_1 > x\}$. Combine (11-58) and (11-59) to obtain the Pollaczek-Khintchine formulas for the steady-state delay and queue-size probabilities.

11-26 Use (11-58) to derive the following formula due to L. Takács:

$$V(x) = 1 - \rho + \rho \int_0^x D(x - y) \frac{G^c(y)}{v} \, dy$$

where $D(\,\cdot\,)$ is the steady-state delay distribution and $v = E(V_1)$.

11-27 Use (11-58) to derive the following formula due to J. Hooke:

$$v(x) = \int_x^\infty \left[\int_0^z d(z - u) \, dG(u) \right] \lambda f(z - x) \, dz \qquad x > 0$$

where $d(z) = dD(z)/dz$ and $f(\,\cdot\,)$ is the density function of the interarrival times. Use the above formula to establish

$$V(x) = P\{(D^* + V_1 - U_e)^+ \le x\}$$

where U_e is an r.v. that is independent of D^* and V_1 and has density function $\lambda F^c(\,\cdot\,)$.

11-4 EFFECTS OF ORDER OF SERVICE ON WAITING TIMES

In this section we investigate the effect of the queue discipline on the distribution of waiting times.

Extremal Properties of FIFO and LIFO

Example 11-7 shows that among those queue disciplines that do not affect L, the asymptotic mean waiting time is the same. Theorem 11-9 below asserts that among those disciplines, the smallest (largest) variance of the asymptotic delay time is achieved by FIFO (LIFO).

Let D_n be the delay of the nth arriving customer, C_n be the number of customers served in the nth busy period,† and $X(t)$ be the number of customers in

---

† A busy period commences when a customer arrives at an empty system and ends at the next departure epoch where the system is empty.

the system at time t. Assume that the arrival and departure epochs cause

1. $E(\sum_{n=1}^{C_1} D_n^2) < \infty$ for any discipline.
2. $\{X(t); t \geq 0\}$ regenerates at the start of each busy period.
3. For any discipline, the d.f. of $X(t)$ is the same as if FIFO were used.

The first assumption makes it meaningful to compare the variances of different queue disciplines over a busy period. The second assumption guarantees that events in different busy periods are independent; it also enables us to employ a limit theorem for regenerative process. The third assumption implies that the order of service does not depend on information about the service times; in particular, a discipline such as "serve next the customer with shortest service time" is ruled out. Another implication of the third assumption is that a server is never idle when customers are in queue. You should check that LIFO and random order of service satisfy the third assumption. For technical reasons that will be apparent, we also assume:

4. The expected number of customers served in each busy period is finite.

Consider a particular busy period, and let T_i denote the (random) arrival epoch of the ith customer to appear during that busy period, Y_i denote the ith epoch at which a customer enters service, and C be the number of customers served during the busy period. The different queue disciplines allowed by the third assumption can all be represented as a permutation of the numbers $\{1, 2, \ldots, C\}$ such that the ith arriving customer enters service at Y_{j_i} with $Y_{j_i} \geq T_i$.

With this notation, the delay of the ith arriving customer is

$$D_i \triangleq Y_{j_i} - T_i \tag{11-60}$$

This is the main result.†

Theorem 11-9 For any queue discipline satisfying assumptions 1 through 4, in each busy period $E(\sum_{i=1}^{C} D_i^2)$ is minimized by FIFO and maximized by LIFO.

PROOF Use (11-60) to obtain

$$\sum_{i=1}^{C} D_i^2 = \sum_{i=1}^{C} Y_{j_i}^2 + \sum_{i=1}^{C} T_i^2 - 2\sum_{i=1}^{C} T_i Y_{j_i} \tag{11-61}$$

Assumption 3 implies that each of the first two sums on the right-side of (11-61) are distributed independently of the queue discipline. Hence, their ex-

---

† The minimization part is due to J. F. C. Kingman, "The Effect of Queue Discipline on Waiting Time Variance," *Proc. Comb. Phil. Soc.* **58**: 163–164 (1962). The maximization part is due to D. G. Tambouratzis, "On a Property of the Variance of the Waiting Time of a Queue," *J. Appl. Prob.* **5**: 702–703 (1968). A generalization of this theorem is given in Oldrich A. Vasicek, "An Inequality for the Variance of Waiting Time under a General Queueing Discipline," *Oper. Res.* **25**:879–884 (1977).

pected values do not depend on the discipline. Let S denote the third sum. We will show that it is maximized by FIFO and minimized by LIFO.

Consider any sample path ω of the queue. Let y_i be the realized value of Y_i, t_i be the realized value of T_i, and $S(\omega)$ be the realized value of S.

The discipline is irrelevant unless at least two customers are present at some y_i. Suppose customers i and $i-j$ $(0 < j < i)$ are present at y_k and $i-j$ is placed into service (this cannot be LIFO). Customer i starts service at some later epoch, y_{k+m} say, with $m > 0$. The contribution to $S(\omega)$ of these two customers is $t_{i-j}y_k + t_i y_{k+m}$, with†

$$t_{i-j} < t_i \le y_k < y_{k+m} \tag{11-62}$$

Observe that

$$t_i y_k + t_{i-j} y_{k+m} - (t_{i-j} y_k + t_i y_{k+m}) = (y_{j+m} - y_k)(t_{i-j} - t_j) < 0 \tag{11-63}$$

where the equality is an algebraic identity and the inequality follows from (11-62). Equation (11-63) implies that starting to serve i before $i-j$ will decrease $S(\omega)$. Thus, any discipline other than LIFO can be changed to achieve a smaller value of $S(\omega)$, and LIFO cannot be improved by pairwise interchanges in the order of service. Since any permutation of a finite set can be turned into any other permutation by a sequence of pairwise interchanges, LIFO minimizes $S(\omega)$ for every ω; therefore it minimizes $E(S)$. Similarly, (11-63) shows that if FIFO were not used, $S(\omega)$ could be made larger, hence $E(S)$ is maximized by FIFO. ☐

Notice that we proved a stronger result than claimed. We showed that $D_1^2(\omega) + \cdots + D_C^2(\omega)$ is minimized by FIFO and maximized by LIFO for every sample path ω. The strategy of establishing a relation among moments by demonstrating the relation for all (perhaps excluding a set of probability zero) sample paths was used in Section 11-3 to prove $L = \lambda W$ and $H = \lambda G$ and is utilized in other proofs in this section.

Corollary 11-9 Let‡ $D^* = \text{dlim}_{i \to \infty} D_i$. Then $\text{Var}(D^*)$ is minimized by FIFO and maximized by LIFO.

PROOF Assumptions 1, 2, and 4 imply that $\{D_i^2 ; i \in I\}$ is a regenerative process satisfying the hypotheses of Theorem 6-10. Hence,

$$E[(D^*)^2] = \lim_{i \to \infty} E(D_i^2) = \frac{E\left(\sum_{i=1}^C D_i^2\right)}{E(C)}$$

Theorem 11-9 asserts $E[(D^*)^2]$ is minimized by FIFO and maximized by LIFO. Assumption 3 implies that the limiting value of the expected queue length does not depend on which discipline is used. Hence, by $L = \lambda W$, $E(D^*)$ is the same for all disciplines, and the assertion of the corollary is established. ☐

---

† Customers who arrive together or start service together are indistinguishable.
‡ The notation $D^* = \text{dlim}_{i \to \infty} D_i$ is defined on p. 182.

Optimality of the Shortest Remaining Processing-Time Discipline

The shortest remaining processing-time (SRPT) discipline places into service at time t that customer present at t whose remaining service time is smallest. This presumes that (1) the service time of a customer is known on arrival, (2) the service time already received by each customer is known, and (3) a newly arrived customer may replace a customer who is in service. We express (3) by saying "preemption of service is allowed."

An important class of queue disciplines is described by this definition.

Definition 11-2 A queue discipline is called *work-conserving* if (a) no servers are free when a customer is in the queue and (b) the discipline does not affect the amount of service time given to a customer or the arrival time of any customer.

Examples of work-conserving disciplines are FIFO, LIFO, and random order of service. For single-server systems, (11-4) shows that the work-in-system process has the same sample path for any work-conserving discipline. In particular, it is the sample path that would be realized if FIFO were used. This might not occur when there is more than one server. (Can you provide an example?)

The SRPT discipline is work-conserving if preemption does not change the time required to serve a customer. This implies that preemption does not induce a setup time and that whenever a previously preempted customer reenters service, service resumes from where it halted. This mode of preemptive behavior is called *preemptive-resume*.

Among all work-conserving disciplines, when there is a single server, it is clear that SRPT always produces, at any time t, the smallest elapsed time to the first departure after t. Since arrival epochs are not affected by the discipline, it should be true (this is the assertion that requires proof) that for any sample path of the system, SRPT produces a pointwise minimum of the number in the system process. Consequently, if it exists, the mean of the asymptotic number in system (L in the notation of Section 11-3) is minimized by SRPT. Applying $L = \lambda W$ shows that SRPT minimizes the expected waiting time in the system.

Theorem 11-10: Optimality of SRPT† Let $X(t)$ be the number of customers in a single-server queueing system at time t. If a work-conserving queue discipline is employed, for each realization of arrival and service times ω and each time t, $X(t; \omega)$ using SRPT is less than or equal to $X(t; \omega)$ using any other discipline.

---

† This proof of the theorem is due to Linus Schrage, "A Proof of the Optimality of the Shortest Remaining Processing Time Discipline," *Oper. Res.* **16**: 687–690 (1968). A different proof has been given by Donald R. Smith, "A New Proof of the Shortest Remaining Processing Time Discipline," *Oper. Res.* **26**: 197–199 (1978).

PROOF We show that for a given sequence of arrival and service times, if SRPT is not used during some interval of time, then $X(t; \omega)$ can be reduced during that interval without being increased at any other time.

Let T_n and V_n be the arrival time and service time of customer n, respectively. Customer n cannot be placed in service before T_n and may be in service at any time greater than or equal to T_n. Let

$$\delta_n(t) = \begin{cases} 1 & \text{if customer } n \text{ is in service at } t \\ 0 & \text{otherwise} \end{cases} \tag{11-64}$$

In particular, $\delta_n(t) = 0$ for $t < T_n$. The server can work on at most one customer at a time, so

$$\sum_{n=1}^{\infty} \delta_n(t) \le 1 \qquad t \ge 0$$

Let R_n be the departure epoch of customer n. In terms of (11-64),

$$R_n \triangleq \min\left\{ x : \int_0^x \delta_n(u) \, du = V_n \right\} \qquad n \in I_+ \tag{11-65}$$

Let $\gamma_n(t)$ be the remaining processing time of job n at time t:

$$\gamma_n(t) \triangleq V_n - \int_0^t \delta_n(u) \, du \qquad t \ge 0 \tag{11-66}$$

From (11-65) and (11-66),

$$R_n = \min\left\{ x : \int_0^x \delta_n(u) \, du = \gamma_n(t) + \int_0^t \delta_n(u) \, du \right\}$$

$$= \min\left\{ x : \int_t^x \delta_n(u) \, du = \gamma_n(t), \quad x \ge t \ge t_n, \quad \gamma_n(t) > 0 \right\} \tag{11-67}$$

Let $J(t)$ be the set of customers present at t:

$$J(t) = \{ n : T_n < t \le R_n \}$$

The SRPT discipline is defined by

$$\gamma_k(t) = \min\{\gamma_n(t) : n \in J(t)\} \Rightarrow \delta_k(t) = 1 \tag{11-68}$$

If more than one k satisfies (11-68), then choose one of these k's arbitrarily.

In what follows, a prime means that some discipline other than SRPT is employed. If SRPT is not used, either all disciplines yield the same system size as SRPT (and the theorem is trivially true) or there is an interval $[t, t + s)$, $s > 0$, such that the customer with the SRPT is not in service and some other customer is. Suppose the latter case prevails; that is, $k \in J'(t)$ and $\gamma_k'(t) < \gamma_j'(t)$ for all $j \in J'(t)$, $j \ne k$, yet $\delta_k'(u) = 0$ for all $u \in [t, t + s)$, and $\delta_j(u) = 1$ for all $u \in [t, t + s)$.

Our task is to show that serving customers j and k according to SRPT would reduce the number of customers present during some part of $[t, t + s)$, and not increase it elsewhere.

Let the superscript o denote the modified discipline that agrees with "prime" except that beginning at time t, work on customer k is completed before customer j enters service. Observe that with "prime" the set $\{u: \delta'_k(u) + \delta'_j(u) = 1\}$ represents the epochs when either k or j is in service; by work conservation, it equals $\{u: \delta^o_k(u) + \delta^o_j(u) = 1\}$. Thus, (11-67) yields

$$\max\{R'_k, R'_j\} = \min\left\{x : \int_t^x [\delta'_k(u) + \delta'_j(u)]\, du = \gamma'_k(t) + \gamma'_j(t)\right\}$$

$$= \min\left\{x : \int_t^x [\delta^o_k(u) + \delta^o_j(u)]\, du = \gamma^o_k(t) + \gamma^o_j(t)\right\}$$

$$= \max\{R^o_k, R^o_j\}$$

It follows from (11-68) that

$$R^o \triangleq \min\{R^o_j, R^o_k\} < \min\{R'_j, R'_k\} \triangleq R'$$

Thus,

$$X^o(u) = \begin{cases} X'(u) & \text{if } u \in (R^o, R') \\ X'(u) - 1 & \text{if } u \in (R^o, R') \end{cases}$$

Hence, any discipline other than SRPT can be improved by making it more consistent with SRPT. $\qquad\square$

Example 11-15 below shows that for an $M/G/1/SRPT$ system where $G(\cdot)$ has a density, the expected waiting time in the steady state is given by

$$E(W^*_{SRPT}) = \lambda \int_0^\infty \frac{\int_0^v xG^c(x)\, dx}{[1 - \rho(v)]^2}\, dG(v) + \int_0^\infty \frac{G^c(v)}{1 - \rho(v)}\, dv \qquad (11\text{-}69)$$

where $G(\cdot)$ is the service-time d.f. and $\rho(x) = \lambda \int_0^x y\, dG(y)$. The numerical results shown in Table 11-1 are given in Schrage and Miller (1966). Table 11-1 indicates that SRPT can achieve a significant reduction in expected waiting time compared to FIFO and that the percentage reduction increases with the traffic intensity.

Notions of Attained Service and Response Time

When the SRPT discipline is employed, at any given time there may be several customers in the system who have received some service. Also, the SRPT discipline distinguishes customers by their service times, so the waiting time of a customer depends on her or his service time. These observations should motivate the introduction of the following random variables.

Table 11-1 Comparison of
$E(W^*_{\text{FIFO}})$ **and** $E(W^*_{\text{SRPT}})$
when $G(t) = 1 - e^{-t}$

λ	$E(W^*_{\text{FIFO}})$	$E(W^*_{\text{SRPT}})$
0.30	$\frac{10}{7}$	1.197
0.50	2	1.425
0.70	$\frac{10}{3}$	1.874
0.80	5	2.352
0.90	10	3.552
0.95	20	5.540
0.98	50	10.494
0.99	100	17.625

With $\delta_n(\,\cdot\,)$ defined by (11-64),

$$\alpha_n(t) \triangleq \int_0^t \delta_n(u)\,du \qquad n \in I_+ \tag{11-70}$$

is the amount of service that customer n has received by time t. We call $\alpha_n(t)$ the *attained service* of customer n at time t. For $0 < x \le V_n$,

$$\tau_n(x) = \min\{t : \alpha_n(t) = x\} - T_n \qquad n \in I_+$$

is the *time to obtain x units of service*. The waiting time of customer n can be written

$$W_n = \tau_n(V_n) \qquad n \in I_+$$

When W_n converges in distribution to a random variable W^* which has a mean, we may write

$$E(W^*) = \int_0^\infty E(W^* \,|\, V = v)\,dG(v) \tag{11-71}$$

For those v such that the event $\{V = v\}$ is possible, the integrand can be interpreted as the expected waiting time of a customer with service time v. Define $\bar{W}(\,\cdot\,)$ by

$$\bar{W}(v) = E(W^* \,|\, V = v) \tag{11-72}$$

We call it the *response time* of a customer with service time v.

The amount of time that customer n is present and not in service is $W_n - V_n$. With disciplines where a customer may rejoin the queue after being in service, e.g. SRPT, $W_n - V_n$ may not be the elapsed time from arrival epoch to start of service (which we call the delay). We call $W_n - V_n$ the *wasted time* of customer n, and the function $\beta(\,\cdot\,)$ defined by

$$\beta(v) \triangleq \bar{W}(v) - v \tag{11-73}$$

is the *wasted-time function*.

For some disciplines, e.g. FIFO, for fixed v and x, when $V_n = v \geq x$, $\tau_n(x)$ does not depend on v. This is not true when SRPT is used (why?). We identify the former case with this definition.

Definition 11-3 A queue discipline is called *service-time-invariant* if given $V_n = v$, $\tau_n(x)$ is independent of v for $0 < x \leq v$.

The measure of congestion $E(W^*)$ does not distinguish among customers having different service times. Queue disciplines such as SRPT give preferred treatment to customers with short service times at the expense of those customers with long service times. Although $E(W^*_{\text{FIFO}}) > E(W^*_{\text{SRPT}})$, SRPT may not be a good discipline to use because $E[W^*_{\text{SRPT}}(v)]$ may be excessively large for some values of v. In the analysis of various queue disciplines for timeshare computer systems, the response time "is generally accepted as the *single most important performance measure*" [Kleinrock (1976, p. 161)].

Example 11-13 For any queueing system using the FIFO discipline,

$$\bar{W}(v) = E(D^*) + v \qquad \text{and} \qquad \beta(v) = E(D^*)$$

For customer n,

$$\alpha_n(t) = \begin{cases} 0 & \text{if } t \leq T_n + D_n \\ x & \text{if } t = T_n + D_n + x \leq T_n + W_n \\ V_n & \text{if } t \geq T_n + W_n \end{cases} \qquad \square$$

Example 11-14 With the LIFO-preemptive-resume discipline, an arriving customer enters service immediately, replacing the customer (if any) in service. Once in service, a customer stays in service until completion or until the next customer arrives, whichever occurs first. At service completion epochs, the most recently preempted customer in the queue resumes service (from the point at which the service was preempted).

Suppose this discipline is used for an M/G/1 queue with arrival rate λ, mean service time v_G, and traffic intensity $\rho = \lambda v_G < 1$. If customer $n + 1$ arrives while customer n is in service, n is in queue while $n + 1$ is served, while all those arrivals that occur when $n + 1$ is in service are served, while all the arrivals that occur when these are in service are served, etc. The expected time for all this to occur is the expected length of a busy period, which is $v_G/(1 - \rho)$ by Proposition 6-2. When $V_n = v$, the expected number of arrivals while customer n is in service is λv. Hence,

$$\beta(v) = \frac{\lambda v v_G}{1 - \rho} = \frac{\rho}{1 - \rho} v$$

and

$$\bar{W}(v) = \frac{v}{1 - \rho}$$

This unusual discipline has these interesting properties:

1. The response time is a linear function of the service time and depends only on ρ.
2. The average waiting time is

$$E(W^*) = \int_0^\infty \bar{W}(v) \, dG(v) = \frac{v_G}{1 - \rho}$$

which does not depend on the form of $G(\,\cdot\,)$.
3. The expected wasted time per unit of service time is

$$\frac{\beta(v)}{v} = \frac{\rho}{1 - \rho}$$

which is independent of v.

The LIFO-preemptive-resume discipline is notable because it leads to simple (and often surprising) analytical solutions. It is not a practical discipline to use. A model of a practical discipline having properties 1 through 3 is described on page 426. $\qquad\square$

Example 11-15: M/G/1/SRPT queue The expected wait, given by (11-69), can be obtained† by first finding the response time and then unconditioning on the length of the service time. For a customer with service time v, let $\bar{C}(v)$ be the expected value of the length of time that starts when the customer first enters the server and ends when service is complete. Let $\bar{D}(v)$ be the expected delay in the steady state. Thus, $\bar{W}(v) = \bar{D}(v) + \bar{C}(v)$; we obtain $\bar{C}(v)$ and then $\bar{D}(v)$.

The arrival rate of customers with service time less than or equal to x is $\lambda(x) \triangleq \lambda G(x)$, and their expected service time is $v_G(x) \triangleq \int_0^x y \, dG(y)/G(x)$. The traffic intensity of these customers is

$$\rho(x) \triangleq \lambda \int_0^x y \, dG(y)$$

Assume that $G(\,\cdot\,)$ has a density $g(\,\cdot\,)$ to avoid having to distinguish among customers having the same service times and avert some technicalities. When the remaining service time of a customer in service is x, service may be preempted by the arrival of customers with service times less than x. The total length of each preemption has the same distribution as the busy period of an M/G/1 queue with mean service time $v_G(x)$ and arrival rate $\lambda(x)$. Hence, the mean length of a preemption is $v_G(x)/[1 - \rho(x)]$.

A customer with remaining service time between x and $x + \Delta x$ will be

---

† The first derivation of (11-69) is given in Schrage and Miller (1966). The derivation given here is new.

preempted with probability $\lambda(x) \Delta x + o(\Delta x)$. Thus,

$$\bar{C}(v) = v + \int_0^v \frac{v_G(x)}{1 - \rho(x)} \lambda(x) \, dx$$

$$= \int_0^v \frac{1}{1 - \rho(x)} \, dx \qquad v \geq 0 \tag{11-74}$$

The delay of a customer with service time v is the sum of the remaining processing times (RPTs) of those customers in the system at its arrival epoch with RPT no larger than v plus the sum of the service times of those customers who arrive during its delay period and have service time smaller than v. Consider the second sum; the expected number of such customers is $\lambda \bar{D}(v)$. The mean service time of each of these customers is $v_G(v)$. Since service times are independent of the arrival epochs, the expected value of the second sum is $\rho(v)\bar{D}(v)$.

The expected value of the first sum is obtained via $L = \lambda W$ as follows. Every customer with service time larger than x eventually has an RPT of x. The arrival rate of these customers is $\lambda G^c(x)$. The expected time that their RPT is between x and $x + \Delta x$ is $\bar{C}(x + \Delta x) - \bar{C}(x)$ (why?). From (11-74),

$$\bar{C}(x + \Delta x) - \bar{C}(x) = \frac{x}{1 - \rho(x)} + o(\Delta x)$$

Customers with service times between x and $x + \Delta x$ arrive at rate $\lambda g(x) \Delta x + o(\Delta x)$. If such a customer has service time y, the expected time that its RPT is y (by definition) is $\bar{D}(y)$. Let $L(x) \Delta x$ be the average number of customers present, in the steady state, whose RPT is between x and $x + \Delta x$. Applying $L = \lambda W$ and ignoring $o(\Delta x)$ terms yield

$$L(x) = \frac{\lambda G^c(x)}{1 - \rho(x)} + \lambda g(x)\bar{D}(x) \qquad x \geq 0$$

Thus, the expectation of the first sum is† $\int_0^v xL(x) \, dx$, and $D(v)$ satisfies

$$\bar{D}(v) = \lambda \int_0^v \frac{xG^c(x)}{1 - \rho(x)} \, dx + \int_0^v \rho'(x)\bar{D}(x) \, dx + \rho(v)\bar{D}(v) \tag{11-75}$$

where $\rho'(x) = d\rho(x)/dx$.

To solve (11-75), differentiate both sides to obtain

$$\bar{D}'(v) - \frac{2\rho'(v)}{1 - \rho(v)} \bar{D}(v) = \frac{\lambda v G^c(v)}{1 - \rho(v)} \tag{11-76}$$

This linear first-order differential equation can be solved by integrating factors. This method works smoothly because $\int \{\rho'(x)/[1 - \rho(x)]\} \, dx = \ln[1 - \rho(x)]$. The solution of (11-76) with $\bar{D}(0) = 0$ is

---

† Observe that we require the result "Poisson arrivals 'see' time averages."

$$\bar{D}(v) = \frac{\lambda}{[1 - \rho(v)]^2} \int_0^v xG^c(x)\, dx \qquad v \geq 0 \tag{11-77}$$

Combining (11-74) and (11-77) yields

$$\bar{W}(v) = \int_0^v \frac{1}{1 - \rho(x)}\, dx + \frac{\lambda}{[1 - \rho(v)]^2} \int_0^v xG^c(x)\, dx \tag{11-78}$$

Substituting (11-78) into (11-71) yields (11-69). □

A Conservation Law for Response Times

When a work-conserving queue discipline is employed and there is a single server, the work-in-system process $\{W(t); t \geq 0\}$ is the same as if FIFO were used. To achieve this property when there is more than one server, we need to restrict attention to a smaller class of queue disciplines.

Definition 11-2a A queue discipline is called *strongly work-conserving* if it is work-conserving and does not distinguish customers by service-time characteristics.

Strongly work-conserving disciplines cannot use any information about service times (e.g., actual values, mean values, or d.f.) in selecting a customer for service. FIFO, LIFO, and random order of service are strongly work-conserving; SRPT is not. The priority disciplines described in the next section are strongly work-conserving when all priority classes have the same service-time d.f.

For emphasis, let $E[W_{\text{FIFO}}(\infty)]$ denote $\lim_{t \to \infty} E[W(t)]$, assuming that the limit exists. The queueing formula $H = \lambda G$ given in (11-51) can be used to obtain a simple relation† among $E[W_{\text{FIFO}}(\infty)]$, $\bar{W}(\cdot)$, and the service-time d.f. for regenerative $GI/G/c$ queues.

Theorem 11-11 In a $GI/G/c$ queue, assume the traffic intensity is less than 1, $P\{U_1 > V_1\} > 0$, a (strongly, when $c > 1$) work-conserving and service-time-invariant queue discipline is employed, and the interarrival times are nonarithmetic. Let λ be the arrival rate, $G(\cdot)$ be the service-time d.f., and $\bar{W}(\cdot)$ be the response time. Then

$$\lambda \int_0^\infty \bar{W}(v)G^c(v)\, dv = E[W_{\text{FIFO}}(\infty)] \tag{11-79}$$

PROOF Let $f_n(t)$ be the remaining service time of customer n at time t; that is, $f_n(t) = \gamma_n(t)$, given by (11-66). A graph of $f_n(\cdot)$ is shown in Figure 11-5a.

---

† This conservation law was first obtained by L. Kleinrock in 1971 for $M/G/1$ queues; see Kleinrock (1976, Sec. 4.9). The extension to $GI/G/1$ queues is given in O'Donovan, "Distribution of Attained and Residual Service in General Queueing Systems," *Oper. Res.* **22**: 570–575 (1974). The extension to $GI/G/c$ is new.

From (11-24),

$$W(t) = \sum_{n=1}^{\infty} f_n(t) \triangleq h(t)$$

By Corollary 6-8b, the steady-state mean of $W(\cdot)$ equals the time average over a single sample path, so

$$H(\omega) \triangleq \lim_{t \to \infty} \frac{1}{t} \int_0^t h(u) \, du = E[W_{\text{FIFO}}(\infty)] \qquad (\text{w.p.1})$$

The time to obtain x units of service is the horizontal distance from T_n to $f_n(V_n - x)$. Use Figure 11-5a and b to obtain

$$g_n = \int_0^{\infty} f_n(t) \, dt = \int_0^{V_n} \tau_n(x) \, dx$$

Since the queue discipline is service-time-invariant,

$$E(g_n) = \int_0^{\infty} \int_0^v E[\tau_n(x)] \, dx \, dG(v)$$

$$= \int_0^{\infty} E[\tau_n(x)] G^c(x) \, dx$$

where the second equality is obtained via integration by parts. Use Theorem 6-10 to equate $\lim_{n \to \infty} E(g_n)$ and the steady-state mean of the g-process, to obtain

$$G(\omega) \triangleq \lim_{n \to \infty} \frac{1}{n} \sum_{i=1}^{n} g_i = \int_0^{\infty} \bar{W}(x) G^c(x) \, dx \qquad (\text{w.p.1})$$

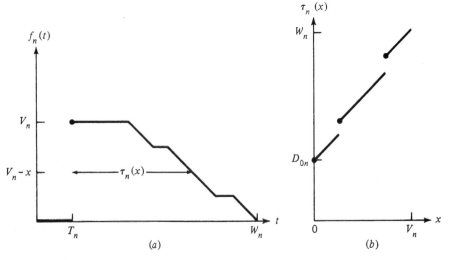

Figure 11-5 Graphs of $f_n(\cdot)$ and $\tau_n(\cdot)$.

Use Theorem 11-8 to write $H = \lambda G$, and (11-79) is obtained. $\qquad\square$

Corollary 11-11a With the hypothesis of Theorem 11-11,

$$\lambda \int_0^\infty \beta(v)G^c(v)\, dv = E[W_{\text{FIFO}}(\infty)] - \frac{\lambda E(V_1^2)}{2} \qquad (11\text{-}80)$$

PROOF Substitute $\beta(v) = \bar{W}(v) - v$ into (11-79). $\qquad\square$

Theorem 11-11 can be sharpened when the arrival process is Poisson.

Corollary 11-11b For an M/G/1 queue,

$$\int_0^\infty \bar{W}(v)G^c(v)\, dv = \frac{E(V_1^2)}{2(1-\rho)} \qquad (11\text{-}81)$$

For an M/M/c queue with mean service time $1/\mu$,

$$\int_0^\infty \bar{W}(v)e^{-\mu v}\, dv = \frac{1}{\mu^2} + \frac{p_0\, a^c}{(c-1)!\,(1-\rho)^2\mu^2} \qquad (11\text{-}82)$$

where p_0 is given by (4-41).

PROOF For the M/G/1 queue, the right-side of (11-79) is given by the Pollaczek-Khintchine equation, (11-52). For the M/M/c queue, the solution of Exercise 11-11 is that the right-side of (11-79) is L/μ, where L is the average number of customers in the system (in the steady state) and is given in Exercise 4-23. $\qquad\square$

Suppose a discipline is changed to provide shorter response times to customers with small service times. Since $G^c(x) \downarrow x$, (11-79) implies that the improvement experienced by the small-service-time customers is less than the detriment placed on some other customers. This does *not* imply that if a discipline satisfies (11-79), then the response time of one segment of customers can be improved only at the expense of another segment; see Exercise 11-31.

Round-Robin Queue Discipline

The SRPT discipline is difficult to implement because it requires that the remaining service times of all customers be known. The primary feature of the SRPT discipline is that customers with a small amount of remaining work are not held in queue while a customer with a large amount of remaining work is served. The round-robin discipline attempts to approximate this feature when the remaining service times are unknown.

In the round-robin discipline, upon entering service, a customer is allocated a time interval of length $\Delta > 0$ (Δ is called the *quantum size*) in the server. Preemption of service is not allowed. If service is completed during this interval, the customer departs at the epoch that service is finished. If service is not completed during the interval, at the end of the quantum the customer joins the tail of the queue (there is one queue for all waiting customers) and waits for another quantum of service. Newly arrived customers also join the tail of the queue.

By breaking up the large service time into many quanta, the round-robin discipline prevents a large service time from delaying, for a long time, many customers with small service times. The round-robin discipline frequently is used to schedule jobs on the central processing unit in a timeshare computer system.

The round-robin discipline is service-time-invariant. When the time required to exchange one customer for another (this is called the *swap time*) is zero, the round-robin discipline is work-conserving. Exercise 11-32 asks you to show that a model with constant swap times can be reduced to a model with zero swap times by modifying the service-time d.f.

Processor Sharing

The expected delay, in the steady state, of an $M/G/1$ queue using the round-robin discipline is obtained in Wolff (1970a).† The result obtained by Wolff is given as the solution of an independent set of linear equations. A more tractable model is obtained by letting $\Delta \downarrow 0$. This provides a good approximation when Δ is small.

When $\Delta \downarrow 0$, the customers in the system go in and out of service very rapidly; in effect, all customers in the system are simultaneously in service. When n customers are in service, the work requirement of each customer is being depleted at rate $1/n$. This model was introduced by L. Kleinrock and is called *processor sharing*.

Theorem 11-12 For a $GI/G/1$ queue using the processor-sharing discipline, assume that $\rho < 1$, the interarrival times are nonarithmetic, and the service-time d.f. has a density function. Then the response time is given by

$$\bar{W}(x) = \frac{2E[W_{\text{FIFO}}(\infty)]}{\lambda E(V_1^2)} x \qquad x \geq 0 \tag{11-83}$$

PROOF Assume the system is in the steady state. Let $L(x)$ be the expected number of customers with attained service time no larger than x. From $L = \lambda W$,

$$L(x) = \int_0^x \lambda G^c(y) \, d\bar{W}(y) \tag{11-84}$$

Let L_B be the average number of customers present when the server is busy. For Δx small, in a time interval of length $L_B \Delta x$ (we ignore terms that go to zero with Δx), the expected amount of service given to each customer is Δx.

---

† A more detailed analysis of the M/M/1 queue is given by I. Adiri and B. Avi-Itzhak, "A Time Sharing Queue," *Manage. Sci.* **15**: 639–657 (1969).

Let $g(\cdot)$ be the density function of the service-time distribution function $G(\cdot)$, and let

$$\mu(x) = \frac{g(x)}{G^c(x)} \qquad x \geq 0$$

The probability that a customer with attained service time x will depart after receiving Δx more units of service time is $\mu(x)\,\Delta x + o(\Delta x)$.

At the end of a time interval of length $L_B\,\Delta x$, if $x > L_B\,\Delta x$, the number of customers with attained service between x and $x + \Delta x$ equals the number of customers present at the start of the interval with attained service between $x - \Delta x$ and x who did not complete service during the interval. Ignoring terms that vanish as $\Delta x \to 0$, we find

$$L(x + \Delta x) - L(x) = [L(x) - L(x - \Delta x)]\left[1 - \frac{G(x) - G(x - \Delta x)}{G^c(x - \Delta x)}\right]$$

Dividing both sides by $(\Delta x)^2$ and letting $\Delta x \to 0$ yield

$$L''(x) = -\mu(x)L'(x) \qquad x > 0$$

where a prime denotes differentiation with respect to x. Thus,

$$\frac{d}{dx}\ln L'(x) = \frac{L''(x)}{L'(x)} = \frac{-g(x)}{G^c(x)} = \frac{d}{dx}\ln G^c(x)$$

and so

$$L'(x) = cG^c(x) \qquad x > 0 \tag{11-85}$$

for some constant c.

Differentiating (11-84) yields

$$L'(x) = \lambda G^c(x)\frac{d\bar{W}(x)}{dx} \qquad x > 0 \tag{11-86}$$

Equating the right-sides of (11-85) and (11-86) and integrating both sides yield [clearly $W(0) = 0$]

$$W(x) = \frac{c}{\lambda}x \qquad x \geq 0 \tag{11-87}$$

The constant c is obtained by substituting (11-87) into (11-79):

$$c\int_0^\infty xG^c(x)\,dx = \frac{cE(V_1^2)}{2} = E[W_{\text{FIFO}}(\infty)]$$

and (11-83) follows directly. $\square$

Corollary 11-12 When the arrival process is Poisson,

$$W(x) = \frac{x}{1 - \rho} \qquad x \geq 0 \tag{11-88}$$

PROOF Use (11-81) to evaluate c in (11-87).   □

Observe that for an $M/G/1$ queue, the processor-sharing discipline has the same response-time function that Example 11-14 shows for the LIFO-preemptive-resume discipline.

EXERCISES

11-28 For the $M/G/1$ queue, compare $\text{Var}(D^*_{\text{FIFO}})$ and $\text{Var}(D^*_{\text{LIFO}})$ for constant and exponentially distributed service times. See Exercise 7-59 and Example 6-10 for relevant formulas.

11-29 Give an example of a multiserver system and of a single-server system, such that the server can be idle when a customer is in queue and the conclusion of Theorem 11-10 is false.

11-30 With the hypothesis of Theorem 11-10, prove that the number in system achieves a pointwise maximum on each sample path when the customers are served according to the rule "longest remaining processing time first."

11-31 Verify that LIFO and LIFO-preemptive-resume satisfy Corollary 11-11*b*.

11-32 Plot the response-time functions for FIFO and LIFO-preemptive-resume in an $M/G/1$ queue. Show that they cross exactly once, and find the intersection point.

11-33 Let† the service times in an $M/G/1$ queue be uniformly distributed on $[\frac{2}{3}, \frac{4}{3}]$. Show that the response time for all customers is smaller with LIFO than with LIFO-preemptive-resume. How do you reconcile this with Corollary 11-11*b*?

11-34 Let θ be the constant swap time and Δ be the quantum size in a queue using the round-robin discipline, and let $G(\cdot)$ be the service-time d.f. Show that the number of customers present at time t in this queue is identical to the same quantity in a similar queue with zero swap time and service-time distribution function $G^{\#}(t) = G(t + i\theta)$ when $(i-1)\Delta < t \leq i\Delta$ for each $i \in I$.

11-35 Let $E(W^*_{PS})$ be the expected value of the waiting time of a customer, in the steady state when the processor-sharing discipline is employed. For the $M/G/1$ queue, show $E(W^*_{PS}) < E(W^*_{\text{FIFO}})$ if, and only if, $\text{Var}(V_1) > [E(V_1)]^2$.

11-5 EFFECTS OF PRIORITIES ON WAITING TIMES

You probably have had the experience of waiting on a line and finding that subsequent arrivals are being served before you because they are VIPs. In this section we examine the effects of partitioning the customers into classes and giving some classes priority over others.

The queue disciplines examined in this section are related to the queue disciplines examined in Section 11-4. In Section 11-4, we focus on queue disciplines that discriminate on a customer-by-customer basis. In this section we focus on queue disciplines that discriminate on the basis of class membership.

Types of Priority Rules

Suppose the customers are divided into J *priority classes*; a member of class j has priority over a member of class i if, and only if, $j < i$. Thus, class 1 has the most

---

† This exercise is provided by A. A. Fredericks.

priority, and class *J* has the least priority. Within each class we assume that customers start service in order of arrival. (What can you say about the effects of this assumption?) A member of class *j* is called a *j-customer.*

Class *j* has *nonpreemptive* (or *head-of-the-line*) priority over class *i* if *j*-customers are served before *i*-customers but a *j*-customer cannot displace an *i*-customer who is in service. Nonpreemptive priorities frequently are used at the queues for devices inside a computer system.

Class *j* has *preemptive* priority over class *i* if *j*-customers are served before *i*-customers and a *j*-customer will displace an *i*-customer who is in service. An example of a model with preemptive priorities is provided by a queueing system where the servers break down from time to time. We can model each breakdown as the arrival of a class 1 customer whose service time is the time to repair the breakdown.

What happens to a customer who is preempted from service? One option is to continue service from the point of interruption; in this case, the discipline is called *preemptive resume.* Another option is to start the service over from the beginning; in this case, the discipline is called *preemptive repeat.* There are two common types of preemptive-repeat disciplines. With the *preemptive-repeat-identical* rule, when service of the preempted customer resumes, a service time of the same *duration* as the interrupted one begins from scratch. With the *preemptive-repeat-different* rule, when service of the preempted customer resumes, a new independent service time begins. The preemptive-resume discipline frequently is appropriate in computer models when preemption is caused by a test program that temporarily seizes a device. In the same setting, the preemptive-repeat-identical discipline might be appropriate when preemption is caused by a critical failure (a machine "crash") because the work required to process a program depends primarily on the properties of the program. The preemptive-repeat-different discipline is appropriate when the variations among service times are due primarily to vagaries of the server.

Since machine breakdowns frequently occur only when the machine is working and arrivals of VIPs may be independent of the number of ordinary customers, it is useful to form two categories of arrival processes with priority 1. Priority 1 arrivals are called *active* if they can arrive only when some member of another priority class is in service. Priority 1 arrivals are called *independent* if they may arrive when the server is either busy or idle.

Conservation Laws for Priority Queues

Recall that in single-server systems, *work-conserving* disciplines result in the same one-dimensional distributions for the work-in-system process as the FIFO discipline. The nonpreemptive and preemptive-resume disciplines are easily seen to be work-conserving. When service times are exponentially distributed, the preemptive-repeat-different and the preemptive-resume disciplines coincide, so preemptive-repeat-different is work-conserving when service times are exponentially distributed. For these disciplines, multiple-server systems are strongly work-conserving when every priority class has the same service-time d.f. We obtain conservation theorems similar in spirit to Theorem 11-11 for these priority rules. The main tool is the queueing formula $H = \lambda G$ established in Theorem 11-8.

The subscript j refers to the jth priority class. Let V_{nj} and D_{nj} be the service time and delay, respectively, of the nth arriving member of class j. Let $W_{\text{FIFO}}(t)$ be the work in the system at time t when the queue discipline is FIFO. In Theorem 11-13 below, the different classes can be priority classes, but they are not required to be (the customers could form a single line and be served FIFO).

Theorem 11-13 Consider a c-server system with J nonpreemptive customer classes. For each class j, assume that the arrival times form a renewal process with rate λ_j and that the service times are i.i.d. random variables. When $c > 1$, assume that all classes have the same service-time distribution function. Let $\rho_j \triangleq \lambda_j E(V_{1j})$. Assume there are numbers d_j^* such that

(a) $\displaystyle \lim_{N \to \infty} \frac{1}{N} \sum_{n=1}^{N} D_{nj} = d_j^*$ (w.p.1), $j = 1, 2, \ldots, J$

Then there exists a number w^* such that

$$\lim_{T \to \infty} \frac{1}{T} \int_0^T W_{\text{FIFO}}(t) \, dt = w^* \qquad \text{w.p.1} \tag{11-89}$$

and

$$\sum_{j=1}^{J} \rho_j d_j^* = w^* - \frac{1}{2} \sum_{j=1}^{J} \lambda_j E(V_{1j}^2) = \text{constant} \tag{11-90}$$

When $\{D_{nj}; n \in I_+\}$ is a regenerative process with a finite mean time between regenerations or a stationary process, d_j^* exists and can be interpreted as the delay in the steady state of a class j customer. Similar remarks apply to w^*. We do not have the tools to establish the existence of d_j^* in the most general case, but the assumption can be verified in an important special case (see Exercise 11-36).

PROOF† Let $f_{nj}(t)$ denote the remaining service time of the nth arriving customer of class j at time t; its graph is the same as the graph of $f_n(\cdot)$ given in Figure 11-4. Repeating the analysis used to derive the Pollaczek-Khintchine formula from $H = \lambda G$ given in Example 11-11 for each class separately, and attaching suitable subscripts, yields

$$G_j = \lim_{N \to \infty} \sum_{n=1}^{N} \int_0^\infty f_{nj}(t)$$

$$= \lim_{N \to \infty} \sum_{n=1}^{N} \left(D_{nj} V_{nj} + \frac{V_{nj}^2}{2} \right)$$

† The first statement and proof of a theorem of this type were given for the $M/G/1$ queue by L. Kleinrock in 1965; see Kleinrock (1976, Sec. 3.4). A proof for the $G/G/1$ queue was given by Linus Schrage, "An Alternative Proof of a Conservation Law for the Queue $G/G/1$," *Oper. Res.* **18**: 185–187 (1970). Our proof is similar to Schrage's. The extension to $GI/G/c$ is new.

Since D_{nj} is independent of V_{nj},

$$G_j = d_j^* E(V_{1j}) + \frac{E(V_{1j}^2)}{2} \qquad j = 1, 2, \ldots, J \qquad (11\text{-}91)$$

Similarly, $h_j(t) = \sum_{n=1}^{\infty} f_{nj}(t)$ is the work in the system at time t of members of class j, so $h(t) \triangleq \sum_{j=1}^{J} h_j(t) = W_{\text{FIFO}}(t)$, where the second equality is due to work conservation.

Assumption (a) implies that hypotheses (a) and (b) of Theorem 11-7 are satisfied. Thus,

$$H_j = \lim_{T \to \infty} \frac{1}{T} \int_0^T h_j(t) \, dt$$

exists for each j. Applying $H = \lambda G$ to each class separately and summing over j yield

$$w^* \triangleq \sum_{j=1}^{T} H_j = \lim_{T \to \infty} \frac{1}{T} \int_0^T h(t) \, dt$$

$$= \lim_{T \to \infty} \frac{1}{T} \int_0^T W_{\text{FIFO}}(t) \, dt$$

exists; moreover,

$$w^* = \sum_{j=1}^{\infty} \lambda_j G_j \qquad (11\text{-}92)$$

Substituting (11-91) into (11-92) and rearranging it yield (11-90). $\qquad \square$

Note that the only use in the proof of the assumption that the arrival processes are renewal processes is that arrival rates for each class exist. If this conclusion were an assumption, the theorem would be true for $G/G/c$ queues.

A particular assertion of (11-90) is that with nonpreemptive priorities, the average delays of the priority classes are constrained by a linear equality. This implies that the benefits of smaller average delays which accrue to those customers with high priority are gained at the expense of increased delays for low-priority customers.

Let us now obtain the analog of Theorem 11-13 for work-conserving preemptive priorities. To ease the notational burden, let us suppress the index j for the class of a customer. Let $f_n(t)$ be the remaining service time of the nth arriving customer in some fixed priority class with priority greater than 1 (we exclude class 1 customers because they do not get preempted, so they are atypical). A sample of the graph of $f_n(\cdot)$ is shown in Figure 11-6.

Let the random variable K_n be the number of times customer n is preempted from service. Each time a customer is preempted we say the customer enters *limbo*. Thus, the delay of customer n is the time spent in queue waiting to start service for the first time, and the total time spent in the system but not in service is the delay plus the time in limbo. For customer n, let the random variable ℓ_{nk} be

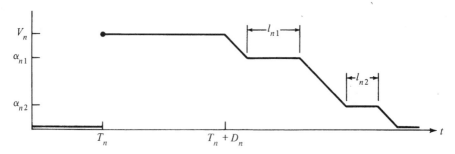

Figure 11-6 A sample graph of $f_n(\cdot)$ for some class greater than or equal to 2.

the length of the kth visit to limbo and the random variable α_{nk} be the attained service at the start of the kth visit to limbo.

Thus,

$$g_n \triangleq \int_0^\infty f_n(t)\, dt = V_n D_n + S_n \qquad n = 1, 2, \ldots \qquad (11\text{-}93)$$

where

$$S_n \triangleq \sum_{k=1}^{K_n} (V_n - \alpha_{kn})\ell_{kn} \qquad (11\text{-}94)$$

Since $V_n - \alpha_{kn}$ is the remaining work required by customer n at the epoch of the ith preemption, S_n is the total "work hours" (when time is measured in hours) spent in limbo by customer n.

Theorem 11-14 Assume that customers arrive at a c-server system according to independent renewal processes and that service times for each class are i.i.d. Let λ_j be the arrival rate of class j, $\rho_j \triangleq \lambda_j E(V_{1j})$, and J be the number of classes. Let the queue discipline be of the work-conserving preemptive type. When $c > 1$, assume that all classes have the same service-time d.f.

If for each class j there are numbers d_j^* and s_j^* such that

(a) $\quad \displaystyle\lim_{n\to\infty} \frac{1}{N} \sum_{n=1}^N D_{nj} = d_j^* \qquad j = 1, 2, \ldots, J$

and

(b) $\quad \displaystyle\lim_{n\to\infty} \frac{1}{N} \sum_{n=1}^N S_{nj} = s_j^* \qquad j = 1, 2, \ldots, J$

then there exists a number w^* such that

$$\lim_{T\to\infty} \frac{1}{T} \int_0^T W_{\text{FIFO}}(t)\, dt = w^* \qquad \text{w.p.1}$$

and

$$\sum_{j=1}^{J} (\rho_j d_j^* + \lambda_j s_j^*) = w^* - \frac{1}{2} \sum_{j=1}^{J} \lambda_j E(V_{1j}^2) = \text{const}$$

PROOF This is left as Exercise 11-37. $\square$

Theorems 11-13 and 11-14 could have been stated as one theorem with $s_j^* \equiv 0$ in the nonpreemptive case.

Expected Waiting Times

Explicit formulas can be obtained for the expected delay, in the steady state, of a class j customer in the M/G/1 and M/M/c models. For priority queues, *Poisson arrivals* means that the arrival processes for the different classes are all Poisson and independent of one another. *Exponential services* means that the service times for each class are i.i.d. with an exponential distribution; the mean service time may depend on the class.

Theorem 11-15 In an M/G/1 queue with nonpreemptive priorities, the mean delay, in the steady state, of a class j customer, $j = 1, 2, \ldots, J$, is given by

$$E(D_j^*) = \frac{\sum_{i=1}^{J} \lambda_i E(V_{1i}^2)}{2(1 - \sum_{i=1}^{j} \rho_i)(1 - \sum_{i=1}^{j-1} \rho_i)} \tag{11-95}$$

when $\sum_{i=1}^{J} \rho_i < 1$.

PROOF You showed in Exercise 11-36 that $E(D_j^*)$ exists and is finite when $\sum_{i=1}^{J} \rho_i < 1$. The delay of a j-customer is

$$\sum_{i=1}^{j} (\text{service times of } i\text{-customers in queue at the } j\text{-customer's arrival epoch})$$

$$+ \sum_{i=1}^{j-1} (\text{service times of } i\text{-customers who arrive while } j\text{-customer is in queue})$$

$+$ remaining service time of customer in service at customer's arrival epoch, if any

Let the random variable Q_i^* be the number of i-customers in queue in the steady state. Since Poisson arrivals "see" time averages, the expected number of i-customers in the queue at the arrival epoch of a j-customer is $E(Q_i^*)$. The expected service time of each of these customers is $E(V_{1i})$, so the expected value of the first sum is

$$\sum_{i=1}^{j} E(Q_i^*)E(V_{1i}) = \sum_{i=1}^{j} \lambda_i E(D_i^*)E(V_{1i}) = \sum_{i=1}^{j} \rho_i E(D_i^*)$$

where $L = \lambda W$ is used to obtain the first equality.

The expected number of i-customers who arrive while a j-customer is in queue is $\lambda_i E(D_i^*)$ (why?). Hence, the expected value of the second sum is

$$\sum_{i=1}^{j-1} \lambda_i E(D_j^*) E(V_{1i}) = E(D_j^*) \sum_{i=1}^{j-1} \rho_i$$

The expected remaining service time of the customer in service (if any) at an arrival epoch of a j-customer is the same for all j (Exercise 11-38). Let η denote the common value. We postpone obtaining an explicit formula for η. We have established

$$E(D_j^*) = \sum_{i=1}^{j} \rho_i E(D_i^*) + E(D_j^*) \sum_{i=1}^{j-1} \rho_i + \eta \qquad j = 1, 2, \ldots, J \quad (11\text{-}96)$$

Solving (11-96) for $E(D_1^*)$, then for $E(D_2^*)$, etc., yields

$$E(D_j^*) = \frac{\eta}{\left(1 - \sum_{i=1}^{j} \rho_i\right)\left(1 - \sum_{i=1}^{j-1} \rho_i\right)} \qquad j = 1, 2, \ldots, J \quad (11\text{-}97)$$

To obtain† η, observe that the memoryless property of the exponential interarrival times implies that for any n, the probability that the nth arriving customer (we consider all classes combined) is a member of class j is $\lambda_j/\sum_{i=1}^{J} \lambda_i$ and this probability is independent of the classes to which the previous $n - 1$ arrivals belonged (why?). Thus, the service-time d.f. of the nth arrival is given by

$$G(t) = \frac{\sum_{j=1}^{J} \lambda_j G_j(t)}{\sum_{i=1}^{J} \lambda_i} \quad (11\text{-}98)$$

where $G_j(\,\cdot\,)$ is the service-time d.f. for members of class j. If the customers were served FIFO, the delays would form the delay process of an $M/G/1$ queue with arrival rate $\lambda = \sum_{j=1}^{J} \lambda_j$ and service-time distribution function $G(\,\cdot\,)$ given by (11-98). The expected delay in the steady state is given by the Pollaczek-Khintchine equation (11-52); since Poisson arrivals "see" time averages, the mean delay is $E[W_{\text{FIFO}}(\infty)]$, which is the number w^* in (11-89):

$$w^* = \frac{\sum_{j=1}^{J} \lambda_j E(V_{1j}^2)}{2\left(1 - \sum_{j=1}^{J} \rho_j\right)} \quad (11\text{-}99)$$

Substituting (11-97) and (11-99) into (11-90) and letting $R_j = \sum_{i=1}^{j} \rho_i$ yield

$$\sum_{j=1}^{J} \rho_j E(D_j^*) = \eta \sum_{j=1}^{J} \frac{\rho_j}{(1 - R_j)(1 - R_{j-1})}$$

$$= \frac{1}{2} \frac{R_J}{1 - R_J} \sum_{j=1}^{J} \lambda_j E(V_{1j}^2)$$

† We give an algebraic derivation. Exercise 11-38 asks you to provide a probabilistic derivation.

The second equality yields

$$\eta = \frac{\sum_{j=1}^{J} \lambda_j E(V_{1j}^2)}{2} \frac{R_J}{1 - R_J \sum_{j=1}^{J} \{\rho_j/[(1 - R_j)(1 - R_{j-1})]\}} \quad (11\text{-}100)$$

Evaluating $R_J/(1 - R_J)$ by long division, you can show that it equals the denominator of the term to its right in (11-100); hence

$$\eta = \frac{1}{2} \sum_{j=1}^{J} \lambda_j E(V_{1j}^2) \quad (11\text{-}101)$$

Substituting (11-101) into (11-97) yields (11-95). □

In the steady state, let Q_j^* be the number of j-customers in the queue. Applying $L = \lambda W$ to each class separately yields

$$E(Q_j^*) = \lambda_j E(D_j^*) \qquad j = 1, 2, \ldots, J \quad (11\text{-}102)$$

Theorem 11-16 Consider an M/M/c queue with nonpreemptive priorities. Let λ_j be the arrival rate of class j and $\lambda \triangleq \sum_{j=1}^{J} \lambda_j$. Assume that the service-time d.f. for all classes is the same exponential distribution with mean $1/\mu$ and that $a \triangleq \lambda/\mu < c$. The mean delay, in the steady state, of a class j customer is given by

$$E(D_j^*) = \frac{C(c, a)/(c\mu)}{(1 - \sum_{i=1}^{j} \rho_i)(1 - \sum_{i=1}^{j-1} \rho_i)} \qquad j = 1, 2, \ldots, J \quad (11\text{-}103)$$

where $\rho_j = \lambda_j/(c\mu_j)$ and $C(c, a)$ is the steady-state probability that all servers are busy and is given by (4-43).

PROOF When all the servers in an M/M/c queue are busy, the mean time between departures is exponentially distributed with mean $1/(c\mu)$. You should check that by setting $E(V_{1j}) = 1/(c\mu)$ in (11-97) and interpreting η as the expected time until a server next becomes free, a valid equation for the M/M/c queue is obtained.

The steady-state probability that an arrival finds all servers busy is $C(c, a)$. The memoryless property of the exponential d.f. implies that for such arrivals, the mean time until the next departure occurs is $1/(c\mu)$. Hence

$$\eta = \frac{C(c, a)}{c\mu} \quad (11\text{-}104)$$

Substituting (11-104) and $E(V_{1j}) = 1/(c\mu)$ in (11-97) yields (11-103). □

The cμ-Rule

Suppose the holding-cost rate of keeping a j-customer in queue is c_j. The asymptotic cost rate, $\mathcal{B}$ say, is (Corollary 6-8b)

$$\mathcal{B} = \sum_{j=1}^{J} c_j E(Q_j^*) = \sum_{j=1}^{J} c_j \lambda_j E(D_j^*) \quad (11\text{-}105)$$

Assume that the customers have been partitioned into J nonpreemptive classes and we have the ability to assign the priority numbers. The priority of a class is fixed once and for all. How should the priorities be assigned to minimize $\mathcal{B}$?

If we give i-customers priority over k-customers, then when one i-customer and one k-customer are in queue at a service completion epoch, the i-customer will enter service and the k-customer will wait for (at least) the service of this i-customer. The expected cost for this avoidable delay to the k-customer (we have the ability to have given it priority) is c_k/μ_i, where $1/\mu_i$ is the mean service time of an i-customer. Similarly, if the k-customer had priority over the i-customer, the expected avoidable cost to the i-customer is c_i/μ_k. The choice of giving priority to the i-customer is better when

$$\frac{c_k}{\mu_i} < \frac{c_i}{\mu_k} \Leftrightarrow c_i \mu_i > c_k \mu_k$$

Definition 11-4 The *$c\mu$-rule* assigns most priority to that class with the largest index $c\mu$; the remaining priorities are assigned accordingly.

The discussion above suggests that the $c\mu$-rule may work well. The next theorem shows that the $c\mu$-rule minimizes (11-105) for the M/G/1 queue.

Theorem 11-17 In an M/G/1 queue with J nonpreemptive priority classes and $E(D_j^*) < \infty$ for each j, the asymptotic cost rate (11-105) is minimized when the priority classes are determined by the $c\mu$-rule.

PROOF† Assume $J \geq 3$ and classes 2 and 3 do not conform to the $c\mu$-rule; i.e., class 2 has priority over class 3, and

$$c_2 \mu_2 < c_3 \mu_3 \tag{11-106}$$

Let $\mathcal{B}$ be the cost rate with this priority assignment. Interchanging the priorities of classes 2 and 3 results in a different cost rate $\mathcal{B}'$. From (11-105) and (11-97),

$$\mathcal{B} - \mathcal{B}' = \frac{(2 - 2\rho_1 - \rho_2 - \rho_3)\rho_2 \rho_3 \eta(c_3 \mu_3 - c_2 \mu_2)}{(1 - \rho_1)(1 - \rho_1 - \rho_2)(1 - \rho_1 - \rho_3)(1 - \rho_1 - \rho_2 - \rho_3)} \tag{11-107}$$

where $\rho_j = \lambda_j/\mu_j$. Combining (11-106) and (11-107) establishes $\mathcal{B}' < \mathcal{B}$.

The above argument applies to priority classes j and $j + 1$ with $j > 1$. To see this, treat all classes with priority over j as a single class having priority 1, adjust the arrival-rate and service-time d.f.'s accordingly (Exercise 11-39 asks you to provide the details), and treat classes j and $j + 1$ as classes 2 and 3. A

† This proof is due to Cox and Smith (1961).

similar (and algebraically simpler) argument handles the case for classes 1 and 2.

Thus, any assignment of priorities that differs from the $c\mu$-rule can be improved by pairwise interchanges, and so the $c\mu$-rule is optimal. $\quad\square$

When $c_1 = c_2 = \cdots = c_J$, Theorem 11-17 asserts that priorities should be assigned in increasing order of mean service time. A special case of this assignment of priorities is explored in Exercise 11-40.

Completion Times

The notion of *completion times* for customers in priority queues allows one to obtain the LST of the steady-state waiting-time d.f. from the theory where there is only one class of arrivals. Completion times were introduced in Gaver (1962). Recall that we assume that service within classes is FIFO.

Definition 11-5 The *completion time* of the nth arriving j-customer is the duration of the interval which starts when that customer enters service and ends at the first epoch where the $(n + 1)$th arriving j-customer may begin (does begin, provided it is present) service.

The completion time of a customer is its service time if no higher-priority customers arrive while the customer is in service. If higher-priority customers do arrive, the completion time is the customer's service time plus the service times of the higher-priority customers who arrive while the original customer is in service, plus the service times of higher-priority customers who arrive while they are in service, etc.

Completion times of a j-customer are unaffected by customers with less priority than j. When the classes arrive according to independent Poisson processes, the classes with priority over j-customers may be grouped into a single class of customers with priority 1, and we may consider the j-customers as having priority 2. For the remainder of this section, we assume Poisson arrival processes and, without loss of generality, two priority classes.

It is convenient to think of the members of class 2 as *customers* and the members of class 1 as *interruptions*. Customers behave in the usual way. Interruptions arrive one at a time; when an interruption is in service, no other interruptions can arrive. Let B_n be the service time of the nth interruption. When an interruption finishes service, the time until the next interruption appears is an r.v. with an exponential d.f. with mean $1/\lambda_1$. Interruptions are a model of server breakdowns, and B_n is the time required to repair the nth breakdown. Interruptions can also represent high-priority customers. Exercise 11-42 asks you to justify interpreting the nth interruption as the nth 1-customer who arrives when no 1-customer is in service, with B_n distributed as the busy period of an $M/G/1$ queue with arrival rate λ_1 and service-time distribution function $G_1(\cdot)$.

Now we present representations of the completion times for the most important priority disciplines. For the nth arriving customer, let V_n be the service time and K_n be the completion time. Let $N(n)$ be the *set* of interruptions who arrive during the service time; i.e., interruption $i \in N(n)$ if it occurs when customer n is in service.

For both nonpreemptive and preemptive-resume disciplines, the elapsed time from when customer n starts service until customer $n + 1$ may enter (does enter, if it has arrived) service is the service time plus the sum of the service times of the interruptions occurring during this service time. Thus

$$K_n = V_n + \sum_{i \in N(n)} B_i \qquad n \in I_+ \tag{11-108}$$

For the preemptive-repeat-identical and preemptive-repeat-different disciplines, a third term representing potential service times that are started but not completed must be added. Hence,

$$K_n = V_n + \sum_{i \in N(n)} B_i + \sum_{i \in N(n)} V'_n(i) \qquad n \in I_+ \tag{11-109}$$

where $V'_n(i)$ is the amount of service received when customer n enters service for the ith time.

The tacit assumption that both $\{V_n\}_{n=1}^{\infty}$ and $\{B_n\}_{n=1}^{\infty}$ are i.i.d. implies that $\{K_n\}_{n=1}^{\infty}$ is a sequence of i.i.d. random variables. The distribution of K_1 is examined in Exercises 11-43 and 11-44.

Waiting-Time Distributions

Now we use the completion times to obtain the steady-state waiting-time distributions for customers and interruptions in an $M/G/1$ setting. Let λ_2 be the arrival rate of customers, $1/\alpha$ be the mean time between occurrences of interruptions, $\tilde{G}_2(\cdot)$ be the LST of the service times of a customer, and v_B and $\tilde{B}(\cdot)$ be the mean and LST, respectively, of the service times of interruptions. When the interruptions represent high-priority customers who arrive according to a Poisson process with rate λ_1 and have service-time distribution function $G_1(\cdot)$, $\tilde{B}(\cdot)$ is the LST of the busy-period d.f. of such an $M/G/1$ queue. Then Proposition 6-2 asserts that $\tilde{B}(\cdot)$ is the root with smallest absolute value of

$$\tilde{B}(s) = \tilde{G}_1[s + \lambda_1 - \lambda_1 \tilde{B}(s)] \tag{11-110}$$

In Theorem 11-18 below, it is not necessary to specify which type of preemptive priority discipline is used. The various types of disciplines lead to different formulas for the LST of the completion time, but do not affect the general formula.

Theorem 11-18 In an $M/G/1$ queue with customers and interruptions, let $\tilde{K}(\cdot)$ and v_K be the LST and mean, respectively, of the completion time of a customer. Assume that $\rho_2 \triangleq \lambda_2 v_K < 1$.

(a) If the interruptions are *active*, then the LST of the steady-state waiting-time d.f. for customers is given by

$$\tilde{W}_{ap}(s) = \frac{(1 - \rho_2)s\tilde{K}(s)}{s - \lambda_2 + \lambda_2 \tilde{K}(s)} \tag{11-111}$$

when interruptions have *preemptive priority*. When interruptions have *non-preemptive priority*, the corresponding LST is

$$\tilde{W}_{an}(s) = \frac{(1 - \rho_2)s\tilde{G}_2(s)}{s - \lambda_2 + \lambda_2 \tilde{K}(s)} \qquad (11\text{-}112)$$

(b) If the interruptions are *independent*, then the LST of the steady-state waiting-time d.f. for customers is given by

$$\tilde{W}_{ip}(s) = \tilde{W}_{ap}(s) \frac{s + \alpha - \alpha\tilde{B}(s)}{s(1 + \alpha v_B)} \qquad (11\text{-}113)$$

when interruptions have *preemptive priority*. When interruptions have *non-preemptive priority*, the corresponding LST is

$$\tilde{W}_{in}(s) = \tilde{W}_{an}(s) \frac{s + \alpha - \alpha\tilde{B}(s)}{s(1 + \alpha v_B)} \qquad (11\text{-}114)$$

PROOF For active preemptive interruptions, the definition of the completion time implies that the waiting-time process for customers is the waiting-time process of an $M/G/1$ queue with arrival rate λ_2 and service times with LST $\tilde{K}(\cdot)$. Thus, $\tilde{W}_{ap}(\cdot)$ is given by the Pollaczek-Khintchine formula for the steady-state *waiting-time* d.f. [Equation (7-66)] with $\tilde{K}(\cdot)$ as the service-time LST, which is (11-111).

For active nonpreemptive interruptions, the delay process for customers corresponds to the delay process of the $M/G/1$ queue described above. Hence, the LST of the steady-state *delay* d.f. is $[(1 - \rho_2)s]/[s - \lambda_2 + \lambda_2 \tilde{K}(s)]$. Since the service time of a customer is not affected by interruptions and is stochastically independent of the delay, $\tilde{W}_{an}(\cdot)$ is the LST of the delay d.f. multiplied by the LST of the service-time d.f., which yields (11-112).

In the case of independent interruptions, observe that both (11-113) and (11-114) are the corresponding LST for the active-interruptions case multiplied by a common factor. This factor represents the effect of interruptions that may be present when a customer arrives when no other customers are present. To obtain this factor, consider what occurs when the interruptions have preemptive priority. Let time 0 denote an epoch when a customer's busy period ends; at this epoch, no customers or interruptions are present. Let the random variable T be the arrival epoch of the next customer. If no interruption is present at T, that customer enters service immediately. If an interruption is present at T, that customer cannot start service until the interruption completes service. Let the random variable $Y(T)$ be the delay of the first customer served in a busy period.

Observe that the process which records the number of customers present at time t corresponds to the number of customers present in an $M/G/1$ queue where the first customer served in each busy period has exceptional service (see Exercise 7-63). Here, the LST of the exceptional service-time d.f. is $\tilde{K}(\,\cdot\,)$ multiplied by the LST of $Y(T)$; the latter LST is obtained in Exercise 5-55. It is left to you to put together these results and show that the generating function of the number of customers present in the steady state, $\hat{\pi}_{ip}(\,\cdot\,)$ say, is

$$\hat{\pi}_{ip}(z) = \tilde{W}_{ap}(x) \frac{x + \alpha - \alpha B(x)}{x(1 + \alpha v_B)}$$

where $x = \lambda - \lambda z$. Since customers are served FIFO, $\tilde{W}_{ip}(s) = \hat{\pi}_{ip}(1 - s/\lambda)$ by the argument in Example 7-19, and (11-113) is obtained.

To obtain (11-114), observe that at the end of the *completion time* of the last customer served in a busy period, there are no interruptions present. Thus, the modification for independent interruptions is identical to the preemptive case. □

The waiting-time d.f. for interruptions with preemptive priority is given by the Pollaczek-Khintchine formula (without any modification) because these interruptions behave as if the customers were not present. The d.f. for nonpreemptive interruptions can be obtained by adapting the derivation of (11-113); see Exercise 11-46.

Differentiating (11-111) and (11-113) with respect to s and letting $s \downarrow 0$ yield

$$E(W_{ap}) = \frac{\lambda v_{2K}}{2(1 - \rho_2)} + v_K \tag{11-115}$$

and

$$E(W_{ip}) = \frac{E(W_{ap})\alpha v_{2B}}{2(1 + \alpha v_B)} \tag{11-116}$$

where v_{2K} and v_{2B} are the second moments of $K(\,\cdot\,)$ and $B(\,\cdot\,)$, respectively.

EXERCISES

11-36 Show that part (*a*) of Theorem 11-13 is valid if (*a*) for each j, the arrivals of j-customers constitute a Poisson process independent of the arrival epochs of other customers and (*b*) $\Sigma_{j=1}^{J} \rho_j < 1$.

11-37 Prove Theorem 11-14.

11-38 Use the assumptions in Theorem 11-15, and prove directly that, in the steady state, the expected remaining service time of the customer in service (if any) at an arrival epoch of a j-customer is $v_{2G}/(2v_{1G}^2)$, where v_{iG} is the ith moment of the distribution function $G(\,\cdot\,)$ given by (11-98). Give a probabilistic proof that the d.f. of the remaining service time defined above is the *equilibrium-excess* distribution of $G(\,\cdot\,)$.

11-39 Consider an M/G/1 model with J priority classes. When priorities are preemptive, explain why j-customers behave as if i-customers did not exist when $i > j$. For both preemptive and nonpreemptive priorities, explain why the effect of all customers in classes $1, 2, \ldots, j$ on members of classes $j + 1, \ldots, J$ can be obtained by treating the former as a single class of top-priority customers with arrival rate $\lambda' = \Sigma_{i=1}^{j} \lambda_i$ and service-time distribution function $\Sigma_{i=1}^{j} \lambda_i G_i(\,\cdot\,)/\lambda'$.

11-40 Assume that the service time of a customer is known at the arrival epoch and that the queue discipline is to serve next the customer in queue with the smallest service time. Preemption is not allowed; that is why this discipline differs from SRPT. This discipline is called *shortest processing time* (SPT). To avoid ties, assume the service-time d.f. has a density $g(\cdot)$. Let $\bar{W}(v)$ be the response time (see page 421 for the definition) of a customer with service time v. For an $M/G/1$ queue with arrival rate λ, show that

$$\bar{W}(v) = v + \frac{\lambda E(V_1^2)/2}{2\left[1 - \lambda \int_0^v xg(x)\, dx\right]^2}$$

Hint: Consider what happens to (11-95) as the number of priority classes becomes infinite.

11-41 Show that Theorem 11-17 is valid for any $GI/G/1$ queue when there are two priority classes.

11-42 In an $M/G/1$ model with two priority classes, explain why the effect of 1-customers on 2-customers is stochastically identical to the effect of interruptions on customers when the service time of an interruption is the busy-period d.f. of an $M/G/1$ queue with arrival rate λ_1 and service-time distribution function $G_1(\cdot)$.

11-43 Show that the LST of the completion time of customers for both nonpreemptive and preemptive-resume priorities is given by

$$\tilde{K}(s) = \tilde{G}_2[s + \alpha - \alpha\tilde{B}(s)]$$

11-44 For preemptive-resume-identical priorities, show that the mean completion time of a customer is

$$v_K = \left(v_B + \frac{1}{\alpha}\right)[\tilde{G}_2(-\alpha) - 1]$$

Show that the corresponding result for preemptive-repeat-different priorities is

$$v_K = \frac{(v_B + 1/\alpha)[1 - \tilde{G}_2(\alpha)]}{\tilde{G}_2(\alpha)}$$

Assume $v_B + 1/\alpha < \infty$. Is $v_K = \infty$ possible? If so, when?

11-45 Suppose interruptions represent 1-customers, as in Exercise 11-39, and consider the four priority schemes described in Exercises 11-42 and 11-43. In Theorem 11-18 it is assumed that $\lambda_2 v_K < 1$. Under what conditions is this assumption valid?

11-46 Describe how to obtain the steady-state waiting-time d.f. for 1-customers who have non-preemptive priority in an $M/G/1$ queue.

11-47 Suppose two independent streams of customers arrive at a single server. Stream 1 is Poisson with rate $\frac{1}{2}$, and the first two moments of the service times are 1 and 4. Stream 2 is Poisson with rate $\frac{1}{36}$, and the first two moments of the service times are 12 and 144. Find the mean waiting time, in the steady state, for customers in each class when the queue discipline is (*a*) FIFO, (*b*) nonpreemptive priority for stream 1, and (*c*) preemptive-resume priority for class 1.

11-6 START-UP AND SHUT-DOWN QUEUEING MODELS

In this section we investigate single-server queueing models where the server is activated and deactivated according to some particular rule. The motivation for these models comes from cost considerations. Suppose that wages, fuel costs, etc. are paid when the server is available and are not charged when the server is unavailable. These costs may be interpreted as the additional costs (above some base costs) incurred when the server is available. These costs provide an incentive for turning off the server when no customers are present.

Suppose a fixed charge R is incurred each time the server is turned on; this cost may represent the effect of power surges, requirements for special personnel, and other one-time costs. If many customers are present when the server is turned on, the cost R is spread over many customers, so there is an incentive for not turning on the server until many customers are present. When, for example, the customers are machines waiting for repair, each unit of time spent in queue or in service represents lost production time. One way to model this effect is to create for each customer a holding cost that is an increasing function of the customer's waiting time. The holding cost provides an incentive for turning on the server before too many customers are present.

In this section, we obtain the expected value of the waiting time in the steady state for two rules for turning on the server and briefly mention a third rule. Our setting is an $M/G/1$ queue with arrival rate λ, mean service time v, and traffic intensity $\rho \triangleq \lambda v < 1$.

The N-Policy

The *N-policy* activates the server when there are N customers waiting for service and deactivates the server when there are no customers in the system. When N is 0 or 1, this model is the same as the ordinary $M/G/1$ model. When the service times have an exponential distribution, the limiting state probabilities are ob-tained in Section 4-6. In Section 7-2 in Volume II, it is shown that the N-policy is optimal with certain cost assumptions.

> **Example 11-16** The N-policy has an interesting interpretation in a production-inventory model. Suppose an inventory starts with S items on hand. Customers arrive according to a Poisson process; each customer requires one item. When the inventory drops to the level $S - N$, the production facility is instructed to start manufacturing. The time required to produce an item is an r.v. with distribution function $G(\cdot)$. Manufacturing continues until the inventory level reaches S, and then the process is repeated. Assume that customers who arrive when no items are available wait as long as required for one to be manufactured. (This assumption is not necessary; see Exercise 11-48.)
>
> Let $X_N(t)$ be the number of customers present at time t in an $M/G/1$ queue operating under the N-policy and $Y_N(t)$ be the number of items in inventory at time t in the production inventory model. Portions of $\{X_N(t); t \geq 0\}$ and $Y_N(t); t \geq 0\}$ are shown in Figure 11-7. In Figure 11-7a, τ_1 is the epoch when the server is turned on, and τ_3 is the epoch when the server is turned off. In Figure 11-7b, τ_1 is the epoch when production starts, and τ_3 is the epoch when production stops. It is easy to see that
>
> $$Y_N(t) = S - X_N(t)$$

so the properties of $\{Y_N(t); \ t \geq 0\}$ can be deduced from the properties of $\{X_N(t); t \geq T\}$. □

It is possible to obtain $\lim_{t \to \infty} P\{X_N(t) = j\}$, but the derivation is tedious and the solution is in the form of a messy generating function. It is more interesting to obtain the mean waiting time in the steady state.

Theorem 11-19 Let $E(W_N^*)$ be the expected waiting time in the steady state of an $M/G/1$ queue operating with the N-policy. Then

$$E(W_N^*) = E(W_1^*) + \frac{N-1}{2\lambda} \qquad N \in I_+ \qquad (11\text{-}117a)$$

and

$$E(W_0^*) = E(W_1^*) \qquad (11\text{-}117b)$$

where $E(W_1^*)$ is given by the Pollaczek-Khintchine equation

$$E(W_1^*) = \frac{\lambda E(V_1^2)}{2(1-\rho)} + E(V_1) \qquad (11\text{-}117c)$$

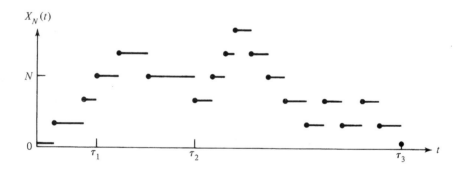

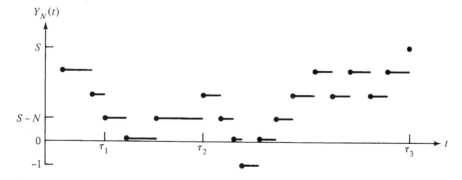

Figure 11-7 (a) The first cycle of the queueing process; (b) first cycle of the inventory process.

PROOF Let W_i be the waiting time of the ith arriving customer. The process $\{W_i ; i \in I_+\}$ regenerates when the server is turned off. Since $\rho < 1$, we may assume, without loss of generality, that the queue is empty when the first customer arrives (why?). Let $C(N)$ be the number of customers served during the first busy period. From Corollary 6-13b,

$$E[C(1)] = \frac{1}{1 - \rho}$$

It is readily established (Exercise 11-49) that

$$E[C(N)] = NE[C(1)] = \frac{N}{1 - \rho} < \infty \qquad N \in I_+ \qquad (11\text{-}118)$$

The discrete version of Theorem 6-8 yields

$$E(W_N^*) = \frac{E(W_1 + \cdots + W_{C(N)})}{E[C(N)]} \qquad (11\text{-}119)$$

Let $E(\mathcal{W}_N)$ denote the numerator in (11-119). To obtain an explicit formula for it, first observe that the ith customer to arrive waits for $N - i - 1$ other customers to arrive before the server is turned on, $1 \le i \le N$. Thus, the contribution to $E(\mathcal{W}_N)$ while the server is turned off [for example, $[0, \tau_1)$ in Figure 11-7a] is

$$\frac{1 + 2 + \cdots + N - 1}{\lambda} = \frac{N(N - 1)}{2\lambda}$$

When the server is turned on, customer 1 enters service. It will not affect $E(\mathcal{W}_N)$ if we assume customers are served in this order: After customer 1 completes service, serve next those customers who arrived while customer 1 was in service (customer 1's descendants). Then serve those customers who arrived while customer 1's descendants were in service, etc. This is the order of service used in the derivation of the $M/G/1$ busy-period d.f. in Section 6-5; hence customers 2, 3, ..., N are each delayed for a random time with this d.f. The contribution to $E(\mathcal{W}_N)$ of customers 2, 3, ..., N during this time interval [for example, $[\tau_1, \tau_2)$ in Figure 11-7a] is

$$(N - 1)E(B_1) = \frac{(N - 1)v}{1 - \rho}$$

where $E(B_1) = v/(1 - \rho)$ is the expected length of a busy period in an $M/G/1$ queue.

To obtain the contribution of customer 1 and customer 1's progeny [e.g., the customers served during $[\tau_1, \tau_2)$ in Figure 11-7a] to $E(\mathcal{W}_N)$, observe that for the $M/G/1$ queue, (11-119) with $N = 1$ implies

$$E(W_1^*) = \frac{E(W_1 + W_2 + \cdots + W_C)}{E(C)}$$

where C is the number of customers served in the first busy period. Since $E(W_1^*)$ is given by the Pollaczek-Khintchine formula (11-117c), the contribution of customer 1 and of customer 1's progeny is explicitly given by

$$E(W_1^*)E[C(1)] = \frac{E(W_1^*)}{1 - \rho}$$

Serve customers in the same fashion when customer 2 enters service. The contribution of customers $3, \dots, N$ to $E(\mathcal{W}_N)$ is $(N - 2)E(B_1)$, and the contribution of customer 2 and customer 2's progeny is $E(W_1^*)E[C(1)]$. Repeating this argument for customers $3, \dots, N$ yields

$$E(\mathcal{W}_N) = \frac{1 + 2 + \cdots + N - 1}{\lambda} + (1 + 2 + \cdots + N - 1)E(B_1)$$

$$+ E(W_1^*)E[C(1)]$$

$$= \frac{N}{1 - \rho}\left[\frac{N - 1}{2\lambda} + E(W_1^*)\right] \qquad (11\text{-}120)$$

Substituting (11-118) and (11-120) into (11-119) yields (11-117a). Since the process $\{W_i; i \in I_+\}$ is the same when $N = 0$ as when $N = 1$, (11-117b) follows directly from (11-117a). $\qquad \square$

The T-Policy

With the T-policy, the server scans the queue T time units after the end of the last busy period to determine whether customers are present. If customers are found, a busy period begins and the server is kept active until the system is empty. If no customers are found, we say a busy period of length 0 occurs. In either case, the next scan is made T units after the end of a busy period.

This model is motivated by the operation of some telephone electronic switching machines, which are stored program computers. In these machines, the central processing unit (CPU) cannot directly "see" whether a customer has arrived at an empty queue, but must periodically instruct a sensor to check the queues that have been idle. This reduces the time the CPU can devote to its other activities. If we model the temporarily diminished capacity of the CPU when it makes a scan by a fixed charge for activating the server and assume that there is a holding charge for waiting customers, we can obtain an optimal scan interval (Exercise 11-51). The N-policy cannot be used in this setting because it assumes that the queue length is monitored continuously when the server is shut down.

Theorem 11-20 Let $E[W^*(T)]$ be the expected waiting time in the steady state of an M/G/1 queue operating with the T-policy. Then

$$E[W^*(T)] = E(W_1^*) + \frac{T}{2} \qquad T \geq 0 \qquad (11\text{-}121)$$

where $E(W_1^*)$ is given by the Pollaczek-Khintchine equation (11-117c).

PROOF Let the random variable N_j be the number of customers present at the start of the jth busy period. The assumption of Poisson arrivals implies $N_1, N_2, \ldots$ are i.i.d. with

$$P\{N_1 = n\} = \frac{e^{-\lambda T}(\lambda T)^n}{n!} \qquad n \in I \tag{11-122}$$

Similarly, all of the busy periods are i.i.d. Let $\mathcal{W}$ be the sum of the waiting times of the customers served in the first busy period, and let C be the number of customers served in that busy period. The regenerative argument leading to (11-119) applies; it and (11-119) imply (why?)

$$
\begin{aligned}
E[W^*(T)] &= \frac{E(\mathcal{W})}{E(C)} \\
&= \frac{\sum_{n=0}^{\infty} E(\mathcal{W} \mid N_1 = n)P\{N_1 = n\}}{\sum_{n=0}^{\infty} E(C \mid N_1 = n)P\{N_1 = n\}}
\end{aligned}
\tag{11-123}
$$

It follows from (11-118), (11-122), and the convention that $N_1 = 0 \Rightarrow C = 0$ that the denominator of (11-123) is $E(N_1)/(1 - \rho) = \lambda T/(1 - \rho)$. Similarly, from (11-120) and (11-122) the numerator of (11-123) is

$$
\begin{aligned}
\sum_{n=1}^{\infty} E(\mathcal{W}_n)P\{N_1 = n\} &= \frac{E(N_1)E(W_1^*) + E(N_1^2 - N_1)/(2\lambda)}{1 - \rho} \\
&= \frac{\lambda T E(W_1^*) + (\lambda T^2/2)}{1 - \rho}
\end{aligned}
$$

Substituting these expressions into (11-123) yields (11-121). $\square$

The D-Policy

The *D-policy* activates the server when the work (i.e., the sum of the service times) of the customers in queue first exceeds the number D. Once turned on, the server is active until the queue is empty, and then the server is turned off.

Recall the cost structure for the operation of a service facility described at the start of this section. It should be apparent that the waiting times of customers and the lengths of busy periods are more accurately controlled by the amount of work present when the server is turned on than by the number of customers present. That is the reason for the introduction of the D-policy. When the holding cost is a linear function of the waiting time, it has been shown[†] that the best D-policy has a smaller asymptotic cost rate than the best N-policy.

The analysis of the $M/G/1$ queue operating under this policy is deferred to Exercise 11-52.

[†] O. J. Boxma, "Note on a Control Problem of Balachandran and Tijms," *Manage. Sci.* **22**: 916–917 (1976).

EXERCISES

11-48 Assume that no backlogs are allowed in the production inventory model in Example 11-16. What is the corresponding effect in the related $M/G/1$ queue?

11-49 Derive (11-118).

11-50 Assume that the cost rate for the server is r when the server is on and 0 when off; there is a fixed charge R for turning on the server; there is no charge for turning off the server; and the cost of holding a customer in the system for time t is ht. Find the asymptotic cost rate when the N-policy is used. What is the best value of N to use? Be sure to take into account that N is an integer.

11-51 Repeat Exercise 11-50 for the T-policy. Show that the best N-policy has a lower cost rate than the best T-policy. Give an intuitive explanation for this fact.

11-52 Suppose successive scans with the T-policy are not of constant length, but are i.i.d. random variables. This is called the *random T-policy*. We can interpret the length of a scan as a vacation time for the server, so this model is also called the *server-vacation model*. What is the expected waiting time in the steady state? Show that the D-policy is a special case of this model, and describe the d.f. of the scan interval. Obtain the expected waiting time in the steady state when the service times have an exponential distribution. How would you do this when the service times have an arbitrary distribution? (The solution in this case is in terms of a renewal function and is messy.)

11-53 Use the random T-policy to obtain the expected waiting time in the steady state of class 2 customers in an $M/G/1$ system with preemptive priorities.

11-7 INSENSITIVITY IN QUEUEING MODELS

Exercise 4-24 asks you to show that, in the $M/M/c/c$ queue, the limiting probability that j customers are present is given by

$$p_j = \frac{a^j/j!}{1 + a + a^2/2 + \cdots + a^c/c!} \qquad j = 0, 1, \ldots, c \qquad (11\text{-}124)$$

In (11-124), $a \triangleq \lambda/\mu$, where λ is the arrival rate and $1/\mu$ is the mean service time. Exercises 8-21 and 8-22 show that (11-124) holds when $c = 2$ and service times have an H_2 or Erlang-2 distribution. In this section we show that (11-124) holds for any service-time d.f.

The $M/G/c/c$ Queue as a Markov Process

Consider a queueing model with Poisson arrivals at rate λ and service times with distribution function $G(\,\cdot\,)$ having mean $1/\mu$. There are c servers and no waiting positions. Let $N(t)$ be the number of customers present at time t. When $G(\,\cdot\,)$ is not the exponential d.f., $\{N(t); t \geq 0\}$ is not a Markov process. A Markov process can be constructed by appending to the state description a vector representing the amount of service received by each customer in the system.

When $N(t) = i > 0$, let us agree to place the customers in servers $1, 2, \ldots, i$. Let $Z_j(t)$ be the amount of service already received by the customer from server j at time t; if server j is idle at time t, $Z_j(t) = 0$. Define the vector

$$\mathbf{Z}(t) = [Z_1(t), \ldots, Z_c(t)]$$

The process $\{[N(t), \mathbf{Z}(t)]; t \geq 0\}$ is easily seen to be a Markov process (why?). Let

$$P_j(\mathbf{x}; t) = P\{N(t) = j, \quad \mathbf{Z}(t) \leq \mathbf{x}\}$$

where $\mathbf{x} = (x_1, \ldots, x_c)$. It is shown below that for any choice of $[N(0), Z(0)]$,

$$\lim_{t \to \infty} P_j(\mathbf{x}; t) \triangleq p_j(\mathbf{x}) = p_j \prod_{i=1}^{j} \mu \int_0^{x_i} G^c(u) \, du \tag{11-125}$$

where an empty product is taken to be 1. Equation (11-124) follows from (11-125) by letting each $x_i \to \infty$.

Generalized Semi-Markov Process Description of a $GI/G/c$ Queue

A Markov process whose states are vectors consisting of one discrete element and at least one supplementary variable is called a *generalized semi-Markov process,* abbreviated GSMP. The process $\{[N(t), \mathbf{Z}(t)]; t \geq 0\}$ described above is a GSMP. Some simplification in notation and nomenclature is achieved if we restrict our discussion of GSMPs to those describing a $GI/G/c$ queue. The general case is discussed in, e.g., Schassberger (1976).

Let $N(t)$ be the number of customers present at time t and 8 be the set of possible values of $N(\cdot)$. Associate with each customer in the system a clock that records the attained service time. When a customer completes service, the attained service time equals the customer's service time, and we say that the customer's clock is *reset.* Similarly, associate a clock with the arrival process that records the elapsed time since the last arrival epoch. This clock is reset at each arrival epoch.

Let clock 0 be the clock for the arrival process. Number the servers from 1 to c and the queue positions from $c + 1$ onward, where the customer in position $c + 1$ is first in line. By *location* k we mean that part of the queueing system which received number k. Location k is *active* when it is occupied and *inactive* otherwise. When location k is active, clock k refers to the clock associated with the customer in location k; when location k is inactive, clock k is *fictitious.*

Observe that $N(\cdot)$ changes its value whenever a clock is reset. *Whenever* $N(\cdot) = i$ and clock k is reset, *assume* that the next value of $N(\cdot)$ is j with probability $p_k(i; j)$. We call $\{p_k(i; j)\}$ the *conditional transition probabilities.*

In general, service need not be given to a customer at unit rate (see Exercise 11-54 for an example). Let $r_{ki} \geq 0$ be the rate at which clock k runs when $N(\cdot) = i$. Let $G_k(\cdot)$ be the d.f. of the total amount of service required by a customer in location k. If t is an epoch where $N(\cdot)$ enters state i with $Z_k(t) = 0$ and $N(u) = i$ for all $u \in (t, t + x)$, then the probability that clock k is not reset during $(t, t + x]$ is $G_k^c(r_{ki} x)$. Clocks associated with exponential service times or interarrival times are called *exponential;* all other clocks are called *nonexponential.*

The GSMP is the continuous-time process formed by appending to the $N(\cdot)$ a vector representing all the clocks. A GSMP is said to be *insensitive* if, for each i,

$\lim_{t \to \infty} P\{N(t) = i\}$ is unaffected when the nonexponential-service or interarrival-time r.v.'s are replaced by exponential r.v.'s with the same mean. Our goal in this section is to show that the GSMP representing the $M/G/c/c$ queue is insensitive. This result is important because we know how to evaluate the limiting probabilities of the $M/M/c/c$ queue.

Example 11-17: $M/G/c/c$ queue Clock 0 is exponential, clocks 1 through c are nonexponential, and clocks with an index greater than c never appear. When i customers are present, clocks $1, \ldots, i$ are active and clocks $i + 1, \ldots, c$ are fictitious. The conditional transition probabilities are given by

$$p_0(i; j) = \begin{cases} 1 & \text{if } j = i + 1 \le c \\ 0 & \text{otherwise} \end{cases} \qquad p_k(i; j) = \begin{cases} 1 & \text{if } j = i - 1 \\ 0 & \text{otherwise} \end{cases}$$

where $1 \le k \le c$. The rates r_{ki} are all 1. $\qquad\qquad\square$

Sufficient Conditions for Insensitivity: An Algorithm

We start with a particular GSMP $\{N(t); t \ge 0\}$ for which we want to establish insensitivity. The *Markov version*, $\{M(t); t \ge 0\}$ say, is formed by changing each nonexponential clock to an exponential clock with the same mean. The process $\{M(t); t \ge 0\}$ is a CTMC. Let $1/\alpha_k$ be the mean of the distribution function $G_k(\cdot)$. In the case of the $M/G/c/c$ queue, α_0 is the arrival rate, and $1/\alpha_1 = 1/\alpha_2 = \cdots = 1/\alpha_c$ is the mean service time. Define

$$q_{ijk} \triangleq \alpha_i \, p_k(i; j) r_{ki}$$

It is the rate at which transitions occur from state i to state j when clock k is reset. The Q-matrix for $\{M(t); t \ge 0\}$ has elements

$$q_{ij} = \sum_{k \in 8} q_{ijk} \qquad i \ne j \qquad q_{ii} = -\sum_{j \ne i} q_{ij}$$

Suppose $\{M(t); t \ge 0\}$ has a limiting distribution $\mathbf{p}$; then (by Theorem 8-6) $\mathbf{p}$ satisfies

$$\mathbf{p}Q = \mathbf{0} \qquad\qquad (11\text{-}126)$$

We call (11-126) the *global balance equation*.

Form a digraph for $\{M(t); t \ge 0\}$ by drawing an arc from i to j whenever $q_{ijk} > 0$. For each pair (i, j) there is one arc from i to j for each k such that $q_{ijk} > 0$. The arc corresponding to a particular q_{ijk} is denoted (i, j_k). The digraph for the $M/G/3/3$ queue is shown in Figure 11-8.

Associate with each arc in the digraph a label which is a vector of 0's and 1's. The size of the vector is the number of nonexponential clocks in the GSMP. Recall the clock numbering scheme given above, and number the nonexponential clocks from smallest to largest in the usual way. If clock k is exponential, the nth element of the label for arc $(i, j)_k$ is 1 if, and only if, the nth nonexponential clock

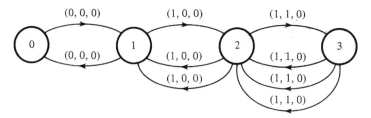

Figure 11-8 Digraph and labels for the M/G/3/3 queue.

is not fictitious in state i. If clock k is nonexponential and n refers to clock k, the nth element of the label for arc $(i, j)_k$ is 0. The labels for the M/G/3/3 queue are shown in Figure 11-8.

Example 11-18: Example 11-17 continued In the M/G/c/c queue, clock 0 is exponential, and clocks $1, \ldots, c$ are nonexponential. The digraph and labels when $c = 3$ are shown in Figure 11-8. $\square$

When there are M nonexponential clocks, there are $2^M - 1$ possible labels. Number the labels in any way you please, and let l_m denote the mth one. Define

$$\delta_m(i, j, k) \triangleq \begin{cases} 1 & \text{if arc } (i, j) \text{ has label } m \\ 0 & \text{otherwise} \end{cases}$$

For each arc label m, it *may* happen that the flow of probability from i to j along arcs with label m equals the flow of probability from j to i along those same arcs. If this balance does occur, we have the *restricted flow equation* for $\{M(t); t \geq 0\}$:

$$p_i \sum_{k \in \mathcal{S}} \delta_m(i, j, k) q_{ijk} = \sum_{k \in \mathcal{S}} p_k \delta_m(j, i, k) q_{jik} \tag{11-127}$$

If (11-127) is valid for all arcs and all labels, then (11-127) implies (11-126). However, (11-127) may be valid for just some (or none) of the arcs and labels. The connection among global balance, the restricted flow equation, and insensitivity is given by this theorem.

Theorem 11-21 For a given GSMP, let $\mathcal{A}(i)$ be the set of active locations whenever $N(\cdot) = i$. If the limiting probability vector of the Markov version satisfies (11-126) and (11-127), then for each $i \in \mathcal{S}$,

$$\lim_{t \to \infty} P\{N(t) = i, Z_0(t) \leq x_0, Z_1(t) \leq x_1, \ldots\}$$

$$= p_i \prod_{i \in \mathcal{A}(i)} \alpha_i \int_0^{x_i} G_i^c(u) \, du \tag{11-128}$$

When $\mathcal{A}(i)$ is empty, the product in (11-128) is 1.

PROOF See David Y. Burman (1982). $\square$

Corollary 11-21 When (11-128) is valid,

$$\lim_{t \to \infty} P\{N(t) = i\} = p_i \qquad i \in S$$

PROOF Send each x_i to infinity. ☐

Example 11-19: Example 11-17 continued In the $M/G/c/c$ queue, let λ be the arrival rate and $1/\mu$ be the mean service time. The global balance equations can be written (via Proposition 4-2)

$$\lambda p_{i-1} = i\mu p_i \qquad i = 1, 2, \ldots, c \qquad (11\text{-}129)$$

By extrapolating Figure 11-8 to arbitrary c, it is apparent that the label on all $(i, i + 1)$ arcs is the same as the label on the $(i + 1, i)$ arc. Hence, the restricted flow equation reduces to (11-129), and condition (*b*) of Theorem 11-21 is verified. Condition (*b*) of Theorem 11-21 is easily validated, and the insensitivity property is established. ☐

EXERCISES

11-54 Suppose the servers in an $M/G/2/2$ system work at different rates. Establish the insensitivity property for this model. Extend the result to an arbitrary number of servers.

11-55 Show that the $M/G/1$ queue operating with the LIFO-preemptive-resume discipline has the insensitivity property.

11-56 Show that the $M/G/1$ queue with the processor-sharing discipline has the insensitivity property.

BIBLIOGRAPHIC GUIDE

The virtual-delay process was introduced by V. Beneš and L. Takács; an account of their work is given in their books [Beneš (1963); Takács (1962)]. The relation between the steady-state d.f.'s of the virtual-delay and delay processes was given by Takács; a nice proof is given in Lemoine (1974).

The first rigorous proof that Poisson arrivals "see" time averages is given in Strauch (1970); the validity of that proof has been questioned, e.g., in Wolff (1970*b*), which contains a proof for single-server systems. The most general proof is in Wolff (1982).

The first general proof of $L = \lambda W$ is given in Little (1961). Prior to Little's paper, intuitive justifications were advanced and the relation was employed in proofs. [See, e.g., Cobham (1954).] A variety of proofs have been offered since 1961; notable ones are Jewell (1967) and Stidham (1974). The idea of extending $L = \lambda W$ appears in Maxwell (1970) and is developed in Brumelle (1971). Applications of $H = \lambda G$ to relations between higher moments are given in Brumelle (1972).

Sections 11-4 and 11-5 were influenced by Wolff (1970*b*), for work conservation, and Kleinrock (1976), for conservation laws and the notion of attained service.

Priorities were first considered in Cobham (1954); Theorems 11-15 and 11-16 are proved there. Preemptive priorities are introduced in Phipps (1956). Completion times and their ramifications were exploited first in Gaver (1962) and then in Avi-Itzhak and Naor (1963).

The first start-up and shut-down queueing model appeared in Yadin and Naor (1963), who considered the N-policy. Further refinements were made in Heyman (1968), Sobel (1969), and Bell (1971). The D-policy is due to Balachandran (1973), and the T-policy is due to Heyman (1977).

The insensitivity property for the $M/G/c/c$ queue was first established in Vaulot (1927). A simpler proof is given by Takács (1969); our proof follows Burman (1982). A general treatment of the generalized semi-Markov process is given in Schassberger (1976) and its references.

Several books on queueing theory complement the material presented in this chapter. Among these are (in addition to those books mentioned above) Cohen (1969, 1976), Cooper (1981), Cox and Smith (1961), Gnedenko and Kovalenko (1968), Gross and Harris (1974), and Kleinrock (1975).

A survey of the first 60 years of queueing research (1907–1967) is given in Bhat (1969). An algorithmic approach to queueing models is given in Neuts (1981).

References

Avi-Itzhak, B., and P. Naor: "Some Queueing Problems with the Service Station Subject to Breakdown," *Oper. Res.* **11**: 303–320 (1963).

Balachandran, K. R.: "Control Policies for a Single Server System," *Manage. Sci.* **19**: 1013–1018 (1973).

Bell, Colin E.: "Characterization and Computation of Optimal Policies for Operating an $M/G/1$ Queueing System with Removable Server," *Oper. Res.* **19**: 208–218 (1971).

Beneš, Václav E.: *General Stochastic Processes in the Theory of Queues*, Addison-Wesley, Reading, Mass. (1963).

Bhat, U. Narayan: "Sixty-Years of Queueing Theory," *Manage. Sci.* **15**: B280–B294 (1969).

Brumelle, Shelby L.: "On the Relation between Customer and Time Averages in Queues," *J. Appl. Prob.* **8**: 508–520 (1971).

———: "A Generalization of $L = \lambda W$ to Moments of Queue Length and Waiting Times," *Oper. Res.* **20**: 1127–1136 (1972).

Burman, David Y.: "Insensitivity in Queueing Systems," *Adv. Appl. Probab.* **14** (1982).

Cobham, Alan: "Priority Assignment in Waiting Line Problems," *Oper. Res.* **2**: 70–76 (1954).

Cohen, J. W.: *The Single Server Queue*, Wiley, New York (1969).

———: *On Regenerative Processes in Queueing Theory*, Springer-Verlag, New York (1976).

Cooper, Robert B.: *Introduction to Queueing Theory*, 2d ed., North-Holland, New York (1981).

Cox, D. R., and Walter L. Smith: *Queues*, Wiley, New York (1961).

Gaver, D. P., Jr.: "A Waiting Line with Interrupted Service, Including Priorities," *J. Roy. Stat. Soc.* **B25**: 73–90 (1962).

Gnedenko, B. V., and I. N. Kovalenko: *Introduction to Queueing Theory*, Israel Program for Scientific Translations, Jerusalem (1968).

Gross, Donald, and Carl M. Harris: *Fundamentals of Queueing Theory*, Wiley, New York (1974).

Heyman, Daniel P.: "Optimal Operating Policies for M/G/1 Queueing Systems," *Oper. Res.* **16**: 362–382 (1968).

———: "The T-Policy for the M/G/1 Queue," *Manage. Sci.* **23**: 775–778 (1977).

———, and Shaler Stidham, Jr.: "A Note on the Relation between Customer and Time Averages in Queues," *Oper. Res.* **28**: 983–944 (1980).

Jewell, William S.: "A Simple Proof of $L = \lambda W$," *Oper. Res.* **15**: 1109–1116 (1967).

Kleinrock, Leonard: *Queueing Systems*, vol. 1: *Theory*, Wiley, New York (1975).

———: *Queueing Systems*, vol. 2: *Computer Applications*, Wiley, New York (1976).

Lemoine, Austin J.: "On Two Stationary Distributions for the Stable GI/G/1 Queue," *J. Appl. Prob.* **11**: 849–852 (1974).

Little, John D. C.: "A Proof for the Queueing Formula: $L = \lambda W$," *Oper. Res.* **9**: 383–387 (1961).

Maxwell, William L.: "On the Generality of the Equation $L = \lambda W$," *Oper. Res.* **18**: 172–174 (1970).

Neuts, Marcel F.: *Matrix-Geometric Solutions in Stochastic Models*, Johns Hopkins, Baltimore, Md. (1981).

Phipps, Thomas E., Jr.: "Machine Repair as a Priority Waiting-Line Problem," *Oper. Res.* **4**: 76–85 (1956).

Schassberger, R.: "On the Equilibrium Distribution of a Class of Finite-State Generalized Semi-Markov Processes," *Math. Oper. Res.* **1**: 395–406 (1976).

Schrage, Linus E., and Louis W. Miller: "The Queue M/G/1 with the Shortest Remaining Processing Time Discipline," *Oper. Res.* **14**: 670–684 (1966).

Sobel, Matthew J.: "Optimal Average-Cost Policy for a Queue with Start-up and Shut-down Costs," *Oper. Res.* **17**: 145–162 (1969).

Stidham, Shaler, Jr.: "A Last Word on $L = \lambda W$," *Oper. Res.* **22**: 417–421 (1974).

Strauch, Ralph E.: "When a Queue Looks the Same to an Arriving Customer as to an Observer," *Manage. Sci.* **17**: 140–141 (1970).

Takács, Lajos: *Introduction to the Theory of Queues*, Oxford University Press, New York (1962).

———: "On Erlang's Formula," *Ann. Math. Stat.* **40**: 71–78 (1969).

Vaulot, A. E.: "Extension des Formules d'Erlang au cas où les durées des conversations suivent une loi quelconque," *Rev. Gén. Electr.* **22**: 1164–1171 (1927).

Wolff, Ronald W.: "Time Sharing with Priorities," *SIAM J. Appl. Math.* **19**: 566–574 (1970*a*).

———: "Work Conserving Priorities," *J. Appl. Prob.***7**: 327–337 (1970*b*).

——— : "Poisson Arrivals See Time Averages," *Oper. Res.* **30** (1982).

Yadin, M., and P. Naor: "Queueing Systems with a Removable Service Station," *Oper. Res. Quart.* **14**: 393–405 (1963).

TWELVE

NETWORKS OF QUEUES

In a network of queues, customers receive service from one or more service stations consecutively. Applications of models of this type appear in many diverse areas. In a production process, the service centers are work stations where fabrication, assembly, or inspection activities occur. A customer (i.e., work order) is routed through these centers until all the required operations have been performed. Example 10-10 is a queueing network model of a multiprogrammed computer, and Example 9-1 is a queueing network model of a portion of a hospital. Many communications systems and portions of vehicular traffic systems have been modeled with networks of queues.

The stochastic processes commonly used to describe the evolution of a network of queues are very complicated. Except for isolated cases, analyses have been restricted to models where interarrival times and service times are exponentially distributed, so that a Markov process is obtained. These are the only types of models we consider.

12-1 JACKSON NETWORKS

A *Jackson network* is a collection of queues with exponential service times in which customers travel from one queue to another according to a Markov chain.

The specific assumptions are as follows.

Assumption 12-1 The *network* consists of N service centers numbered 1, 2, ..., N. Service center i contains c_i identical servers.

Assumption 12-2 Customers from *outside the network* (called *exogenous* customers) arrive at service center i according to a Poisson process with rate λ_i, $i = 1, 2, \ldots, N$.

Assumption 12-3 After receiving service at center i, a customer leaves the network with probability $p_{i0} > 0$ or goes *instantaneously* to service center j with probability† p_{ij}, $\sum_{j=0}^{N} p_{ij} = 1$. These probabilities are independent of the history of the network.

Assumption 12-4 Customers arriving at service center i (from inside or outside the network) are served FIFO. The service times are i.i.d. exponential r.v.'s with mean $1/\mu_i$, $i = 1, 2, \ldots, N$.

Assumption 12-3 states that the successive service centers visited by a customer form a Markov chain with transition matrix $P = (p_{ij})$, where $p_{00} \triangleq 1$. We call P the *routing matrix* of the network, and Assumption 12-3 is called *Markov routing*.

Traffic Equation

Let α_i be the asymptotic arrival rate at service center i. The rate at which exogenous customers arrive at service center i is λ_i, from Assumption 12-2. The rate at which customers arrive at service center i from service center j is easily seen to be $\alpha_j p_{ji}$. Adding these rates yields

$$\alpha_i = \lambda_i + \sum_{j=1}^{N} \alpha_j p_{ji} \qquad i = 1, 2, \ldots, N \tag{12-1}$$

This is called the *traffic equation*.

Summing both sides of (12-1) over i and using $1 - p_{j0} = \sum_{j=1}^{N} p_{ji}$, we obtain

$$\sum_{j=1}^{N} p_{j0} \alpha_j = \sum_{j=1}^{N} \lambda_j \tag{12-2}$$

The left-side of (12-2) is the asymptotic rate at which customers leave the network, so (12-2) expresses the conservation law "what goes in must come out."

To verify that (12-1) has a unique solution, write it in matrix form as

$$\boldsymbol{\alpha} = \boldsymbol{\lambda} + \boldsymbol{\alpha} P_0 \tag{12-3}$$

where $\boldsymbol{\alpha} = (\alpha_1, \ldots, \alpha_N)$, $\boldsymbol{\lambda} = (\lambda_1, \ldots, \lambda_N)$, and $P_0 = (p_{ij})$ with $1 \leq i, j \leq N$. The ith row of P_0 sums to $1 - p_{i0} < 1$, hence

$$\lim_{n \to \infty} P_0^n = 0 \tag{12-4}$$

From Theorem 7-3, (12-4) implies $(I - P_0)^{-1}$ exists and consists of nonnegative elements. Thus, (12-3) has the unique solution

$$\boldsymbol{\alpha} = \boldsymbol{\lambda}(I - P_0)^{-1} \tag{12-5}$$

which is nonnegative.

---

† We allow $p_{ii} > 0$. How do you interpret it?

Limiting Probabilities

J. R. Jackson obtained the fascinating theorem that, in the steady state, the distribution of the number of customers in each service center can be computed *as if* the service centers were independent $M/M/c$ queues with center i having arrival rate α_i. The computational implications of this theorem are that we can decompose a network with N service centers into N $M/M/c$ queues and achieve considerable computational savings. We introduce the following notation in order to state the theorem precisely.

Let $X_i(t)$ be the number of customers in service center i at time t, $\mathbf{s} = (s_1, \ldots, s_N)$, where $s_i \in I$,

$$P(\mathbf{s}; t) = P\{X_i(t) = s_i, i = 1, \ldots, N\} \tag{12-6}$$

and, assuming it exists,

$$p(\mathbf{s}) = \lim_{t \to \infty} P(\mathbf{s}; t) \tag{12-7}$$

In (12-6) and (12-7) we allow $P(\mathbf{s}; t)$ to depend on some initial values for each $X_i(0)$; the explicit dependence is suppressed for notational ease. The left-side of (12-7) is independent of the initial values.

Theorem 12-1: J. R. Jackson's theorem In a Jackson network, assume $a_i \triangleq \alpha_i/\mu_i < c_i$ for each service center. Then the limiting probabilities $p(\cdot)$ exist and

$$p(\mathbf{s}) = \prod_{i=1}^{N} \psi_i(s_i) \tag{12-8}$$

where for each $n \in I$ and $i = 1, 2, \ldots, N$,

$$\psi_i(n) = \begin{cases} \dfrac{\psi_i(0)a_i^n}{n!} & \text{if } n \leq c_i \\[3mm] \dfrac{\psi_i(0)\rho_i^{n-c_i}a_i^{c_i}}{(c_i)!} & \text{if } n \geq c_i \end{cases} \tag{12-9}$$

with $\rho_i \triangleq a_i/c_i$, $n \in I$, and $\psi_i(0)$ is a normalizing constant that yields $\sum_{n=0}^{\infty} \psi_i(n) = 1$.

From (4-42) we recognize (12-9) as the limiting probability distribution of a queue with c_i servers, Poisson arrivals at rate α_i, and exponential service times with mean $1/\mu_i$.

PROOF Assumptions 12-1 to 12-4 imply that $\{[X_1(t), \ldots, X_N(t)]; t \geq 0\}$ is a continuous-time Markov chain.† Let $q(\mathbf{s}; \mathbf{s}')$ be the transition rate from state

---

† To conform to the requirement that a CTMC have a state space in I, the vector-valued process must be mapped into a scalar-valued process. This is inconvenient and unnecessary.

s to state s′. It is clear that this CTMC is irreducible; so by Theorem 8-6, $p(\cdot)$ satisfies (if it exists)

$$\sum_{s' \neq s} p(s')q(s'; s) = p(s) \sum_{s' \neq s} q(s; s')$$ (12-10)

We show that (12-8) satisfies (12-10) by substituting it into (12-10) and obtaining an identity.

Let $\mathbf{1}_i$ be a row vector of N elements with a 1 in the ith position and 0's elsewhere. If we neglect $o(\Delta t)$ terms, a transition from s to $s + \mathbf{1}_i$ occurs when an exogenous customer arrives at service center i. Thus,

$$q(s; s + \mathbf{1}_i) = \lambda_i \qquad i = 1, \ldots, N$$ (12-11a)

Similarly, a transition from s to $s - \mathbf{1}_i$ occurs when a customer leaves the network after completing service at i. Let $s_i \wedge c_i$ denote the smaller of s_i and c_i; then $(s_i \wedge c_i)\mu_i(\Delta t)p_{i0} + o(\Delta t)$ is the probability that such a transition occurs during $(t, t + \Delta t)$. Thus,

$$q(s; s - \mathbf{1}_i) = (s_i \wedge c_i)\mu_i p_{i0}$$ (12-11b)

If a customer finishes service at i and is routed to j, a transition from s to $s - \mathbf{1}_i + \mathbf{1}_j$ occurs. Reasoning as above yields

$$q(s; s - \mathbf{1}_i + \mathbf{1}_j) = (s_i \wedge c_i)\mu_i p_{ij}$$ (12-11c)

Since all other transitions out of s during $(t, t + \Delta t)$ occur with probability $o(\Delta t)$,

$$q(s; s') = 0 \qquad \text{if} \qquad s' \notin \{s + \mathbf{1}_i, s - \mathbf{1}_i, s - \mathbf{1}_i + \mathbf{1}_j\}$$ (12-11d)

Instead of substituting (12-11) into (12-10) directly, considerable algebraic effort is saved by an observation due to P. Whittle and called the *method of partial balance*. Suppose the set $\{s' \neq s\}$ is partitioned into a finite number of disjoint subsets, $S_1, \ldots, S_K$ say. If the *partial-balance equations*

$$\sum_{s' \in S_k} p(s')q(s'; s) = p(s) \sum_{s' \in S_k} q(s; s')$$ (12-12)

hold for each $k = 1, \ldots, K$, then (12-10) is true. By choosing the S_k judiciously, the effort in verifying (12-10) can be simplified. [The method of partial balance has the drawback that for some CTMCs and some choices of the subsets, (12-12) may not hold for some k even though (12-10) is true. At present, there is no known way to check a priori that a set of partial-balance equations is consistent.]

For the problem at hand, choose $S_1 = \{s + \mathbf{1}_i, i = 1, \ldots, N\}$ and S_2 as the complementary set. When $k = 1$, (12-12) is

$$\sum_{i=1}^{N} p(s + \mathbf{1}_i)q(s + \mathbf{1}_i; s) = p(s) \sum_{i=1}^{N} q(s; s + \mathbf{1}_i)$$

From (12-11a) and (12-11b) we obtain

$$\sum_{i=1}^{N} p(\mathbf{s} + \mathbf{1}_i)[(s_i + 1) \wedge c_i]\mu_i p_{i0} = p(\mathbf{s}) \sum_{i=1}^{N} \lambda_i \qquad (12\text{-}13a)$$

Similarly, for $k = 2$, (12-12) yields

$$p(\mathbf{s} - \mathbf{1}_i)\lambda_i + \sum_{j \neq i} p(\mathbf{s} - \mathbf{1}_i + \mathbf{1}_j)[(s_j + 1) \wedge c_j]\mu_j p_{ji} = p(\mathbf{s})(s_i \wedge c_i)\mu_i \quad (12\text{-}13b)$$

for $i = 1, \ldots, N$. Exercise 12-5 asks you to provide the interpretation of (12-13a) and (12-13b).

Substituting (12-8) into (12-13) yields (after some algebra that is left to you) (12-1), so (12-10) is valid. We have now established that $p(\,\cdot\,)$ is a stationary distribution of the CTMC.

According to Theorem 8-6, if

$$\sum_{\mathbf{s}} p(\mathbf{s}) \sum_{\mathbf{s}'} q(\mathbf{s}; \mathbf{s}') < \infty \qquad (12\text{-}14)$$

then $p(\,\cdot\,)$ is a limiting distribution. If $c_i < \infty$ for $i = 1, \ldots, N$, then (12-11) shows that the inner sum is a bounded function of $\mathbf{s}$; hence (12-14) is valid. The proof when some $c_i = \infty$ is left as Exercise 12-6. $\square$

The product form of (12-8) implies that in the steady state, $X_1(t), \ldots, X_N(t)$ are mutually independent for each t. It does not imply that $X_1(t)$ is independent of $X_2(t')$ for $t \neq t'$.

Internal Flows in the Network

Theorem 12-1 asserts that each service center in a Jackson network behaves, in the steady state, as if it received Poisson arrivals at rate α_i. We now present an example† of a Jackson network where the total arrival stream (consisting of the exogenous arrivals and the arrivals from within the network) at a service center is not a Poisson process.

Suppose there is one service center consisting of a single server. Let λ be the arrival rate of exogenous customers, $1/\mu$ be the mean service time, and p be the probability that a customer joins the end of the queue after leaving the server. (In the notation of Theorem 12-1, $p = p_{11}$ and $1 - p = p_{10}$.) Customers who rejoin the queue are said to be *fed back*.

It follows from Theorem 12-1 that the limiting probability that n customers are present, ψ_n say, is given by

$$\psi_n = (1 - \rho)\rho^n \qquad n \in I \qquad (12\text{-}15)$$

where $\rho = \lambda/[(1 - p)\mu]$.

Proposition 12-1 In the steady state, the probability that a fed-back customer joins a queue containing n other customers is ψ_n for all $n \in I$.

---

† This example is due to P. J. Burke, "Proof of a Conjecture on the Interarrival-Time Distribution in an M/M/1 Queue with Feedback," *IEEE Transactions on Communications*, **24**: 575–576 (1976).

PROOF The number in the network is easily seen to be a birth-and-death process, so in the steady state, the number of customers present at departure (not feedback) epochs has the same distribution as the number of customers present at arrival epochs of exogenous customers (Exercise 4-50). Since customers are fed back (or depart) in a way that is independent of the number of customers in the queue, the distribution "seen" by fed-back customers is the same as the distribution "left behind" by departing customers. Thus, the fed-back customers "see" the same distribution of customers as is "seen" by the exogenous arrivals. Since the latter form a Poisson process, they "see" time averages, and the proposition follows. □

As a consequence of this proposition the steady-state probability that n customers are present just before a customer (exogenous or fed-back) joins the queue is ψ_n. The next proposition asserts that the times between arrivals (exogenous and fed-back) are not exponentially distributed, so the arrival process cannot be Poisson.

Proposition 12-2 In the steady state, let T be an epoch where a customer (exogenous or fed-back) joins the queue, and let $T + U$ be the next such epoch. Then U is independent of T, and

$$P\{U > t\} = \beta e^{-\mu t} + (1 - \beta)e^{-\lambda t} \tag{12-16}$$

where $\beta = p\mu/(\mu - \lambda)$.

PROOF Let $H(t)$ be the probability that none of the customers present at T (there is at least one by the definition of T) are fed back prior to $T + t$. The memoryless property of the exponential service and (exogenous) interarrival times implies that U is independent of T and that

$$P\{U > t\} = e^{-\lambda t}H(t) \tag{12-17}$$

Toward obtaining $H(\cdot)$, let $H_n(t)$ be the corresponding conditional probability when it is given that n customers are present just before T (do not count the arrival at T). Let H_0 be the probability that none of the customers present at T is fed back. Then

$$H(t) = \sum_{n=0}^{\infty} \psi_n H_n(t) + H_0$$

When n customers are present at T, the jth one in line is the first one to be fed back if, and only if, it is fed back and the first $j - 1$ customers are not fed back. This occurs with probability $(1 - p)^{j-1}p$. The elapsed time from T until the jth customer completes service is the sum of j independent exponential r.v.'s, so it has a gamma density with shape parameter j. Hence,

$$h_n(t) \triangleq -\frac{d}{dt} H_n(t) = \sum_{j=1}^{n+1} \frac{(1 - p)^{j-1}p\mu(\mu t)^{j-1}e^{-\mu t}}{(j - 1)!} \tag{12-18}$$

Substituting (12-15) and (12-18) into

$$h(t) \triangleq -\frac{d}{dt} H(t) = \sum_{n=0}^{\infty} \psi_n h_n(t)$$

yields

$$h(t) = (1 - \rho)p\mu e^{-\mu t} \sum_{n=0}^{\infty} \rho^n \sum_{j=1}^{n+1} \frac{[(1 - p)\mu t]^{j-1}}{(j - 1)!}$$

$$= (1 - \rho)p\mu e^{-\mu t} \sum_{j=1}^{\infty} \frac{[(1 - p)\mu t]^{j-1}}{(j - 1)!} \sum_{n=j-1}^{\infty} \rho^n \qquad (12\text{-}19)$$

$$= p\mu e^{-(\mu - \lambda)t}$$

The probability that none of the customers present at T (count the arrival at T) is fed back is

$$H_0 = \sum_{n=0}^{\infty} \psi_n (1 - p)^{n+1} = \frac{(1 - p)\mu - \lambda}{\mu - \lambda} \triangleq 1 - \beta \qquad (12\text{-}20)$$

Consequently,

$$H(t) = H_0 - \int_0^t h(x) \, dx = 1 - \beta + \beta e^{-(\mu - \lambda)t} \qquad (12\text{-}21)$$

and substituting (12-21) into (12-17) yields (12-16). $\qquad\qquad\square$

There are processes inside some Jackson networks that are Poisson. When $c_1 = c_2 = \cdots = c_N = 1$, it can be shown† that if $p_{ij} > 0$ and the probability of being routed from j to i in n steps is zero for every n, then the customers leaving service center i and going directly to service center j form a Poisson process with rate $\alpha_i p_{ij}$.

Closed Jackson Networks

In a *closed network*, a fixed number of customers circulates through the network. There are no exogenous customers and no departures. (A network with exogenous arrivals is called *open*.) A *closed Jackson network* satisfies Assumptions 12-1, 12-3, and 12-4 with Assumption 12-3 changed so that $p_{i0} = 0$ for each i. We let M denote the number of customers in the network.

Since there are no exogenous arrivals, the traffic equation for the closed network is obtained from (12-1) by setting each $\lambda_i = 0$:

$$\alpha_i = \sum_{j=1}^{N} \alpha_j p_{ji} \qquad i = 1, 2, \ldots, N \qquad (12\text{-}22)$$

The matrix $P = (p_{ij})$, $1 \le i, j \le N$, is stochastic. If P is reducible, the customers in

---

† Frederick J. Beutler and Benjamin Melamed, "Decomposition and Customer Streams of Feedback Networks of Queues in Equilibrium," *Oper. Res.* **26**: 1059–1072 (1978).

distinct communicating classes will form independent closed Jackson networks, each having an irreducible routing matrix. So, without loss of generality, we may assume that P is irreducible. With this assumption, it is evident from Markov chain theory that (12-22) has a nonnegative solution different from the zero vector which is unique up to multiplication by a scalar.

The essential features of Theorem 12-1 for open Jackson networks carry over to closed Jackson networks.

Theorem 12-2: Jackson-Gordon-Newell theorem† In a closed Jackson network with an irreducible routing matrix, let $\alpha = (\alpha_1, \ldots, \alpha_N)$ be any nonnegative and nonzero solution of (12-22),

$$a_i = \frac{\alpha_i}{\mu_i} \quad \text{and} \quad \rho_i = \frac{a_i}{c_i} \quad i = 1, \ldots, N$$

Then the limiting probabilities $p(\,\cdot\,)$ exist and are given by

$$p(\mathbf{s}) = K \prod_{i=1}^{N} \psi_i(s_i) \tag{12-23}$$

where for each i,

$$\psi_i(n) = \begin{cases} \dfrac{a_i^n}{n!} & \text{if } n \le c_i \\[2mm] \dfrac{\rho_i^{n-c_i} a_i^{c_i}}{(c_i)!} & \text{if } n > c_i \end{cases} \tag{12-24}$$

and K is the normalizing constant that yields $\sum p(\mathbf{s}) = 1$ when the sum is taken over all the possible state vectors $\mathbf{s}$.

PROOF The proof is similar to the proof of Theorem 12-1 and is left as Exercise 12-7. □

The major difficulty in computing $p(\mathbf{s})$ is obtaining the value of K. The number of different states is the number of ways M customers can be placed in N service centers, which is $\binom{N+M-1}{M}$. This may be a very large number. For example, when $M = 100$ and $N = 10$, there are roughly 4.25×10^{12} states of the network. Algorithms for obtaining K are available.‡ These algorithms (whose descriptions are lengthy) are recursive in the number of customers. A potential advantage of computing K in this way is illustrated in Example 12-1.

---

† This theorem was first published in Jackson (1963). A more detailed version appears in Gordon and Newell (1967).

‡ See J. Buzen, "Computational Algorithms for Closed Networks with Exponential Servers," *Comm. Assoc. Comput. Mach.* **16**: 527–531 (1973), or A. C. Williams and R. A. Bhandiwad, "A Generating Function Approach to Queueing Network Analysis of Multiprogrammed Computers," *Networks* **6**: 1–12 (1976).

Example 12-1: Multiprogrammed computer system A specially structured closed Jackson network is used as a model of a multiprogrammed computer system in Example 10-10. The special structure consists of the particular routing probabilities

$$p_{1j} = \pi_j \quad \text{and} \quad p_{j1} = 1 \quad j = 2, \ldots, N$$

with $\sum_{j=2}^{N} p_{1j} = 1$. Let a_i be the average number of busy servers at i; in Example 10-10 it is shown that

$$a_i = \gamma_i a_1 \tag{12-25}$$

where $\gamma_i \triangleq \mu_1 \pi_i / \mu_i$. Thus, the average occupancy of every service station can be obtained from a_1.

To obtain a_1, let X_1^* be the number of customers in service center 1 in the steady state. The constant K appearing in (12-23) depends on M, so let us emphasize that dependence by writing the constant as $K(M)$. It is left to you (Exercise 12-7) to establish

$$P\{X_1^* = k\} = \frac{\psi_1(k)K(M)}{K(M-k)} \tag{12-26}$$

and so

$$a_1 = K(M) \left[\sum_{k=1}^{c_1} \frac{k\psi_1(k)}{K(M-k)} + c_1 \sum_{k=c_1+1}^{M} \frac{\psi_i(k)}{K(M-k)} \right] \qquad \square$$

Generalizations

The assumptions of the Jackson network are relaxed somewhat in Jackson (1963). Assumption 12-2 is replaced by this assumption.

Assumption 12-2a When $\sum_{i=1}^{N} X_i(t) = n$, an exogenous customer arrives during $(t, t + \Delta t]$ with probability $\lambda_n \Delta t + o(\Delta t)$. An exogenous customer is placed in service center i with probability r_i which is independent of the assignment of all previous exogenous arrivals.

The condition $p_{i0} > 0$ for all i in Assumption 12-3 is replaced by the hypothesis that (12-4) is valid. That is, each customer leaves the network after a finite number of transitions among the service centers (w.p.1). Assumption 12-4 is replaced by this assumption.

Assumption 12-4a When $X_i(t) = n$, a customer at service center i will complete service during $(t, t + \Delta t]$ with probability $\mu_n \Delta t + o(\Delta t)$, with $\mu_0 = 0$.

With these assumptions, it is shown that the limiting† probabilities (12-7)

---

† Jackson neglects to check that the appropriate version of (12-14) is satisfied, so he proves results for only the stationary distribution. By considering the trivial case of $N = 1$ and recalling Theorem 4-4, it is clear that some condition similar to (4-39) is required. It is sufficient to assume that $\{\lambda_n\}$ and $\{\mu_n\}$ are bounded.

have a product solution similar to (12-8) in which $\psi_i(\cdot)$ has the form of the limiting probabilities in a birth-and-death process with state-dependent birth rates and death rates [Equation (4-37)]. The proof is similar to the proof of Theorem 12-1.

Several extensions of the Jackson network are given in Baskett, Chandy, Muntz, and Palacios (1975). They make either Assumption 12-4a or one of the following assumptions: (1) There is a single server, and the queue discipline is processor sharing (page 426). (2) There are infinitely many servers, and the service-time d.f. has a rational Laplace transform (i.e., the service-time d.f. is a random mixture of gamma d.f.'s). (3) There is a single server, the queue discipline is LIFO-preemptive-resume (page 422), and the service-time d.f. has a rational Laplace transform. Not all the service centers have to satisfy the same assumption. Customers may be grouped into different classes, and a customer may change classes in a Markov fashion while moving from one service center to another. The mode of service at a service center may be different for various classes of customers. Assumption 12-2a is also invoked. Networks of this type are called BCMP networks.

The main result concerning BCMP networks is that the limiting probabilities (12-7) have a product form similar to (12-8). The method of proof parallels the proof of Theorem 12-1 and is much more intricate. A BCMP network appears to be a good model for multiprogrammed and timeshare computer systems. It permits, for example, the central processing unit to be a single server with processor sharing and the input-output devices to be parallel servers with customers served FIFO. The different customer classes may be used to distinguish batch jobs from timeshare jobs or to reflect the status of a job.

EXERCISES

12-1 Assumption 12-3 requires $p_{i0} > 0$ for each i. This is a stronger requirement than is needed for the proof of Theorem 12-1 to be valid. What weaker condition is sufficient for the given proof to be valid?

12-2 Suppose travel times from service center i to service center j are i.i.d. exponential r.v.'s. The network may be open or closed. Construct an equivalent Jackson network in which these travel times are zero. How are the limiting probabilities affected by the nonzero travel times? What changes if the travel times have a density with a rational Laplace transform?

12-3 Assumption 12-4 requires that customers be served FIFO at each service center. This is not necessary for the given proof of Theorem 12-1. What property of the queue discipline is required for that proof to be valid?

12-4 Provide interpretations of (12-13a) and (12-13b) that enable you to write them directly.

12-5 Show that the left-side of (12-4) equals $\sum_{i=1}^{N} \lambda_i$ for any choice of $c_1, \ldots, c_N$; $c_i = \infty$ is allowed. *Hint:* Recall (4-43).

12-6 Prove Theorem 12-2. *Hint:* Use the notion of partial balance, and establish that in verifying that, (12-23) satisfies

$$(c_i \wedge s_i)\mu_i \, p(\mathbf{s}) = \sum_{j=1}^{N} [c_j \wedge (s_j + 1)]\mu_j \, p_{ji} \, p(\mathbf{s} + \mathbf{1}_j - \mathbf{1}_i)$$

for each i constitutes a proof.

12-7 Establish (12-26).

12-8 In Example 12-1, let $N = 2$, $c_1 = c_2 = 1$, and $\mu_1 = \mu_2 = \mu$. Let ψ_M be the mean time elapsing between arrival epochs of a particular job at station 1. Show that

$$\psi_M = \frac{M + 1}{\mu}$$

You may want to refer to Exercise 11-19; in that exercise, ψ_M is denoted by $E(C)$.

12-2 WAITING TIMES IN TANDEM QUEUES

In this section we consider a special case of an open Jackson network in which the departures from service station i all go to service station $i + 1$, $i < N$, and the departures from service station N leave the network. This type of network is called a *tandem*. The special traffic structure of a tandem allows us to obtain conveniently used formulas for waiting-time d.f.'s in certain cases. Tandem queueing models have been employed to model diverse phenomena such as assembly-line production and vehicular traffic flow.

Independence of Waiting Times for Single-Server Work Centers

The assumptions for an open Jackson network made in Section 12-1 are still in force. The routing matrix for a tandem is described by

$$p_{i, i+1} = 1 \qquad i = 1, \ldots, N \qquad \text{and} \qquad p_{N0} = 1$$

Hence, the solution of the traffic equation (12-1) is simply

$$\alpha_i = \lambda \qquad i = 1, \ldots, N$$

Our goal is to show that when each work center consists of a single server, the waiting time of a customer at work center i is independent of his or her waiting time at work center j for any $j \neq i$. This will enable us to establish that the waiting time in work center i has the same d.f. as in an M/M/1 queue in isolation with arrival rate λ and mean service time $1/\mu_i$.

> **Theorem 12-3** Let W_{ki} be the waiting time of the kth arriving customer in the ith service center of a tandem queueing network that is in the steady state at time 0. At each service center, assume that customers are served FIFO. If $c_i = 1$ and $\rho_i = \lambda/\mu_i < 1$ for each i, then $W_{k1}, \ldots, W_{kN}$ are mutually independent.

PROOF† Let

$$
\begin{aligned}
V_{ki} &= \text{service time of customer } k \text{ in center } i \\
T_k &= \text{arrival time of customer } k \text{ at center 1} \\
R_k &= \text{departure time of customer } k \text{ from center 1} \\
\tau_k &= R_k - R_{k-1} \\
X_i(t) &= \text{number of customers at center } i \text{ at time } t
\end{aligned}
$$

† This proof is due to Edgar Reich, "Note on Queues in Tandem," *Ann. Math. Stat.* **34**: 338–341 (1963).

The assumption that customers are served FIFO implies that $X_1(R_k^+)$ equals the number of arrivals during $(T_k, T_k + W_{k1}]$. Hence,

$$P\{X_1(R_k^+) = n \mid W_{k1}, \ldots, W_{kN}\} = \frac{e^{-\lambda W_{k1}}(\lambda W_{k1})^n}{n!} \qquad n \in I \qquad (12\text{-}27)$$

Multiplying both sides of (12-27) by z^n ($|z| \leq 1$) and summing over $n \in I$ yield

$$\sum_{n=0}^{\infty} z^n P\{X_1(R_k^+) = n \mid W_{k1}, \ldots, W_{kN}\} = e^{-\lambda W_{k1}(1-z)}$$

Unconditioning on W_{k1} yields

$$\begin{aligned}
\hat{W}_{k1}(z) &\triangleq \sum_{n=0}^{\infty} z^n P\{X_1(R_k^+) = n \mid W_{k2}, \ldots, W_{kN}\} \\
&= \int_0^{\infty} e^{-\lambda x(1-z)} \, dP\{W_{k1} \leq x \mid W_{k2}, \ldots, W_{kN}\}
\end{aligned} \qquad (12\text{-}28)$$

Recall that in Exercise 8-30 you showed that $X_1(R_k^+)$ is independent of the previous departure epochs, so it is independent of $\tau_k, \tau_{k-1}, \ldots$. Of course, $X_1(R_k^+)$ is independent of those customers who have not yet left center 1 and of the service times in center $j \geq 2$ of those customers who have previously left center 1. Since these r.v.'s completely determine $W_{k2}, \ldots, W_{kN}$, $X_1(R_k^+)$ is independent of $(W_{k2}, \ldots, W_{kN})$; so, then, is the right-side of (12-28). This has the further implication that

$$P\{W_{k1} \leq x \mid W_{k2}, \ldots, W_{kN}\} = P\{W_{k1} \leq x\}$$

Repeating this argument on $W_{k2}, \ldots, W_{k,N-1}$, in turn, will establish that W_{ki} is independent of $(W_{k,i+1}, \ldots, W_{kN})$, and so

$$P\{W_{ki} \leq x_i; \ i = 1, \ldots, N\} = \prod_{i=1}^{N} P\{W_{ki} \leq x_i\} \qquad k \in I_+ \qquad \square$$

Corollary 12-3 For $i = 1, \ldots, N$,

$$P\{W_{ki} \leq t\} = 1 - e^{-(\mu_k - \lambda)t} \qquad k \in I_+ \qquad (12\text{-}29)$$

PROOF It is clear that $P\{W_{k1} \leq t\}$ is given by the steady-state waiting-time d.f. of an M/M/1 queue, which is (12-29) (from Example 4-12). Since the departures from service center 1 are the departures from an M/M/1 queue, they form a Poisson process (Theorem 8-9, Burke's output theorem). Thus, the argument above applies to W_{k2} because it is independent of W_{k1}. Proceeding in this fashion for $i = 3, \ldots, N$ completes the proof. $\square$

The requirement that each service center have but one server can be relaxed for the first and last centers of the tandem.

Theorem 12-4 Theorem 12-3 remains valid if $c_i = 1$ for all $i \in [2, N-1]$.

PROOF See P. J. Burke, "The Output Process of a Stationary M/M/s Queueing System," *Ann. Math. Stat.* **39**: 1144–1152 (1968). $\square$

Examples of Strange Behavior

A consequence of Theorem 12-4 is that for any choice of c_i and c_{i+1}, W_{ki} and $W_{k,i+1}$ are independent (Exercise 12-10). The following example shows that if $N = 3$ and $c_2 \geq 2$, then W_{k1} and W_{k3} are dependent. This means that the condition in Theorem 12-4 that the intermediate service stations contain only one server cannot be weakened.

Example 12-2† Consider a tandem with three service centers, where $c_1 = c_3 = 1$ and $c_2 \geq 2$. The mean service time at each service center is the same, $1/\mu$ say. Let λ be the arrival rate of the exogenous customers, with $\lambda < \mu$.

Corollary 12-3 implies that in the steady state,

$$E(W_{k3}) = \frac{1}{\mu - \lambda} \qquad k \in I_+$$

By producing a customer index l such that

$$E(W_{l3} \mid W_{l1} \text{ is large}) \neq \frac{1}{\mu - \lambda} \tag{12-30}$$

we will establish that W_{k3} and W_{k1} are not, in general, independent. Burke's way of establishing (12-4) is to produce a lower bound on the probability that customer l is delayed at service center 3 given that W_{l1} is large.

Consider what occurs in the steady state. From Corollary 12-3, for any given $\epsilon > 0$, there is a smallest customer index $l \in I_+$ where $W_{l1} = w_{l1}$ which is so large that $e^{-\lambda w_{l1}} < \epsilon$; this is the probability that no customers are present at the epoch when customer l leaves service center 1. Thus, we can assume there is a customer, customer j say, that starts service at center 1 at the epoch at which customer l leaves center 1. We denote this epoch by R_l. Let D_{li} be the delay (i.e., the time spent in queue) of customer l at service center i. Since the service times in all service centers are i.i.d.,

$$P\{j \text{ enters center 2 before } l \text{ leaves center 2} \mid D_{l2} = 0\} = \tfrac{1}{2}$$

If service centers 2 and 3 are both empty at R_l, then $D_{l3} > 0$ if (but not only if) customer j enters center 2 before customer l leaves it, customer j completes service at center 2 before customer l does, and customer j is still in service at center 3 when customer l leaves center 2. Since all service times are

---

† This example is due to P. J. Burke, "The Dependence of Sojourn Times in Tandem M/M/s Queues," *Oper. Res.* **17**: 754–755 (1969). An example that shows dependent waiting times in a network that is not a tandem is given in B. Simon and R. Foley, "Some Results on Sojourn Times in Acyclic Networks," *Manage. Sci.* **25**: 1027–1035 (1979).

i.i.d. and exponential, we obtain

$$P\{D_l > 0 \,|\, X_2(R_l^-) = X_3(R_l^-) = 0\} \geq \tfrac{1}{8} - \epsilon \qquad (12\text{-}31)$$

[The ϵ occurs because $\{X_1(R_l^-) \geq 0\}$ occurs with positive probability that is smaller than ϵ.]

By similar arguments you can show that (Exercise 12-11)

$$P\{D_{l3} > 0 \,|\, X_2(R_l^-) = 0\} \geq \tfrac{1}{4} \qquad (12\text{-}32)$$

and

$$P\{D_{l3} > 0 \,|\, X_3(R_l^-) = 0\} \geq \tfrac{1}{4} \qquad (12\text{-}33)$$

From (12-31), (12-32), and (12-33) and the definition of customer l, we obtain

$$P\{D_{l3} > 0 \,|\, W_{l1} \geq -(\log \epsilon)/\lambda\} \geq \tfrac{1}{8} - \epsilon \qquad (12\text{-}34)$$

Since $\{D_{l3} > 0\}$ implies that the server at center 3 is occupied when customer l arrives there,

$$E[D_{l3} \,|\, W_{l1} \geq -(\log \epsilon)/\lambda] \geq \frac{7}{8\mu} + \frac{\epsilon}{\mu} \qquad (12\text{-}35)$$

Since λ does not appear in (12-35), an appropriate choice of ϵ yields (12-30).

□

In the proof of Theorem 12-3, we argued that those customers in service center i at the epoch when customer k left there all arrived during customer k's wait in center i. That argument is always valid for single-server centers where customers are served FIFO. In Example 12-2, $c_2 \geq 2$ and customers need not leave center 2 in the same order in which they start service there. The ability of one customer to pass another in a multiple-server center was exploited to produce a counterexample to the conjecture that waiting times are independent.

Delays at service center 1 tend to be small when few customers are there at an arrival epoch, which suggests that delays in successive service centers are dependent (and, moreover, are positively correlated). On the other hand, write

$$W_{k1} = D_{k1} + V_{k1}$$

and

$$W_{k2} = D_{k2} + V_{k2}$$

Theorem 12-4 asserts that W_{k1} is independent of W_{k2}, and V_{k1} is independent of V_{k2} by assumption, so it is plausible that D_{k1} is independent of D_{k2}. The following example settles the issue in favor of dependence.

Example 12-3† We consider a tandem with two service centers, each containing one server. The arrival rate is λ, and the mean service time at center i

---

† This example is due to P. J. Burke, "The Dependence of Delays in Tandem Queues," *Ann. Math. Stat.* **35**: 874–875 (1964).

is $1/\mu_i$. In the steady state, the delay distributions of all customers are the same, so distributional results for customer 1 apply to all customers. Let D_i be the delay of customer 1 at service center i. We show that in the steady state,

$$P\{D_2 = 0\} < P\{D_2 = 0 \mid D_1 = 0\} \tag{12-36}$$

From Theorem 12-1, we obtain

$$P\{D_2 = 0\} = 1 - \rho_2 \tag{12-37}$$

where $\rho_2 = \lambda/\mu_2$. Let T be the epoch when customer 1 arrives; then $\{D_1 = 0\}$ is equivalent to $\{X_1(T) = 0\}$. Since Poisson arrivals "see" time averages and Theorem 12-1 implies $X_1(T)$ is independent of $X_2(T)$,

$$P\{N_2(T) = n \mid D_1 = 0\} = (1 - \rho_2)\rho_2^n \qquad n \in I \tag{12-38}$$

Let $W(T)$ be the work present at center 2 at epoch T; then (12-38) implies

$$P\{W(T) = 0\} = 1 - \rho_2 \quad \text{and}$$
$$\text{Pd}\{W(T) = x\} = \rho_2(\mu_2 - \lambda)e^{-(\mu_2 - \lambda)x} \qquad x > 0 \tag{12-39}$$

where $\text{Pd}\{\,\cdot\,\}$ stands for probability density.

Let V be the service time of customer 1. The event $\{D_2 = 0\}$ occurs if, and only if, the event $\{V > W(T)\}$ occurs. Thus

$$P\{D_2 = 0 \mid D_1 = 0, \quad W(T) = x\} = e^{-\mu_1 x} \qquad x \geq 0$$

and unconditioning on $W(T)$ via (12-39) yields

$$P\{D_2 = 0 \mid D_1 = 0\} = 1 - \rho_2 + \rho_2 \int_0^\infty e^{-\mu_1 x} e^{-(\mu_2 - \lambda)x} \, dx$$

$$= 1 - \rho_2 + \frac{\rho_2(\mu_2 - \lambda)}{\mu_1 + \mu_2 - \lambda}$$

$$> 1 - \rho_2 = P\{D_2 = 0\}$$

and (12-36) is established. $\qquad\qquad\qquad\qquad\qquad\qquad\qquad\qquad\square$

EXERCISES

12-9 Where does the proof of Theorem 12-3 break down if some $c_i > 1$ or if service is not FIFO?

12-10 Let c_i be the number of exponential servers in the ith service center of a tandem. It is not necessary that all c_i servers have the same mean service time. The exogenous arrival process is Poisson. Show that Theorem 12-4 implies that W_{ki} is independent of $W_{k,i+1}$ for each $k \in I_+$, $i = 1, \ldots, N - 1$.

12-11 Derive (12-32) and (12-33). *Hint*: There is no need to consider customer j.

12-12 In Example 12-3, derive

$$P\{D_2 > x \mid D_1 = 0\} = \left(\frac{\rho_2 \mu_1}{\mu_1 + \mu_2} - \lambda\right)e^{-(\mu_2 - \lambda)x} \quad x > 0$$

What conclusions can you draw from this?

BIBLIOGRAPHIC GUIDE

The first paper to treat networks of queues was by R. R. P. Jackson (1954). In that paper, Theorem 12-1 was established for a tandem model consisting of single-server service centers. Theorem 12-1 was established in J. R. Jackson (1957) and extended in a variety of ways in J. R. Jackson (1963). Theorem 12-2 is a special case of a theorem in J. R. Jackson (1963) which was rediscovered in Gordon and Newell (1967). The latter paper initiated many applications of closed Jackson network models of computer systems. The extension of the Jackson network to the BCMP network appears in Baskett, Chandy, Muntz, and Palacios (1975). A comprehensive treatment of these topics appears in Kelley (1979).

Closed networks of $M/M/c$ queues with arbitrarily distributed transit times (which may depend on the pair of service centers involved) are considered in Posner and Bernholtz (1968a). The analogous model with one server at each service center and several classes of customers is described in Posner and Bernholtz (1968b).

Section 12-2 is based on the work of P. J. Burke and E. Reich. In addition to the papers referred to there, see the survey paper Burke (1972). Papers which survey other aspects of networks of queues include Disney (1975) and Lemoine (1977, 1978).

References

Baskett, F.; K. M. Chandy; R. R. Muntz; and F. G. Palacios: "Open, Closed, and Mixed Networks of Queues with Different Classes of Customers," *J. Assoc. Comput. Mach.* **22**: 248–260 (1975).

Burke, P. J.: "Output Processes and Tandem Queues," *Proc. Symp. Computer Communications Networks and Teletraffic*, Polytechnic Institute of Brooklyn Press, New York (1972).

Disney, Ralph L.: "Random Flow in Queueing Networks: A Review and Critique," *Trans. AIIE* **7**: 268–288 (1975).

Gordon, William J., and Gordon F. Newell: "Closed Queueing Systems with Exponential Servers," *Oper. Res.* **15**: 254–265 (1967).

Jackson, James R.: "Networks of Waiting Lines," *Oper. Res.* **5**: 518–521 (1957).

————: "Jobshop-Like Queueing Systems," *Manage. Sci.* **10**: 131–142 (1963).

Jackson, R. R. P.: "Queueing Systems with Phase-Type Service," *Oper. Res. Quart.* **5**: 109–120 (1954).

Kelley, F. P.: *Reversibility and Stochastic Networks*, Wiley, New York (1979).

Lemoine, Austin J.: "Networks of Queues—A Survey of Equilibrium Analysis," *Manage. Sci.* **24**: 464–481 (1977).

————: "Networks of Queues—A Survey of Weak Convergence Results," *Manage. Sci.* **24**: 1175–1193 (1978).

Posner, M., and B. Bernholtz: "Closed Finite Queueing Networks with Time Lags," *Oper. Res.* **16**: 962–976 (1968a).

———— and ————: "Closed Finite Queueing Networks with Time Lags and with Several Classes of Units," *Oper. Res.* **16**: 977–985 (1968b).

THIRTEEN

BOUNDS AND APPROXIMATIONS

Some of the exact solutions to queueing models have been found to be too complicated for practical use. For example, the steady-state delay distribution for the $GI/G/1$ queue is described by the integral equation (9-92). This equation has not been found to be convenient for doing numerical calculations in many practical problems. For the $GI/G/c$ (and even the $M/G/c$) queues, the analytical solutions that have been obtained are intractable and are not presented in this book.

For these reasons, bounds and approximations for the operating characteristics of various queueing models have been sought. Since the quantities of interest (e.g., the mean delay) can be computed in principle but not in practice, the bounds and approximations should be easy to compute if they are to be useful. Of course, they should also be accurate.

13-1 BOUNDS FOR QUEUEING MODELS

In many applications, the steady-state delay is the operating characteristic of primary interest. In this section we concentrate our efforts on bounding it. Throughout the section we use the notation

$$T_n = \text{arrival epoch of customer } n$$
$$U_n = T_n - T_{n-1} \text{ with } U_1 = T_1$$
$$1/\lambda = E(U_1) \text{ and } \sigma_u^2 = \text{Var } (U_1)$$
$$V_n = \text{service time of customer } n$$
$$1/\mu = E(V_1) \text{ and } \sigma_v^2 = \text{Var } (V_1)$$
$$D_n = \text{delay of customer } n$$

The random variable D^* will represent the steady-state delay (see Section 6-4). When bounding $E(D^*)$, we assume that $\lambda/\mu < c$ and $\sigma_v^2 < \infty$ so that $E(D^*) < \infty$ (by Theorem 11-6).

Bounds for the Delay in the *GI/G/*1 Queue

Upper and lower bounds for $E(D^*)$ in the *GI/G/*1 queue can be constructed from the following representation of $E(D^*)$.

Theorem 13-1 In a *GI/G/*1 queue, let

$$Y_n = V_{n-1} - U_n \qquad n \in I_+$$

and†

$$Z_n = (D_{n-1} + Y_n)^- \quad n \in I_+$$

Then

$$E(D^*) = \frac{\lambda[\sigma_u^2 + \sigma_v^2 - \text{Var}\,(Z^*)]}{2(1 - \rho)} \tag{13-1}$$

where $\text{Var}\,(Z^*) = \lim_{n \to \infty} \text{Var}\,(Z_n)$ and $\rho = \lambda/\mu$.

PROOF Recall (9-65) which states that

$$D_n = (D_{n-1} + Y_n)^+ \qquad n \in I_+ \tag{13-2}$$

when customers are served FIFO. The order of service does affect $E(D^*)$. From (13-2) and the definition of Z_n, we obtain

$$D_{n-1} + Y_n = (D_{n-1} + Y_n)^+ - (D_{n-1} + Y_n)^- = D_n - Z_n \tag{13-3}$$

Taking expectations of the outer terms yields

$$E(Z_n) = E(D_n) - E(D_{n-1}) - E(Y_n) \tag{13-4}$$

Since $E(Y_n) = 1/\mu - 1/\lambda$ for each $n \in I_+$, taking limits as $n \to \infty$ in (13-4) yields

$$\lim_{n \to \infty} E(Z_n) = \frac{1 - \rho}{\lambda} \tag{13-5}$$

For any random variable A, $A = A^+ - A^-$ and $A^2 = (A^+)^2 + (A^-)^2$. Thus,

$$\text{Var}\,(A) = E(A^2) - [E(A)]^2 = \text{Var}\,(A^+) + \text{Var}\,(A^-) + 2E(A^+)E(A^-) \tag{13-6}$$

Use (13-6) to take variances on both sides of the first equality in (13-3);

† Recall the notation $a^+ = \max\{a, 0\}$ and $a^- = \max\{-a, 0\}$. Observe that if n_i is the ith index for which $Z_n > 0$, then Z_{n_i} is the length of the ith idle period.

recalling (13-2) and the independence of D_{n-1} and Y_n yield

$$\text{Var}(D_{n-1}) + \text{Var}(Y_n) = \text{Var}(D_n) + \text{Var}(Z_n) + 2E(D_n)E(Z_n) \quad (13\text{-}7)$$

Take limits as $n \to \infty$ on both sides of (13-7), and then use (13-5) and $\text{Var}(Y_n) = \text{Var}(Y_1) = \sigma_u^2 + \sigma_v^2$ to obtain

$$\sigma_u^2 + \sigma_v^2 = \text{Var}(Z^*) + \frac{2E(D^*)(1-\rho)}{\lambda} \quad (13\text{-}8)$$

Rearranging (13-8) yields (13-1). $\qquad\qquad\qquad\qquad\qquad \square$

From (13-1), upper and lower bounds on $\text{Var}(Z^*)$ will produce bounds on $E(D^*)$.

Corollary 13-1 In a $GI/G/1$ queue with traffic intensity $\rho < 1$,

$$\frac{\lambda E[(Y_1^+)^2]}{2(1-\rho)} \le E(D^*) \le \frac{\lambda(\sigma_u^2 + \sigma_v^2)}{2(1-\rho)} \quad (13\text{-}9)$$

PROOF The upper bound is obtained from the crude bound $\text{Var}(Z^*) \ge 0$. The lower bound is obtained from observing that $D_{n-1} \ge 0$ implies

$$Z_n^2 = [(D_{n-1} + Y_n)^-]^2 \le (Y_n^-)^2 \qquad n \in I_+$$

Combining this with (13-7) yields

$$\begin{aligned}
\text{Var}(Z_n) &\le E[(Y_n^-)^2] - [E(Z_n)]^2 \\
&\le E[(Y_n)^2 - (Y_n^+)^2] - \{E(Y_n^2) - 2E(Y_n)[E(D_n) - E(D_{n-1})]\} \\
&= \text{Var}(Y_n) - E[(Y_n^+)^2] + 2E(Y_n)[E(D_n) - E(D_{n-1})] \qquad n \in I_+
\end{aligned}$$
$$(13\text{-}10)$$

Since $\text{Var}(Y_n) = \text{Var}(Y_1) = \sigma_u^2 + \sigma_v^2$ and $E[(Y_n^+)^2] = E[(Y_1^+)^2]$, taking limits as $n \to \infty$ in (13-10) yields

$$\text{Var}(Z^*) \le \sigma_u^2 + \sigma_v^2 - E[(Y_1^+)^2] \quad (13\text{-}11)$$

Substituting (13-11) in (13-1) produces the lower bound. $\qquad\qquad \square$

The upper bound in (13-9) depends on only the mean and variance of the interarrival and service times, so it is easy to calculate. The derivation suggests that the bound will work best when D^* tends to be large, i.e., when ρ is close to 1. It can be shown that this bound becomes sharp as $\rho \uparrow 1$. The term $E[(Y_1^+)^2]$ in the lower bound may depend on the complete specification of the interarrival- and service-time d.f.'s and may not have a closed-form expression (Exercise 13-1). It would be desirable to obtain a lower bound that depends on only mean and variance information, but such bounds are likely to be very loose and often provide no more information than the trite bound $E(D^*) \ge 0$ (see Exercises 13-2 and 13-3).

In many problems, one is interested in the probability of long delays. Obtaining the steady-state delay distribution may require considerable computation. An upper bound on this quantity provides a conservative estimate.

Theorem 13-2 Let $D(\,\cdot\,)$ be the steady-state delay d.f. of a $GI/G/1$ queue and $H(\,\cdot\,)$ be the d.f. of Y_1. Define θ by†

$$\theta = \sup\left\{t > 0 : \int_{-\infty}^{\infty} e^{ty} \, dH(y) < 1\right\} \tag{13-12}$$

Then

$$D^c(t) \triangleq 1 - D(t) \leq e^{-\theta t} \qquad t > 0 \tag{13-13}$$

PROOF Observe that when $t \geq 0$ and $\int_{-\infty}^{\infty} e^{ty} \, dH(y) \leq 1$,

$$1 - H(t) + \int_{-\infty}^{t} e^{-\theta(t-y)} \, dH(y) \leq e^{-\theta t} \int_{-\infty}^{\infty} e^{\theta y} \, dH(y) \leq e^{-\theta t}$$

We prove that any nonincreasing function on $(0, \infty)$, say $\eta(\,\cdot\,)$, with $0 \leq \eta(t) \leq 1$ for all $t \in (0, \infty)$, and which satisfies

$$\eta(t) \geq 1 - H(t) + \int_{-\infty}^{t} \eta(t - y) \, dH(y) \qquad t \geq 0 \tag{13-14}$$

is an upper bound on $D^c(t)$. Since $e^{-\theta t}$ is such a function, (13-13) will be validated.

The proof of the assertion is by induction. Let $\eta(\,\cdot\,)$ satisfy (13-14). Choose $D_0 = 0$ (this does not affect the limiting d.f.). Clearly $P\{D_0 \geq t\} \leq \eta(t)$. Make the inductive assumption that $P\{D_n \geq t\} \leq \eta(t)$ is valid for some n. Then (13-2) yields

$$P\{D_{n+1} \geq t\} = P\{D_n + Y_n \geq t\} = \int_{-\infty}^{\infty} P\{D_n \geq t - y\} \, dH(y)$$

$$= \int_{-\infty}^{t} P\{D_n \geq t - y\} \, dH(y) + 1 - H(t)$$

$$\leq \int_{-\infty}^{t} \eta(t - y) \, dH(y) + 1 - H(t)$$

$$\leq \eta(t) \qquad \text{[from (13-14)]}$$

and the induction succeeds. ☐

Bounds for the Expected Delay in the $GI/G/c$ Queue

Tight bounds which are easy to apply have not been found for the $GI/G/c$ queue. The best available bounds are given below.

† If the set in (13-12) is empty, set $\theta = 0$.

We assume, without loss of generality, that customers enter service in *order of arrival*. If customers are assigned cyclically to the servers (customers k, $c + k$, $2c + k$, ... go to server k), this certainly increases the mean delay in the steady state (Exercise 13-5). Then $E(D^*)$ is bounded above by the expected delay in a $GI/G/1$ queue with the same service-time d.f. and where the mean and variance of the interarrival times are c times as large. From Corollary 13-1 we obtain

$$E(D^*) \leq \frac{\lambda(\sigma_v^2 + c\sigma_u^2)}{2(c - a)} \tag{13-15}$$

where $a = \lambda/\mu$.

Recall (11-55) and write it in the form

$$E(D^*) = \frac{E[W^*(\infty)]}{a} - \frac{E(V_1^2)}{2E(V_1)} \tag{13-16}$$

Thus, a lower bound for $E(D^*)$ can be obtained from a lower bound for $E[W^*(\infty)]$. Let $W(t)$ be the work in the system at time t and T_n be the arrival epoch of customer n. The maximum work that can be done during (T_{n-1}, T_n) is $c(T_n - T_{n-1})$, so

$$W(T_n) \geq [W(T_{n-1}) + V_n - cU_n]^+ \qquad n \in I_+ \tag{13-17}$$

By comparing (13-17) with (13-2), it is apparent that $W(T_n)$ is no smaller than the delay of customer n in a $GI/G/1$ queue with the same service times and with all the interarrival times multiplied by c. Let $E(D_1^*)$ denote the mean delay in the steady state of this $GI/G/1$ queue. Then $E[W^*(\infty)] \geq E(D_1^*)$, and substitution into (13-16) yields†

$$E(D^*) \geq \frac{E(D_1^*)}{a} - \frac{\mu E(V_1^2)}{2} \tag{13-18}$$

The lower bound for $E(D_1^*)$ given in Corollary 13-1 may be combined with (13-18).

A particularly important special case is the $M/G/c$ queue. Then the Pollaczek-Khintchine equation (11-52) gives $E(D_1^*)$ exactly:

$$E(D_1^*) = \frac{\lambda E(V_1^2)}{2(c - a)}$$

Substitution into (13-18) yields

$$E(D^*) \geq \frac{\mu E(V_1^2)\,[1/(c - a) - 1]}{2} \tag{13-19}$$

The right-side of (13-19) is positive if, and only if, $a < c - 1$, so the bound is nontrivial only when $\rho \triangleq a/c > (c - 1)/c$.

---

† This bound is believed to be new. It is better than the bound given by (40) in Kingman (1970) if, and only if, ρ is less than the coefficient of variation of the interarrival times.

Comments

Since the bounds developed above hold for all interarrival- and service-time d.f.'s, it is unreasonable to expect them to be close to the exact value in any particular case. It is apparent from (13-1) that the mean delay is not determined by only the means and variances, so bounds based on only these parameters cannot be too tight. The performance of the bounds described in this section is left to Exercises 13-6 and 13-7.

One way to improve the bounds is to restrict the d.f. of the interarrival times. In Marshall (1968), improved bounds for single-server systems are obtained by assuming that the d.f. of the interarrival times has a monotone hazard rate. This idea is extended to many-server systems in Brumelle (1971).

EXERCISES

13-1 Express the d.f. of Y_1^+ in terms of the interarrival- and service-time d.f.'s. How would you compute $E[Y_1^+)^2]$? Calculate the bounds in (13-9) for the M/M/1 queue; how good are they?

13-2 Exhibit interarrival- and service-time d.f.'s such that $P\{V_n < [E(V_n) + E(U_n)]/2 < U_n\} = 1$ and the variance of U_n is as large as desired. What does this exercise imply about a lower bound for $E(D^*)$ that depends on only the first two moments of the interarrival- and service-time d.f.'s?

13-3 Let u and v be nonnegative numbers. Show that $[(v - u)^+]^2 \geq (v - u)^2 - u^2$. Use this inequality to establish

$$E(D^*) \geq \frac{\lambda^2 \operatorname{Var}(V_1) + \rho^2 - 2\rho}{2\lambda(1 - \rho)}$$

When is this bound trivial? How well does it work for M/G/1 queues?

13-4 Find conditions on U_1 and V_1 that make the set in (13-12) nonempty.

13-5 Give a rigorous derivation of (13-15). *Hint*: Explain why the workload in the c single-server systems is never smaller than the workload in the c-server system, and use (13-16).

13-6 Evaluate the upper bound for $E(D^*)$ in an $E_5/M/1$ queue, and evaluate the upper and lower bounds for an M/M/5 queue. How well do the bounds perform? The following table will assist your computations.

$\rho = \dfrac{\lambda}{c}$	$E(D^*)$	
	M/M/5	$E_5/M/1$
0.1	0.00004	0.04
0.5	0.052	0.40
0.75	0.37	1.56
0.90	1.52	5.14
0.95	3.51	11.14

In this table, $\mu = 1$.

13-7 Evaluate the bound given by (13-13) for the M/M/1 and $E_2/M/1$ queues. The exact probabilities are given in the following table, where $\mu = 1$.

	$P\{D^* > 2\}$	
λ	M/M/1	E_2/M/1
0.1	0.165	0.14
0.5	0.368	0.29
0.75	0.607	0.52
0.90	0.819	0.77
0.95	0.905	0.88

How would you evaluate the bound for the E_5/M/1 queue?

13-2 DIFFUSION PROCESS APPROXIMATIONS

The formula for the probability that j customers are present at time t in an M/M/1 queue is given by (4-126). This formula is hard to work with, and its generalization for the M/G/1 queue is not known. In this section, we use a diffusion process to approximate the behavior of the GI/G/1 queue and obtain approximate formulas for the transient and steady-state probabilities. We also employ a diffusion process to approximate the delay process. Tractable and exact steady-state solutions for the GI/G/c queue are not known; we present a diffusion process that can be used to approximate the steady-state distribution of the number of customers present.

Review of Reflected Brownian Motion

Let $\{B(t); t \geq 0\}$ be a Brownian motion (abbreviated BM) process on the line, and set $B(0) = x_0$. Recall from Section 9-4 that $F(t, x; x_0) \triangleq P\{B(t) \leq x \mid B(0) = x_0\}$ satisfies

$$\frac{\partial}{\partial t} F(t, x; x_0) = -c \frac{\partial}{\partial x} F(t, x; x_0) + \frac{D^2}{2} \frac{\partial^2}{\partial x^2} F(t, x; x_0) \qquad t \geq 0 \quad (13\text{-}20)$$

where

$$c \, \Delta t = E[B(t + \Delta t) - B(t)] + o(\Delta t)$$

and

$$D^2(\Delta t) = \text{Var} \, [B(t + \Delta t) - B(t)] + o(\Delta t)$$

We call c the infinitesimal mean (or drift) and D^2 the infinitesimal variance (or diffusion coefficient). The initial condition $B(0) = x_0$ is represented by

$$F(0, x; x_0) = \begin{cases} 0 & \text{if } x < x_0 \\ 1 & \text{if } x \geq x_0 \end{cases} \qquad (13\text{-}21)$$

When a reflecting barrier for the BM is placed on the x axis, the boundary condition

$$F(t, 0; x_0) = 0 \qquad x_0 > 0, \qquad t > 0 \qquad (13\text{-}22)$$

is obtained. The solution of (13-20) subject to (13-21) and (13-22) is

$$F(x, t; x_0) = \Phi\left(\frac{x - x_0 - ct}{D\sqrt{t}}\right) - e^{2xc/D^2}\Phi\left(\frac{-x - x_0 - ct}{D\sqrt{t}}\right) \qquad (13\text{-}23)$$

where

$$\Phi(z) = \frac{1}{\sqrt{2\pi}} \int_{-\infty}^{z} e^{-u^2/2} \, du$$

is the normal d.f. When $c < 0$,

$$F(x) = \lim_{t \to \infty} F(x, t; x_0) = 1 - e^{-2x(-c)/D^2} \qquad (13\text{-}24a)$$

and

$$\lim_{t \to \infty} E[B(t) \mid B(0) = x_0] = \frac{D^2}{-2c} \qquad (13\text{-}24b)$$

Both limits are independent of x_0 and represent steady-state values.

An Approximation for the Number Present in an M/M/1 Queue

In order to motivate the diffusion model employed in approximating the $GI/G/1$ queue, and to indicate its efficacy for the $M/M/1$ queue in particular, we first develop an approximation for the number present in an $M/M/1$ queue. The complicated nature of the exact transient solution is displayed in Section 4-10; a potential use of this approximation is found in transient analyses.

In the $M/M/1$ queue, let λ be the arrival rate, $1/\mu$ be the mean service time, and $\rho = \lambda/\mu$. Let $X(t)$ be the number of customers in the queue at time t, and

$$p(t, n; n_0) = P\{X(t) = n \mid X(0) = n_0\} \qquad n \in I, \qquad t \geq 0$$

For $t > 0$ and $n \in I_+$, $p(t, n; n_0)$ satisfies

$$\frac{d}{dt} p(t, n; n_0) = \lambda p(t, n - 1; n_0) + \mu p(t, n + 1; n_0) - (\lambda + \mu)p(t, n; n_0) \qquad (13\text{-}25)$$

and

$$\frac{d}{dt} p(t, 0; n_0) = \mu p(t, 1; n_0) - \lambda p(t, 0; n_0) \qquad (13\text{-}25')$$

The initial condition is

$$p(0, n; n_0) = \begin{cases} 1 & \text{if } n = n_0 \\ 0 & \text{if } n \neq n_0 \end{cases} \qquad (13\text{-}26)$$

and the boundary condition is

$$p(t, n; n_0) = 0 \qquad n < 0, \, t \geq 0 \qquad (13\text{-}27)$$

The idea of the approximation is to replace (13-25) by a partial differential equation that is easier to solve. We do this by replacing the discrete variable n by the continuous variable x, or $p(t, n; n_0)$ by $f(t, x; x_0)$ in (13-25). Then we expand (13-25) in a Taylor's series about the point $(t, x; x_0)$, keeping only first- and second-order terms (Exercise 9-42). This yields (we denote partial differentiation by a subscript and delete the argument)

$$f_t = -(\lambda - \mu)f_x + \frac{\lambda + \mu}{2} f_{xx} \qquad x, t > 0 \qquad (13\text{-}28)$$

If we define

$$F(t, x; x_0) = \int_{-\infty}^{x} f(t, y; x_0)\, dy$$

it can be easily shown that $F(\,\cdot\,,\,\cdot\,;\,\cdot\,)$ also satisfies (13-28). We take

$$F(0, x; x_0) = \begin{cases} 0 & \text{if } x < x_0 \\ 1 & \text{if } x \geq x_0 \end{cases} \qquad (13\text{-}29)$$

and

$$F(t, 0; x_0) = 0 \qquad t > 0 \qquad (13\text{-}30)$$

as natural replacements for (13-26) and (13-27). This system of equations is identical in form to (13-20), (13-21), and (13-22), so $F(t, x; x_0)$ is given by the right-side of (13-23) with $c = \lambda - \mu$ and $D^2 = \lambda + \mu$.

Consider now the asymptotic behavior of $p(\,\cdot\,)$ and $F(\,\cdot\,)$. From the theory of the M/M/1 queue, we have (Exercise 4-22) $p_n = \lim_{t \to \infty} p(t, n; n_0) = (1 - \rho)\rho^{n-1}$ for $n \in I_+$, and from (13-24a) we obtain

$$F(t) = \lim_{t \to \infty} F(t, x; x_0) = 1 - \exp \frac{-2x(1 - \rho)}{1 + \rho}$$

If we approximate p_n by

$$\hat{p}_n = \int_{n-1}^{n} dF(x) \qquad n \in I_+$$

we obtain

$$\hat{p}_n = (1 - e^{-2(1 - \rho)/(1 + \rho)})e^{-2(n-1)(1 - \rho)/(1 + \rho)}$$

$$= (1 - \alpha)\alpha^{n-1} \qquad n \in I_+$$

where $\alpha = \exp[-2(1 - \rho)/(1 + \rho)]$. Thus $\hat{p}_n$ has the same form as p_n. If ρ is close to 1, then $\alpha \doteq \rho$ and hence

$$\hat{p}_n \doteq p_n \qquad \rho \doteq 1$$

so $\hat{p}_n$ is a good approximation of p_n when ρ is slightly less than 1.

Approximate System Length for a GI/G/1 Queue

We consider first a heuristic derivation of (13-28) for the M/M/1 queue. During the interval $(t, t + \Delta t]$, the number of customers in the system changes by the number of arrivals minus the number of service completions, and when $X(t) = n > 0$, this change has expectation $(\lambda - \mu)\,\Delta t + o(\Delta t)$ and variance $(\lambda + \mu)\,\Delta t + o(\Delta t)$. To approximate $\{X(t); t \geq 0\}$ by a diffusion process with the same infinitesimal mean and variance, set $c = \lambda - \mu$ and $D^2 = \lambda + \mu$ in (13-20), which yields (13-28). This suggests that an appropriate choice of c and D^2 will yield a good approximation for the queue length of the $GI/G/1$ queue.

For a $GI/G/1$ queue, let $1/\lambda$ and σ_u^2 be the mean and variance, respectively, of the interarrival times and $1/\mu$ and σ_v^2 be the mean and variance, respectively, of the service times. Let $X(t)$ be the number of customers present at time t and $A(t)$ and $R(t)$ be the number of arrivals and departures, respectively, during $(0, t]$. Then

$$X(t) = X(0) + A(t) - R(t) \qquad t > 0 \tag{13-31}$$

In Section 5-3 it is shown that for a renewal process $\{N(t); t \geq 0\}$, where the interrenewal times have mean v and variance σ^2, for large values of t

$$E[N(t)] \approx \frac{t}{v} \tag{13-32}$$

and

$$\text{Var}\,[N(t)] \approx \frac{\sigma^2 t}{v^3} \tag{13-33}$$

By hypothesis, $\{A(t); t \geq 0\}$ is a renewal process, so from (13-32) and (13-33) we obtain

$$E[A(t)] \approx \lambda t \qquad \text{Var}\,[A(t)] \approx \lambda^3 \sigma_u^2 t \tag{13-34}$$

The process $\{R(t); t \geq 0\}$ is not a renewal process, nor is it independent of the arrival process. But in heavy traffic (ρ close to 1), the server will be occupied most of the time, so we approximate $R(t)$ by $\hat{R}(t)$, where

$$E[\hat{R}(t)] \approx \mu t \qquad \text{Var}\,[\hat{R}(t)] \approx \mu^3 \sigma_v^2 t \tag{13-35}$$

and we treat the arrival and departure processes as if they were independent. Substituting $\hat{R}(t)$ for $R(t)$ in (13-31) and using (13-34) and (13-35), we obtain the approximate results

$$\lim_{t \to \infty} \frac{E[X(t)]}{t} \doteq \lambda - \mu \tag{13-36}$$

and

$$\lim_{t \to \infty} \frac{\text{Var}\,[X(t)]}{t} \doteq \lambda^3 \sigma_u^2 + \mu^3 \sigma_v^2 \tag{13-37}$$

This suggests that we approximate $X(t)$ by a diffusion process, $\hat{X}(t)$ say, with infinitesimal mean and variance given by

$$c = \lambda - \mu \tag{13-38}$$

and

$$D^2 = \lambda^3 \sigma_u^2 + \mu^3 \sigma_v^2 \tag{13-39}$$

respectively. If we let $F(t, x; x_0) = P\{\hat{X}(t) \le x \,|\, \hat{X}(0) = x_0\}$, then $F(\,\cdot\,, \cdot\,;\cdot\,)$ satisfies (13-20), (13-21), and (13-22); hence it is given by (13-23), with c and D^2 as above.

As a partial check on the efficacy of this approximation, consider what happens as $t \to \infty$. Substituting (13-38) and (13-39) into (13-24b) yields

$$\lim_{t \to \infty} E[\hat{X}(t)] = \frac{\lambda^2(\sigma_v^2/\rho + \sigma_u^2 \rho^2)/\rho}{2(1 - \rho)} \tag{13-40}$$

For the $M/G/1$ queue, the Pollaczek-Khintchine formula (11-52), $E(W^*) = E(D^*) + 1/\mu$, and $L = \lambda W$ yield

$$\lim_{t \to \infty} E[X(t)] = \frac{\lambda^2(\sigma_v^2 + 1/\mu^2)}{2(1 - \rho)} \tag{13-41}$$

Since $\sigma_u^2 = 1/\lambda^2$ in this case, (13-40) becomes

$$\lim_{t \to \infty} E[\hat{X}(t)] = \frac{\lambda^2(\sigma_v^2/\rho + 1/\mu^2)/\rho}{2(1 - \rho)} \tag{13-40a}$$

The right-sides of (13-40a) and (13-41) are close when ρ is close to 1.

Approximate Virtual Delay for a $GI/G/1$ Queue

Let $S(t)$ be the work that has arrived by time t. Recall (11-1):

$$S(t) = V_1 + \cdots + V_{A(t)}$$

From (13-34) we obtain, for large values of t,

$$E[S(t)] = E(V_1)E[A(t)] \approx \rho t \tag{13-42}$$

Using the conditional variance relationship Var $[S(t)] = E(\text{Var } [S(t) \,|\, A(t)]) + \text{Var } [E(S(t) \,|\, A(t))]$ and (13-34), we obtain, for large values of t,

$$\text{Var } [S(t)] \approx \lambda(\sigma_v^2 + \rho^2 \sigma_u^2)t \tag{13-43}$$

The sample paths of the virtual-delay process are sawtooth functions with a jump of size V_i at the arrival epoch of the ith customer followed by a decline of slope -1; the process has an impenetrable boundary at the axis of abscissas (see Figure 11-1). Assume that (13-42) and (13-43) hold for all t, so that

$$\lim_{\Delta t \to 0} \frac{E[S(t + \Delta t)] - E[S(t)]}{\Delta t} = \rho \tag{13-44}$$

and

$$\lim_{\Delta t \to 0} \frac{\text{Var } [S(t + \Delta t)] - \text{Var } [S(t)]}{\Delta t} = \lambda(\sigma_v^2 + \rho^2 \sigma_u^2) \tag{13-45}$$

Let $W(t)$ be the virtual delay at time t. When $W(t) \geq \Delta t$,

$$W(t + \Delta t) - W(t) = S(t + \Delta t) - S(t) - \Delta t \qquad (13\text{-}46)$$

From (13-44) and (13-45), the mean and variance of increments in $W(\cdot)$ are proportional to Δt when Δt is small, which is a property that BM possesses. This makes it plausible to approximate the virtual-delay process by a BM with a reflecting barrier at the x axis, $\{\hat{W}(t); t \geq 0\}$ say. The parameters for this process,

$$c = \rho - 1 \quad \text{and} \quad D^2 = \lambda(\sigma_v^2 + \rho^2\sigma_u^2) \qquad (13\text{-}47)$$

are obtained from (13-44) through (13-46). The d.f. of $\hat{W}(t)$ is given by (13-23); the asymptotic mean is obtained from (13-24b) and is

$$\lim_{t \to \infty} E[\hat{W}(t)] = \frac{\lambda(\sigma_v^2 + \rho^2\sigma_u^2)}{2(1 - \rho)} \qquad (13\text{-}48)$$

For the M/G/1 queue, $\sigma_u^2 = 1/\lambda^2$ and (13-48) agrees exactly with the Pollaczek-Khintchine formula.

Substituting (13-48) into the exact relationship between the mean delay and the mean virtual delay (13-16) yields the approximation

$$E(D^*) \doteq \frac{\lambda(\sigma_v^2 + \rho\sigma_u^2)}{2(1 - \rho)} - \frac{1}{2\mu} \qquad (13\text{-}49)$$

When the arrivals are not Poisson, the steady-state delay (D^* say) and the steady-state virtual delay [$W(\infty)$ say] typically have different d.f.'s. However, it can be shown† that as $\rho \uparrow 1$, the d.f. of D^* tends to the d.f. of $W(\infty)$. From (13-24), an approximation for the d.f. of D^* is an exponential distribution with a mean given by (13-48).

A Numerical Example

We test the diffusion approximation for $X(\cdot)$ by comparing $E[\hat{X}(\cdot)]$ to a simulation. Consider a $GI/G/1$ queue that is empty at $t = 0$, where the interarrival times are uniformly distributed from 0 to 20 minutes and where the service times are uniformly distributed from 0 to 19 minutes. Thus

$$\frac{1}{\lambda} = 10 \qquad \sigma_u^2 = \frac{100}{3} \qquad 1/\mu = 9.5 \qquad \sigma_v^2 = \frac{(9.5)^2}{3}$$

and so $\rho = 0.95$.

Let X^* denote the number present in the steady state, and let $\hat{X}$ be its approximation. From (13-24a) and (13-24b), we have

$$E(\hat{X}) = \text{Var}(\hat{X}) = 6.50$$

The simulation consists of n (to be chosen) simulated sample paths of $X(\cdot)$,

---

† J. A. Hooke, "Some Limit Theorems for Priority Queues," Tech. Rep. 91, Cornell University, Ithaca, N.Y. (1969).

Table 13-1 Comparison of diffusion approximation and simulation results for the mean number present

					τ					
$E[\hat{x}(\tau)]$	0.1	0.2	0.3	0.4	0.5	0.6	0.7	0.8	0.9	1.0
Diffusion	2.68	3.50	4.02	4.40	4.68	4.91	5.11	5.27	5.41	5.52
Simulation	2.25	2.91	3.40	3.71	4.16	4.39	4.61	5.05	5.18	5.13

and $E[X(t)]$ is estimated by averaging the simulated sample paths at time t. Call this estimate $\hat{x}(t)$. Assume that the simulation program produces i.i.d. sample paths, and approximate $E[X(t)]$ and Var $[X(t)]$ by $E(\hat{X})$ and Var $(\hat{X})$, respectively, when t is large. Then standard sampling theory indicates that when n is sufficiently large, $\hat{x}(t)$ has a normal distribution with mean 6.5 and variance $6.5/n$. Hence, given $\alpha > 0$, the number of simulated sample paths required to have

$$P\{|\hat{x}(t) - E[X(t)]| \leq \alpha E[X(t)]\} \geq 0.99$$

is the smallest integer at least as large as $(2.575)^2/\alpha$. Choosing $\alpha = 0.05$ yields $n = 134$.

From this analysis, the choice of $n = 150$ emerged. One of the potential uses of a diffusion approximation is as an aid in assessing the accuracy of a simulation experiment.

In Exercise 9-38, you showed that if one makes the change of variables $\zeta = -(c/D^2)x$ and $\tau = (c^2/D^2)t$, (13-20) becomes

$$F_\tau = -F_\zeta + \frac{F_{\zeta\zeta}}{2}$$

This can be solved once and for all, and the solution for any c and D^2 can be recovered by scaling. For our example, $-c/D^2 = 13.00$ and $D^2/c^2 = 2472.2$ minutes.

The results of the simulation experiment are shown in Table 13-1. In the table, the diffusion process consistently overestimates the mean number of customers in the system.

Theoretical Justification of Diffusion Approximations

The diffusion approximations presented above are based on heuristic arguments. Their efficacy as approximations would seem to depend on the plausibility of the derivation and numerical evidence. However, there are theorems which show that when the traffic intensity is slightly less than 1, the queue-length and virtual-delay processes are approximately BM processes with a reflecting barrier at the x axis. The limit theorem is established for the $GI/G/c$ queue in Iglehart and Whitt (1970). Now we sketch that limit theorem when $c = 1$.

Consider a sequence of $GI/G/1$ systems, and index them by $n \in I_+$. For the

nth system, let λ_n and $1/\mu_n$ denote the arrival rate and mean service time, respectively, and let $X_n(t)$ be the number in system at time t. For the above quantities, let the absence of a subscript denote the limit with respect to n; for example, $\lambda = \lim_{n \to \infty} \lambda_n$. For each n, let σ_{un}^2 and σ_{vn}^2 be as before.

Let D^2 be given by (13-39). In Sec. 3 of Iglehart and Whitt (1970), it is shown that if $\lim_{n \to \infty} (\lambda_n - \mu_n)\sqrt{n} \to d$, where d is some finite constant, then as $n \to \infty$,

$$\frac{X_n(nt')}{D\sqrt{n}} \xrightarrow{w} B(t', d/D) \tag{13-50}$$

where $B(\cdot, d/D)$ is the BM process with a negative drift d/D, infinitesimal variance 1, and a reflecting barrier at the x axis. The "w" over the arrow in (13-50) denotes "weak convergence," which is a mode of convergence used for stochastic processes. (We do not assume that you know what weak convergence is. See the footnote on p. 354 that indicates why convergence theory for stochastic processes is beyond the scope of this book.)

For our purposes it is sufficient to interpret (13-50) as stating that when n is large, both sides of (13-50) are "close." For large n (that is, $\rho_n = \lambda_n/\mu_n$ is close to 1), d is approximately $(\lambda - \mu)\sqrt{n}$. By setting $t = nt'$, (13-50) asserts that $X(t)$ is approximately $D\sqrt{n}\, B(t/n, d/D)$; that is, it is approximately reflected BM with infinitesimal mean

$$\frac{D\sqrt{n}\,(d/D)}{n} = \lambda - \mu$$

and infinitesimal variance

$$\frac{D^2 n}{n} = D^2$$

This is the diffusion approximation for $\{X(t); t \geq 0\}$ given above.

Steady-State Probabilities for the $GI/G/c$ Queue

The idea behind the $GI/G/1$ diffusion model can be adapted to multiserver queues. In principle, an approximation for the transient behavior is obtained, but a tractable, closed-form solution to the partial differential equation is not available. We content ourselves with approximating the steady-state probability that n customers are present.

The number of servers does not affect the arrival process, so (13-34) is valid when there are c servers. When $n < c$ customers are present, there are n busy servers, departures occur at rate $n\mu$, and (13-33) suggests that we approximate the variance of the times between departures by $n\mu^3\sigma_v^2$. Similarly, when $n \geq c$ customers are present, there are c busy servers, departures occur at rate $c\mu$, and (13-33) suggests that we approximate the variance of the times between departures by $c\mu^3\sigma_v^2$. Repeating the arguments used in the single-server case, we obtain an approximating diffusion process with a state-dependent infinitesimal mean and variance given by

$$\alpha(x) = \lambda - \min\{x, c\} \mu \tag{13-51}$$

and

$$\beta(x) = \lambda^3 \sigma_u^2 + \min\{x, c\} \mu^3 \sigma_v^2 \tag{13-52}$$

respectively.

The forward Kolmogorov equation of this process is obtained by substituting (13-51) and (13-52) into (9-129). In the steady state, the partial derivative of the transition density is zero, and we obtain

$$\frac{d}{dx} [\alpha(x)f(x)] = \frac{d^2}{dx^2} \frac{\beta(x)f(x)}{2} \tag{13-53}$$

The boundary condition for (13-53) can be obtained by generalizing (9-121):

$$2\alpha(0)f(0) = \frac{d}{dx} [\beta(x)f(x)]_{x=0} \tag{13-54}$$

You can easily verify, by direct substitution, that

$$f(x) = \frac{K}{\beta(x)} \exp\left[2 \int_0^x \frac{\alpha(y)}{\beta(y)} dy\right] \tag{13-55}$$

is the most general solution of (13-53) subject to (13-54).

Since $\alpha(x)$ and $\beta(x)$ are discontinuous at $x = c$, we *choose* to write $f(\cdot)$ as the sum of two solutions of (13-53):

$$f(x) = \begin{cases} K_1 f_1(x) & \text{if } 0 \le x \le c \\ K_2 f_2(x) & \text{if } x > c \end{cases} \tag{13-56}$$

where $f_1(\cdot)$ is given by (13-55) with $\alpha(x) = \lambda - x\mu$ and $\beta(x) = \lambda^3 \sigma_u^2 + x\mu^3 \sigma_v^2$ and $f_2(\cdot)$ is given by (13-55) with $\alpha(x) \equiv \alpha(c)$ and $\beta(x) \equiv \beta(c)$. The unknown constant K is absorbed in K_1 and K_2, both of which remain to be determined.

By evaluating the integrals in (13-55), (13-56) can be written as

$$f(x) = K_1 (\lambda^3 \sigma_u^2 + x\mu^3 \sigma_v^2)^{k-1} \exp\left(\frac{-2x}{\mu^2 \sigma_v^2}\right) \qquad 0 \le x \le c \tag{13-57a}$$

where

$$k = \frac{2\lambda}{\mu^3 \sigma_v^2} \left[\left(\frac{\lambda \sigma_u^2}{\mu \sigma_v^2}\right)^2 + 1\right]$$

and

$$f(x) = K_2 \exp\left[\frac{-2x(c\mu - \lambda)}{c\mu^3 \sigma_v^2 + \lambda^3 \sigma_u^2}\right] \qquad x \ge c \tag{13-57b}$$

Table 13-2 Comparison of exact and diffusion results for M/M/c queue

c		$\rho = 0.75$		$\rho = 0.85$		$\rho = 0.95$	
		Mean	Variance	Mean	Variance	Mean	Variance
3	exact	3.95	3.54	6.69	6.19	20.08	19.47
	approximate	4.11	3.55	6.83	6.20	20.20	19.47
5	exact	5.14	3.64	7.96	6.25	21.42	19.49
	approximate	5.20	3.65	8.02	6.26	21.47	19.49
7	exact	6.41	3.75	9.33	6.31	22.88	19.50
	approximate	6.45	3.77	9.36	6.32	22.91	19.51
9	exact	7.74	3.87	10.77	6.37	24.40	19.52
	approximate	7.76	3.89	10.79	6.39	24.42	19.50

The constants K_1 and K_2 are determined as follows. We should certainly require

$$1 = \int_0^\infty f(x)\,dx = K_1 \int_0^c f_1(x)\,dx + K_2 \int_c^\infty f_2(x)\,dx \tag{13-58}$$

In Halachami and Franta (1978), it is argued that the steady-state probability that n customers are present should be approximated by

$$\hat{p}_n = \int_{n-1/2}^{n+1/2} f(x)\,dx \qquad n \in I_+$$

Since $f(\,\cdot\,)$ changes form at $x = c$, it seems reasonable to require that both forms provide the same estimate. Thus, we require

$$K_1 \int_{c-1/2}^{c+1/2} f_1(x)\,dx = K_2 \int_{c-1/2}^{c+1/2} f_2(x)\,dx \tag{13-59}$$

to hold. Solving (13-58) and (13-59) simultaneously yields K_1 and K_2.

The exact values for the mean and variance of the number of customers present in the steady state can be computed easily for the M/M/c queue. These are compared to their approximations obtained from the diffusion process in Table 13-2.†

In Table 13-2, the diffusion approximation always overestimates the mean; this property was also observed in the single-server model. It is not known if this relationship holds in general.

Diffusion Approximations for Networks

Exponential distributions are essential in the analysis of queueing networks given in Chapter 12. It is desirable to develop approximations for more general models,

† This table is reprinted by permission from Baruch Halachami and W. R. Franta, "Erratum to 'A Diffusion Approximation to the Multi-Server Queue,'" *Manage. Sci.* **24**: 1448 (1978).

and one might try a diffusion approximation. The natural analog of the general form of the diffusion equation (13-20) is readily obtained, but the analog of the boundary equation (13-22) is known only in some special cases with two service centers. Some progress toward a tractable diffusion approximation is given in Harrison (1973).

EXERCISES

13-8 For the M/G/1 queue, compare $\lim_{t \to \infty} \text{Var} [\hat{X}(t)]$ with its exact value. What happens as $\rho \uparrow 1$?

13-9 Suppose a GI/G/1 queue is initially empty, so the first busy period has the same distribution as all subsequent busy periods. When $V_1 = x$, the length of the first busy period is the first-passage time of $W(\cdot)$ from x to 0. Use a diffusion process to approximate the d.f., mean, and variance of the busy period. For the M/G/1 queue, compare the mean and variance you obtain to the exact results.

13-3 SYSTEM APPROXIMATIONS

A method for constructing approximations in complicated models is to posit a simpler model whose operating characteristics are believed to approximate those of the given model. The simpler model is solved, and its operating characteristics are taken as approximations of the true operating characteristics. We call these simpler models *system approximations* because the approximation is obtained from an exact analysis of an approximate system.

System approximations are obtained in an ad hoc manner. The choice of an approximate system should be justified by plausibility arguments, and the quality of the approximation should be tested by comparing it to either known solutions in particular cases or simulations. Since there is usually no known bound on the error of the approximation, one can never be sure that the approximation will be sufficiently accurate in a particular situation.

In this section we present two system approximations. One is for the M/G/c queue, and the other is for a queueing model of a multiprogrammed computer.

An Approximation for the Steady-State Expected Delay in an M/G/c Queue

The diffusion approximation for the GI/G/c queue given in Section 13-2 does not take advantage of the special properties of the Poisson process when it is the arrival process. This suggests that a sharper approximation is available for the M/G/c queue. We present a system approximation† that is exact when service times are exponential, so it is better than the diffusion process approximation in that situation (see Table 13-2).

---

† This approximation appears in A. M. Lee and P. A. Longton, "Queueing Processes Associated with Airline Passenger Check-in," *Oper. Res. Quart.* **10**: 56–71 (1959). The analysis given here is due to Shirley A. Nozaki and Sheldon M. Ross, "Approximations in Finite-Capacity Multi-Server Queues with Poisson Arrivals," *J. Appl. Prob.* **15**: 826–834 (1978).

We continue to use the notation employed in Sections 13-1 and 13-2, and we assume $E(V_1^2) < \infty$ and $a \triangleq \lambda/\mu < c$, so $E(D^*)$ exists and is finite (by Theorem 11-6). For simplicity, we assume that customers enter service FIFO, but any work-conserving and service-time-invariant discipline could be used.

Recall (13-16):

$$E(D^*) = \frac{E[W^*(\infty)]}{a} - \frac{E(V_1^2)}{2E(V_1)} \qquad (13\text{-}60)$$

We approximate $E(D^*)$ by substituting an approximation for $E[W^*(\infty)]$ into (13-60). This is the first system approximation.

System Approximation 13-1 In the steady state, if it is given that k servers are busy at an arrival epoch or at an epoch when a customer enters service, then the expected work (i.e., the sum of the remaining service times) in those k servers is approximated by $kE(V_1^2)/[2E(V_1)]$.

Notice that $E(V_1^2)/[2E(V_1)]$ is the expected value of the equilibrium excess of a service time; see (5-47a). This approximation is exact when the service times are exponential (by the memoryless property) and at arrival epochs when $k = 1$ (by Theorem 6-11 and Poisson arrivals "see" time averages). It is also exact for the $M/G/c/c$ queue (see Section 11-7).

The amount of work in the system when a customer arrives is the amount of work in those servers other than her or his own when the customer enters service plus the amount of work processed while she or he was in queue. The expected value of the latter is $cE(D^*)$ because all servers are busy when a customer is in queue. From System Approximation 13-1, the expected value of the former is $m_b E(V_1^2)/[2E(V_1)]$, where m_b is the mean number of busy servers at the start of service epochs and needs to be determined.

Let p_j be the limiting probability that j customers are in the system. Since Poisson arrivals see time averages, p_j is the steady-state probability that j customers are present at an arrival epoch. When $j < c$ customers are present at an arrival epoch, the arriving customer enters service immediately; when $j \geq c$ customers are present, that customer is delayed and enters service at some epoch where $c - 1$ other customers are in service. Thus,

$$m_b = \sum_{j=0}^{c-1} j p_j + (c-1) \sum_{j=c}^{\infty} p_j$$

In Example 11-8, it is shown that

$$\sum_{j=0}^{c-1} j p_j + c \sum_{j=c}^{\infty} p_j = a$$

Since $\sum_{j=c}^{\infty} p_j \triangleq P_D$ is the probability that an arriving customer is delayed, we may write

$$m_b = a - P_D$$

Thus,

$$E[W(\infty)] \doteq cE(D^*) + \frac{(a - P_D)E(V_1^2)}{2E(V_1)} \tag{13-61}$$

Substituting (13-61) into (13-60) yields

$$E(D^*) \doteq \frac{P_D E(V_1^2)}{2E(V_1)(c - a)} \tag{13-62}$$

Now we must approximate P_D.

System Approximation 13-2 Let P_D be approximated by the probability of delay in an M/M/c queue with the same arrival rate, mean service time, and number of servers.

System Approximation 13-2 states that P_D is approximated by $C(c, a)$, given by (4-43). Several numerical investigations of special cases have shown this approximation to be reasonable. Substituting (4-43) into (13-62) yields the approximation

$$E(D^*) = \frac{C(c, a)E(V_1^2)}{2E(V_1)(c - a)} \tag{13-63}$$

It can be shown (Exercise 13-12) that (13-63) is a good approximation when a is slightly smaller than c, that is, in heavy traffic.

From $L = \lambda W$, multiplying both sides of (13-63) by λ yields an approximation for the average number of customers in the queue; the latter has been tabulated for Erlang service times. Table 13-3 shows that (13-63) is close to the exact values for two Erlang distributions.

Table 13-3 Comparison of approximate and exact average queue size for Erlang-k service times

	$c = 3$		$c = 4$		$c = 10$	
ρ	$k = 2$	$k = 3$	$k = 2$	$k = 3$	$k = 2$	
0.3	0.023	0.020	0.012	0.011	0.00037	approximate
	0.024	0.022	0.013	0.012	0.00043	exact
0.7	0.862	0.766	0.750	0.667	0.388	approximate
	0.878	0.787	0.769	0.692	0.408	exact
0.9	5.515	4.902	5.317	4.727	4.514	approximate
	5.544	4.941	5.354	4.775	4.576	exact
0.99	72.852	64.757	72.609	64.542	71.555	approximate
	72.887	64.809	72.656	64.603	71.647	exact

An Approximate Model for a Multiprogramming Computer System

A closed queueing network that may be used to describe a multiprogrammed computer is described in Example 10-10. In Exercise 11-18, you showed that the expectation of the elapsed time between an epoch when a job joins the queue at station 1 (which represents the central processor) and the next such epoch (this is called the cycle time) is given by

$$E(C) = \frac{M}{a_1 \mu_1} \tag{13-64}$$

where M is the number of customers in the network, a_1 is the average number of busy servers in station 1, and μ_1 is the reciprocal of the mean service time at station 1. A formula for a_1 is given in Example 12-1.

In the model of Example 10-10, the number of jobs in the computer is fixed, and no new jobs arrive. Now we change this model in three ways. First, the number of jobs is not fixed, but is a nonnegative, integer-valued r.v. that is bounded above by the maximum number of jobs that can simultaneously reside in the computer. This number is called the *degree of multiprogramming* and is given by the computer system's design; it is denoted by J. Second, jobs arrive at the computer from an external source, which may be a collection of remote-access devices or batch loadings. A new job generated by the external source enters the computer only if the number of jobs residing there is less than J; otherwise, the new job joins a queue. Third, once a job enters the computer, it has to complete a random number of cycles, Y say, and then it departs. At each departure epoch, if the queue is not empty, the first in line enters the computer; if the queue is empty, the number of jobs in the computer decreases by 1. A schematic diagram of this model is shown in Figure 13-1.

Let $X(t)$ be the number of jobs that are in the queue or in the computer at time t. Assume that when $X(t) = n$, the arrival rate is λ_n, $n \in I$. When j jobs are in the computer, we assume the computer behaves as the closed network model in Example 10-10. When j jobs are in the computer, let ψ_j be the mean cycle time and ξ_j be the mean number of job completions per unit time. We obtain ψ_j from (13-64), and ξ_j is given by (why?)

$$\xi_j = \frac{j}{\psi_j E(Y)} \tag{13-65}$$

This is the system approximation.

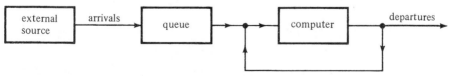

Figure 13-1 Schematic diagram of the model.

System Approximation The process $\{X(t); t \geq 0\}$ is approximated by a birth-and-death process with birth rate λ_n in state $n \in I$, death rate ξ_n in state $n < J$, and death rate ξ_J in state $n \geq J$.

The system approximation is based on the following considerations. In a large computer system, jobs are divided into smaller units which are processed sequentially by the computer. If the distribution of the number of these units per job is geometric with a large mean, and if the number of jobs in the computer does not change rapidly (this should hold when the computer is heavily loaded), then the probability of a departure in an interval of length Δt should be close to $\xi_j \, \Delta t$ when the state is j.

The properties of the approximate model are easily obtained from the birth-and-death process theorems in Chapter 4. From Corollary 4-4a, the limiting probabilities, p_n say, are given by (4-35) and (4-20'), namely

$$p_n = p_0 \prod_{i=0}^{n-1} \frac{\lambda_i}{\xi_{i+1}} \qquad n \in I_+ \tag{13-66a}$$

where p_0 is chosen so that $\sum_{n=0}^{\infty} p_n = 1$. These probabilities exist when

$$\sum_{n=J}^{\infty} \frac{\lambda_J \cdots \lambda_n}{\xi_J^n} < \infty \qquad \text{and} \qquad \sum_{n=J}^{\infty} \frac{1}{\lambda_n} = \infty \tag{13-66b}$$

In the special case where $\lambda_n \equiv \lambda$, (13-66b) is simply

$$\lambda < \xi_J \tag{13-66c}$$

From Theorem 4-9, we deduce that

$$\frac{\sum_{n=0}^{J-1} \lambda_n p_n}{\sum_{n=0}^{\infty} \lambda_n p_n}$$

is the limiting probability that a job is delayed in the queue and that

$$\frac{\sum_{n=J}^{\infty} \lambda_n p_n [(\xi_J t)^{n-J}/(n-J)!] \xi_J e^{-\xi_J t}}{\sum_{i=0}^{\infty} \lambda_i p_i}$$

is the density function, evaluated at $t > 0$, of the steady-state delay in queue.

Comparisons of the approximate model with some exact analyses are left to Exercises 13-13, 13-14, and 13-15.

EXERCISES

13-10 Show that the approximations in (13-62) and (13-63) obey the lower and upper bounds in Section 13-1. *Hint*: Use $L = \lambda W$ to prove that $0 \leq \rho - P_D \leq (1 - \rho)(c - 1)$ for every M/G/c queue with $\rho < 1$.

13-11 One way to estimate $E(D^*)$ for the M/G/c queue is the following. Let r be the ratio of $E(D^*)$ for the M/G/1 queue to $E(D^*)$ for the M/M/1 queue with the same arrival rate and traffic intensity. This can be interpreted as the effect on $E(D^*)$ of nonexponential service times. Approximate $E(D^*)$ for the

$M/G/c$ queue by r multiplied by the expected delay for the $M/M/c$ queue with the same arrival rate, number of servers, and traffic intensity.

Compare this approximation to the approximation given by (13-63). You may be surprised by your answer.†

13-12 From a result in Iglehart and Whitt (1970), it can be shown that for a $GI/G/c$ queue,

$$\lim_{\rho \uparrow 1} (1 - \rho)E(D^*) = \frac{c\lambda(\sigma_u^2 + \sigma_v^2)}{2}$$

Use this and your answer to Exercise 13-11 to prove that (13-63) is a good approximation when ρ is slightly less than 1. What does it suggest as an approximation for the expected delay in a $GI/G/c$ queue?

13-13 Suppose a computer consists of a single central processing unit and a single input-output device, both having exponential service times with mean $1/\mu$. In Exercise 12-8 you showed that this implies $\psi_j = (j + 1)/\mu$, $j \in I_+$. Assume that the random variable Y obeys $P\{Y = i\} = \theta(1 - \theta)^{i-1}$, $i \in I_+$, and that the arrivals are Poisson. Find the explicit formulas for p_n, $n \in I$, and the expected delay in the queue for the appropriate model.

13-14 Solve the model in Exercise 13-13 *exactly* for p_n, $n \in I$, when $J = 1$ and when $J = \infty$. When $J = \infty$, you should get the same answer you obtained in Exercise 13-13. When $J = 1$, let $E(D^*)$ be the expected delay in the queue and $\hat{d}$ be its approximation. Show that

$$0 \le \frac{\hat{d} - E(D^*)}{E(D^*)} \le \frac{\theta}{4 - \theta}$$

13-15 Show that the approximate state probabilities are, in fact, exact when $J = \infty$ and the number of devices in the computer (N in Example 10-10) is arbitrary. Maintain all the other assumptions in Exercise 13-13.

BIBLIOGRAPHIC GUIDE

The first bounds for the $GI/G/1$ queue were published by J. F. C. Kingman in 1962. Our treatment follows Kingman (1970). The upper bound in Corollary 13-1 was also obtained in Marshall (1968). Marshall found a lower bound too; for those simple cases where comparisons have been made, Marshall's bound performs better than the lower bound in (13-9) when ρ is close to 0 and performs worse when ρ is close to 1. Marshall initiated the use of nonparametric families of interarrival times, based on the failure rate, to improve the quality of the bounds. This idea was extended to multiple-server systems in Brumelle (1971). A survey of bounds and approximations is given in Stoyan (1977).

The first paper which shows that a queueing process can be approximated by a diffusion process is Iglehart (1965). In Iglehart and Whitt (1970), heavy traffic limit theorems are obtained for processes associated with the $GI/G/c$ queue. Whitt (1974) surveys convergence theorems and their application to queues. Our presentation of the single-server system is inspired by Gaver (1968) and Newell (1968); the details are taken from Heyman (1975). The approximation for the multiserver system is due to Halachami and Franta (1978). Diffusion approximations of this sort are given in Newell (1971, 1973, and 1979) and Fischer (1977). Diffusion approximations for communication systems have been explored by

---

† The relation between the two approximations was told to us by D. R. Smith.

D. P. Gaver and J. P. Lehoczky; see, e.g., Gaver and Lehoczky (1979) and the references there.

Since diffusion processes exhibit small, rapid upward and downward fluctuations, they cannot be used as approximations for inventory models with occasional large infusions of stock, such as the (S, s) model. When there are returns and demands, such as in the model in Example 9-8, a diffusion process approximation may be reasonable [see Heyman (1977)].

System approximations for queues are surveyed in Stoyan (1977). Some more recent approximations are by Hokstad (1978) and Boxma, Cohen, and Huffels (1979).

References

Boxma, O. J.; J. W. Cohen; and N. Huffels: "Approximations of the Mean Waiting Time in an $M/G/s$ Queueing System," *Oper. Res.* **27**: 1115–1127 (1979).

Brumelle, Shelby L.: "Some Inequalities for Parallel-Server Queues," *Oper. Res.* **19**: 402–413 (1971).

Fischer, M. J.: "An Approximation to Queueing Systems with Interruptions," *Manage. Sci.* **24**: 338–344 (1977).

Gaver, D. P.: "Diffusion Approximation and Models for Certain Congestion Problems," *J. Appl. Prob.* **5**: 607–623 (1968).

Gaver, Donald P., and John A. Lehoczky: "A Diffusion Approximation Model for a Communication System Allowing Message Interference," *IEEE Trans. Commun.* **27**: 1190–1199 (1979).

Halachami, Baruch, and W. R. Franta: "A Diffusion Approximation to the Multi-server Queue," *Manage. Sci.* **24**: 522–529 (1978).

Harrison, J. Michael: "The Heavy Traffic Approximation for Single Server Queues in Series," *J. Appl. Prob.* **10**: 613–629 (1973).

Heyman, D. P.: "A Diffusion Model Approximation for the $GI/G/1$ Queue in Heavy Traffic," *Bell Syst. Tech. J.* **54**: 1637–1646 (1975).

———: "Optimal Disposal Policies for a Single-Item Inventory System with Returns," *Nav. Res. Log. Quart.* **24**: 385–405 (1977).

Hokstad, Per: "Approximations for the $M/G/m$ Queue," *Oper. Res.* **26**: 510–523 (1978).

Iglehart, D. L.: "Limit Diffusion Approximations for the Many-Server Queue and the Repairman Problem," *J. Appl. Prob.* **2**: 429–441 (1965).

———, and W. Whitt: "Multiple Channel Queues in Heavy Traffic II," *Adv. Appl. Probab.* **2**: 355–369 (1970).

Kingman, J. F. C.: "Inequalities in the Theory of Queues," *J. Roy. Stat. Soc.* **B32**: 102–110 (1970).

Marshall, K. T.: "Some Inequalities in Queueing," *Oper. Res.* **16**: 651–665 (1968).

Newell, Gordon F.: "Queues with Time-Dependent Arrival Rates I: The Transition through Saturation," *J. Appl. Prob.* **5**: 436–451 (1968).

———: *Applications of Queueing Theory*, Chapman & Hall, London (1971).

———: *Approximate Stochastic Behavior of n-Server Service Systems with Large n*, Lecture Notes in Economics and Mathematical Systems No. 87, Springer-Verlag, New York (1973).

———: *Approximate Behavior of Tandem Queues*, Lecture Notes in Economics and Mathematical Systems No. 171, Springer-Verlag, New York (1979).

Stoyan, Dietrich: "Bounds and Approximations in Queueing through Monotonicity and Continuity," *Oper. Res.* **25**: 851–863 (1977).

Whitt, Ward: "Heavy Traffic Limit Theorems for Queues: A Survey," *Mathematical Methods in Queueing Theory*, A. B. Clarke (ed.), Lecture Notes in Economics and Mathematical Systems No. 98, Springer-Verlag, New York (1974).

BACKGROUND MATERIAL

The factual information necessary to understand the material in this book usually is presented in an undergraduate calculus sequence and a calculus-based course on probability. Not all courses cover and emphasize the particular facts utilized in this book, so this appendix is designed to be a ready reference. Most of the material on probability theory in Section A-1 should be familiar. The concepts of conditional probability and conditional expectation are used repeatedly in studying stochastic processes, and you should become comfortable with them. The simple, but crucial, properties of the exponential distribution given in Section A-2 are not brought out in some probability courses and may be new to you.

Certain manipulations in the text are simplified by the use of Laplace transforms and generating functions. These methods may well have been absent from your previous studies. Section A-4 is a short list of transforms that suffices for inverting the transforms encountered in this book. Section A-5 contains a review of facts from advanced calculus, such as the definition of lim sup, conditions for interchanging the order of integration, and "little oh" notation.

A-1 RUDIMENTS OF PROBABILITY THEORY

The basic concepts of probability theory include sample space, event, probability measure (or, simply, probability), random variable, and distribution function.

Sample Spaces and Probability Measures

A *sample space*, Ω say, is the set of all possible outcomes of a potential experiment. An event is a subset of the sample space; i.e., it is a collection of outcomes.

If A is an event and the outcome of the experiment is the member ω of Ω, then A occurs if $\omega \in A$. Events are the objects to which probabilities are assigned.

In any application, we are free to choose Ω and the events in any way we please. Our choice should be related to the phenomenon we are modeling. In discussing the time to failure of a piece of equipment, for example, a convenient choice for Ω might be $[0, \infty)$; if we knew that the item always lasted at least a but never more than b, we might choose $\Omega = [a, b]$. It would not be a good idea to choose $\Omega = (-b, a/2)$. The events have some more natural structure to them. We would like to ensure that Ω is an event because it is a "sure thing." To continue with our example, if A_1 is the event "the time to a failure is no more than 1," that is,

$$\Omega = [0, \infty) \quad \text{and} \quad A_1 = \{\omega : \omega \leq 1\}$$

then the set

$$A_1^c = \{\omega : \omega > 1\}$$

represents the time to failure being greater than 1 (the *complement* of A_1). We would like to have A_1^c be an event, so that both the occurrence and the nonoccurrence of something substantial are events.

If A_1 and A_2 both are events, then their union $A_1 \cup A_2$ is the set $\{\omega : \omega \in A_1$ and/or $\omega \in A_2\}$. With A_1 as above and $A_2 = \{\omega : \frac{1}{2} \leq \omega \leq 2\}$, $A_1 \cup A_2 = \{\omega : 0 \leq \omega \leq 2\}$; that is, $A_1 \cup A_2$ is the set of outcomes such that the time to failure is between 0 and 2. We would like $A_1 \cup A_2$ to be an event too. Continuing in this way, we are led to want $A_1 \cup A_2 \cup A_3 \cdots$ to be an event when each A_i is an event.

Thus, a *family of events*, $\mathcal{F}$ say, is constructed such that

$$\Omega \in \mathcal{F} \tag{A-1}$$

$$A \in \mathcal{F} \Rightarrow A^c \in \mathcal{F} \quad \text{where} \quad A^c \triangleq \Omega - A \tag{A-2}$$

and $\quad$ if $A_1, A_2, \ldots$ are in $\mathcal{F}$, then so is their union. $\tag{A-3}$

The technical name for a family set satisfying (A-1) through (A-3) is a *σ-field*.

We write $\bigcup_{i=1}^{n} A_i$ for the union of $A_1, A_2, \ldots, A_n$. With these axioms, we can readily conclude that (1) the empty set, $\varnothing$ say, is in $\mathcal{F}$; (2) if A_1 and A_2 are in $\mathcal{F}$, then so is $A_1 \cap A_2$ (the intersection, consisting of points common to A_1 and A_2); and (3) the union of a finite number of sets is in $\mathcal{F}$.

Now let us look at the axioms that define a probability measure. Before we start, it is appropriate to mention that there are ways other than the axiomatic method to define a probability measure, and more than one set of axioms has been proposed by the proponents of the axiomatic method. The axiomatic method is the most common way to define probability, and the axioms presented here are almost universally accepted.

Definition A-1 Given a sample space Ω and a family of events $\mathcal{F}$, a *probability measure* (or *probability*) is a function, $P\{ \cdot \}$ say, which assigns a real number to each event A in $\mathcal{F}$ such that

(a) For any $A \in \mathcal{F}$, $0 \leq P\{A\} \leq 1$

(b) $P\{\Omega\} = 1$

(c) For any infinite sequence of disjoint events $A_1, A_2, \ldots$ (that is, $A_i \cap A_j = \varnothing$ whenever $i \neq j$),

$$P\left\{\bigcup_{i=1}^{\infty} A_i\right\} = \sum_{i=1}^{\infty} P\{A_i\}$$

From this definition it is readily established that (1) $P\{\varnothing\} = 0$; (2) if $A_1, \ldots, A_n$ is a finite sequence of disjoint events, then $P\{\bigcup_{i=1}^{n} A_n\} = \sum_{i=1}^{n} P\{A_i\}$; (3) $P\{A^c\} = 1 - P\{A\}$; (4) if $A_1 \subset A_2$ (A_1 is a subset of A_2), then $P\{A_2\} \geq P\{A_1\}$; and (5) for any two events A_1 and A_2, $P\{A_1 \cup A_2\} = P\{A_1\} + P\{A_2\} - P\{A_1 \cap A_2\}$.

In every probability model, there are a sample space Ω, a family of events $\mathcal{F}$, and a probability measure $P\{\cdot\}$, which are chosen in that order. The triple $\{\Omega, \mathcal{F}, P\}$ is called a *probability space*. The sample space and probability measure arise naturally in applications, but usually the family $\mathcal{F}$ is not mentioned. When Ω is countable, $\mathcal{F}$ is typically the set of all subsets, and it need not be mentioned explicitly. When Ω is uncountable, $\mathcal{F}$ can be very complicated. Suppose $\Omega = [0, \infty)$ and we want to assign probabilities to all intervals of the form $(a, b]$. Then typically $\mathcal{F}$ is chosen to be the smallest σ-field that includes all such intervals; the family of all subsets of $[0, \infty)$ is not chosen because that leads to technical difficulties. Thus, there are subsets of Ω whose probability is not defined; these sets are very hard to find and are of no practical importance. The important thing to remember is that the foundation of every probability model is a probability space, and only sets in $\mathcal{F}$ have probability.

When Ω is uncountable, usually there is no way to construct $\mathcal{F}$ without including some nonempty sets that will be assigned probability zero if we wish to include all those sets which we think deserve positive probability. Often one finds interesting statements that are true except when one of these events of probability zero occurs, in which case we say the statement is true "with probability 1," abbreviated w.p.1. This is a fine point that is mentioned to explain why "w.p.1" is written after some statements.

Random Variables and Distribution Functions

A random variable is, as its name suggests, a variable whose value is determined by a random mechanism. In mathematics generally, things whose values are determined by some mechanism are functions, so a random variable is a function. A real-valued random variable is a real-valued function, $X(\cdot)$ say, defined on Ω such that if the outcome $\omega \in \Omega$ occurs, the random variable assumes the value $X(\omega)$.

To fix these ideas firmly, let us examine a very simple random variable. A single flip of a coin can result in either a head or a tail showing. Denote these outcomes by H and T, respectively, so the sample space for a single flip of the coin is $\Omega = \{H, T\}$. If a dollar is won when the coin lands heads and lost when the coin lands tails, then the payoff from a flip of the coin is a random variable. This random variable is described by

$$X(\omega) = \begin{cases} 1 & \text{if } \omega = H \\ -1 & \text{if } \omega = T \end{cases}$$

Some functions from Ω to the real numbers cannot be random variables, but any function we are likely to choose will be a random variable.† We frequently abbreviate random variable as r.v.

An r.v. that achieves a countable number of values is called *discrete*; all other r.v.'s are called *continuous*.

Just as the function $h(\cdot) = g[f(\cdot)]$ denotes that function which assumes the value $g[f(t)]$ at the point t, the random variable $Y(\cdot) = g[X(\cdot)]$ is the r.v. which assumes the value $g[X(\omega)]$ when outcome ω occurs. We typically drop the arguments when denoting an r.v., writing X for $X(\cdot)$ and $Y = g(X)$ for $Y(\cdot) = g[X(\cdot)]$. The ω's are used occasionally for emphasis.

The probability measure for events is employed to attach probabilities to values (or *realizations*) of r.v.'s in this manner. For a given set of real numbers B, let $A = \{\omega : X(\omega) \in B\}$. Then

$$P\{X \in B\} = P\{A\}$$

In order for the left-side to be defined, we must have $A \in \mathcal{F}$. This is where restrictions on $X(\cdot)$ appear.

Associated with every random variable X is a unique function, $F(\cdot)$ say, called its *distribution function*, which is abbreviated by d.f. This function is defined by

$$F(x) = P\{X \le x\} \qquad -\infty < x < \infty$$

If‡ $F(\infty) = 1$, then $F(\cdot)$ is called *honest* (or *proper*); otherwise, it is *defective*, and $1 - F(\infty)$ is called the *defect*. The interval $[a, b]$ is the *support* of $F(\cdot)$ if it is the smallest interval such that $F(x) = 0$ for all $x < a$ and $F(x) = 1$ for all $x > b$. The d.f. contains all the probabilistic information about the random variable. For example,

$$P\{a < X \le b\} = P\{X \le b\} - P\{X \le a\}$$

$$= F(b) - F(a) \qquad a < b$$

When X is discrete and takes the values $x_1 < x_2 < x_3 < \cdots$, say, then

$$P\{X = x_i\} = F(x_i) - F(x_{i-1})$$

---

† For those readers familiar with measure theory, a random variable is a measurable function. In this book, only measurable functions are allowed.

‡ A distribution function $F(\cdot)$ is a bounded monotone function, so $\lim_{x \to \infty} F(x)$ necessarily exists. We write this limit as $F(\infty)$ although $F(\cdot)$ is defined on $(-\infty, \infty)$ which does not include $\pm\infty$. Similarly, we write $F(\infty, x_2, \ldots, x_n)$ for $\lim_{x \to \infty} F(x, x_2, \ldots, x_n)$.

When X is continuous, often $F(\cdot)$ has a derivative; this derivative is called the (*probability*) *density function* of X, and is abbreviated p.d.f. Let $f(\cdot)$ be the derivative of $F(\cdot)$. When it is notationally inconvenient to denote the p.d.f. explicitly, we write $\mathrm{Pd}\{X = x\}$ for the density function of the random variable X evaluated at x. By definition,

$$f(x) = \lim_{\Delta x \to 0} \frac{F(x + \Delta x) - F(x)}{\Delta x}$$

so that for small† positive values of Δx,

$$f(x)\,\Delta x = F(x + \Delta x) - F(x) + o(\Delta x)$$

Since

$$F(x + \Delta x) - F(x) = P\{x < X \leq x + \Delta x\}$$

we have

$$f(x)\,\Delta x \approx P\{x < X \leq x + \Delta x\}$$

where "$\approx$" means equal except for terms of order Δx or smaller [i.e., except for $o(\Delta x)$]. Notice that $f(x)$ by itself is not the probability on the right, and $f(x)\,\Delta x$ is not the probability that $X = x$. In fact, the latter probability is zero because

$$P\{X = x\} = \lim_{\Delta x \to 0} [F(x + \Delta x) - F(x - \Delta x)] = F(x) - F(x) = 0$$

whenever $F(\cdot)$ is continuous at x. In this case, $X \neq x$ is an example of a statement that holds w.p.1.

These ideas can be extended to cover more than one random variable at a time. Let

$$F(x_1, x_2, \ldots, x_n) = P\{X_1 \leq x_1, X_2 \leq x_2, \ldots, X_n \leq x_n\}$$

for any nonempty set of random variables $X_1, X_2, \ldots, X_n$ and real numbers $x_1, x_2, \ldots, x_n$. Then $F(\cdot, \cdot, \ldots, \cdot)$ is the *joint distribution function* of the random variables $X_1, X_2, \ldots, X_n$. If $P\{X_1 < \infty\} = 1$, then

$$F(\infty, x_2, \ldots, x_n) = P\{X_2 \leq x_2, \ldots, X_n \leq x_n\}$$

and similar statements hold when various other x's are set at $+\infty$. When all the x's but one, x_i say, are set at $+\infty$, the resulting function of x_i alone is called the *marginal distribution function* of X_i. The derivative of the marginal d.f. (when it exists) is called the *marginal density function*. The function (assuming it exists)

$$f(x_1, \ldots, x_n) = \frac{\partial^n}{\partial x_1 \cdots \partial x_n} F(x_1, \ldots, x_n)$$

is called the *joint density function* of $X_1, \ldots, X_n$. The marginal density function of

---

† The notation $o(x)$ stands for any function of the real variable x, say $g(x)$, such that $\lim_{x \to 0} [g(x)/x] = 0$. The beginning of Section A-5 contains a discussion of this notation.

X_i is obtained from the joint density function by integrating the joint density function, with respect to x_j for each $j \neq i$, from $-\infty$ to $+\infty$. For example, let $f_1(\cdot)$ be the marginal density of X_1. Then

$$f_1(x_1) = \int_{-\infty}^{\infty} \cdots \int_{-\infty}^{\infty} f(x_1, x_2, \ldots, x_n) \, dx_2 \cdots dx_n$$

A very important concept is *stochastic independence*, commonly referred to simply as *independence*.

Definition A-2 The random variables X_1 and X_2 with joint distribution function $F(\cdot, \cdot)$ are said to be *independent* if for all x_1 and x_2,

$$F(x_1, x_2) = F(x_1, \infty)F(\infty, x_2)$$

That is, a pair of r.v.'s is independent if their joint distribution function can be expressed as the product of the two marginal distributions. This definition is extended in the obvious way to any finite number of random variables. Observe that when X_1 and X_2 are independent, the joint probability density function satisfies

$$f(x_1, x_2) = f_1(x_1)f_2(x_2)$$

for all x_1 and x_2.

An infinite collection of random variables is independent if every finite set of at least two of them consists of independent random variables. An important consideration when there are three or more random variables is *mutual independence*. It may happen that X_1 and X_2 and X_2 and X_3 are two pairs of independent random variables, but X_1 and X_3 are not independent (in which case we say that they are *dependent*). If every subset of a collection of (at least two) random variables (i.e., all possible pairs, triplets, and so forth) contains only independent random variables, the collection consists of mutually independent random variables.

It is also useful to define independent events. The events A and B are independent if

$$P\{A \cap B\} = P\{A\}P\{B\}$$

When $X_1, X_2, \ldots, X_n$ are mutually independent and have the same d.f., we say they are *independent and identically distributed*, which is abbreviated i.i.d.

Expectations

The intuitive idea behind the expected value of a random variable is that it is a weighted sum of its potential values, where the weight associated with each value is the probability of obtaining that value. The symbol $E(X)$ is commonly used to denote the expected value of the random variable X.

Definition A-3 The *expected value* of an r.v. with distribution function $F(\cdot)$ is

$$E(X) = \int_{-\infty}^{\infty} x \, dF(x)$$

provided the integral† is absolutely convergent.

Other common names for $E(X)$ are *mean* and *first moment*.

When $F(\cdot)$ has a density function $f(\cdot)$, $E(X)$ is

$$E(X) = \int_{-\infty}^{\infty} xf(x) \, dx$$

When X is discrete, we obtain

$$E(X) = \sum_{i=-\infty}^{\infty} x_i \, P\{X = x_i\}$$

which conforms exactly to our intuitive notion of $E(X)$.

When X can assume both positive and negative values,

$$\int_{-\infty}^{\infty} x \, dF(x) = \int_{-\infty}^{0} x \, dF(x) + \int_{0}^{\infty} x \, dF(x)$$

and the integral on the left may not exist because both integrals on the right may diverge, the first to $-\infty$ and the second to $+\infty$. In this case, we say that $E(X)$ does not exist. However, when $X \geq 0$, the contribution of the first integral on the right is zero, so $E(X)$ either exists or is represented by an integral that diverges to $+\infty$. In the latter situation, we say $E(X) = +\infty$. Similar remarks apply when $X \leq 0$.

When X is a random variable and $g(\cdot)$ is a reasonable‡ function, $Y = g(X)$ also is a random variable and may have an expected value. Suppose that X and Y are discrete. Then $Y = y_i$ whenever $X \in S_i = \{x_j : g(x_j) = y_i\}$, and so

$$E(Y) \triangleq \sum_{i=-\infty}^{\infty} y_i \, P\{Y = y_i\} = \sum_{i=-\infty}^{\infty} y_i \, P\{X \in S_i\} = \sum_{j=-\infty}^{\infty} g(x_j) P\{X = x_j\}$$

This analysis can be extended to arbitrary random variables, and the formula

$$E(Y) = E[g(X)] = \int_{-\infty}^{\infty} g(x) \, dF(x) \tag{A-4}$$

is always valid when Y is a bona fide random variable. Let us look at some examples of this result (assume the integrals exist):

1. When $g(X) = X^n$, we obtain

$$E(X^n) = \int_{-\infty}^{\infty} x^n \, dF(x)$$

This is called the *n*th *moment* of X.

---

† See Section A-5 if you are not familiar with this notation for an integral.
‡ For those readers familiar with measure theory, reasonable means measurable.

2. When $g(X) = cX$ for some constant c,

$$E(cX) = c \int_{-\infty}^{\infty} x \, dF(x) = cE(X)$$

3. When $g(X) = [X - E(X)]^2 = Y$,

$$E(Y) = \int_{-\infty}^{\infty} [x - E(X)]^2 \, dF(x) = E(X^2) - [E(X)]^2$$

and $E(Y)$ is called the *variance* of X, denoted by Var (X) [often the notation $V(X)$ or σ_X^2 is used].

Since the integrand is nonnegative, Var $(X) \geq 0$. The positive square root of Var (X) is called the *standard deviation* of X and is typically denoted by σ_X. Using (2) and (3), we deduce that

$$\text{Var } (cX) = E(c^2 X^2) - [cE(X)]^2 = c^2 \text{ Var } (X)$$

for any constant c.

In the next four examples we assume that $F(\cdot, \cdot)$ has a p.d.f. $f(\cdot, \cdot)$; this simplifies the notation. The conclusions obtained are true for any d.f.

4. When $g(X_1, X_2) = X_1 + X_2$,

$$E(X_1 + X_2) = \int_{-\infty}^{\infty} \int_{-\infty}^{\infty} (x_1 + x_2) f(x_1, x_2) \, dx_1 \, dx_2$$

$$= \int_{-\infty}^{\infty} x_1 \int_{-\infty}^{\infty} f(x_1, x_2) \, dx_2 \, dx_1$$

$$+ \int_{-\infty}^{\infty} x_2 \int_{-\infty}^{\infty} f(x_1, x_2) \, dx_1 \, dx_2$$

$$= \int_{-\infty}^{\infty} x_1 f_1(x_1) \, dx_1 + \int_{-\infty}^{\infty} x_2 f_2(x_2) \, dx_2$$

$$= E(X_1) + E(X_2)$$

Combining (2) and (4) and using mathematical induction, we obtain

$$E\left(\sum_{i=1}^{n} c_i X_i \right) = \sum_{i=1}^{n} c_i E(X_i)$$

for any finite and positive integer n.

5. When $g(X_1, X_2) = X_1 X_2$ and X_1 and X_2 are independent,

$$E(X_1 X_2) = \int_{-\infty}^{\infty} \int_{-\infty}^{\infty} x_1 x_2 f(x_1, x_2) \, dx_1 \, dx_2$$

$$= \int_{-\infty}^{\infty} \int_{-\infty}^{\infty} x_1 f_1(x_1) x_2 f_2(x_2) \, dx_1 \, dx_2$$

$$= E(X_1) E(X_2)$$

where $f_1(\cdot)$ and $f_2(\cdot)$ are the p.d.f.'s of X_1 and X_2, respectively.

6. When $g(X_1, X_2) = [X_1 - E(X_1)][X_2 - E(X_2)] = Y$, $E(Y)$ is called the *covariance* of X_1 and X_2, denoted by Cov (X_1, X_2) (sometimes the notation σ_{X_1, X_2} is used).

From (2),

$$\text{Cov} (X_1, X_2) = \int_{-\infty}^{\infty} \int_{-\infty}^{\infty} [x_1 x_2 - x_1 E(X_2) - x_2 E(X_1) - E(X_1) E(X_2)]$$

$$\times f(x_1, x_2) \, dx_1 \, dx_2$$

$$= E(X_1 X_2) - 2E(X_1) E(X_2) + E(X_1) E(X_2)$$

$$= E(X_1 X_2) - E(X_1) E(X_2) = \text{Cov} (X_2, X_1)$$

When X_1 and X_2 are independent, (5) yields

$$\text{Cov} (X_1, X_2) = E(X_1) E(X_2) - E(X_1) E(X_2) = 0$$

However, it is not necessarily true that X_1 and X_2 are independent if Cov $(X_1, X_2) = 0$.

7. When $g(X_1, X_2) = [c_1 X_1 + c_2 X_2 - E(c_1 X_1 + c_2 X_2)]^2$, where c_1 and c_2 are constants, use (2), (3), (4), and (5) to obtain

$$E[g(x)] = \text{Var} (c_1 X_1 + c_2 X_2)$$

$$= E(c_1 [X_1 - E(X_1)])^2 + E(c_2 [X_2 - E(X_2)])^2$$

$$+ 2c_1 c_2 E[X_1 - E(X_1)] E[X_2 - E(X_2)]$$

$$= c_1^2 \text{ Var} (X_1) + c_2^2 \text{ Var} (X_2) + 2c_1 c_2 \text{ Cov} (X_1, X_2)$$

When X_1 and X_2 are independent, Cov $(X_1, X_2) = 0$, so

$$\text{Var} (c_1 X_1 + c_2 X_2) = c_1^2 \text{ Var} (X_1) + c_2^2 \text{ Var} (X_2)$$

8. When $g(X) = e^{-sX}$ and $X \geq 0$, we obtain

$$E(e^{-sX}) = \int_0^{\infty} e^{-sx} \, dF(x)$$

This is the *Laplace-Stieltjes transform* (abbreviated LST) of $F(\cdot)$. When $F(\cdot)$

has p.d.f. $f(\cdot)$,

$$E(e^{-sX}) = \int_0^\infty e^{-sx} f(x)\, dx$$

which is the *Laplace transform*† of $f(\cdot)$.

The rudiments of Laplace transform theory are given in Section A-3.

9. Let $v = E(X)$ and $\sigma^2 = \text{Var}\,(X)$, and choose $Y = (X - v)^2$. For any number $z > 0$,

$$\sigma^2 = E(Y) \geq \int_{y=z^2}^\infty y\, dF(y) \geq z^2 \int_{y=z^2}^\infty dF(y) = z^2 P\{Y \geq z^2\}$$

hence

$$P\{|X - v| \geq z\} \leq \frac{\sigma^2}{z^2}$$

This is called *Chebychev's inequality.*

Conditional Probability

The basic idea of conditional probability can be easily illustrated by the roll of a fair die. If X is the number of pips showing after a roll, then $P\{X = i\} = \frac{1}{6}$ for $i = 1, 2, \ldots, 6$. But if we are told that an even number of pips is showing, we would naturally say that $P\{X = i\} = \frac{1}{3}$ when i is 2, 4, or 6 and $P\{X = i\} = 0$ otherwise. What has happened is that the set of possible outcomes has been reduced and the assignments of probabilities have been adjusted accordingly. We now want to do this in a formal way so that conditional probability calculations can be made rigorously and systematically.

Definition A-4 For any two events A and B with $P\{B\} > 0$, the *conditional probability* of A given B is denoted by $P\{A \mid B\}$ and is defined by

$$P\{A \mid B\} = \frac{P\{A \cap B\}}{P\{B\}}$$

When $P\{B\} = 0$, $P\{A \mid B\}$ is not defined. It is often convenient to adopt the *convention* that $P\{A \mid B\} = 0$ when $P\{B\} = 0$.

The first consequence of this definition is a rule for finding $P\{A \cap B\}$. In many applications, we naturally have $P\{B\}$ and $P\{A \mid B\}$. Then

$$P\{A \cap B\} = P\{A \mid B\} P\{B\}$$

When A and B are independent, $P\{A \cap B\} = P\{A\} P\{B\}$, and so

$$P\{A \mid B\} = P\{A\}$$

This is the intuitive idea behind the use of the word independence.

---

† We also call $E(e^{-sX})$ the Laplace (or Laplace-Stieltjes) transform of X.

If $A_1, A_2, \ldots$ are disjoint events with $\bigcup_{i=1}^{\infty} A_i = \Omega$, it is easily shown that

$$P\{B\} = \sum_{i=1}^{\infty} P\{B \cap A_i\}$$

Since $P\{B \cap A_i\} = P\{B \mid A_i\} P\{A_i\}$,

$$P\{B\} = \sum_{i=1}^{\infty} P\{B \mid A_i\} P\{A_i\}$$

which is called the *theorem of total probability*. Substituting this formula for $P\{B\}$ into the definition of $P\{A \mid B\}$, we obtain *Bayes' formula*:

$$P\{A_j \mid B\} = \frac{P\{A_j \cap B\}}{P\{B\}}$$

$$= \frac{P\{B \mid A_j\} P\{A_j\}}{\sum_{i=1}^{\infty} P\{B \mid A_i\} P\{A_i\}}$$

This notion of conditional probability can be expressed in terms of random variables and distribution functions, and we often do so. Let X and Y be random variables with joint distribution function $F(\cdot, \cdot)$, and let the marginal d.f.'s of X and Y be $F_1(\cdot)$ and $F_2(\cdot)$, respectively. From the definition of conditional probability,

$$P\{X \leq x \mid Y \leq y\} = \frac{P\{X \leq x, Y \leq y\}}{P\{Y \leq y\}}$$

Denoting the left-side by $F_1(x \mid Y \leq y)$, which is called the *conditional d.f.* of X, we obtain

$$F_1(x \mid Y \leq y) = \frac{F(x, y)}{F_2(y)}$$

When the appropriate derivatives exist, we can write the *conditional density function*

$$f_1(x \mid Y \leq y) = \frac{\partial}{\partial x} \frac{F(x, y)}{F_2(y)}$$

When X and Y are discrete r.v.'s and $P\{Y = y\} > 0$,

$$P\{X \leq x \mid Y = y\} = \frac{P\{X \leq x, Y = y\}}{P\{Y = y\}}$$

is a direct consequence of Definition A-4. When Y is not discrete, $P\{Y = y\} > 0$ is the exception, so we have to offer a different definition. If the theorem of total probability *were* valid when Y is continuous with distribution function $G(\cdot)$, it would state

$$P\{X \leq x\} = \int_{-\infty}^{\infty} P\{X \leq x \mid Y = y\} \, dG(y)$$

We *define* $P\{X \leq x \mid Y = y\}$ as *any* solution (there is always one, but there may be many) of this equation. We usually use $P\{X \leq x \mid Y = y\}$ to obtain $P\{X \leq x\}$ from the definition of the conditional probability, so nonuniqueness is not a major issue. The important fact is that it is permissible to condition on the value of an r.v., say $Y = y$, when $P\{Y = y\} = 0$.

Conditional Expectation

Conditional probabilities can be viewed as establishing a new sample space B and new probabilities for events. Random variables defined over the original sample space Ω remain random variables when conditioned on the event B, but have a smaller domain of definition. They may still have expectations, and our intuitive idea of what an expectation represents remains the same. To be both intuitive and precise, we use discrete random variables to introduce the conditional expectation. The ideas and results are applicable to continuous random variables if we replace sums by integrals.

For any event B and discrete random variable X, the *conditional expectation of X given B* is defined by

$$E(X \mid B) = \sum_{i=-\infty}^{\infty} iP\{X = i \mid B\}$$

and it is a number. If we take the event B to be $\{Y = b\}$ for some discrete random variable Y, we obtain

$$E(Y \mid Y = b) = \sum_{i=-\infty}^{\infty} iP\{X = i \mid Y = b\}$$

Different values of b may produce different values of $E(X \mid Y = b)$. We usually write the *function* $E(X \mid Y = \cdot)$ as $E(X \mid Y)$ and express it as "the expected value of X given Y." Since $E(X \mid Y)$ is function whose value is determined by the value taken on by Y, *it is a random variable.*

One of the main reasons we study conditional expectations is that they frequently provide a convenient way to obtain an unconditional (i.e., "ordinary") expectation. Since $E(X \mid Y)$ is a random variable, form its expectation, $E[E(X \mid Y)]$; then

$$E[E(X \mid Y)] = \sum_{b=-\infty}^{\infty} E(X \mid Y = b)P\{Y = b\}$$

$$= \sum_{b=-\infty}^{\infty} \sum_{i=-\infty}^{\infty} iP\{X = i \mid Y = b\}P\{Y = b\}$$

$$= \sum_{i=-\infty}^{\infty} \sum_{b=-\infty}^{\infty} P\{X = i \mid Y = b\}P\{Y = b\}$$

$$= \sum_{i=-\infty}^{\infty} iP\{X = i\} = E(X)$$

This procedure for obtaining $E(X)$ is employed frequently in the study of stochastic processes. To illustrate its use, let X be the number of pips showing on a fairly rolled die, and define Y to be 1 if X is even and 0 if X is odd. Then

$$E(X \mid Y = 1) = 4 \qquad E(X \mid Y = 0) = 3$$

$$P\{Y = 1\} = P\{Y = 0\} = \tfrac{1}{2}$$

and so

$$E(X) = E[E(X \mid Y)] = \frac{4 + 3}{2}$$

Since $E(X \mid Y)$ is a random variable, it has all the endowments of random variables we have studied previously, such as a variance, a d.f., and a Laplace transform.

Sums of Random Variables

Let X_1 and X_2 be random variables, and define the random variable S by $S = X_1 + X_2$. When X_1 and X_2 are discrete, it is easy to see that

$$P\{S = k\} = \sum_{i=-\infty}^{\infty} P\{X_2 = k - i \mid X_1 = i\} P\{X_1 = i\}$$

When X_1 and X_2 have distribution functions $F_1(\cdot)$ and $F_2(\cdot)$, respectively, then the d.f. of S, say $G(\cdot)$, is given by

$$G(s) = \int_{-\infty}^{\infty} F_2(s - x \mid X_1 = x) \, dF_1(x)$$

If either X_1 or X_2 possesses a density function, then so does S.

When X_1 and X_2 are independent and nonnegative, then the above formulas reduce to

$$P\{S = k\} = \sum_{i=0}^{k} P\{X_2 = k - i\} P\{X_1 = i\}$$

and

$$G(s) = \int_0^s F_2(s - x) \, dF_1(x) = \int_0^s F_1(x) \, dF_2(s - x)$$

The expressions on the right-side of the last two equations are called *convolutions* and are studied more extensively in Section A-3.

Once we know how to find the distribution of the sum of two random variables, in principle, we can find the distribution of a sum of any finite number of them. We can find the distribution of $X_1 + X_2$, then of $X_3 + (X_1 + X_2)$, then of $X_4 + (X_1 + X_2 + X_3)$, etc.

Now we consider sums of a large number of random variables and the famous limit theorems for them. To this end, define $S_0 = 0$ and

$$S_n = X_1 + X_2 + \cdots + X_n \qquad n \in I_+$$

If each X_i has mean μ, then

$$E(S_n) = n\mu \qquad n \in I$$

Furthermore, if the X_i's are independent and have the common variance σ^2, then

$$\text{Var } (S_n) = n\sigma^2$$

and

$$\text{Var } \left(\frac{S_n}{n}\right) = \text{Var } \left(\frac{X_1}{n} + \frac{X_2}{n} + \cdots + \frac{X_n}{n}\right)$$

$$= \frac{n\sigma^2}{n^2} = \frac{\sigma^2}{n}$$

Thus, $E(S_n/n) = \mu$, which is a constant, and Var $(S_n/n) \to 0$ as $n \to \infty$, which suggests that S_n/n, which is a random variable, approaches a constant as $n \to \infty$. This is a typical interpretation of "statistical regularity" or a "law of averages," but it is not a precise mathematical statement. The precise statements are the laws of large numbers given now.

Theorem A-1: Weak law of large numbers Let $X_1, X_2, \ldots$ be mutually independent random variables, and let $v_k = E(X_k)$ and $\sigma_k^2 = \text{Var } (X_k)$ for each k. Set $S_n = X_1 + \cdots + X_n$ so that $E(S_n) \triangleq m_n = \sum_1^n v_k$ and Var $(S_n) \triangleq s_n^2 = \sum_1^n \sigma_k^2$. If $s_n^2/n \to 0$, then for any $\varepsilon > 0$,

$$\lim_{n \to \infty} P\left\{\frac{|S_n - m_n|}{n} < \varepsilon\right\} = 1$$

This version of the weak law of large numbers requires that the random variables possess a variance. This restriction is not necessary when $X_1, X_2, \ldots$ have the same distribution.

The weak law states that when all the v_k's are the same, v say, given any $\varepsilon > 0$, there exists a $\delta > 0$ (possibly depending on ε) and an N (possibly depending on ε and δ) such that for any $n > N$,

$$P\left\{\left|\frac{S_n}{n} - v\right| < \varepsilon\right\} > 1 - \delta$$

That is, for all sufficiently large n, S_n/n is close to v with probability close to 1. What the weak law does *not* say is that if $|S_n/n - v|$ is small for one large value of n, it will remain small for all larger values of n. To draw that conclusion, we need to take the limit inside the probability statement, and that stronger result is the strong law of large numbers.

Theorem A-2: Strong law of large numbers With the notation of Theorem A-1, assume that either

(a) $\displaystyle\sum_{k=1}^{\infty} \frac{\sigma_k^2}{k^2} < \infty$

or

(b) $X_1, X_2, \ldots$ are i.i.d. with $E(X_1) < \infty$

Then

$$\lim_{n \to \infty} \frac{S_n - m_n}{n} = 1 \qquad \text{(w.p.1)} \qquad \text{(A-5)}$$

Equivalently, for every pair $\varepsilon > 0$ and $\delta > 0$, there is an N such that for every $r \in I$,

$$P\left\{ \frac{|S_N - m_N|}{N} < \varepsilon, \ldots, \frac{|S_{N+r} - m_{N+r}|}{N+r} < \varepsilon \right\} > 1 - \delta$$

In case (b), $m_n = n\mu$ and $\lim_{n \to \infty} (S_n/n) = \mu$. Note that the left-side of (A-5) is an r.v. and the right-side is a constant.

The strong law justifies the interpretation of probability measure as the long-run frequency of occurrence. To see this, consider many independent replications of an experiment performed under identical circumstances, and let ω_n be the outcome of the nth experiment. For each $n = 1, 2, \ldots$ and any event A, define the random variable X_n by

$$X_n = \begin{cases} 1 & \text{if } \omega_n \in A \\ 0 & \text{if } \omega_n \notin A \end{cases}$$

Hence $E(X_n) = P\{A\}$. The strong law implies

$$\frac{1}{n} \sum_{i=1}^{n} X_j \to P\{A\} \qquad \text{(w.p.1)}$$

and the left-side is the relative frequency of occurrences of A.

The weak and strong laws both state that the difference between S_n and its expectation is small compared to n, when n is large. In absolute terms, S_n may be (in fact, usually is) far from m_n. To determine how far S_n is from m_n, divide $|S_n - m_n|$ by a function of n that goes to infinity more slowly than n does.

Theorem A-3: Law of the iterated logarithm When $X_1, X_2, \ldots$ are i.i.d. with mean 0 and variance σ^2,

$$\limsup_{n \to \infty} \frac{|S_n|}{\sigma \sqrt{2n \ln(\ln n)}} = 1 \qquad \text{(w.p.1)}$$

This theorem states that, for $\lambda > 1$,

$$|S_n| > \lambda \sigma \sqrt{2n \ln(\ln n)}$$

holds w.p.1 for only finitely many values of n, and for $0 < \lambda < 1$, it holds for infinitely many values of n.

The next limit theorem is perhaps the most widely known, used, and abused theorem in probability theory. The simple version given here suffices for our purposes.

> **Theorem A-4: Central-limit theorem** When $X_1, X_2, \ldots$ are i.i.d. with mean v and variance σ^2, then for every fixed z,
>
> $$\lim_{n \to \infty} P\left\{ \frac{S_n - nv}{\sigma\sqrt{n}} < z \right\} = \Phi(z)$$
>
> where $\Phi(z)$ is the standard normal d.f. given by
>
> $$\Phi(x) = \frac{1}{\sqrt{2\pi}} \int_{-\infty}^{x} e^{-u^2/2} \, du$$

This theorem asserts that if one adds a large number of i.i.d. random variables and *normalizes* (subtracts from S_n its mean nv and divides by the standard deviation $\sigma\sqrt{n}$) so the resulting r.v. has mean 0 and variance 1, then this latter r.v. has a d.f. that is close (pointwise) to the distribution function $\Phi(\cdot)$. An important feature of the central-limit theorem to remember is that the conclusion is typically known to be true only when the summands are mutually independent.

A-2 EXPONENTIAL FAMILY OF DISTRIBUTIONS

The exponential distribution is very important in applied probability models. This distribution often fits data on times to complete a telephone conversation, times between failures of complex equipment, and times between job arrivals at a computer. It is also useful as a building block. By combining exponential distributions, a wide variety of nonexponential distributions can be obtained.

Exponential Distribution

> **Definition A-5** The distribution function $F(\cdot)$ is *exponential* with parameter $\alpha > 0$ if
>
> $$F(x) = \begin{cases} 1 - e^{-\alpha x} & x \geq 0 \\ 0 & x < 0 \end{cases}$$

An r.v. with an exponential d.f. is called *exponential*.

The following properties of the exponential d.f. follow immediately from the definition.

> **Proposition A-1** If X is exponential with parameter α, then

(a) X has a density function $f(\,\cdot\,)$ given by

$$f(x) = \begin{cases} \alpha e^{-\alpha x} & x \geq 0 \\ 0 & x < 0 \end{cases}$$

(b) $F^c(x) \triangleq P\{X > x\} = \begin{cases} 1 & x < 0 \\ e^{-\alpha x} & x \geq 0 \end{cases}$

(c) $E(X^n) = \alpha \int_0^\infty x^n e^{-\alpha x}\, dx = \dfrac{n!}{\alpha^n}$ $n \in I$;   in particular, $E(X) = 1/\alpha$

(d) $\text{Var}\,(X) = \dfrac{1}{\alpha^2}$

(e) $E(e^{-sx}) = \alpha \int_0^\infty e^{-sx} e^{-\alpha x}\, dx = \dfrac{\alpha}{s + \alpha}$

PROOF This is left to you. □

Observe that (c) and (d) imply that the ratio of the standard deviation to the mean is 1.

The importance of the exponential distribution is due to its unique memoryless property. If X is exponentially distributed with parameter α, then

$$P\{X > t + s \mid X > s\} = \frac{e^{-\alpha(t+s)}}{e^{-\alpha s}}$$

$$= e^{-\alpha t} = P\{X > t\} \qquad t, s \geq 0 \tag{A-6}$$

Suppose X represents the time to failure of some device. Equation (A-6) asserts that if the device has lasted until time s, the distribution of the remaining life is independent of s; in particular, it is the same as the time to failure of a new device. In other words, when failure times are exponentially distributed, a used item is as good as new. We call the property expressed by (A-6) the *memoryless property* of the exponential distribution.

Here is another version of the memoryless property. Suppose $F(\,\cdot\,)$ is the d.f. of a nonnegative random variable X and $F(\,\cdot\,)$ has a density $f(\,\cdot\,)$. The function

$$h(t) \triangleq \frac{f(t)}{F^c(t)} \qquad t \geq 0$$

is the *hazard rate* (or *failure rate*) of X [of $F(\,\cdot\,)$]. If X represents the time to failure of a device, then the probability that the device fails during $(t, t + \Delta t]$ is $h(t)\,\Delta t + o(\Delta t)$. When $F(t) = 1 - e^{-\alpha t}$,

$$h(t) = \frac{\alpha e^{-\alpha t}}{e^{-\alpha t}} = \alpha \qquad t \geq 0 \tag{A-7}$$

i.e., the exponential distribution has a constant hazard rate. In other words, no matter how long the device has been working, the probability that it will fail during an interval of length Δt is $\alpha \, \Delta t + o(\Delta t)$.

The exponential distribution is singled out for so much attention because it is the only d.f. with a density that is memoryless.

Proposition A-2 Let X be a nonnegative r.v. with distribution function $F(\,\cdot\,)$ possessing a density $f(\,\cdot\,)$, and let $E(X) = 1/\alpha$, $0 < \alpha < \infty$. Then the following three statements are equivalent:

(a) X has the exponential distribution

$$F(t) = 1 - e^{-\alpha t}$$

(b) X has the *memoryless property*

$$P\{X > t + s \,|\, X > s\} = P\{X > t\} \qquad \text{for all } t, s > 0$$

(c) X has the *constant-hazard-rate property*

$$\frac{f(t)}{F^c(t)} = \text{const} \qquad \text{for all } t \geq 0$$

PROOF We showed above that $(a) \Rightarrow (b)$. We now show $(b) \Rightarrow (c) \Rightarrow (a)$. To show $(b) \Rightarrow (c)$, observe that (b) implies

$$F(t + s) - F(t) = P\{t < X \leq t + s\} = P\{X \leq t + s \,|\, X > t\}P\{X > t\}$$
$$= P\{X \leq s\}P\{X > t\} = F(s)F^c(t)$$

Thus,

$$f(t) = \lim_{s\downarrow 0} \frac{F(t + s) - F(t)}{s}$$
$$= \lim_{s\downarrow 0} \frac{F(s)}{s} F^c(t) = f(0)F^c(t)$$

and hence $f(t)/F^c(t)$ equals the constant $f(0)$. To show (c) implies (a), let the constant in (c) be β. Since $f(t) = -\,dF^c(t)/dt$, (c) implies

$$\beta = \frac{-1}{F^c(t)} \frac{dF^c(t)}{dt} = -\frac{d}{dt}[\ln F^c(t)]$$

Hence

$$\int d\,[\ln F^c(t)] = -\beta \int dt + \ln \gamma$$

where $\ln \gamma$ is the constant of integration. Thus,

$$F^c(t) = \gamma e^{-\beta t} \qquad t \geq 0$$

Since $1 = F^c(0) = \gamma, f(t) = \beta e^{-\beta t}$. Since $E(X) = 1/\alpha$,

$$\frac{1}{\alpha} = \int_0^\infty t\beta e^{-\beta t} \, dt = \frac{1}{\beta}$$

and the demonstration is complete. $\qquad\qquad\qquad\qquad\qquad\qquad\qquad$ $\square$

The condition that $F(\cdot)$ have a density is crucial for the "only if" part of Proposition A-1. When X is a discrete r.v., the geometric distribution is memoryless (see Exercises A-1 through A-3).

The next proposition is used frequently in the text.

Proposition A-3 Let $X_1, X_2, \ldots, X_n$ be independent r.v.'s with $P\{X_i \le t\} = 1 - e^{-\alpha_i t}$, $i = 1, 2, \ldots, n$. Let $M = \min\{X_1, X_2, \ldots, X_n\}$. Then

$$P\{M \le t\} = 1 - e^{-(\alpha_1 + \cdots + \alpha_n)t} \qquad t \ge 0$$

PROOF Observe that $\{M > t\}$ occurs if, and only if, each $X_i > t$. Using the independence assumption yields

$$P\{M > t\} = \prod_{i=1}^n P\{X_i > t\} = \prod_{i=1}^n e^{-\alpha_i t} = e^{-(\alpha_1 + \cdots + \alpha_n)t}$$

and the desired result follows immediately. $\qquad\qquad\qquad\qquad\qquad\qquad$ $\square$

Gamma Distribution

Let X_1 and X_2 be i.i.d. exponential r.v.'s with mean $1/\lambda$. The random variable $S_2 = X_1 + X_2$ has density

$$f_2(t) = \int_0^t \lambda e^{-\lambda(t-x)} \lambda e^{-\lambda x} \, dx = \lambda^2 t e^{-\lambda t} \qquad t \ge 0$$

When $X_1, X_2, \ldots, X_k$ are i.i.d. exponential r.v.'s with mean $1/\lambda$, a simple induction will show that the random variable $S_k = X_1 + \cdots + X_k$ has density

$$f_k(t) = \frac{\lambda^k t^{k-1} e^{-\lambda t}}{(k-1)!} \qquad k \in I_+ \qquad\qquad (A-8)$$

We call $f_k(t)$ the *gamma density with shape parameter k and scale parameter λ.* There is no reason to restrict k to being an integer. Suppose we want a density function of the form

$$f_\kappa(t) = ct^{\kappa-1} e^{-\lambda t} \qquad t \ge 0$$

where $\kappa > 0$, $\lambda > 0$, and c is chosen so that $f_\kappa(\cdot)$ integrates to 1. Setting $x = \lambda t$ yields

$$1 = c \int_0^\infty t^{\kappa-1} e^{-\lambda t} \, dt = \frac{c}{\lambda^{\kappa-1}} \int_0^\infty x^{\kappa-1} e^{-x} \, dx$$

The last integral is the gamma function

$$\Gamma(\kappa) = \int_0^\infty x^{\kappa-1} e^{-x} \, dx$$

Integration by parts establishes

$$\Gamma(\kappa) = (\kappa - 1)\Gamma(\kappa - 1)$$

and $\Gamma(k) = (k - 1)!$ for $k \in I_+$.

Thus,

$$f_\kappa(t) = \frac{\lambda(\lambda t)^{\kappa-1} e^{-\lambda t}}{\Gamma(\kappa)} \qquad t > 0 \tag{A-9}$$

is a bona fide density function for any $\kappa > 0$; it is called the *gamma density with shape parameter κ and scale parameter λ.*

The reason why κ is called the shape parameter and λ is called the scale parameter can be seen in Figures A-1 and A-2. For fixed λ, changes in κ lead to qualitative changes in the shape of the density, while for fixed κ, changes in λ lead primarily to changes in the scale of the function.

The properties of the gamma density that we use are given in this proposition.

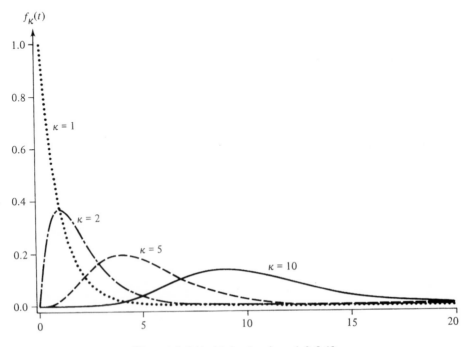

Figure A-1 $f_\kappa(t)$ with $\lambda = 1$ and $\kappa = 1, 2, 5, 10$.

Proposition A-4 Let X be an r.v. whose density function is given by (A-9). Then

(i) $E(X) = \dfrac{\kappa}{\lambda}$

(ii) $\text{Var}(X) = \dfrac{\kappa}{\lambda^2}$

(iii) $E(e^{-sX}) = \dfrac{\lambda^\kappa}{(s + \lambda)^\kappa}$

PROOF This is left as Exercise A-4. □

Observe that $+\sqrt{\text{Var}(X)}/E(X) = 1/\sqrt{\kappa}$, which is less than 1 when $\kappa > 1$ and larger than 1 when $\kappa < 1$. Thus, large values of κ lead to distributions that are "more regular" than the exponential, and small values of κ correspond to distributions that are "more irregular" than the exponential.

By putting $\kappa = k \in I_+$ and $\lambda = k\alpha$ in (A-8), a family of distributions with mean $1/\alpha$ and variance $(k\alpha^2)^{-1}$ is achieved. This special form of the gamma density is called the *Erlang* density and denoted by E_k. The Erlang density frequently is used in queueing models to describe r.v.'s where the data indicate that the exponential d.f. is not appropriate because the ratio of the standard deviation to the mean is much smaller than 1. Since the Erlang distribution

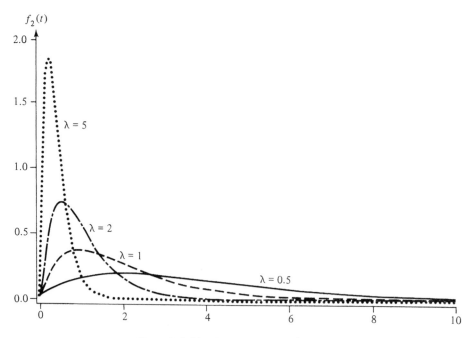

Figure A-2 $f_\kappa(t)$ with $\kappa = 2$ and $\lambda = \frac{1}{2}, 1, 2, 5$.

represents the sum of i.i.d. exponential r.v.'s, some properties of the exponential d.f. can be used in the subsequent analysis of models with Erlang distributions (see Example 8-5).

Hyperexponential Distribution

Let X_1 and X_2 be independent exponential r.v.'s with $E(X_i) = 1/\alpha_i$. Define the random variable Y by

$$Y = \begin{cases} X_1 & \text{with probability } p \\ X_2 & \text{with probability } 1 - p \end{cases}$$

The density function of Y is

$$f_2(t) = p\alpha_1 e^{-\alpha_1 t} + (1 - p)\alpha_2 e^{-\alpha_2 t} \qquad t \ge 0 \qquad \text{(A-10)}$$

This is called the *hyperexponential density of order 2* and denoted by H_2. A physical model that leads to the H_2 density for failure times (say) occurs when there are two manufacturing plants and plant i produces objects which have exponentially distributed failure times with mean $1/\alpha_i$. Plant 1 produces $100p$ percent of the total objects. When the objects are prepared for sale to the public, the outputs from both plants are mixed. The failure time of a randomly chosen object is then given by the random variable Y defined above.

The density in (A-10) has the obvious generalization

$$f_k(t) = \sum_{i=1}^{k} p_i \alpha_i e^{-\alpha_i t} \qquad t \ge 0 \qquad \text{(A-11)}$$

where $p_i \ge 0$, $\sum_{i=1}^{k} p_i = 1$, and each $\alpha_i > 0$. This is the hyperexponential density of order k and is denoted by H_k. The salient features of the H_k density are requested in Exercise A-5.

EXERCISES

A-1 A nonnegative discrete random variable N is memoryless if $P\{N = m + n \,|\, N \ge m\} = P\{N = n\}$ for all $m, n \in I$. The hazard rate at n is $P\{N = n\}/P\{N \ge n\}$. Suppose $P\{N = n\} = (1 - \pi)\pi^n$, $n \in I$, for some π between 0 and 1. Is the failure rate a constant? Is the r.v. memoryless?

A-2 (Continuation) Show that $(1 - \pi)\pi^{n-1}$, $n \in I_+$, is the unique discrete memoryless distribution.

A-3 (Continuation) Prove that when $P\{N = 0\} = 0$, the memoryless property holds if, and only if, the constant-hazard-rate property holds.

A-4 Prove Proposition A-4.

A-5 Let the random variable Y have an H_k density. Find $E(Y)$, Var (Y), $P\{Y \le y\}$ and $E(e^{-sY})$.

A-3 LAPLACE TRANSFORMS AND GENERATING FUNCTIONS

This section contains a brief introduction to Laplace transforms, Laplace-Stieltjes transforms, and generating functions. Laplace transforms are applied to functions defined on the nonnegative real numbers, and generating functions are applied to functions defined on the nonnegative integers. Formal proofs of the results usually are absent.

Laplace Transforms

Definition A-6 Let $f(\cdot)$ be a real-valued function with domain $[0, \infty)$. The *Laplace transform of* $f(\cdot)$, which may be denoted by $\mathcal{L}\{f(\cdot)\}$ or $\tilde{f}(\cdot)$, is defined by

$$\mathcal{L}\{f(\cdot)\} \triangleq \tilde{f}(s) \triangleq \int_0^\infty e^{-st} f(t) \, dt \qquad (A\text{-}12)$$

where s is allowed to be complex. We say that $\mathcal{L}\{f(\cdot)\}$ exists if (A-12) converges for at least one value of s.

Since (A-12) is a Riemann integral, $f(\cdot)$ must be a Riemann-integrable function. We almost always treat s as a real number. A sufficient (but not necessary) condition for $\mathcal{L}\{f(\cdot)\}$ to exist is

$$|f(t)| < Me^{\gamma t} \qquad t > T, s > \gamma \qquad (A\text{-}13a)$$

for some positive constants M and T and any constant γ. A sufficient condition for $\tilde{f}(s)$ to exist for all $s \geq 0$ is

$$|f(t)| < Mt^m \qquad t > T \qquad (A\text{-}13b)$$

for some $m > 0$. From now on we consider only functions where $\mathcal{L}\{f(\cdot)\}$ exists.

The early importance of Laplace transforms came from the way they simplified differential and integral equations and their limit theorems. In probability theory, transforms are especially useful in dealing with sums of independent random variables. First we need the fundamental result that will establish that Laplace transforms are unique.

Theorem A-5 If $\tilde{f}(s) = 0$ for all s, then $f(t) \equiv 0$.

Corollary A-5 If $\tilde{f}(s) = \mathcal{L}\{f(\cdot)\}$ and $\tilde{g}(s) = \mathcal{L}\{g(\cdot)\}$, then

$$\tilde{f}(s) \equiv \tilde{g}(s)$$

if, and only if,

$$f(t) \equiv g(t)$$

From now on, a tilde always denotes the Laplace transform of the function to which it is attached, as in (A-12). Because of Corollary A-5, for any transform $\tilde{f}(s)$, there is a unique function $f(\cdot)$ for which $\mathcal{L}\{f(\cdot)\} = \tilde{f}(s)$, and we may write

$$f(t) = \mathcal{L}^{-1}\{\tilde{f}(s)\}$$

where $\mathcal{L}^{-1}\{\cdot\}$ refers to the *inverse transform*. Equation (A-14) below expresses $f(t)$ in terms of $\tilde{f}(s)$ and is called the *inversion formula*. The formula is not used in the text. It is displayed so that you have some appreciation of how a Laplace transform can be inverted.

Theorem A-6: Inversion formula If $\tilde{f}(s) = \mathcal{L}\{f(\,\cdot\,)\}$, then

$$f(t) = \mathcal{L}^{-1}\{\tilde{f}(\,\cdot\,)\} = \frac{1}{2\pi i} \int_{\gamma - i\infty}^{\gamma + i\infty} e^{st}\tilde{f}(s) \, ds \qquad t > 0 \qquad \text{(A-14)}$$

and $f(t) = 0$ for $t < 0$. The integration in (A-14) is to be taken along the line $s = \gamma$ (γ is real) in the complex plane, where γ is chosen such that all the singularities of $\tilde{f}(s)$ lie to the left of $s = \gamma$. Here, $i = \sqrt{-1}$.

Note that this theorem regards s as complex.

We usually invert Laplace transforms by inspection. Properties A-1 through A-4 can be obtained directly from (A-12).

Property A-1: Linearity For any two constants c_1 and c_2,

$$\mathcal{L}\{c_1 f_1(\,\cdot\,) + c_2 f_2(\,\cdot\,)\} = c_1 \tilde{f}_1(s) + c_2 \tilde{f}_2(s)$$

Property A-2 For any constant c,

$$\mathcal{L}\{e^{ct}f(t)\} = \tilde{f}(s - c)$$

Property A-3: Translation If $g(t)$ is defined by

$$g(t) = \begin{cases} f(t - c) & t > c \\ 0 & t \le c \end{cases}$$

for some constant $c > 0$, then

$$\tilde{g}(s) = e^{-sc}\tilde{f}(s)$$

Property A-4: Change of scale For $c > 0$,

$$\mathcal{L}\{f(ct)\} = \frac{\tilde{f}(s/c)}{c}$$

Integrating (A-12) by parts yields

$$\tilde{f}(s) = \frac{1}{s} \int_0^\infty e^{-st}\dot{f}(t) \, dt + \frac{1}{s} \lim_{t \downarrow 0} f(t) \qquad \text{(A-15)}$$

where

$$\dot{f}(t) = \frac{d}{dt} f(t)$$

Properties A-5 through A-8 are consequences of (A-15).

Property A-5: Differentiation If $\dot{f}(\,\cdot\,)$ is the derivative of $f(\,\cdot\,)$ and $\dot{f}(\,\cdot\,)$ is bounded, then

$$\mathcal{L}\{\dot{f}(\,\cdot\,)\} = s\tilde{f}(s) - \lim_{t\downarrow 0} f(t)$$

If $f(0)$ is defined, the above limit is $f(0)$. Applying Property A-5 several times and assuming all derivatives are bounded and continuous at zero generalize Property A-5 to

$$\mathcal{L}\{f^{(n)}(\,\cdot\,)\} = s^n\tilde{f}(s) - s^{n-1}f(0) - s^{n-2}\dot{f}(0) - \cdots - f^{(n-1)}(0)$$

where $f^{(n)}(\,\cdot\,)$ is the nth derivative of $f(\,\cdot\,)$. Interchanging the roles of $f(\,\cdot\,)$ and $\dot{f}(\,\cdot\,)$ yields the next property.

Property A-6: Integration Let

$$F(t) = \int_0^t f(x)\,dx$$

Then

$$\tilde{F}(s) = \frac{1}{s}\tilde{f}(s)$$

Notice that we can deduce that $F(\,\cdot\,)$ is Laplace-transformable since $f(\,\cdot\,)$ is, which implies that $\lim_{t\to\infty} e^{-st}F(t) = 0$.

Property A-7: Initial-value theorem When the indicated limits both exist, then

$$\lim_{t\downarrow 0} f(t) = \lim_{s\to\infty} s\tilde{f}(s)$$

Property A-8: Final-value theorem When the indicated limits both exist, then

$$\lim_{t\to\infty} f(t) = \lim_{s\downarrow 0} s\tilde{f}(s)$$

When Properties A-7 and A-8 are used to obtain information about $\tilde{f}(\,\cdot\,)$ from $f(\,\cdot\,)$, these are called *Abelian* theorems; when they are employed to find the limits of $f(\,\cdot\,)$ from limits of $\tilde{f}(\,\cdot\,)$, they are called *Tauberian* theorems.

Observing that (Proposition A-13, Leibnitz' rule, explains how to take the derivative)

$$\frac{d}{ds}\tilde{f}(s) = \frac{d}{ds}\int_0^\infty e^{-st}f(t)\,dt = -\int_0^\infty e^{-st}tf(t)\,dt$$

will establish the next two properties.

Property A-9: Multiplication by t^n

$$\mathcal{L}\{t^n f(t)\} = (-1)^n \frac{d^n}{ds^n}\tilde{f}(s)$$

Property A-10: Division by t

$$\mathcal{L}\left\{\frac{f(t)}{t}\right\} = \int_s^\infty \tilde{f}(x) \, dx$$

Define $g(t) = f(t)/t$; one of the consequences of Property A-10 is that

$$-d\tilde{g}(s) = \tilde{f}(s) \, ds$$

Combining this with the final-value theorem yields this property.

Property A-11: Asymptotic-rate theorem When the indicated limits both exist, then

$$\lim_{t \to \infty} \frac{f(t)}{t} = \lim_{s \downarrow 0} s^2 \tilde{f}(s)$$

Perhaps the most important reason Laplace transforms are used in probability theory is due to the following property.

Property A-12: Convolution theorem The *convolution* of the two functions $f(\,\cdot\,)$ and $g(\,\cdot\,)$, say $h(\,\cdot\,)$, is defined by

$$h(t) = \int_0^t f(t-u)g(u) \, du = \int_0^t g(t-u)f(u) \, du \tag{A-16}$$

and denoted by

$$h(t) = f * g(t)$$

If both $f(\,\cdot\,)$ and $g(\,\cdot\,)$ have a Laplace transform, then so does $h(\,\cdot\,)$, and

$$\tilde{h}(s) = \tilde{f}(s)\tilde{g}(s)$$

These properties are used in a variety of ways in the text. One is to invert Laplace transforms by inspection. Table A-1 can be prepared by direct evaluation of (A-12).

**Table A-1
Elementary
Laplace transforms**

$f(t)$	$\mathcal{L}\{f(t)\}$
1	$\dfrac{1}{s}$
e^{at}	$\dfrac{1}{s-a} \quad s > a$
$\sin at$	$\dfrac{a}{s^2 + a^2}$

Suppose we wish to invert $1/s^2$. By writing

$$\frac{1}{s^2} = \frac{1}{s}\frac{1}{s}$$

and using Property A-6, we get

$$\mathcal{L}^{-1}\left\{\frac{1}{s^2}\right\} = \int_0^t 1\, dx = t$$

Similarly,

$$\mathcal{L}^{-1}\left\{\frac{1}{s^3}\right\} = \int_0^t x\, dx = \frac{t^2}{2}$$

and, in general,

$$\mathcal{L}^{-1}\left\{\frac{1}{s^{n+1}}\right\} = \frac{t^n}{n!}$$

Now consider

$$\tilde{f}(s) = \left(\frac{\lambda}{s+\lambda}\right)^2 = \frac{\lambda}{s+\lambda}\frac{\lambda}{s+\lambda}$$

From Property A-12,

$$\mathcal{L}^{-1}\left\{\frac{\lambda^2}{(s+\lambda)^2}\right\} = \lambda^2 \int_0^t e^{-\lambda(t-x)}e^{\lambda x}\, dx = \lambda^2 t e^{-\lambda t}$$

When $f(\,\cdot\,)$ is a p.d.f. of a nonnegative random variable, X say, $\tilde{f}(s)$ has some special properties and interpretations. First we observe that $\tilde{f}(s)$ always exists for $s \geq 0$.

Property A-13: Existence and bounds When $f(\,\cdot\,)$ is a p.d.f., $\tilde{f}(s)$ exists for $s \geq 0$ (Re $s \geq 0$ when† s is complex) and

$$0 \leq \tilde{f}(s) \leq 1$$

PROOF Since the distribution function

$$F(t) = \int_0^t f(x)\, dx$$

exists, $f(\,\cdot\,)$ is integrable, so (A-12) makes sense as an integral. Choose $s = 0$; then

$$\tilde{f}(0) = \int_0^\infty f(x)\, dx = P\{X < \infty\} \leq 1 \tag{A-17}$$

This shows $\tilde{f}(s)$ exists and proves the upper bound: Since $f(x) \geq 0$ for all x

† Re s denotes the real part of the complex number; i.e., when $s = a + bi$, Re $s = a$.

and e^{-sx} is a decreasing function of s, $\tilde{f}(s) \leq \tilde{f}(0)$ for $s \geq 0$, so $\tilde{f}(s)$ exists for all $s \geq 0$. Since $e^{-st} \geq 0$,

$$\tilde{f}(s) = \int_0^\infty e^{-st} f(t) \, dt \geq 0 \qquad \square$$

A very important feature of the Laplace transform of a p.d.f. is that it can be used to obtain moments.

Property A-14: Moment evaluation When $f(\cdot)$ is the p.d.f. of a nonnegative random variable X,

$$E(X^n) = (-1)^n \frac{d^n}{ds^n} \tilde{f}(s)\Big|_{s=0}$$

PROOF Since $d^n e^{-st}/ds^n = (-1)^n t^n e^{-st}$, Leibnitz' rule yields

$$\frac{d^n}{ds^n} \tilde{f}(s) = (-1)^n \int_0^\infty e^{-st} t^n f(t) \, dt$$

Proposition A-10 (the dominated convergence theorem) allows us to set $s = 0$ and completes the proof. $\qquad \square$

Properties A-5 and A-6 link Laplace transforms and differential equations, particularly those with constant coefficients.

Example A-1 For a given function $g(\cdot)$ and number a, solve

$$\dot{f}(t) = af(t) + g(t) \qquad \text{(A-18)}$$

for $f(t)$ with the boundary condition $f(0) = \alpha$. We can do this by equating the Laplace transforms of both sides. By Property A-5, the Laplace transform of the left-side is

$$s\tilde{f}(s) - \alpha$$

and by Property A-1, the Laplace transform of the right-side is

$$a\tilde{f}(s) + \tilde{g}(s)$$

Thus,

$$s\tilde{f}(s) - \alpha = a\tilde{f}(s) + \tilde{g}(s)$$

or

$$\tilde{f}(s) = \frac{\alpha + \tilde{g}(s)}{s - a} = \frac{\alpha}{s - a} + \tilde{g}(s)(s - a)^{-1}$$

This can be inverted term by term by inspection. Since $(s - a)^{-1} = \mathcal{L}^{-1}\{e^{at}\}$,

$$f(t) = \alpha e^{at} + \int_0^t e^{a(t-x)} g(x) \, dx$$

solves (A-18); the last term is obtained from the convolution property. Since Laplace transforms are unique, this is the only solution of (A-18). □

Laplace-Stieltjes Transforms

Laplace-Stieltjes transforms (LSTs) are a generalization of Laplace transforms to functions that are not necessarily Riemann-integrable. This generalization is desirable when we are dealing with random variables that have a concentration of probability at a point. An example is the waiting time in a queue. With positive probability the waiting time is zero, hence the waiting-time distribution does not have a density function. A large class of examples is described by discrete distributions. In this case, the LST provides a unifying link between the Laplace transforms and probability generating functions.

Definition A-7 Let $F(\,\cdot\,)$ be a real-valued function with domain $[0, \infty)$. The *Laplace-Stieltjes transform* of $F(\,\cdot\,)$, denoted by $\tilde{F}(s)$, is defined by

$$\tilde{F}(s) = \int_0^\infty e^{-st}\, dF(t) \qquad (A\text{-}19)$$

where s may be complex. We say that $\tilde{F}(\,\cdot\,)$ *exists* if (A-19) converges for at least one value of s.

It is common practice to use the same notation for LST and Laplace transforms and to distinguish them by context. This usually does not cause any difficulties.

If $F(\,\cdot\,)$ has derivative $f(\,\cdot\,)$, then

$$\tilde{F}(s) = \int_0^\infty e^{-st} f(t)\, dt = \mathcal{L}\{f(\,\cdot\,)\} \qquad (A\text{-}20)$$

This suggests that the properties of Laplace transforms will carry over to the LST, and that is essentially correct. Except for Properties A-7, A-8, A-11, and A-12, all the Laplace transform results apply to LSTs without change. The slight changes needed in the excepted properties are listed below; the "a" denotes the LST form of the property.

Property A-7a: Initial-value theorem When the indicated limits both exist, then

$$\lim_{t\downarrow 0} F(t) = \lim_{s\to\infty} \tilde{F}(s)$$

Property A-8a: Final-value theorem When the indicated limits both exist, then

$$\lim_{t\to\infty} F(t) = \lim_{s\downarrow 0} \tilde{F}(s)$$

Property A-11a: Asymptotic-rate theorem When the indicated limits both exist,

$$\lim_{t \to \infty} \frac{F(t)}{t} = \lim_{s \downarrow 0} s\tilde{F}(s)$$

Property A-12a: Convolution theorem Let $H(\cdot)$ be defined by

$$H(t) = \int_0^t F(t - u) \, dG(u) \tag{A-21}$$

If both $F(\cdot)$ and $G(\cdot)$ have an LST, then so does $H(\cdot)$, and

$$\tilde{H}(s) = \tilde{F}(s)\tilde{G}(s)$$

The following theorem is often handy for proving limit theorems.

Property A-15a: Continuity theorem Let $F_n(\cdot)$ be a d.f. for each $n \in I$ and F be a possibly defective d.f. Then

$$\lim_{n \to \infty} F_n(t) = F(t) \qquad \text{for all } t \geq 0$$

if, and only if,

$$\lim_{n \to \infty} \tilde{F}_n(s) = \tilde{F}(s) \qquad \text{for all } s > 0$$

Generating Functions

If we think of LSTs as transforms of distribution functions of nonnegative random variables and of Laplace transforms as the special case of LSTs when there is a density function, then generating functions can be viewed as the special case of LSTs for discrete nonnegative random variables.

Definition A-8 The *generating function* of a sequence $\{a_0, a_1, \ldots\}$, denoted $\hat{A}(\cdot)$, is defined by

$$\hat{A}(z) = \sum_{n=0}^{\infty} z^n a_n \tag{A-22}$$

We say the generating function *exists* if the sum in (A-22) converges for at least one value of z. The set of z's for which the sum is convergent is called the *region of convergence*. The variable z may be either real or complex.

We usually take z to be real. The sum in (A-22) converges if, for some N,

$$|a_n| \leq M^{\alpha n} \qquad \text{for all } n > N \qquad \text{and} \qquad |z| < \frac{1}{\alpha}$$

for some positive constants M and α. A sufficient condition for $\hat{A}(z)$ to exist for all

z's such that $|z| < 1$ is the existence of N and M such that

$$|a_n| < Mn^m \qquad \text{for all } n > N$$

and some $m > 0$.

We can view generating functions as a special case of LSTs where

$$z = e^{-s} \tag{A-23}$$

and the function $F(\,\cdot\,)$ is constant except for jumps of size a_n at each n. This means that all theorems proved for LSTs must hold for generating functions, when properly modified to account for (A-23). This immediately implies that generating functions are unique and have Property A-1 (linearity). You are invited to develop the analogous statements of Properties A-2 through A-4.

The remaining properties of generating functions are obtained easily by manipulating the sum in (A-22). Denote the generating-function form of a Laplace transform property by "b."

Property A-5b: Differences Let $b_0 = a_0$ and $b_n = a_n - a_{n-1}$, $n = 1, 2, \ldots$. Then

$$\hat{B}(z) \triangleq \sum_{n=0}^{\infty} z^n b_n = (1 - z)\hat{A}(z)$$

Property A-6b: Partial sums Let $b_n = \sum_{k=0}^{n} a_k$. Then

$$\hat{B}(z) = \frac{\hat{A}(z)}{1 - z}$$

The above two properties can be checked by comparing the coefficients of like powers of z.

Property A-7b: Initial-value theorem If $\hat{A}(z) = \sum_{n=0}^{\infty} z^n a_n$, then

$$a_0 = \lim_{z \downarrow 0} \hat{A}(z)$$

The next property follows from LST results and the observation that (A-23) implies

$$z = 1 - s + o(s)$$

for s near zero.

Property A-8b: Final-value theorem When the indicated limits both exist,

$$\lim_{n \to \infty} a_n = \lim_{z \uparrow 1} (1 - z)\hat{A}(z)$$

Property A-9b: Multiplication by n^k Let

$$\hat{A}_k(z) \triangleq \sum_{n=0}^{\infty} z^n n^k a_n \qquad k = 0, 1, \ldots$$

Then

$$\hat{A}_k(z) = z \frac{d}{dz} \hat{A}_{k-1}(z) \qquad k = 1, 2, \ldots$$

where $\hat{A}_0(z) = \hat{A}(z)$.

Property A-10b: Division by n

$$\sum_{n=0}^{\infty} \left| \frac{z^n a_n}{n} \right| = \int_0^z \frac{\hat{A}(x)}{x} \, dx$$

Property A-11b: Asymptotic-rate theorem When the indicated limits both exist, then

$$\lim_{n \to \infty} \frac{a_n}{n} = \lim_{z \uparrow 1} (1 - z)^2 \hat{A}(z)$$

Property A-12b: Convolution theorem The convolution of the two discrete functions a and b, say c, is defined by

$$c_n = \sum_{k=0}^{n} a_{n-k} b_k$$

If a and b each have a generating function, then so does c, and

$$\hat{C}(z) = \hat{A}(z)\hat{B}(z)$$

When $\{a_n\}$ is a probability function of a nonnegative random variable, $\hat{A}(z)$ is called a *probability generating function* (PGF). Since this means $0 \le a_n \le 1$, it is straightforward to establish this property.

Property A-13b: Existence and bounds For a probability function of a nonnegative random variable, $\hat{A}(z)$ exists for $|z| \le 1$ and

$$0 \le \hat{A}(z) \le 1$$

When $\sum_{n=0}^{\infty} a_n = 1$,

$$\hat{A}(1) = 1$$

Property A-14b When $\hat{A}(z)$ is a PGF of a random variable X,

$$E[X(X - 1) \cdots (X - k + 1)] = \frac{d^k}{dz^k} \hat{A}(z) \Big|_{z=1}$$

The proof follows immediately from the observation that

$$\frac{d^k}{dz^k} z^n \Big|_{z=1} = n(n-1) \cdots (n - k + 1) \qquad n > k$$

We call $E[X(X-1) \cdots (X-k+1)]$ the kth *factorial moment* of X. A corollary to Property A-14b is

$$\text{Var}\,(X) = \frac{d^2}{dz^2}\,\hat{A}(z)\Big|_{z=1} + \frac{d}{dz}\,\hat{A}(z)\Big|_{z=1} - \left[\frac{d}{dz}\,\hat{A}(z)\Big|_{z=1}\right]^2$$

Property A-15b: Continuity theorem Let $\hat{A}_n(\,\cdot\,)$ be the PGF of the random variable X_n, $n \in I_+$, and $\hat{A}(\,\cdot\,)$ be the PGF of the random variable X. Then

$$\lim_{n \to \infty} P\{X_n = k\} = P\{X = k\} \qquad k \in I_+$$

if, and only if,

$$\lim_{n \to \infty} \hat{A}_n(z) = \hat{A}(z) \qquad z \neq 1$$

Transforms and Discounted Costs

Money received in the future frequently is considered to be less valuable than the same amount received in the present because of interest considerations. If you had a dollar now, you could put it in the bank and have more than a dollar a year from now, so a dollar today is more valuable than a dollar a year from today.†

A common way to model this change in value is to use *continuous discounting*, which says that a dollar at time t from now is worth $e^{-\beta t}$ dollars now, where $\beta > 0$ is the continuous interest rate. The quantity $e^{-\beta t}$ is called the *present value* of the dollar received at time t.

Property A-15 When a dollar is received at time T which is a random variable with distribution function $F(\,\cdot\,)$, the expected present value of the dollar is

$$E(e^{-\beta T}) = \tilde{F}(\beta)$$

Thus, $\tilde{F}(\beta)$ can be interpreted as an expected present value.

If a *cost rate* of r is incurred up to the random time T, the total cost will be rT. By the above arguments, the present value of the cost incurred during $(t, t + \Delta t)$ is

$$re^{-\beta t}\,\Delta t + o(\Delta t) \qquad t < T$$

and zero otherwise. Thus, we can establish this property.

Property A-16 When costs are incurred at unit rate until the random time T,

† An axiomatic justification for discounting is presented in Chapter 2 of Volume II.

the expected present value of the total cost is

$$E\left(\int_0^T e^{-\beta t}\, dt\right) = \frac{1}{\beta}\, E(1 - e^{-\beta T}) = \frac{1}{\beta}\,[1 - \tilde{F}(\beta)]$$

In discrete-time models, the value of a dollar n time periods from now is taken to be $(1 + \alpha)^{-n}$, where α is the rate of interest per period. The number $\delta = (1 + \alpha)^{-1}$ is the *discount factor* and is analogous to $e^{-\beta}$ in the continuous-time case. When a dollar is received at the end of period N, where N is a random variable with $a_n = P\{N = n\}$, the expected present value of the dollar at interest rate α per period is

$$E[(1 + \alpha)^{-N}] = \sum_{n=0}^{\infty} \delta^n a_n = \hat{A}(\delta)$$

When a dollar is received at the end of each of N periods, where N is a random variable with $a_n = P\{N = n\}$, the expected present value of the dollars received at interest rate α per period is

$$E\left(\sum_{n=0}^{N} (1 + \alpha)^{-n} - 1\right) = \frac{\delta}{1 - \delta}\, E(1 - \delta^N) = \frac{1}{\alpha}\,[1 - \hat{A}(\delta)]$$

In some discounted-cost models, the quantity sought is the transform, not the distribution.

EXERCISES

A-6 Show that the solution of

$$f(t) = 1 - e^{-\mu t} + r\int_0^t f(t - x)\mu e^{-\mu x}\, dx \qquad t \geq 0$$

with r a given constant between 0 and 1, is

$$f(t) = \frac{1 - e^{-\mu(1 - r)t}}{1 - r} \qquad t \geq 0$$

A-7 Let $G(t)$ be a given d.f. of a nonnegative r.v. and r be a given constant between 0 and 1. Let $F(\cdot)$ satisfy

$$F(t) = G(t) + r\int_0^t F(t - x)\, dG(x) \qquad t \geq 0$$

When it exists, show that

$$\lim_{t \to \infty} F(t) = \frac{1}{1 - r}$$

A-8 Let X and Y be independent nonnegative r.v.'s with LSTs $\tilde{F}(\cdot)$ and $\tilde{G}(\cdot)$, respectively. Find the LST of the random variable $Z \triangleq X - Y$.

A-9 Let X be a discrete nonnegative r.v. with $f_n = P\{X = n\}$, $n \in I$. Find the generating function of the functions $F_n = P\{X \leq n\}$ and $F_n^c = P\{X > n\}$.

A-10 Consider the sequence defined by $a_0 = a_1 = 1$,

$$a_n = a_{n-1} + a_{n-2} \qquad n = 2, 3, \ldots$$

Find the generating function, and display the first eight terms of the sequence.

A-11 Let f_n, $n \in I_+$, be given, with $\sum_{n=1}^{\infty} n f_n = \nu$. Define m_n, $n \in I_+$, by

$$m_n = f_n + \sum_{k=1}^{n} f_{n-k} m_k$$

Show that when it exists, $\lim_{n \to \infty} m_n = \nu$.

A-4 A SHORT LIST OF LAPLACE TRANSFORMS AND GENERATING-FUNCTION PAIRS

Table A-2 Some Laplace transform pairs

$f(t)$	$\tilde{f}(s)$	
1	$\dfrac{1}{s}$	
t	$\dfrac{1}{s^2}$	
t^k	$\dfrac{k!}{s^{k+1}}$	
$e^{\lambda t}$	$(s - \lambda)^{-1}$	$s > \lambda$
$\lambda e^{-\lambda t}$	$\dfrac{\lambda}{s + \lambda}$	
$\lambda \dfrac{(\lambda t)^{n-1}}{(n-1)!} e^{-\lambda t}$	$\dfrac{\lambda^n}{(s + \lambda)^n}$	
$\sin \theta t$	$\dfrac{\theta}{s^2 + \theta^2}$	
$o(t)$	$o\left(\dfrac{1}{s}\right)$	

Table A-3 Some generating-function pairs

a_n	$\hat{A}(z)$
1	$(1 - z)^{-1}$
n	$z(1 - z)^{-2}$
$n\lambda^n$	$\lambda z(1 - \lambda z)^{-2}$
λ^n	$(1 - z\lambda)^{-1} \qquad |\lambda z| < 1$
$(1 - p)p^n \qquad 0 \le p \le 1$	$(1 - p)(1 - zp)^{-1}$
$\begin{cases} 0 & 0 \le n \le r - 1 \\ \dbinom{n-1}{n-r} p^{r-1}(1-p)^{n-r} & n \ge r \end{cases}$	$(1 - p)^r(1 - zp)^{-r}$
$\dbinom{r}{n} p^n q^{r-n} \qquad 0 \le n \le r$	$(q + pz)^r$
a_{n-1}	$z\hat{A}(z)$
a_{n+1}	$z^{-1}[\hat{A}(z) - a_0]$

A-5 FACTS FROM MATHEMATICAL ANALYSIS

In the study of stochastic processes frequently we use limiting operations. This section contains statements that are utilized to justify various formal operations, such as interchanging the order of integration and taking the derivative of an infinite sum of functions.

Big-oh, Little-oh Notation

The function $f(\cdot)$ is $o[g(\cdot)]$ (read "f is little-oh of g") as $x \to a$ if $\lim_{x \to a} [f(x)/g(x)] = 0$. When $g(x) = x$, we write $f(\cdot)$ is $o(\cdot)$ and say "$f(\cdot)$ is little-oh of x." The most common circumstance is for $a = 0$ and $g(x) = x$; then $f(\cdot)$ is $o(\cdot)$ means $\lim_{x \to 0} [f(x)/x] = 0$. The purpose of this notation is to describe an essential property of a function without specifying the function in detail. For example, let X be an r.v. with density function $f(\cdot)$. We may write

$$P\{x < X < x + \Delta x\} = f(x)\,\Delta x + o(\Delta x)$$

when Δx is small. The most common example of an $o(\cdot)$ function that we encounter is $x^{1+\delta}$ for some $\delta > 0$, with $x \to 0$.

There are two other important special cases. Choosing $g(x) \equiv 1$ gives $f(\cdot)$ is $o(1)$ if $\lim_{x \to a} f(x) = 0$. Choosing $g(x) = 1/x$ gives $f(\cdot)$ is $o(1/x)$ if $\lim_{x \to a} x f(x) = 0$.

It follows directly from the definition that the sum of two (and hence of any *finite* number) of little-oh functions is a little-oh function, and multiplying a little-oh function by a constant yields another little-oh function.

For $a < \infty$, we write $f(\cdot)$ is $O[g(\cdot)]$ (read "f is big-oh of g") if $f(x)/g(x)$ is bounded for all x in a neighborhood of a. For $a = \infty$, $f(\cdot)$ is $O[g(\cdot)]$ if $f(x)/g(x)$ is bounded for all sufficiently large x. In particular, $f(\cdot)$ is $O(x)$ if $\lim_{x \to \infty} [f(x)/x]$ is finite. Big-oh notation is not used extensively in this book.

Convergence Concepts

The *supremum* (abbreviated *sup*) of a set is its least upper bound, and the *infinum* (abbreviated *inf*) of a set is its greatest lower bound. The sup and inf need not be members of the set; if they are members of the set, they are called the *maximum* and *minimum*, respectively. For example, the set $(0, 1]$ has an inf of 0 and a maximum of 1.

Let $f(x)$ be a function that is defined for some values of x near the point a. Let $\delta > 0$ be chosen arbitrarily, and define the functions

$$\phi_a(\delta) \triangleq \sup\{f(x) : 0 < |x - a| < \delta\}$$

and

$$\psi_a(\delta) \triangleq \inf\{f(x) : 0 < |x - a| < \delta\}$$

Observe that these are monotone in δ, so they have limits (possibly infinite) as

$\delta \downarrow 0$. The *limit superior* (abbreviated *lim sup*) and the *limit inferior* (abbreviated *lim inf*) of $f(\cdot)$ at point a are defined by

$$\limsup_{x \to a} f(x) \triangleq \lim_{\delta \to 0^+} \phi_a(\delta)$$

and

$$\liminf_{x \to a} f(x) \triangleq \lim_{\delta \to 0^+} \psi_a(\delta)$$

respectively.

The limit superior is clearly at least as large as the limit inferior. When equality holds, their common value is $\lim_{x \to a} f(x)$; when equality does not hold, the limit does not exist. A useful way to prove $\lim_{x \to a} f(x) = L$ is to establish

$$\liminf_{x \to a} f(x) \geq L \geq \limsup_{x \to a} f(x)$$

A sequence of functions $\{f_n(\cdot)\}$, each defined on a set A, *converges pointwise* to the function $f(\cdot)$ if $\lim_{n \to \infty} f_n(x) = f(x)$ for each $x \in A$. Sometimes a stronger mode of convergence of functions is required. The sequence $\{f_n(\cdot)\}$ *converges uniformly* on A to $f(\cdot)$ if for any $\epsilon > 0$ there exists a number $N = N(\epsilon)$ (which does not depend on x) such that $n > N$ implies $|f_n(x) - f(x)| < \epsilon$ for all $x \in A$. Alternatively, if $\sup\{|f_n(x) - f(x)| : x \in A\} < \epsilon$ for all $n > N$, then $f_n(\cdot)$ converges to $f(\cdot)$ uniformly.

To establish that a sum of functions converges uniformly, one may use this proposition.

Proposition A-5 If $|f_n(x)| \leq a_n$ for all $x \in A$ and $\sum_{n=0}^{\infty} a_n < \infty$, then $\sum_{n=1}^{\infty} f_n(x)$ converges uniformly (and absolutely) on A.

Uniform convergence is connected to integration and differentiation of infinite sums.

Proposition A-6 If $f_n(\cdot)$ is integrable on $[a, b]$ for each $n \in I_+$ and $f_n(\cdot)$ converges uniformly to $f(\cdot)$ on $[a, b]$, then $f(\cdot)$ is integrable on $[a, b]$ and

$$\lim_{n \to \infty} \int_a^b f_n(x)\, dx = \int_a^b \lim_{n \to \infty} f_n(x)\, dx = \int_a^b f(x)\, dx$$

In particular, choosing $f_n(x) = \sum_{i=1}^n g_i(x)$ yields

$$\int_a^b \sum_{i=1}^{\infty} g_i(x)\, dx = \sum_{i=1}^{\infty} \int_a^b g_i(x)\, dx \qquad (A\text{-}24)$$

Proposition A-7 If (a) $f_n(\cdot)$ is differentiable on $[a, b]$ for each $n \in I_+$; (b) for some $x_0 \in [a, b]$, $\lim_{n \to \infty} f_n(x_0)$ exists; and (c) the first derivatives of $f_n(\cdot)$ converge uniformly on $[a, b]$, then $f_n(\cdot)$ converges uniformly on $[a, b]$ to some differentiable function $f(\cdot)$ and

$$\frac{d}{dx} f(x) = \lim_{n \to \infty} \frac{d}{dx} f_n(x)$$

Choosing $f_n(x) = \sum_{i=1}^{n} g_i(x)$ yields

$$\frac{d}{dx} \sum_{i=0}^{\infty} g_i(x) = \sum_{i=0}^{\infty} \frac{d}{dx} g_i(x)$$

whenever $f_n(\cdot)$ satisfies Proposition A-7.

Integration Theorems

The purpose of uniform convergence in Proposition A-6 is to guarantee that $f(\cdot)$ is integrable. If Proposition A-6 were relied on to justify bringing limits inside integrals, considerable effort would be expended in establishing uniform convergence. We can avoid these tasks by using a notion of integral that encompasses a larger class of functions. The Lebesgue integral is the appropriate integral for this task. Those readers familiar with Lebesgue integration can interpret the integrals in this book in that sense. *This book does not require a knowledge of Lebesgue integration.*

The integrals used in this book are of the *Riemann-Stieltjes* type. This is their definition.

Definition A-9: Riemann-Stieltjes integral Let $f(\cdot)$ and $g(\cdot)$ be defined and bounded on a closed interval $[a, b]$. Then $f(\cdot)$ is *Riemann-Stieltjes-integrable with respect to* $g(\cdot)$ *on* $[a, b]$, with *Riemann-Stieltjes integral* $\mathfrak{I}$, if for any $\epsilon > 0$ there exists $\delta > 0$ such that whenever

(a) $$a = a_0 < a_1 < \cdots < a_n = b$$

with $d \triangleq \max\{a_i - a_{i-1}; \quad i = 1, 2, \ldots, n\} < \delta$ and (b) $a_{i-1} \leq x_i \leq a_i$, $i = 1, 2, \ldots, n$, hold then

(c) $$\left| \sum_{i=1}^{n} f(x_i)[g(a_i) - g(a_{i-1})] - \mathfrak{I} \right| < \epsilon$$

Letting $\Delta g_i = g(a_i) - g(a_{i-1})$, we write (c) in limit notation as

$$\mathfrak{I} \triangleq \int_a^b f(x) \, dg(x) = \lim_{d \to 0} \sum_{i=1}^{n} f(x_i) \, \Delta g_i$$

Riemann-Stieltjes integrals where a or b is infinite are obtained by taking limits as $a \to -\infty$ or $b \to +\infty$. The following are the properties of Riemann-Stieltjes integrals we employ.

Proposition A-8 Riemann-Stieltjes integrals possess these properties:
(a) If $\int_a^b f(x) \, dg(x)$ exists and $g(\cdot)$ has derivative $g'(\cdot)$, then

$$\int_a^b f(x) \, dg(x) = \int_a^b f(x) g'(x) \, dx$$

(b) If $f(\,\cdot\,)$ is integrable with respect to $g(\,\cdot\,)$ on $[a, b]$, then $g(\,\cdot\,)$ is integrable with respect to $f(\,\cdot\,)$ on $[a, b]$, and the *integration-by-parts formula*

$$\int_a^b f(x)\, dg(x) + \int_a^b g(x)\, df(x) = f(b)g(b) - f(a)g(a)$$

is valid.

(c) If, on the interval $[a, b]$, one of the functions $f(\,\cdot\,)$ and $g(\,\cdot\,)$ is continuous and the other is monotonic, then $\int_a^b f(x)\, dg(x)$ exists.

(d) If $f(\,\cdot\,)$ is continuous and $g(\,\cdot\,)$ is monotonic on $[a, b]$, then there exists a point y, $a \le y \le b$, such that

$$\int_a^b f(x)\, dg(x) = f(y)[g(b) - g(a)]$$

This is called the *first mean-value theorem for Riemann-Stieltjes integrals*.

The problem with using Riemann-Stieltjes integrals is that $\int_a^b f(x)\, dg(x)$ is not defined if $f(\,\cdot\,)$ and $g(\,\cdot\,)$ both have a jump at the same point. To circumvent this problem, we adopt the following convention throughout this book.

Convention If $f(\,\cdot\,)$ and $g(\,\cdot\,)$ both have a jump at y, $a < y < b$, then

$$\int_a^b f(x)\, dg(x) = \int_a^{y^-} f(x)\, dg(x) + f(y)[g(y^+) - g(y^-)]$$

$$+ \int_{y^+}^b f(x)\, dg(x) \tag{A-25}$$

The notations $\int_a^{y^-} f(x)\, dg(x)$ and $\int_{y^+}^b f(x)\, dg(x)$ mean that the contribution at y is excluded. More precisely, in part c of Proposition A-8, we take $g(a_n) = \lim_{u \uparrow y} g(u)$ in the former integral and $g(a_0) = \lim_{u \downarrow y} g(u)$ in the latter integral. The term $g(y^+) - g(y^-)$ is the value of the jump at y.

The convention is extended to a countable number of jumps by iteration. With this convention, if $f(\,\cdot\,)$ and $g(\,\cdot\,)$ are step functions with common jump points, the integral becomes a sum. This means that we do not have to distinguish sums from integrals. More important, if $g(\,\cdot\,)$ is nonnegative and nondecreasing (e.g., if it is a d.f.), then the integral in (A-25) agrees with the Lebesgue integral, and the powerful theorems of that theory apply.

Proposition A-9: Monotone convergence theorem Assume (a) $f_n(\,\cdot\,) \ge 0$ for each $n \in I_+$, (b) $f_n(x) \uparrow f(x)$ pointwise as $n \to \infty$, and (c) $g(\,\cdot\,) \ge 0$ and nondecreasing. Then

$$\lim_{n \to \infty} \int_a^b f_n(x)\, dg(x) = \int_a^b f(x)\, dg(x)$$

whenever $0 \le a \le b$.

Choosing $f_n(x) = \sum_{i=1}^{n} g_i(x)$ and $g(x) = x$ establishes (A-24) for nonnegative functions.

Here is another theorem of this type.

Proposition A-10: Dominated convergence theorem Suppose that for each†
$x \in [a, b]$, $\lim_{t \to s} f(t, x) = f(s, x)$ and $|f(t, x)| \leq h(x)$ for each x and some function $h(\cdot)$. Then if $\int_a^b h(x) \, dx$ exists and $g(\cdot) \geq 0$ and is nondecreasing, then

$$\int_a^b f(t, x) \, dg(x) \to \int_a^b f(s, x) \, dg(x)$$

In particular, if $\partial f(t, x)/\partial t$ exists at $t = s$ and is bounded by some function $h(x)$ which is integrable, then

$$\frac{d}{dt} \int_a^b f(t, x) \, dx = \int_a^b \frac{\partial f(t, x)}{\partial t} \, dx$$

When the roles of $f(\cdot, \cdot)$ and $g(\cdot)$ are interchanged, this is the corresponding theorem.

Proposition A-11: Helley-Bray theorem Let $g(\cdot)$ be continuous and $F_n(\cdot)$ be a d.f. for each $n \in I$. If $F_n(\cdot) \to F(\cdot)$ pointwise and $F(\cdot)$ is a d.f., then

$$\lim_{n \to \infty} \int_a^b g(x) \, dF_n(x) = \int_a^b g(x) \, dF(x)$$

whenever $-\infty < a < b < \infty$. If $g(\cdot)$ is bounded, then

$$\lim_{n \to \infty} \int_{-\infty}^{\infty} g(x) \, dF_n(x) = \int_{-\infty}^{\infty} g(x) \, dF(x)$$

The interchange of the order of integration is justified by the next proposition.

Proposition A-12: Fubini's theorem If $f(x, y) \geq 0$ for all (x, y) in its domain and $g_1(\cdot)$ and $g_2(\cdot)$ are nonnegative and nondecreasing, then

$$\int_{-\infty}^{\infty} \left[\int_{-\infty}^{\infty} f(x, y) \, dg_2(y) \right] dg_1(x) = \int_{-\infty}^{\infty} \left[\int_{-\infty}^{\infty} f(x, y) \, dg_1(x) \right] dg_2(y)$$

$$\text{(A-26)}$$

If $f(x, y) < 0$ for some (x, y), (A-26) holds if the integrals are absolutely convergent.

† This theorem remains true for the intervals $(-\infty, b]$, $[a, \infty)$, and $(-\infty, \infty)$.

In many applications of Fubini's theorem, $g_1(\cdot)$ and $g_2(\cdot)$ are d.f.'s; then (A-26) holds whenever $f(\cdot, \cdot)$ is bounded.

This is the formula for differentiating a definite integral.

Proposition A-13: Leibnitz' rule Let

$$F(x) = \int_{a(x)}^{b(x)} f(x, y) \, dy$$

If $f(x, y)$ has a continuous derivative with respect to x in the region $c \le x \le d$, if $a(x) \le y \le b(x)$, and if $a(\cdot)$ and $b(\cdot)$ are differentiable, then

$$\frac{d}{dx} F(x) = \int_{a(x)}^{b(x)} \frac{\partial f(x, y)}{\partial x} \, dy + f[x, b(x)] \frac{db(x)}{dx} - f[x, a(x)] \frac{da(x)}{dx}$$

whenever $c \le x \le d$.

In particular,

$$\frac{d}{dx} \int_a^b f(x, y) \, dy = \int_a^b \frac{\partial f(x, y)}{\partial x} \, dy$$

This equation is valid when $b = \infty$ or $a = -\infty$ provided the right-side is finite.

Taylor's Series

A Taylor's series expansion of a function is often useful in obtaining asymptotic behavior.

Proposition A-14: Taylor's theorem Let $f(y)$ be a continuous function possessing $n + 1$ continuous derivatives for all y between a and x inclusive. Then

$$f(x) = f(a) + f^{(1)}(a)(x - a) + \frac{f^{(2)}(a)(x - a)^2}{2!}$$

$$+ \cdots + \frac{f^{(n)}(x)(x - a)^n}{n!} + R_n(x, a) \qquad \text{(A-27)}$$

where

$$R_n(x, a) = \int_a^x \frac{(x - y)^n}{n!} f^{(n+1)}(y) \, dy$$

and

$$f^{(n)}(y) \triangleq \frac{d^n}{dy^n} f(y)$$

The mean-value theorem for integrals may be employed to obtain Lagrange's form of the remainder:

$$R_n(x, a) = \frac{f^{(n+1)}(t)(x-a)^{n+1}}{(n+1)!} \tag{A-28}$$

for some t between a and x.

When $a = 0$, (A-27) is called a *Maclaurin series*. Here is a typical example of the use we have for Proposition A-14. Let $F(\cdot)$ be the d.f. of a nonnegative random variable X, $\tilde{F}(\cdot)$ be the LST of $F(\cdot)$, and $v_n = E(X^n)$. Then $\tilde{F}(s)$ has the Maclaurin's series:

$$\tilde{F}(s) = 1 - vs + v_2 \frac{s^2}{2} + \cdots + v_n \frac{s^n}{n!} + o(s) \tag{A-29}$$

whenever $v_{n+1} < \infty$. A proof of (A-29) can be constructed from the propositions presented in this appendix; doing so is left to the interested reader.

BIBLIOGRAPHIC GUIDE

References

There are many excellent books on probability theory. Included among those that do not use measure theory are the following:

Clark, A. Bruce, and Ralph L. Disney: *Probability and Random Processes for Engineers and Scientists*, Wiley, New York (1970).
Feller, William: *An Introduction to Probability Theory and Its Applications*, vol. 1, 3d ed., Wiley, New York (1968).
Neuts, Marcel F.: *Probability*, Allyn and Bacon, Boston (1973).
Parzen, Emanuel: *Modern Probability Theory and Its Applications*, Wiley, New York (1960).
Ross, Sheldon M.: *A First Course in Probability*, Macmillan, New York (1976).

Among the probability books that use measure theory are:

Breiman, Leo: *Probability*, Addison-Wesley, Reading, Mass. (1968).
Chung, Kai Lai: *A Course in Probability Theory*, 2d ed., Academic, New York (1974).
Feller, William: *An Introduction to Probability Theory and Its Applications*, vol. 2, 2d ed., Wiley, New York (1971).
Loève, Michel: *Probability Theory*, 3d ed., Van Nostrand, Princeton, N.J. (1963).

The classical work on Laplace transforms is:

Widder, D. V.: *The Laplace Transform*, Princeton University Press, Princeton, N.J. (1946).

Techniques for numerically inverting Laplace transforms are given in the following:

Jagerman, D. L. "An Inversion Technique for the Laplace Transform with Application to Approximation, *Bell Syst. Tech. J.* **57**:669–710 (1978).
Weeks, W. T., "Numerical Inversion of Laplace Transforms," *J. Assoc. Comput. Mach.* **13**:419–429 (1966).

Among the many tables of Laplace transforms are:

Churchill, R. V.: *Operational Mathematics*, 3d ed., McGraw-Hill, New York (1972).

Doetsch, G.: *Introduction to the Theory and Applications of the Laplace Transformation*, Springer-Verlag, New York (1974).

Nixon, E. E.: *Handbook of Laplace Transformation: Fundamentals, Applications, Tables, and Examples*, 2d ed., Prentice-Hall, Englewood Cliffs, N.J. (1965).

Oberhettinger, F., and L. Badii: *Tables of Laplace Transforms*, Springer-Verlag, New York (1973).

Roberts, G. E., and H. Kaufman: *Table of Laplace Transforms*, Saunders, Philadelphia (1966).

Spiegel, M. R.: *Theory and Problems of Laplace Transforms*, Schaum's Outline Series, McGraw-Hill, New York (1965).

Books on mathematical analysis that do not treat measure theory include:

Apostle, Tom M.: *Mathematical Analysis*, 2d ed., Addison-Wesley, Reading, Mass. (1974).

Bartle, Robert G.: *Elements of Real Analysis*, 2d ed., Wiley, New York (1976).

Olmsted, John M. H.: *Advanced Calculus*, Appleton-Century-Crofts, New York (1961).

Rudin, Walter: *Principles of Mathematical Analysis*, 2d ed., McGraw-Hill, New York (1964).

Books on measure theory and integration include:

Bartle, Robert G.: *The Elements of Integration*, Wiley, New York (1966).

Halmos, Paul R.: *Measure Theory*, Van Nostrand, Princeton, N.J. (1950).

Royden, H. L.: *Real Analysis*, Macmillan, New York (1963).

AUTHOR INDEX

SUBJECT INDEX